OPTICAL FIBER
TELECOMMUNICATIONS

Contributors

BRIAN G. BAGLEY

LEE L. BLYLER, JR.

CHARLES A. BURRUS

H. CRAIG CASEY, JR.

ALLEN H. CHERIN

ALAN G. CHYNOWETH

LEONARD G. COHEN

JACK COOK

BERNARD R. EICHENBAUM

WILLIAM G. FRENCH

WILLIAM B. GARDNER

DETLEF GLOGE

RAYMOND E. JAEGER

PETER KAISER

DAVID KALISH

IVAN P. KAMINOW

RAYMOND A. KEMPF

P. LELAND KEY

CHARLES R. KURKJIAN

PAUL D. LAZAY

TIEN PEI LEE

TINGYE LI

JOHN B. MacCHESNEY

ENRIQUE A. J. MARCATILI

DIETRICH MARCUSE

CALVIN M. MILLER

STEWART E. MILLER

JAMES W. MITCHELL

SUZANNE R. NAGEL

KURT NASSAU

A. DAVID PEARSON

STEWART D. PERSONICK

GEORGE E. PETERSON

HERMAN M. PRESBY

PETER K. RUNGE

HAROLD SCHONHORN

MORTON I. SCHWARTZ

PETER W. SMITH

ROGERS H. STOLEN

BASANT K. TARIYAL

ARTHUR R. TYNES

TSUEY TANG WANG

JOHN C. WILLIAMS

OPTICAL FIBER
TELECOMMUNICATIONS

Edited by
STEWART E. MILLER
Bell Laboratories
Crawford Hill Laboratory
Holmdel, New Jersey

ALAN G. CHYNOWETH
Bell Laboratories
Murray Hill, New Jersey

ACADEMIC PRESS   New York   San Francisco   London   1979
A Subsidiary of Harcourt Brace Jovanovich, Publishers

ACADEMIC PRESS, INC.
111 Fifth Avenue, New York, New York 10003

United Kingdom Edition published by
ACADEMIC PRESS, INC. (LONDON) LTD.
24/28 Oval Road, London NW1 7DX

Library of Congress Cataloging in Publication Data

Main entry under title:

Optical fiber telecommunications.

    Includes bibliographies.
    1. Optical communications. 2. Fiber optics.
I. Miller, Stewart E. II. Chynoweth, A. G.
TK5103.59.068   621.38'0414   78-20046
ISBN 0-12-497350-7

PRINTED IN THE UNITED STATES OF AMERICA

80 81 82   9 8 7 6 5 4 3 2

Contents

Chapter 4   Dispersion Properties of Fibers

Detlef Gloge, Enrique A. J. Marcatili, Dietrich Marcuse,
and Stewart D. Personick

Chapter 5   Nonlinear Properties of Optical Fibers

Rogers H. Stolen

Chapter 6   Fiber Design Considerations

Detlef Gloge and William B. Gardner

Chapter 7 Materials, Properties, and Choices

*Brian G. Bagley, Charles R. Kurkjian, James W. Mitchell,
George E. Peterson, and Arthur R. Tynes*

Chapter 8 Fiber Preform Preparation

*William G. French, Raymond E. Jaeger, John B. MacChesney,
Suzanne R. Nagel, Kurt Nassau, and A. David Pearson*

Chapter 9 Fiber Drawing and Control

*Raymond E. Jaeger, A. David Pearson, John C. Williams,
and Herman M. Presby*

Chapter 10   Coatings and Jackets

Lee L. Blyler, Jr., Bernard R. Eichenbaum, and Harold Schonhorn

Chapter 11   Fiber Characterization

*Leonard G. Cohen, Peter Kaiser, Paul D. Lazay, and
Herman M. Presby*

Chapter 12   Fiber Characterization — Mechanical

*David Kalish, P. Leland Key, Charles R. Kurkjian,
Basant K. Tariyal, and Tsuey Tang Wang*

Chapter 13   Optical Cable Design

Morton I. Schwartz, Detlef Gloge, and Raymond A. Kempf

Chapter 14   Fiber Splicing

Detlef Gloge, Allen H. Cherin, Calvin M. Miller, and
Peter W. Smith

Chapter 15   Optical Fiber Connectors

Jack Cook and Peter K. Runge

Chapter 16   Optical Sources

Charles A. Burrus, H. Craig Casey, Jr., and Tingye Li

Chapter 17   Modulation Techniques

Ivan P. Kaminow and Tingye Li

Chapter 18   Photodetectors

Tien Pei Lee and Tingye Li

Chapter 19   Receiver Design

Stewart D. Personick

Chapter 20   Transmission System Design

Stewart E. Miller

Chapter 21   Potential Applications

Stewart E. Miller

List of Contributors

Numbers in parentheses indicate the pages on which the authors' contributions begin.

Brian G. Bagley (167), Bell Laboratories, Murray Hill, New Jersey 07974

Lee L. Blyler, Jr. (299), Bell Laboratories, Murray Hill, New Jersey 07974

Charles A. Burrus (499), Bell Laboratories, Crawford Hill Laboratory, Holmdel, New Jersey 07733

H. Craig Casey, Jr. (499), Bell Laboratories, Murray Hill, New Jersey 07974

Allen H. Cherin (455), Bell Laboratories, Norcross, Georgia 30071

Alan G. Chynoweth (1), Bell Laboratories, Murray Hill, New Jersey 07974

Leonard G. Cohen (343), Bell Laboratories, Crawford Hill Laboratory, Holmdel, New Jersey 07733

Jack Cook (483), Bell Laboratories, Holmdel, New Jersey 07733

Bernard R. Eichenbaum (299), Bell Laboratories, Norcross, Georgia 30071

William G. French (233), Bell Laboratories, Murray Hill, New Jersey 07974

William B. Gardner (151), Bell Laboratories, Norcross, Georgia 30071

Detlef Gloge (37, 101, 151, 435, 455), Bell Laboratories, Crawford Hill Laboratory, Holmdel, New Jersey 07733

Raymond E. Jaeger* (233, 263), Bell Laboratories, Murray Hill, New Jersey 07974

Peter Kaiser (343), Bell Laboratories, Crawford Hill Laboratory, Holmdel, New Jersey 07733

David Kalish (401), Bell Laboratories, Norcross, Georgia 30071

* Present Address: Galileo Electro-Optics Corporation, Sturbridge, Massachusetts 01518

Ivan P. Kaminow (557), Bell Laboratories, Crawford Hill Laboratory, Holmdel, New Jersey 07733

Raymond A. Kempf (435)* Bell Laboratories, Norcross, Georgia 30071

P. Leland Key (401), Bell Laboratories, Murray Hill, New Jersey 07974

Charles R. Kurkjian (167, 401), Bell Laboratories, Murray Hill, New Jersey 07974

Paul D. Lazay (343), Bell Laboratories, Murray Hill, New Jersey 07974

Tien Pei Lee (593), Bell Laboratories, Crawford Hill Laboratory, Holmdel, New Jersey 07733

Tingye Li (499, 557, 593), Bell Laboratories, Crawford Hill Laboratory, Holmdel, New Jersey 07733

John B. MacChesney (233), Bell Laboratories, Murray Hill, New Jersey 07974

Enrique A. J. Marcatili (17, 37, 101), Bell Laboratories, Crawford Hill Laboratory, Holmdel, New Jersey 07733

Dietrich Marcuse (37, 101), Bell Laboratories, Crawford Hill Laboratory, Holmdel, New Jersey 07733

Calvin M. Miller (455), Bell Laboratories, Norcross, Georgia 30071

Stewart E. Miller (1, 653, 675), Bell Laboratories, Crawford Hill Laboratory, Holmdel, New Jersey 07733

James W. Mitchell (167), Bell Laboratories, Murray Hill, New Jersey 07974

Suzanne R. Nagel (233), Bell Laboratories, Murray Hill, New Jersey 07974

Kurt Nassau (233), Bell Laboratories, Murray Hill, New Jersey 07974

A. David Pearson (233, 263), Bell Laboratories, Murray Hill, New Jersey 07974

Stewart D. Personick** (101, 627), Bell Laboratories, Holmdel, New Jersey 07733

George E. Peterson (167), Bell Laboratories, Murray Hill, New Jersey 07974

Herman M. Presby (263, 343), Bell Laboratories, Crawford Hill Laboratory, Holmdel, New Jersey 07733

Peter K. Runge (483), Bell Laboratories, Crawford Hill Laboratory, Holmdel, New Jersey 07733

---

* Present Address: 5042 Vernon Oaks Drive, Dunwoody, Georgia 30338
** Present Address: TRW-Vidar Corporation, Mountain View, California 94050

Harold Schonhorn (299), Bell Laboratories, Murray Hill, New Jersey 07974

Morton I. Schwartz (435), Bell Laboratories, Norcross, Georgia 30071

Peter W. Smith (455), Bell Laboratories, Crawford Hill Laboratory, Holmdel, New Jersey 07733

Rogers H. Stolen (125), Bell Laboratories, Crawford Hill Laboratory, Holmdel, New Jersey 07733

Basant K. Tariyal (401), Bell Laboratories, Norcross, Georgia 30071

Arthur R. Tynes (167), Bell Laboratories, Crawford Hill Laboratory, Holmdel, New Jersey 07733

Tsuey Tang Wang (401), Bell Laboratories, Murray Hill, New Jersey 07974

John C. Williams (263), Bell Laboratories, Murray Hill, New Jersey 07974

Foreword

Technological advance is usually the result of incremental innovations that move existing art forward in relatively small steps. Major break-throughs and entirely new technologies come much less frequently. These offer not just a better way of doing things already being done, but pro-vide a basis for new classes of products and services to serve society's needs. Often they are the result of scientific discoveries that open the way to previously unforeseen applications. In other cases, the applications may have been foreseen, but the scientific discoveries were needed to realize them. The laser, first described in a publication by Townes and Schawlow in 1958, was a scientific discovery that fitted both descriptions. In these twenty years following the advent of the laser, there has not only been an astounding surge of scientific progress in the understanding of optical phenomena and in the use of optical tools in many branches of science, but there have also been created new technologies with enormous po-tential for useful application. Among these, optoelectronic technology for communications stands out.

Communications using light is far from a new idea. Alexander Graham Bell, a century ago, sought to send speech over visible light beams, and his Photophone was capable of transmitting speech information over distances of several hundred meters. By today's standards his schemes for modulating the light beam and detecting the signals were crude, and his system proved impractical. Nevertheless, the basic elements for lightwave communication were there.

In principle, the whole electromagnetic spectrum has been available as a medium for communications since the epochal discoveries in the middle of the 19th century of James Clerk Maxwell and Michael Faraday. Their understanding of the nature of electromagnetic radiation led, in turn, to the discovery by Heinrich Hertz in 1887 of long wavelength radia-tion and to its use in radio by Guglielmo Marconi in 1895. Since then, the history of the transmission of communications signals has largely been a record of advance to progressively shorter wavelengths, eventually to microwaves. This orderly exploitation of higher frequencies and the band-width increases they offer led farsighted engineers like Hartley (1945) and Tyrrell (1951) to look beyond the microwave era that began with World

War II and to set forth* the circumstances that would permit the next logical step—use of still higher, optical frequencies for communications. Thus a background of theory and enlightened speculation was available in 1960, when the laser came along to offer what was almost immediately seen to be a key element in a lightwave communications system.

Inspired by the advent of the first working laser, built by Maiman[†] in 1960, research on lightwave communications began in earnest in the early 1960s; Kompfner and Miller and their associates at Bell Laboratories' Crawford Hill laboratory undertook a broad program of research to explore the fundamentals of lightwave transmission, as well as the elements of a lightwave communications system. At the same time, an intense research effort was launched by solid state physicists and chemists on previously unknown or unexplored optical properties of solids, and on the discovery and development of new laser, light-emitting diode, and nonlinear optical materials. This research effort provided the basis for the new optoelectronic technology and brought lightwave communications within reach.

But the question of the transmission medium remained unsettled. From the beginning, glass fibers had been an appealing choice, but losses were far too high to make them appear to be a practical possibility. Following a proposal in 1966 by Kao,[‡] a British engineer, that this loss could be reduced, research programs on glass fibers were launched by several organizations. These culminated in the 1970 achievement by the Corning Glass Works of losses below 20 dB/km. Advances by Bell Laboratories and Corning in the years immediately following brought superior materials and processes for the production of fibers and led to losses of 1 dB/km and below. By this time, semiconductor lasers capable of continuous, room-temperature operation were available. With these successes, all of the elements for a practical communications system were finally in hand, even though much engineering work remained to be done. By the middle of the decade, experimental versions of commercial systems appeared.

Throughout the period between the conception of the laser and the first practical lightwave communications systems, progress has been the result of an interdisciplinary effort involving electrical engineers, physicists, and materials scientists and engineers. Each group has stimulated and strengthened the work of the others; without this close collaboration, the accomplishment of workable systems simply would not have happened. While this kind of interdisciplinary research and development has characterized semiconductor electronics—and there are some other examples

---

* Internal publications, Bell Laboratories.
† Hughes Aircraft Company.
‡ Standard Telecommunications Laboratory.

—it is not common in many other areas of technology. For these, solid state electronics serves as a model, and lightwave communications have demonstrated this strikingly, just as semiconductor electronics did before it. The wide range of disciplinary talents available in an industrial laboratory environment has fostered this interdisciplinary exploration of the new optical science and technology, and as a result, no essential aspect of either the science or the technology was left unattended to in this remarkably swift advance to a major new systems technology.

What can be said of the future for lightwave communications? It is too early to answer that question with certainty. It can be said with considerable assurance that lightwave transmission systems will have a place, in competition with longer wavelength radio systems, waveguides, and electrons in copper wires, but it cannot yet be said how far they will go in superseding or displacing these established transmission technologies. Already it can safely be predicted that lightwave systems will compete successfully, technically, and economically for short-haul high-capacity communications links, such as data links and metropolitan area telecommunications trunking. It is likely that they will prove technically feasible for long-haul communications, both on land and underseas, and for local distribution networks, but the economics is less certain, for the present, in these cases. The past history of electronics technologies—especially semiconductor electronics—offers ample reason for optimism, however. In the early stages of radically new technologies it is possible to see only the beginnings of systems opportunities, but it is not possible to predict how many more will develop or what will result from the learning experience furnished by the first systems to be developed. Lightwave systems are in this stage now; and it remains to be seen whether such advanced research concepts as integrated optics will flourish to the point of enabling electronics engineers to go beyond transmission and to develop a whole spectrum of electronics components and systems based on photons rather than electrons. One thing is certain, and that is that the inherent excitement of such possibilities will provide abundant incentives for research scientists and exploratory development engineers for years to come.

N. B. HANNAY
Bell Laboratories

Preface

In the manner described by N. B. Hannay in this book's Foreword, optical fiber telecommunications have become a reality. Around the world, thousands of scientists and engineers are engaged in designing first-generation commercial systems and in exploring new concepts potentially applicable to second-generation commercial systems. Both types of activity will be facilitated by a comprehensive treatment of the fundamentals. This field, like many others in the electronics industry, is characterized by rapid publication in the scientific journals. Major review articles began appearing about five years ago. This book serves as a basic text, providing scientific and fundamental engineering principles for all parts of a modern lightwave telecommunication system: the fiber, including cabling and splicing; the lightwave sources, both lasers and light-emitting diodes; the detectors; the transmitters and receivers; the system design principles; and the potential applications. Extensive references are made to the original scientific literature so that readers can go back and reexamine the stepping stones and alternatives to the currently preferred approaches which are here described in detail.

As a basic text, this book should prove useful to students, scientists, and engineers in academic, industrial, and other institutions. Specialists in one of the components will find the book useful to review their own field and to gain perspective on how their work relates to other specialties. Design engineers will find it useful in that it brings together all facets of the communication system in one book.

Optical fiber telecommunications bring together the collaborative efforts of workers from the system area of telecommunications, the glass industry, and the semiconductor electronics industry, which currently produces the most attractive lightwave sources and detectors, as well as the transistors and integrated circuits used in transmitters and receivers. Workers in all these fields should find this work a useful reference.

The book is organized so that it is understandable to a reader on the graduate level with no specialized knowledge of lightwave communication and yet provides a comprehensive treatment. The first two chapters give historical background, outline the detailed chapter organization, and lead the reader through the evolution of the new transmission medium, the glass fiber. After the first two chapters, the in-depth treatments of various specialties can be used selectively without reading earlier chapters. At the cost of some repetition, Chapters 3 to 21 are essentially stand-alone treatments, much as review papers in scientific journals are. One consistent set of symbols for important quantities is used throughout.

This book differs from earlier ones in the comprehensive nature of the treatment of all subjects. Furthermore, all the authors are drawn from an organization that is itself engaged in all aspects of the development of optical telecommunications technology and systems. Each chapter is written by one or more researchers in his or her own specialty—over forty contributing authors in all. Emphasis is on fundamental science and design principles of enduring value, not on specific design choices or standardization.

Chapter 1

Evolution of Optical Communications

ALAN G. CHYNOWETH
STEWART E. MILLER

Telecommunications via optical fiber waveguides has many roots. Alexander Graham Bell (1880) transmitted speech several hundred meters over a beam of light shortly after his invention of the telephone. Little came of it, for the art could not support the novel concept. Several additional probes into the potential of visible wavelengths for telecommunications were made in the 1940s and 1950s at Bell Laboratories, the last of which called for coherent sources of light in order to permit efficient information transmission (Kompfner, 1972).

All forms of optical telecommunications utilize (1) a visible or near-infrared source which is modulated by the information-bearing signal, (2) a transmission medium, and (3) a detector which recovers the modulation as a baseband signal practically identical to that taken as the input to the system.

1.1 SOURCES AND DETECTORS

The invention and experimental demonstration of the laser brought new life to the prospect for optical telecommunications and triggered a broad effort to gain the science and technology necessary to make it a reality. The laser (Schawlow and Townes, 1958) is a coherent source which, in principle, makes feasible in the optical wavelength region all of the communication techniques employed in the microwave region. The laboratory demonstrations of laser action, first in ruby (Maiman, 1960) and then in gases (Javan *et al.*, 1961), provided immediate evidence of feasibility. For several years the helium–neon gaseous laser was the workhorse of optical

1

communications research; without it the art could not have moved along as it did (Gordon and White, 1964). Nevertheless, it was the demonstration of laser operations in semiconductor devices (Hall *et al.*, 1962; Nathan *et al.*, 1962) that gave the first hint of practical, low-cost optical communications with the potential for miniaturization and improved reliability generally associated with solid-state components.

The first semiconductor lasers, based on simple p–n junctions formed in gallium arsenide, were logical and relatively modest extensions of the technology that was being widely pursued to arrive at efficient electroluminescent devices (light-emitting diodes), but they worked only for a short while, and even so, only at liquid nitrogen temperatures and under pulsed excitation. Considerable further work had to be done to arrive at lasers that would run continuously at normal working temperatures.

Crucial first steps toward meeting these requirements were the concepts and technology surrounding the fabrication of heterojunction structures. In conventional p–n homojunctions, the same host semiconductor material (e.g., GaAs) is used for both sides of the junction. In a heterojunction, the host material on the n-side differs from that on the p-side. It was early recognized (Alferov and Kazarinov, 1963; Kroemer, 1963) that the potential barriers of heterojunction structures offered the possibility of confining and concentrating the injected electron and hole populations while, at the same time, providing a rudimentary optical cavity structure that would facilitate the moding necessary for laser action.

Proof of the correctness of the heterojunction approach first came from the demonstration of reduced threshold currents needed to obtain laser action in single heterojunction structures in the GaAlAs material system (Hayashi *et al.*, 1969; Kressel and Nelson, 1969). Later, using double heterojunctions in which the carriers are confined in the potential well formed by sandwiching a thin, narrower energy-gap material between somewhat higher energy-gap n- and p-type materials, "continuous" operation at room temperature was demonstrated (Hayashi *et al.*, 1970; Alferov *et al.*, 1971).

For the early demonstrations, an hour or so of continuous operation served to prove the point. But for practical communications systems, mean lifetimes to failure of a million hours or more would be necessary. It took intensive investigation on the part of many scientists and engineers to identify defects and other factors that limited laser lifetime; a particularly important breakthrough occurred when "dark-line defects" (so-called because of their appearance under the microscope using an infrared image converter) were observed to be migrating and spreading into the active region of laser diodes, eventually quenching their lasing action (De-Loach *et al.*, 1973; Petroff and Hartman, 1974). These dark-line defects appear to be related to strain-induced dislocations while their migration could be accounted for by a newly discovered phenomenon, enhancement of impurity diffusion rates by the local, radiationless dumping of electron-hole recombination energy at the impurities (Kimerling *et al.*, 1976). Developing fabrication methods that minimize strain in the diodes has so far been the principal factor in achieving million-hour lasers (Hartman *et al.*, 1977).

Primary attractions of lasers as light sources in optical communications are the narrow linewidths of their emissions, and the relative ease of achieving good coupling efficiency into the transmitting medium because of their directionality. These properties make lasers particularly attractive for the longer distances between source and receiver. But in many situations these distances can be quite modest; it was early recognized that light-emitting diodes could be used as light sources in these situations. The potential attractions of LEDs are that they may be somewhat less costly to fabricate than lasers, and long life may be easier to attain, although it is still too early to be sure that these advantages will materialize. While the simplest junction structures can serve in short optical links, within switching machines, for example, diode structures which exhibit much higher radiance and diode-fiber coupling efficiency have been developed for medium distances of up to a kilometer or so, depending on the modulation frequency (Burrus and Miller, 1971).

While semiconductor lasers and light-emitting diodes are the sources currently under intensive development as optical communications sources, exploratory work with other possible sources also continues. Such sources include optically pumped lasers, particularly based on neodymium ions in various glassy or crystalline hosts. Neodymium lasers offer the attractions of very narrow linewidth and emission in the region of the spectrum, 1.1 to 1.3 μm, where glass fibers have low attenuation and dispersion. Furthermore, these lasers can be pumped by GaAlAs LEDs, thus making an all solid-state device possible. Miniature Nd-doped lasers have been demonstrated using end-pumping into a crystalline fiber of Nd–yttrium aluminum garnet (Stone and Burrus, 1978).

The obvious attractiveness of devices that operate in the infrared wavelength region where glass fibers show lowest loss and dispersion underscores the recent demonstration of lasing in the junctions made in the semiconductor material system, gallium indium arsenide phosphide (Shen *et al.*, 1977). It is too early to know whether efficient, reliable, and cost-effective lasers and LEDs can be made in this system, but early results are encouraging. More of a challenge may be the development of suitable detector devices, perhaps using the same basic material system, to operate at the same wavelengths, a region where silicon devices are no longer effective and the most obvious present alternative, germanium, exhibits lower sensitivity on account of its background dark current.

As just noted, silicon photodetector devices are suitable for use at the wavelengths emitted by GaAlAs devices. Indeed, the relatively mature state of silicon technology has meant that suitable detectors for this region have been comparatively straightforward to develop. The detector devices are of two types: (i) p–i–n diodes, which yield a signal approximately equivalent to one electron per incident photon, and (ii) avalanche photo-diodes, operated in reverse bias close to breakdown so as to provide internal multiplication, yielding a signal equivalent to perhaps 10 to 100 electrons per incident photon. The p–i–n diodes can be used for short optical links where maximum sensitivity is not required, but in many other situations the advantages of the avalanche photodiode make it the preferred choice. A key requirement in avalanche diodes is the avoidance of structural and other defects which can give rise to microplasmas, and hence, electrical instability. This has been no mean task in silicon, and it is yet to be seen what difficulties of this sort may have to be overcome in developing detectors for the longer infrared wavelengths.

1.2 EVOLUTION OF THE TRANSMISSION MEDIUM—FIBER LIGHTGUIDES

The transmission system concepts which are attractive depend on the nature of the available sources, transmission media, and receivers. It is not surprising, therefore, that system research has tracked the evolutionary and revolutionary changes in these elements. Naturally there has been a two-way interaction, with system concepts stimulating new components as well as the reverse.

Immediately after the laboratory demonstrations of laser action, rather broadly based research was initiated at Bell Labs (1) on beam waveguide transmission media (Kompfner, 1965) and (2) on radiolike transmission through the atmosphere. Borrowing from the ancient art of astronomy and inventing new structures better suited to the telecommunications application, researchers were able to confine the transmitted optical energy into beams which could be received almost in their entirety at the receiver—leading to loss in the order of 1 dB/km for very broad visible or near-infrared bands. A coherent carrier wave in the visible or near-infrared can in theory be modulated with an enormous amount of information, which served as the goal. However, realization was accompanied by costs. For atmospheric transmission, severe losses resulted from fog and snow all the way from 0.63- to 10-μm wavelength (Chu and Hogg, 1968). Consequently, special-purpose links for length 100–300 m or "fair weather" communication applications seem to be the only places where atmospheric optical transmission is feasible.

Beam waveguide transmission through lenses or periscopic mirror pairs spaced about 100 m apart proved quite practical. The lens guide could be enclosed in an underground conduit less than 15 cm in diameter, providing independence of weather. The gas lens, an innovation stimulated by the telecommunications interest (Berreman, 1964; Miller, 1970), made it possible to guide laser beams around curves a few hundred meters in radius, thus providing potential relief from the straight-line installation requirements of glass-lens guides. However, all forms of beam waveguide involved expensive installations which appeared attractive economically only when the equivalent of a million or more voice circuits were to be carried. By using many laser beams in a one-lens sequence a vast communication potential could be provided at excellent economy. This art remains on tap for future use, should the need arise.

In parallel with research on optical beam guides, alternative laser-waveguiding techniques were sought in laboratories in many countries. A specific root for fiber transmission as we now know it is to be found in work at Standard Telecommunications Laboratories in England (Kao and Hockham, 1966). Although the best existing fibers had losses greater than 1000 dB/km, Kao and Hockham speculated that losses as low as 20 dB/km should be achievable and they further suggested that such fibers would be useful in telecommunications. No materials expert is on record as predicting that losses as low as 20 dB/km would be achieved, but many groups began research to explore this potentiality. Both multicomponent glass and high-silica glass systems were explored. Multicomponent glasses offered the attractions of lower working temperatures, thereby lessening the chances of contamination, but purification of the starting materials and homogeneity of the glass proved to be major technical obstacles. Nevertheless, impressive progress was made, especially in Japan, England, and the United States. In Japan, the Nippon Sheet Glass Co. and Nippon Electric Co. joined forces to develop the fiber they later called "Selfoc." It was the first communications fiber, one with a graded index profile but with losses above 100 dB/km (Uchida *et al.*, 1969). In England, work was coordinated by the British Post Office and included work in universities as well as in industry. Painstaking efforts in all three countries eventually led to multicomponent glass fibers with losses in the range 4–7 dB/km, truly remarkable achievements (see Chapter 8).

Meanwhile, in the United States, Corning Glass Works was investigating high-silica glasses for optical fibers, and Bell Laboratories started paying more attention to this system after the difficulties of working with the multicomponent glasses became more apparent. Corning was the first to announce a low-loss fiber, produced by a group working with R. D. Maurer, in 1970 (Kapron *et al.*, 1970); the fiber was single-mode, hundreds of meters long, and had losses under 20 dB/km. Later, this group succeeded in reducing the losses to less than 4 dB/km. In brief, the method developed by Corning, the so-called "soot method," was to form a pure or doped high silica powdery glass deposit around a cylindrical mandrel by flame pyrolysis of the constituent gases. This deposit was then consolidated into a structurally homogeneous glass by subjecting it to a carefully controlled time and temperature cycle. After removal of the central mandrel, this consolidated glass preform could be collapsed into a solid rod and drawn into a fiber. Since the composition of the constituent gases could be varied at will with time during the deposition, a radial variation in the refractive index of the deposit could be introduced. Starting with germania-doped silica deposits and gradually reducing the germania content resulted in a decreasing refractive index outward along the radius from the central axis of the preform. It is one of the kindnesses of nature, however, that as the preform is drawn into a fiber, the cylindrical symmetry of the preform is maintained so that the radial distribution of the refractive index in the fiber is a reproduction in miniature of that deliberately introduced into the preform.

The work in high-silica glasses at Bell Laboratories got its first big impetus from the discovery that suitably heat-treated, boron-doped silica could have a refractive index less than that of pure silica alone (van Uitert *et al.*, 1973). It had been recognized, following Kao, that on account of it having the lowest known optical loss, pure silica was potentially a very attractive material for use as the core of an optical waveguide since the core carries most of the optical energy. What was needed was a cladding material with lower refractive index, a material that resulted from the borosilicate discovery. The scientists at Bell Labs explored various embodiments of the combination of pure silica and boron-doped silica, but, working independently, focused on an approach somewhat different from that which Corning had developed. Starting with a commercial quality fused silica tube, the constituent gases of the desired glass deposit were fed into the inside of the tube where they were reacted at a suitable temperature to deposit the glass. As in the Corning process, the gas constituents could be programmed so as to introduce any desired radial distribution in the refractive index of the glass deposit. After collapsing the tube into a solid rod, a fiber could be drawn. A major breakthrough in making this process more practical occurred when MacChesney speeded up the pyrolytic deposition process by traversing an external flame along the tube—the modified chemical vapor deposition (MCVD) process (MacChesney *et al.*, 1974a,b).

Using the MCVD process, fibers that had transmission losses equal to, or better than, those obtained by Corning were fabricated (French *et al.*, 1974), while more recently, losses under 1 dB/km have been obtained by a Japanese group (Horiguchi and Osanai, 1976). The MCVD process has been adopted, in fact, by many other laboratories, but exploration of other processes continues.

Thus, in the exploration and race for practical and economic optical fibers, we see a sequence of international flavor, with contributions from many people in many places, building on previous work in the best scientific tradition. It is of interest to pause in this introductory narrative to summarize pictorially the progress that has been made in two key pacing items in the early phase of optical communications development. Though oversimplified, Fig. 1.1 gives some idea of the advances that were made in reducing the optical loss of fibers and in increasing the mean time to failure of semiconductor lasers. The approximate levels that had to be reached before optical communications systems could be taken seriously are also indicated. Fiber losses of less than a few decibels per kilometer are necessary for repeaterless runs of a few kilometers or more, while the mean time to failure for lasers has to be of the order of 10^6 hours if the actual failure rate due to the statistical spread of individual lifetimes is to be tolerable.

Low attenuation is not the only requirement put on a fiber lightguide—it must have low dispersion as well if a light pulse is not to be broadened out of all recognition in a relatively short transmission distance. Such broadening, which limits the bandwidth of the transmission medium, can be minimized by careful control of the radial distribution of the refractive index, a capability of both the Corning and Bell processes noted above.

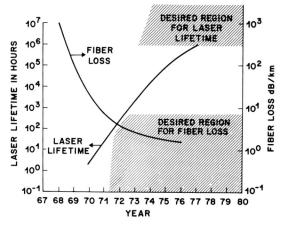

Fig. 1.1 Curves illustrating the progress that has been made regarding two key pacing items: optical loss in the fiber and semiconductor laser lifetime. Without fiber losses less than a few decibels per kilometer, and laser lifetimes of a million hours or more, optical communications systems would hardly be competitive with conventional transmission systems. The fiber target was reached in the early 1970s, followed about 5 years later by achievement of the laser target.

The drawn fiber must also have adequate mechanical strength. Immediately after they are drawn, fibers have very high strength, but their strength rapidly deteriorates if their surfaces are left unprotected. Much remains unknown about the detailed mechanisms of glass fiber strength degradation, but it is found that the strength deteriorates much more slowly if the surfaces are coated with suitable substances immediately after drawing. Intensive studies at Bell Labs and elsewhere have led to substances, and to methods of applying them in-line during fiber drawing, that preserve the high strengths needed for the subsequent mechanical integrity and robustness of practical optical cables.

The successful demonstration of a low-loss optical fiber in the laboratory is a far cry from the practical implementation of optical cables in the field. It is essential that practical methods be found for splicing fibers and cables, methods that can be used under the extremely adverse physical conditions not uncommon in the working communications system. Thus, an important segment in recounting the development phase of optical fiber systems concerns splicing techniques. Work at various laboratories has led to solutions to these problems, as well as to designs for fiber and cable connectors. It remains to be seen which of these various designs will prove the most efficient, convenient, and cost-effective, but that connections and splices showing losses of only about 0.1 dB can be repeatedly made in the laboratory seems no mean feat.

Equally important is the physical design of the cables themselves. Cables containing up to 144 optical fibers have been developed at Bell Labs, for example, with all the fibers packed in plastic ribbons and housed within a reinforced plastic sheath in such a way that the fibers are not strained either when the cable is being pulled through conduits or after installation. The communications capacity of such a cable is equivalent to about 45,000 two-way voice channels with the electronics initially developed, but it has a potential for greater capacity. Some idea of the enormous traffic potential of optical fiber systems is given in Fig. 1.2 which compares the optical cable with various other Bell System transmission media with equivalent capacity.

1.3 OPTICAL FIBER COMMUNICATIONS SYSTEMS

Once one has available a low-loss optical fiber waveguide—now frequently abbreviated "lightguide"—one considers its usefulness wherever wire pairs or coaxial cables are used. Systems were initially conceived as using single-mode, high-capacity fibers in conjunction with laser sources. It soon was recognized, however, that highly multimode fibers in combination with LED (or laser) sources had an important place also. The single-mode form offers the largest bandwidth capability (a bit-rate repeater-spacing product near 50,000 Mbit. km), but requires a single-

COMPARISON OF WESTERN ELECTRIC COPPER WIRE CABLES WITH LIGHTGUIDE CABLE
Digital Alternatives for Metropolitan Trunking 2-way Voice Circuits

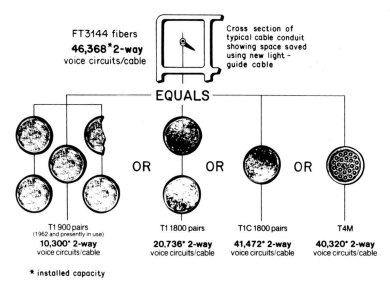

Fig. 1.2 This illustration gives some idea of the enormous communications capacity of optical cables compared with copper wire pair or coaxial cables. For definitions of the Bell System FT3144, T1, T1C, and T4M designations, see Dorros (1977).

mode laser and necessitates micron-scale precision in making splices. Multimode fibers are required when LEDs are used as sources in order to transmit significant power. They require far less exacting splicing tolerances and, by sophisticated design of the transverse index profile, theoretically can be given a bandwidth capability of up to 14,000 Mbit. km, of which 1000 Mbit. km has been realized in the laboratory.

Compared with wire-line alternatives, the advantages of fibers extend well beyond lower losses and wider bandwidths. Their small size (hair-like) can be important in crowded city ducts. Their freedom from ground-loops, inductive pickup, electromagnetic radiation, cross talk, and low-temperature sensitivity are general assets that can be advantageous in "special purpose" applications.

This variety of virtues leads to serious consideration of lightwave transmission systems for many parts of the telecommunications plant. These include the connection from subscriber to nearby central office, trunking facilities in metropolitan areas, intercity links, and undersea intercontinental cables. Further use will appear internal to other electronic equipment—as internal links in switching machines or in computers, for example. "Wiring" for communication within a building may one day be done with fibers. All of these potential applications bring a breadth to optical-fiber telecommunications far greater than the early optical communication concepts. Fibers are expected to be used with modest (i.e., ~1 Mbit/sec) information rate per fiber as well as with large information rates. This breadth makes lightguide transmission a topic of importance as well as of interest to a great many people.

As a natural step in the transition from research to commercial use of this new technology, Bell Labs has planned and carried out a system experiment in Atlanta, and now (as this is written) is in the process of evaluating a lightwave system carrying customers' voice, data, and video signals in the Chicago business area.

The Atlanta experiment (Jacobs, 1976) used an exploratory design of a system aimed at the metropolitan area trunking; it used a 44.7 Mbit/sec pulse rate to carry 672 voice circuits on a single fiber about 100 μm in diameter. The cable, containing 144 fibers, was fabricated by the Western Electric Co. in a 1-km length and cut to 650 m to fit the length of a standard underground plastic duct located adjacent to the Bell Labs building. By series hookup of several fibers, error-free transmission over a span between repeaters of 10.9 km was demonstrated, and the component performances justified a system design span of about 7 km. These results were so encouraging that a similar cable was fabricated, and the same lightwave repeater equipment was installed in Chicago to get further environmental and field-handling experience with operating company people. The 1.5 miles of 24-fiber cable was successfully installed and is now in operation (Mullins, 1977).

Other systems trials are currently under way or being planned by other organizations, both in the United States and abroad.

In California the General Telephone and Electronics Corporation installed a cable containing six fibers and a number of wire pairs from a telephone center in Long Beach to another telephone office in Artesia, a distance of about 9 km (Basch and Beaudette, 1977). Two intermediate repeaters were employed. The fibers in this system were manufactured by Corning Glass Works and were fabricated into the cable by General Cable Corporation. Two of the fibers carry commercial traffic in the form of a standard digital carrier signal at 1.544 Mbit/sec. This installation, like the Bell System installation in Chicago, was carried out by regular operating company personnel, and was intended to provide experience and demonstrate feasibility. Neither the Chicago nor the Long Beach system can be considered to be a prototype of a system ready for large-scale manufacture.

In England the British Post Office sponsored a "feasibility trial" using a 2-fiber cable, 13 km long in 1-km sections, installed in ducts and operating at a line rate of 8.448 Mbit/sec (Brace and Cameron, 1977). The cable was fabricated by British Insulated Callender's Cables Limited using Corning

Glass Works fibers. Also in England Standard Telecommunications Laboratories carried out a "field demonstration" over a 9-km total span with two intermediate repeaters located in telephone buildings; the line rate was 140 Mbit/sec and the cable was fabricated by Standard Telecommunications and Cables Corporation using fibers manufactured by Corning Glass Works (Hill *et al.*, 1977).

In Japan a vigorous research and development program is also going forward on lightwave communications via fiberguide cables (Kawahata, 1977). Several experimental systems were installed to explore the advantages of lightguides for communication along high voltage power transmission lines. In addition the Japanese government has funded a program called HI-OVIS (Higachi-Ikoma Optical-Visual Information System). The plan is to bring two-way video, data, and voice communications into a series of homes via fiber lightguides. This installation was completed in 1978 and is being used experimentally; up to the present the "customers" use the service without charge.

In Germany, Siemens and Halske has installed a 2.1-km-long cable including 10 Corning fibers and has operated it for over a year carrying digital speech (2.048 Mbit/sec) or video (FM on a 20-MHz subcarrier) (Bark *et al.*, 1977).

This sampling of early experience carries one consistent theme: serious efforts to utilize lightguide cables in a field environment have met with success and revealed no serious problems.

1.4 OTHER ASPECTS OF THE INNOVATION OF OPTICAL COMMUNICATIONS SYSTEMS

Generally, it is not sufficient simply to demonstrate that information can be transmitted reliably at adequate rates over optical systems in order for investment in such systems to continue. It must be demonstrated also that they are compatible with existing electrical communications systems with which they must interconnect. They should operate at compatible frequencies, voltages, insertion losses, and delays.

But in addition, optical systems must offer economic advantages. In view of the great reduction in material involved, as evident in Fig. 1.2, the optical cable itself might cost considerably less to make than its metallic counterparts for comparable channel capacity. But Fig. 1.2 also dramatizes one of the underlying dilemmas in the contemporary development of electronics technology; namely, the information processing and carrying capacity per unit volume of modern electronics technology is now so vast that unless the *demands* for such capacity are correspondingly large, a relatively small production volume of electronics devices suffices. For example, if *all* of the Bell System's trunk network, which at present uses microwave radio, coaxial cables, and wire pairs, were to be converted to optical systems, the traffic could, in principle, be handled by about 10 million dollars' worth of lasers, a market certainly far too small to justify the laser development costs alone. Instead, the development costs of the optical devices and the cables are more appropriately viewed from the perspective of overall system economics, particularly the savings that accrue to the communications system by not having to install so much conventional equipment, and by the income from new services that optical communications make practical.

But these kinds of problems are familiar; they seem to occur repeatedly, whenever new technology is being introduced that supplants the old. There is growing confidence that the birth pains will pass and that optical communications will become established as a major and healthy technological force. And once established, optical technology is unlikely to remain as just another means of transmitting signals from here to there using sources, fibers, and detectors. Instead, the optical domain can be expected to expand so that more and more of the signal processing can be done optically if the technical needs and economics warrant. Already, exploratory research has indicated some of the directions this expansion might follow, directions where functions previously performed electrically might be done optically. These functions might include amplification, signal regeneration, modulation and mixing, switching, memory, and so on. And perhaps paralleling the path along which electrical semiconductor electronics has evolved, we may see more and more of these functions integrated on to a common substrate, perhaps even performed on single "chips" as integrated optical circuits. Research on individual elements of such circuits is already being actively and widely pursued, but this subject is somewhat beyond the scope of the present book.

In the following paragraphs we provide the reader with an expanded index to the book. It will be useful for many readers to read selectively among the chapters, and we hope to facilitate this by providing some commentary here.

Chapter 2 provides a technically oriented introduction to telecommunications fibers. The various fiber types are identified, and the key performance criteria are explained. The evolution of ideas which the research process yielded are reviewed.

Chapter 3 gives an in-depth theoretical treatment of electromagnetic wave propagation in fibers, within the bounds of linearity with respect to wave amplitude. This serves as a foundation for later chapters, which treat important properties from an engineering or design viewpoint.

Chapter 4 covers the delay-distortion properties of the important fiber types, and quantitatively shows how the dispersion is influenced by the material characteristics and fiber structure.

Chapter 5 describes the nonlinear wave propagation which occurs at large wave amplitude and gives the power threshold at which stimulated Brillouin, stimulated Raman, and self-phase modulation will occur.

Chapter 6 brings together the engineering judgments necessary to arrive at a fiber design to meet a certain requirement.

Chapter 7 reviews the pertinent basic understanding of the optical properties of glasses, the causes of optical absorption and scattering losses, the impurities that must be guarded against and the chemical methods for detecting them, and methods of measuring the optical properties of ultralow loss bulk samples of glass.

Chapter 8 describes the various methods that have been developed for preparing glass preforms prior to drawing fibers. Special emphasis is given to the chemical vapor deposition processes which have proved so adaptable and successful.

Chapter 9 discusses the various techniques that have been used for drawing fibers and for achieving the required dimensional tolerances.

Chapter 10 is concerned with the theory and practice of applying protective polymeric coatings to fibers as they are drawn in order to preserve their strength and minimize environmental attack.

Chapter 11 describes in some detail the techniques which were originated to optically characterize communication fibers. New highly precise measurements have made possible the remarkable advance in fiber transmission technology by quantitatively showing the results of changes in materials, process, or structure of the fiber.

Chapter 12 provides information on the mechanical strength and fatigue properties of fibers drawn under various conditions and with different protective coatings.

Chapter 13 gives some of the engineering and design considerations involved in arriving at a lightguide cable design. Several designs, for a few fibers or for a hundred, are illustrated.

Chapter 14 concerns itself with splicing groups of fibers, the type of problem encountered in joining two sections of multifiber cable.

Chapter 15 concerns itself with the demountable optical connector which is an essential element of a repeater station. These connectors are the terminals of rearrangeable optical patch cords or of plug-in electronics boards containing lasers or optical detectors.

Chapter 16 describes the lightwave generators which are used in transmitters for lightguide transmission links. The properties of LEDs and lasers that relate closely to the telecommunications function are given in some detail.

Chapter 17 describes the state of the art on techniques for modulating lightwave generators with telecommunications signals.

Chapter 18 covers photodetectors used for lightwave transmission. The choice of detector materials is discussed. Device characteristics which are important to the transmission system are described, and device design interrelations are identified.

Chapter 19 describes quantitatively the device performance interactions which determine the receiver signal-to-noise performance.

Chapter 20 relates to the system interactions between the various components, and describes some of the trade-offs the system designer makes in arriving at an effective overall design.

Chapter 21 identifies in broad terms the numerous potential applications for lightguide transmission, ranging from short on-premises links to undersea systems for intercontinental communication. A brief indication is given of the current status.

REFERENCES

Alferov, Zh. I., and Kazarinov, R. F. (1963). Author's certificate No. 181737, Claim no. 950840 of March 30.

Alferov, Zh. I., Andreev, V. M., Garbuzov, Yu. V., Zhilyaev, E. P., Morozov, E. P., Portnoi, E. L., and Triofim, V. G. (1971). Investigation of the influence of the AlAs–GaAs heterostructure parameters on the laser threshold current and realization of continuous emission at room temperature. *Sov. Phys.—Semicond. (Engl. Transl.)* **4,** 1573; translated from *Fiz. Tekh. Poluprovodn.* **4,** 1826 (1970).

Bark, P., Boscher, G., Gier, J., Goldmann, H., and Zeidler, G. (1977). Installation of an experimental optical cable link and experiences obtained with the transmission of TV- and telephone signals. *Conf. Proc. Eur. Conf. Opt. Commun., 3rd, 1977* p. 243.

Basch, E. E., and Beaudette, R. A. (1977). The GTE fiber optic system. *Nat. Telecommun. Conf. Rec., 1977* p. 14:2.

Bell, A. G. (1880). Selenium and the photophone. *Electrician* **5,** 214.

Berreman, D. W. (1964). A lens or light guide using convectively distorted thermal gradients in gases. *Bell Syst. Tech. J.* **43,** 1469.

Brace, D., and Cameron, K. (1977). BPO 8448 Kbit/s optical cable feasibility trial. *Conf. Proc. Eur. Conf. Opt. Commun., 3rd, 1977* p. 237.

Burrus, C. A., and Miller, B. I. (1971). Small-area, double-heterostructure aluminum-gallium arsenide electroluminescent diode sources for optical-fiber transmission lines. *Opt. Commun.* **4,** 307.

Chu, T. S., and Hogg, D. C. (1968). Effects of precipitation on propagation at 0.63, 3.5, and 1.6 microns. *Bell Syst. Tech. J.* **47,** 723.

DeLoach, B. C., Hakki, B. W., Hartman, R. L., and D'Asaro, L. A. (1973). Degradation of CW GaAs double-heterojunction lasers at 300 K. *Proc. IEEE* **61,** 1042.

Dorros, I. (1977). "Engineering and Operations in the Bell System." Bell Laboratories.

French, W. G., MacChesney, J. B., O'Connor, P. B., and Tasker, G. W. (1974). Optical waveguides with very low losses. *Bell Syst. Tech. J.* **53,** 951.

Gordon, E. I., and White, A. D. (1964). Single frequency gas lasers at 6328 Å. *Proc. IEEE* **52,** 206.

Hall, R. N., Fenner, G. E., Kingsley, J. D., Soltys, T. J., and Carlson, R. O. (1962). Coherent light emission from GaAs junctions. *Phys. Rev. Lett.* **9,** 366.

Hartman, R. L., Schumaker, N. E., and Dixon, R. W. (1977). Continuously operated (Al, Ga) as double-heterostructure lasers with 70°C lifetimes as long as two years. *Appl. Phys. Lett.* **31,** 756.

Hayashi, I., Panish, M. B., and Foy, P. W. (1969). A low-threshold room temperature injection laser. *IEEE J. Quantum Electron.* **QE-5**, 211.

Hayashi, I., Panish, M. B., Foy, P. W., and Sumski, S. (1970). Junction lasers which operate continuously at room temperature. *Appl. Phys. Lett.* **17**, 109.

Hill, D. R., Jessop, A., and Howard, P. J. (1977). A 140 Mbit/s field demonstration system. *Conf. Proc. Eur. Conf. Opt. Commun., 3rd, 1977* p. 240.

Horiguchi, M., and Osanai, H. (1976). Spectral losses of low-OH-content optical fibers. *Electron. Lett.* **12**, 310.

Jacobs, I. (1976). Lightwave communications passes its first test. *Bell Lab. Rec.* **54**, 290.

Javan, A., Bennett, W. R., Jr., and Herriott, D. R. (1961). Gas optical lasers. *Phys. Rev. Lett.* **6**, 106.

Kao, K. C., and Hockham, G. A. (1966). Dielectric-fiber surface waveguides for optical frequencies. *Proc. IEEE* **133**, 1151.

Kapron, F. P., Keck, D. B., and Maurer, R. D. (1970). Radiation losses in glass optical waveguides. *Appl. Phys. Lett.* **17**, 423.

Kawahata, M. (1977). Fiber optics application to full two-way CATV system—Hi-OVIS. *Natl. Telecommun. Conf. Rec. 1977* p. 14:4.

Kimerling, L. C., Petroff, P. M., and Leamy, H. J. (1976). Injection-stimulated dislocation motion in semiconductors. *Appl. Phys. Lett.* **28**, 297.

Kompfner, R. (1965). Optical communications. *Science* **150**, 149.

Kompfner, R. (1972). Optics at Bell Laboratories—optical communications. *Appl. Opt.* **11**, 2412.

Kressel, H., and Nelson, H. (1969). Close confinement gallium arsenide p–n junction lasers with reduced optical loss at room temperature. *RCA Rev.* **30**, 106.

Kroemer, H. (1963). A proposed class of heterojunction injection lasers. *Proc. IEEE* **51**, 1783.

MacChesney, J. B., O'Connor, P. B., DiMarcello, F. V., Simpson, J. R., and Lazay, P. D. (1974a). Preparation of low loss optical fibers using simultaneous vapor deposition and fusion. *Proc. Int. Congr. Glass, 10th, 1974* Vol. 6, p. 40.

MacChesney, J. B., O'Connor, P. B., and Presby, H. M. (1974b). A new technique for the preparation of low-loss and graded-index optical fibers. *Proc. IEEE* **62**, 1280.

Maiman, T. H. (1960). Stimulated optical radiation in ruby. *Nature (London)* **6**, 106.

Miller, S. E. (1970). Optical communications research progress. *Science* **170**, 685.

Mullins, J. H. (1977). A Bell system optical fiber system—Chicago installation. *Nat. Telecommun. Conf. Rec. 1977* p. 14:1.

Nathan, M. I., Dumke, W. P., Burns, G., Dill, F. H., Jr., and Lasher, G. (1962). Stimulated emission of radiation from GaAs p–n junctions. *Appl. Phys. Lett.* **1**, 62.

Petroff, P., and Hartman, R. L. (1974). Rapid degradation phenomenon in heterojunction GaAlAs-GaAs lasers. *J. Appl. Phys.* **45**, 3899.

Schawlow, A. L., and Townes, C. H. (1958). Infrared and optical masers. *Phys. Rev.* **112**, 1940.

Shen, C. C., Hsieh, J. J., and Lind, T. A. (1977). 1500-hr. Continuous CW operation of double-heterostructure GaInAsP/InP lasers. *Appl. Phys. Lett.* **30**, 353.

Stone, J., and Burrus, C. A. (1978). Nd:YAG self-contained LED-pumped single-crystal fiber laser. *Fiber Integrated Opt.* **2**, 19.

Uchida, T., Furukawa, M., Kitano, I., Koizumi, K., and Matsamura, H. (1969). A light-focusing guide. *IEEE J. Quantum Electron.* **qe-5**, 331.

van Uitert, L. G., Pinnow, D. A., Williams, J. C., Rich, T. C., Jaeger, R. E., and Grodkiewicz, W. H. (1973). Borosilicate glasses for fiber optical waveguides. *Mater. Res. Bull.* **8**, 469.

Chapter 2

Objectives of Early Fibers: Evolution of Fiber Types

ENRIQUE A. J. MARCATILI

2.1 PURPOSE OF THE CHAPTER AND RELATION TO THE REST OF THE BOOK

One purpose of this chapter is to answer qualitatively the following questions. What are optical fibers? How do they work? What are the limitations induced by loss and dispersion? Another purpose, and the most important one, is to show how those evolving limitations have controlled the conception and development of some fiber types as well as the history of fiber objectives.

The chapter is directed to the beginner in fiber optics who may find that a light-touch view of the evolution of fibers for optical communication is a useful introduction to the quantitative and far more detailed treatment to be found in Chapters 3–6. Pertinent references can be found in later chapters.

In Section 2.2 different types of fibers and basic ideas about guidance are introduced; loss and dispersion, the most important characteristics of fiber transmission, are the subjects of the two following sections; all of this information serves as ground work to Section 2.5 where the objectives and evolution of fiber types are described.

2.2 GUIDANCE IN OPTICAL FIBERS

2.2.1 Cross Sections and Index Profiles of Fibers

Fibers for optical communication are long flexible filaments of small cross section, comparable to a human hair, and are made of transparent

17

dielectrics whose function is to confine and guide visible and infrared light over long distances.

Guides with many cross sections and refractive-index profiles have been investigated; the most important are shown in Fig. 2.1 together with typical dimensions. A plastic and lossy jacket, commonly applied to the outside of the fiber to prevent cross talk with other guides and to keep the fiber strong by preventing chemical and abrasive attack on its surface, is not shown in these cross sections.

The step-index fiber (Fig. 2.1a) consists of a core of uniform refractive index n_1 made either of a highly transparent solid material such as high-silica content glass, multicomponent glass, or a low-loss liquid such as tetrachloroethylene or hexachlorobuta-1,3-diene. The cladding intimately surrounding the core is also a dielectric of slightly smaller index $n_1(1 - \Delta)$. However, since most of the electromagnetic power is bound to travel along the core, the cladding material, unlike that of the core, may be lossy, hundreds of decibels per kilometer, without substantially increasing the attenuation of the transmitted signal. The cladding is made of high-silica glass, multicomponent glass, or plastic.

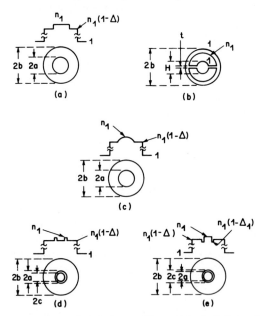

Fig. 2.1 Fiber cross sections, their index profiles and typical dimensions (a, b, c, t, H are in microns): (a) Step-index fiber. Single mode: $a \cong 5$, $b \cong 50$, $\Delta \cong 0.002$. Multimode: $a \cong 25, b \cong 50, \Delta \cong 0.01$. (b) Single-material fiber. Single mode: $b \cong 50, t \cong 5, H \cong 10$. Multimode: $b \cong 50$, $t \cong 3$, $H \cong 30$. (c) Graded-index fiber. Single mode: $c \cong 5$, $b \cong 50$, $\Delta \cong 0.002$. Multimode: $a = 25, b \cong 50, \Delta \cong 0.01$. (d) Ring fiber. Single mode: $a \cong 10, b \cong 50$, $c \cong 7$, $\Delta \cong 0.002$. (e) W fiber. Single mode: $a \cong 9$, $b \cong 50$, $c \cong 19$, $\Delta \cong 0.001$, $\Delta_1 \cong 0.005$.

Even if it does not look like one, the single-material fiber (Fig. 2.1b) is equivalent to a step-index fiber because the thin slab and air surrounding the central core behave like a uniform cladding of index slightly smaller than n_1. Core, slab, and protective external tube have been made exclusively of fused silica.

In graded-index fibers (Fig. 2.1c) the refractive index of the core gradually decreases from the center toward the core–cladding interface, and later we will see why this guiding medium is so important for wide-band transmission. Either high-silica content or multicomponent glasses are used to make these guides.

The number of guided modes in fibers is proportional to the core cross section and to the index difference $n_1\Delta$ between axis and cladding. Therefore, by reducing their product sufficiently, the fiber can be made to guide a single mode. Typical dimensions and values of $n_1\Delta$ are shown in the captions of Figs. 2.1a, b, and c.

Ring fibers and W fibers (Figs. 2.1d and e) have three regions of uniform index. In the ring fiber the middle region has the largest index while in the W fiber the highest and lowest indices are in the inner and middle regions, respectively.

2.2.2 Guidance in Single and Multimode Fibers

The how and the why guidance occurs in different fiber types, together with a qualitative discussion of the parameters that influence their guiding properties, are the objects of this subsection.

All dielectric fibers guide because of total internal reflection. Consider first the step-index fiber Fig. 2.1a, made of lossless dielectrics and having a uniform core index n_1 and a smaller cladding index $n_1(1 - \Delta)$. For most practical fibers the index difference is small; in fact Δ varies between a few parts per thousand and a few parts per hundred. In the meridional cutaway view of such a fiber, Fig. 2.2, consider a ray traveling in the plane of the drawing and making an angle θ with the axis. If $\theta \leq (2\Delta)^{1/2}$, the ray experiences total internal reflection as it grazes the core–cladding interface, and zigzags repeatedly along the fiber without attenuation, always maintaining the same angle θ with the axis. On the other hand a meridional ray making an angle larger than $(2\Delta)^{1/2}$ with the axis (dotted ray) is partially reflected at the core–cladding interface, and the rest is refracted

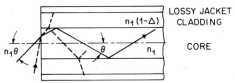

Fig. 2.2 Meridional ray trajectories in a multimode step-index fiber.

and absorbed in the lossy jacket. For $\Delta = 0.01$, the critical angle $(2\Delta)^{1/2}$ that separates guided from unguided rays is 0.14 rad.

At the square-cut ends of the fiber the guided rays refract and make an angle $n_1\theta$ with the axis. Conversely, meridional rays impinging on the core at the end of a fiber and making an angle smaller than $n_1(2\Delta)^{1/2}$ with the axis are guided by the fiber. It turns out that this result applies not only to meridional rays but to all rays; therefore the angle $n_1(2\Delta)^{1/2}$ is a measure of the light-collecting ability of the guide and is called the numerical aperture (NA) of the fiber.

Most rays, though, do not follow a meridional path but rather a helical path (Fig. 2.3). The projection of the trajectory of one of these skewed guided rays on a plane perpendicular to the axis is also shown.

Three types of rays can be distinguished from the point of view of containment within the core: guided, leaky, and unguided. To characterize them let us first remember that a ray is completely defined by the angle θ that the ray makes with the guide axis and by the angle of incidence ϕ at the core–cladding interface. Guided rays are totally reflected at the core–cladding interface. Rays with $\theta \le (2\Delta)^{1/2}$ belong to this class. Rays with $\theta \ge (2\Delta)^{1/2}$ and $\phi \ge (\pi/2) - (2\Delta)^{1/2}$ are leaky rays. If the core–cladding interface were flat these rays would be totally reflected but since it is not these rays are only partially reflected. Some of these rays may have small losses, much smaller than those of the third kind, unguided rays, for which $\phi \le (\pi/2) - (2\Delta)^{1/2}$.

For long-distance transmission only guided rays are important. They are responsible for most of the transmission phenomena since the others are substantially attenuated. However, in the process of measuring short, laboratory-size fibers, leaky modes, if ignored, may lead to errors.

The guided ray depicted in Fig. 2.3 is completely contained within two cylindrical coaxial surfaces, the so-called caustic surfaces where the total reflection of the ray takes place. One of them is the core–cladding interface, the other depends on the values of the angles θ and ϕ. The projections of these surfaces on a plane perpendicular to the fiber axis are the circles A and B.

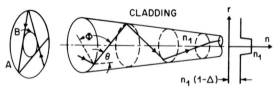

Fig. 2.3 Skew ray in a step-index fiber; Φ is its angle of incidence; θ is its angle with the axis. A and B are end views of caustic surfaces. Dash-circles in external caustic surface help to visualize ray trajectory.

All the results discussed in this subsection are independent of the core radius a, of the cladding thickness t, and of the free space wavelength λ. This lack of influence stems from the fact that the ray picture is strictly valid for $\lambda = 0$. For nonzero λ the ray picture becomes approximate but still extremely useful when the number of guided modes $(2\pi n_1 a/\lambda)^2 \, 2\Delta$ is large compared to unity.

The exact modal or wave picture results from solving Maxwell's equations. Let us review some discrepancies between the two approaches.

While the ray picture says that a guided ray can make any angle smaller than $(2\Delta)^{1/2}$ with the fiber's axis, the wave picture says that only a discrete number of modes (or their equivalent rays) are indeed permitted. This discrepancy becomes quite significant when the number of guided modes is small. In fact, for $a < 0.27\lambda/n \, \Delta^{1/2}$ the wave approach predicts that the fiber can guide a single mode in each polarization. For $\Delta = 0.001, n = 1.46$ and $\lambda = 1 \ \mu m$ the fiber is monomode if $a < 5.85 \ \mu m$.

The modal analysis also shows that between the caustic surfaces the electromagnetic field varies in an oscillatory manner along the radial, azimuthal, and axial directions; outside the caustics the field components decay almost exponentially in the radial direction but keep on varying periodically in the other two. The speed with which field decays inside the internal caustic and outside the external one varies with the mode. It is because of this decay that the cladding can have a finite thickness without affecting substantially either the field or the attenuation of most of the guide modes. Indeed, only modes close to cutoff $[\theta \cong (2\Delta)^{1/2}]$ have a slow decay in the cladding and consequently are attenuated by the lossy jacket.

Another discrepancy between ray and wave optics appears when the fiber is bent uniformly with a constant radius of curvature. Ray optics predicts erroneously that in spite of the curvature some rays can experience total internal reflection and consequently remain guided without any loss. Wave optics, on the other hand, correctly predicts that every mode of the curved guide exhibits some radiation loss. Qualitatively, this loss can be understood from the fact that the wave front of a mode moves with constant angular velocity and consequently, far away from the center of curvature, there are portions of the decaying field in the cladding that would be traveling faster than the speed of light. This is impossible and consequently that decaying field is radiated.

Typical fibers can negotiate radii of curvature of the order of 1 cm with negligible radiation loss. A more serious problem, which will be considered later, is that of random bends of the fiber axis.

Single-material fibers (Fig. 2.1b) are also step-index fibers. The central core has constant index n_1 and is surrounded by the low-index air and the supporting slabs. Only modes whose field components in the slabs decay

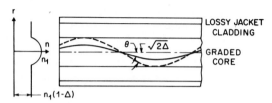

Fig. 2.4 Trajectories of meridional rays in a graded-index fiber.

exponentially away from the core remain guided. The slab and the air behave like an equivalent uniform cladding of index $n_1(1 - \Delta_e)$, where

$$\Delta_e = \tfrac{1}{8}(\lambda/n_1 t)^2.$$

Independently of the core size, single mode guidance is achieved if the dimensions t and H of the fiber (see Fig. 2.1b) satisfy the condition

$$t/H \geq 0.5.$$

What are the ray trajectories in graded-index fibers? Unlike the case of
step-index fiber, meridional rays do not follow zigzag paths with all
turning points at the core–cladding interface, but rather quasi-sinusoidal
paths (Fig. 2.4), with turning points that depend on the angle θ and the
index profile. As in the case of the step-index fibers, skewed rays in
graded-index fibers follow quasi-helical paths (Fig. 2.5). They are also
contained within two cylindrical caustic surfaces but for most rays the external caustic does not coincide with the core–cladding interface.

In the graded-index fibers there are also three kinds of rays, but again
the guided ones are by far the most important.

To discuss the NA and the number of guided modes of these fibers it is
convenient to single out a particular family of index profiles whose importance will become apparent later on when dispersion will be considered.
The index decays monotonically in the radial direction following the
power law

$$n = n_1[1 - \Delta(r/a)^g],$$

where g is a number that defines the profile. For $g = \infty$ the fiber has a
step-index profile. For $g = 2$ the index is parabolic.

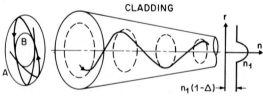

Fig. 2.5 Skew ray in a graded-index fiber. A and B are end views of caustic surfaces.
Dash-circles on external caustic surface help to visualize ray trajectory.

The NA of a fiber with this graded index is $n_1(2\Delta)^{1/2}$ independent of g. The number of guided modes is

$$N = \frac{g\,\Delta}{2+g}\left(\frac{2\pi n_1 a}{\lambda}\right)^2.$$

A step-index fiber ($g = \infty$) carries only twice as many modes as a parabolic index fiber ($g = 2$) if both fibers have the same NA and core diameter.

The ring fiber (Fig. 2.1d) guides because the middle ring has a larger index than the surrounding medium and the electromagnetic power is mostly concentrated in it.

The W fiber (Fig. 2.1e) guides very much like a step-index fiber with the same core size and NA. However, modes that in the step-index fiber are leaky or unguided, in the W fiber must tunnel first through the lower index ring, and consequently they attenuate more slowly.

Like the single-material fiber and the ring fiber, the W fiber can be dimensioned to guide a single mode with a beam size larger than that of the step-index guide with the same NA. This may relieve the required splicing tolerance.

Up to now we have seen how light is guided within a fiber. In the next two subsections we will see how that light is either deliberately coupled to and from fiber-end devices or is coupled to other fibers introducing unwanted cross talk.

2.2.3 Coupling to Sources and Detectors

One important function of the fiber is to transfer as much light as possible from the source to the detector. This is achieved by minimizing the attenuation in the fiber and by optimizing the coupling between the fiber and the source, as well as the coupling between the fiber and the detector. We consider in this subsection the coupling optimizations only.

How do sources radiate in free space? Lasers operating with a single transverse mode can be made to radiate a single mode in free space. Other lasers, not so well behaved, may radiate several modes simultaneously, and spatially incoherent sources such as light-emitting diodes radiate a very large number of modes.

No matter how sophisticated the matching optics between a source and a fiber, the number of free-space modes emitted by the source that can be coupled to the fiber can not exceed the number of fiber modes. Therefore, for good coupling efficiency, incoherent sources must be used with large-core, large-NA multimode fibers, while single-transverse-mode lasers can be used either with single or multimode fibers. Similarly, multimode fibers can only be efficiently coupled to large area detectors.

2.2.4 Cross Talk between Fibers

Fibers packed in tapes or cables follow parallel paths over long distances. This directional coupler-like structure requires a very small coupling per unit length between degenerate modes (modes with identical propagation constant) to avoid intolerable cross talk. However, the coupling level can be easily kept to a safe level by increasing the separation of the fibers to decrease the intensity of the exponentially decaying overlapping field.

A second coupling mechanism between fibers may occur via a double scattering process. Imperfections in one fiber such as bending of the axis, diameter variations, bubbles, etc., can scatter power that is channeled into a second fiber via another set of scattering imperfections. This coupling mechanism can be avoided by the presence of the lossy-plastic jacket surrounding each fiber.

A more-difficult-to-control source of cross talk among fibers exists at multifiber splices, where a group of closely spaced fibers at a cable end must be aligned and joined to its counterpart in another cable end. Spacings, offsets, tilts, fiber-end-break defects, bubbles in matching fluid, etc., produce cross talk and there is no simple overall solution to these scattering mechanisms short of close tolerance control.

2.3 LOSS MECHANISMS

Transmission loss is the most important of the optical properties of a fiber. It determines to a large extent the separation between the repeaters that regenerate the transmitted signals. Consequently, the system cost is significantly controlled by loss.

Three mechanisms are responsible for transmission loss in fibers: material absorption, linear scattering, and nonlinear scattering. A description of these mechanisms, their relative importance, and techniques to reduce them are discussed in this section.

2.3.1 Material Absorption

Material absorption is a loss mechanism by which part of the transmitted power is dissipated as heat in the guide. In absolutely pure glasses the absorption in the visible and near-infrared regions of the spectrum, up to about 1.1 μm, is negligible and is mostly due to the tails of absorption peaks centered in the ultraviolet and far infrared. Above 1.1 μm, the tails of the far-infrared absorption peaks become significant and account for most of the pure glass losses [see Fig. 2.6; Origuchi and Osanai (1976)]. Glasses, however, are not pure. They frequently contain impurity ions with electronic transitions in the visible and near infrared that cause ab-

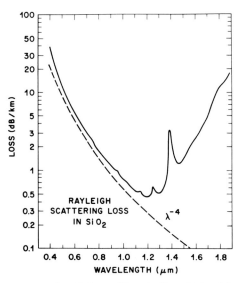

Fig. 2.6   Loss versus wavelength in multimode fiber (Origuchi and Osanai, 1976).

sorption. Typically one part per billion in weight of any of the ubiquitous ions of chromium, copper, iron, manganese, etc., in their worst valence state is enough to introduce a peak attenuation of about 1 dB/km. Another absorbing ion, one that has proven very difficult to eradicate, and the only one grossly responsible for absorption peaks in the loss versus wavelength curve of a recent low loss fiber (Fig. 2.6) is hydroxyl (OH). The fundamental stretching vibration peak of this ion occurs at approximately 2.72 μm. Its overtones appear almost harmonically at 1.38, 0.95, and 0.72 μm. Combinations between those overtones and the fundamental vibration of the SiO_2 tetrahedron appear at 1.24, 1.13, and 0.88 μm. The first five of these six "water" absorption peaks are easily seen in Fig. 2.6. Typically one part per million in weight of hydroxyl causes an attenuation of 1 dB/km at 0.95 μm. Consequently this fiber has about one part per ten million of OH. No doubt, further processing improvements will reduce this water content to even lower levels.

2.3.2 Loss Due to Linear Scattering

Linear scattering occurs when part of the power carried by one mode of the fiber is transferred linearly (proportionally to that power) into another mode. As in all linear processes no change of frequency is involved.

This coupling among modes occurs because the guide is not a mathematically perfect cylindrical structure. Imagine, for example, that a parameter of the fiber varies sinusoidally with period Λ along the fiber. This parameter can be, for example, the axis departure from a straight line, or in a straight fiber the index departure from constant along a line parallel to the axis, etc.

Then, a guided mode with period λ_g couples strongly to another mode of period along the axis λ_{gc} only if

$$\Lambda = |(1/\lambda_g) - (1/\lambda_{gc})|^{-1},$$

that is, if Λ is equal to the beat wavelength of the two modes.

The relation between Λ and the direction in which light is coupled (or scattered) can be deduced from the formula above. For light coupled in the forward direction making a small angle with the axis $\lambda_{gc} \sim \lambda_g$ and consequently Λ is large compared to λ_g. In fact, if the angle is smaller than $(2\Delta)^{1/2}$ the scattered power remains trapped within the core. The scattering occurs between guided modes, and as will be seen in the next section, the main effect of this mode mixing is to influence the dispersive properties of the fibers. On the other hand, if the scattering angle is bigger than $(2\Delta)^{1/2}$ the coupled mode is unguided, and the fiber exhibits scattering losses over and above the absorption losses.

In any multimode fiber the mechanical periodicity that separates mode mixing from scattering radiation is approximately

$$\Lambda_0 = 4a/\Delta^{1/2}.$$

This period is 1 mm for $a = 25$ μm and $\Delta = 0.01$.

Fibers in general do not exhibit a single mechanical wavelength Λ but rather a spectrum of them, and each component scatters according to the previous description. Let us consider now what types of imperfections contribute to that spectrum.

Index fluctuations of a random nature occurring on a small scale compared with the wavelength λ/n_1 have a wide spectrum and consequently scatter light almost omnidirectionally. This is called Rayleigh scattering and produces an attenuation proportional to $1/\lambda^4$. Therefore, Rayleigh scattering is strongly reduced by operating at the longest possible wavelength.

Some of these small scale dielectric fluctuations may be due to compositional variations, phase separations, strains, small bubbles, etc., that can be reduced by improving the fabrication technique. However, there are small-scale index fluctuations which are the result of freezing-in density inhomogenities due to thermal agitation at the time the fiber solidifies. This Rayleigh scattering is fundamental, cannot be eliminated, and is the limiting source of scattering loss in fibers. Theory and experiments show that for fused silica this Rayleigh limit is about 4.8, 0.8, and 0.3 dB/km at 0.63, 1, and 1.3 μm, respectively (see Fig. 2.6). Dopants such as titanium and germanium increase those losses.

Inhomogeneities having spatial periodicities in the range $\lambda/n < \Lambda < 1$ mm scatter mostly in the forward direction, and this type of radiation is called Mie scattering. Sources of these types of imperfections are, for example, core–cladding irregularities, strain, bubbles, diameter fluctuations, and most importantly, axis meandering. Scattering losses introduced by the axis meandering, are called microbending losses. They can easily be generated by the introduction of minute, uncontrolled, lateral forces on the fiber during the cabling process and may account for several decibels per kilometer of excess attenuation. These losses can be reduced by following design criteria that essentially tend to (a) increase guidance by increasing Δ, (b) stiffen the fiber by increasing the fiber's cross section, and (c) filter out the high-frequency components of the mechanical spectrum by housing each fiber in soft plastic (low Young's modulus) and surrounding it with stiffening members such as rigid (high Young's modulus) plastics, graphite, metal, etc.

2.3.3 Loss Due to Nonlinear Scattering

If the field intensity in a fiber is very high, nonlinear phenomena set in, and power from a mode can be transferred to the same or other modes traveling either in the forward or backward direction but shifted in frequency. Unlike linear scattering where the coupling coefficients are independent of the powers in the modes involved, the coupling coefficients in nonlinear scattering are functions of those powers.

Stimulated Brillouin and Raman scatterings are of the nonlinear type and as in any stimulated phenomena they have a threshold. The minimum threshold, and consequently the most critical for optical communication, occurs in a single-mode high-silica fiber transmitting monochromatic light, where Brillouin scattering (mostly in the backward direction) sets in at a threshold power

$$P_T = 8 \times 10^{-5}\alpha/w^2 \quad \text{W.}$$

In this expression α, in decibels per kilometer, is the linear attenuation coefficient of the fiber and w is the beam's full width at half-power density in microns. For a realistic single mode fiber having $\alpha = 2$ dB/km and $w = 10 \ \mu m$, the threshold power P_T is 16 mW.

Raman scattering occurs in the forward direction; for the same fiber its threshold is about three orders of magnitude higher.

For a broad-spectrum source and a multimode fiber the Brillouin and Raman thresholds are even larger.

The net result is that losses introduced by nonlinear scattering can be avoided by the judicious choice of core diameter and signal level.

2.4 DISPERSION IN FIBERS

As light pulses travel along a fiber each of them broadens and eventually overlaps with its neighbors, thus increasing the number of errors at the receiver output. The dispersive properties of the fiber are responsible for this phenomenon and they limit the information-carrying capacity of the guide.

Three mechanisms are responsible for the pulse broadening in fibers: material dispersion, waveguide dispersion and modal dispersion. However, waveguide dispersion seldom contributes significantly.

2.4.1 Material Dispersion

A material of index n is called dispersive if $d^2n/d\lambda^2 \neq 0$. Physically this implies that the phase velocity of a plane wave traveling in this dielectric varies nonlinearly with wavelength and consequently, a light pulse will broaden as it travels through it.

In a fiber of length L the pulse broadening due to material dispersion is practically the same for all rays and equal to

$$\tau_m = (L/c)\lambda\delta\lambda(d^2n/d\lambda^2),$$

where c is the speed of light in free space and $\delta\lambda/\lambda$ is the relative spectral width of the source between $1/e$ points. τ_m/L, in nanoseconds per kilometer, versus wavelenth is plotted in Fig. 2.7 assuming a pure silica-core fiber and an Al–Ga–As light-emitting diode (LED), for which typically $\delta\lambda/\lambda = 0.04$. At $\lambda = 0.8$ μm, the pulse broadening is 4 nsec/km. The substitution of the LED by an Al–Ga–As injection laser ($\delta\lambda/\lambda = 0.002$) reduces the pulse broadening by a factor of 20.

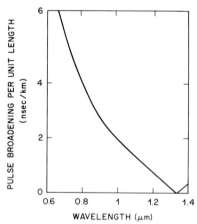

Fig. 2.7 Pulse spread due to material dispersion in silica. Relative width of source $\delta\lambda/\lambda = 0.04$.

Pulse broadening due to material dispersion can be drastically reduced if the carrier wavelength is increased to about 1.3.

The main effect of doping silica with different elements is to shift somewhat the wavelength at which material dispersion is negligible.

To take advantage of low-material dispersion and low-transmission loss (Fig. 2.6) sources and detectors operating in the 1.1 to 1.6-μm wavelength region are actively investigated.

2.4.2 Waveguide Dispersion

In a multimode fiber made of dispersionless material consider a single mode whose propagation constant is β. The group velocity of that mode varies with wavelength; the waveguide is dispersive and broadens a transmitted pulse if $d^2\beta/d\lambda^2 \neq 0$.

Equivalently, in the geometric optics picture, if the angle between the ray representing the mode and the fiber's axis varies with wavelength, the ray trajectory and its flight time also varies. However, the pulse broadening due to waveguide dispersion is in general negligible compared to that caused by material dispersion.

2.4.3 Modal Dispersion

In a multimode fiber the modes have different group velocities. Neglecting material and waveguide dispersion, an impulse equipartitioned among those modes at the input of the fiber evolves into a collection of debunching impulses as they travel along the fiber. If unresolved, the output is a broad single pulse whose duration is equal to the difference between the flight times of the slowest mode and of the fastest.

Like the wave picture above, the ray picture yields the same result. In the step-index fiber (Fig. 2.2) rays with different angles θ have different path lengths. The pulse broadening due to modal dispersion

$$\tau_M = T\Delta$$

is the difference between the flight time $T_{max} = T(1 + \Delta)$ along the longest path $[\theta = (2\Delta)^{1/2}]$ and the flight time $T = (L/c)n_1$ along the shortest path ($\theta = 0$). For silica ($n_1 = 1.458$) and $\Delta = 0.01$, the pulse broadening due to modal dispersion is $\tau_M/L = 48.6$ nsec/km.

There are several mechanisms that reduce modal dispersion. The most effective is the grading of the core index so that its profile is almost a parabola with its maximum on axis. The quasi-equalization of flight times is easy to understand with the help of the ray picture shown in Fig. 2.4. Meridional rays follow almost sinusoidal trajectories of different lengths, but the local group velocity is inversely proportional to the local index, therefore longer sinusoidal paths are compensated by higher speed away from the axis. Though skew rays follow more complicated helical paths the mechanism for modal equalization is the same.

There is no index profile capable of strictly equalizing the group velocity of all modes in a multimode circular symmetric fiber. However, in a fiber whose core index follows the power law

$$n = n_1[1 - 2\Delta(r/a)^g],$$

and whose cladding's index is $n_1(1 - \Delta)$, the modal dispersion is remarkably small if

$$g = g_{opt} \cong 2\left(1 - \frac{\lambda}{\Delta}\frac{d\Delta}{d\lambda}\right).$$

For currently made fibers, $(\lambda/\Delta)(d\Delta/d\lambda)$ is only a few tenths; therefore, the optimized profiles are almost parabolic.

The pulse broadening in this optimized fiber is

$$\tau_{M_{\min}} = T\Delta^2/8$$

and consequently, is $\Delta/8$ times narrower than that of a step-index fiber with the same Δ. For $\Delta = 0.01$ the pulse broadening per unit length is

$$\tau_{M_{\min}}/L = 0.061 \text{ nsec/km},$$

which represent about three orders of magnitude improvement over the step-index profile fiber.

Modal equalization, though effective, is quite critical. For a small departure of g from g_{opt}, say for example that $g = g_{opt}(1 \pm \Delta)$, the pulse broadening is about nine times larger than for $g = g_{opt}$.

Other mechanisms, different from index grading, also reduce modal dispersion. One is the differential attenuation of modes. In step-index fibers, for example, higher order modes having larger field penetration in the cladding than the lower order ones are more sensitive to higher loss in the cladding, to the lossy external jacket and to the core–cladding irregularities. The consequences are differential absorption and scattering losses that tend to eliminate the higher order modes (slower modes) thus reducing the NA of the fiber to an effective value $n_1(2\Delta_{\text{eff}})^{1/2}$ and the pulse broadening to $T\Delta_{\text{eff}}$.

Still another mechanism that reduces modal dispersion is mode mixing. Coupling among guided modes forces power from a slow mode to speed up when transferred to a fast one and vice versa; thus power tends to travel at a speed averaged among all modes.

The ray picture helps also to understand mode mixing. In the step-index fiber (Fig. 2.2) a ray maintains the same angle θ with the axis throughout. Assume now, for example, that the core–cladding interface is irregular because of gentle diameter variations; after every bounce a ray will change its angle with the axis and consequently all rays tend toward the same path length.

In step-index fibers mode mixing among all guided modes reduces the pulse broadening from $(n_1\Delta/c)L$ to $(n_1\Delta/c)(LL_c)^{1/2}$. Instead of growing linearly with the length L of the fiber the pulse width grows at a slower pace, proportionally to $L^{1/2}$. The pulse width is also proportional to the square root of L_c, a characteristic length which is inversely proportional to the coupling strength; the stronger the coupling the smaller the modal dispersion.

Mode mixing, if indiscriminate, may impose a penalty—attenuation. Indeed, as explained in Section 2.3.2, coupling to leaky or unguided modes attenuates the transmitted signal. What is needed is coupling among only the guided modes, or, in terms of the ray picture, one looks for waveguide deformations that vary the angle of the ray with the guide axis without exceeding the critical angle. These equivalent purposes are achieved if the spectral distribution of the coupling deformation (axis meandering, fiber diameter, index variations along the axis, etc.) has significant amplitude only for the spatial periods equal to the beat wavelengths among guided modes. That spectrum must cut off all high-frequency components starting at about $\Delta^{1/2}/4a$. Typically, periods shorter than 1 mm must be avoided.

2.5 FIBER TYPES AND THEIR EVOLUTION

Nature probably produced the earliest bundle of fibers: it is a mineral called ulexite in which capillary needles within a matrix provide guidance.

However, documented experiments about guidance in dielectric waveguides go back to the last century when John Tyndall demonstrated that light follows the curved path of a free-falling water stream.

Forty years later, in 1910, a quantitative analysis of the step-index fiber (Hondros and Debye, 1910) was carried on. However, it was not until the late 1950s that the first all-glass fibers were fabricated (Kapany, 1967).

Many of these fibers, aligned and grouped to form "coherent bundles" capable of transmitting high-resolution images were used to make face plates for cathode-ray tubes, coupling plates for image intensifiers, image scopes for the inspection of inaccessible places, image directors, medical instruments such as endoscopes, etc. Simultaneously groups of unaligned fibers or "incoherent bundles" made to transmit light rather than images, became extensively used in photoelectronics, data processing, photocopy, photography, light distribution, etc.

Depending on the application the most important requirements on these bundles were large light-collecting ability (large NA) and/or high resolution. Thus, multicomponent glasses were used to make either incoherent bundles with NA as large as unity ($\Delta \sim 0.2$) or coherent bundles with individual fibers having an overall diameter of a few microns (Kapany, 1967).

In general, loss in the bundled fibers was not an important consideration since most of the devices were short—a few meters at most. Absorption losses of the order of 1 dB/m in the best optical glasses available were quite tolerable.

Long distance communication through fibers though could not be tackled: their attenuation of about 1000 dB/km was too high. Indeed repeater spacing of about 20 m is economically unthinkable.

The first requirement to make fibers attractive for long-distance transmission was then to reduce the transmission loss. The honor of making the first low-loss fiber compatible with long-distance transmission belongs to Corning Glass Works (Kapron et al., 1970), made with a soot deposition technique. The fiber was essentially single mode and had the then-remarkable low loss of 20 dB/km at 0.63 μm.

Further research was strongly influenced by the fact that in those days, the only source in existence that was efficient, potentially inexpensive, and compatible with fibers was the light-emitting diode. The injection laser had very short life and was multimode. Both sources required low-loss, multimode fibers with core–cladding index difference about 0.01 and core diameter in the order of 50 to 100 μm. Handling experience dictated an external diameter in the 100- to 150-μm region.

For some time, efforts to make these low-loss multimode fibers were hampered by material contamination and fabrication difficulties. In the meantime, two alternatives, the liquid-core fiber and the single-material fiber were discovered.

Silica tubes, drawn to capillary dimensions with the aid of oxy-hydrogen flames and then pressure-filled with tetrachloroethylene were the first liquid-core fibers reported almost simultaneously by two groups (Stone, 1972; Ogilvie et al., 1972; Ogilvie, 1971). The NA of the fiber was approximately 0.34 and the loss, mostly due to Rayleigh scattering and OH absorption, was about 10 dB/km at 1.06 μm.

Later, that loss was reduced to about 7 dB/km by the use of a capillary tube made of Chance–Pilkington ME1 glass filled with hexachlorobuta-1,3-diene (Gambling et al., 1972).

Currently Australia maintains interest in liquid-core fibers for communication.

Single-material fibers, Fig. 2.1b, on the other hand, were invented to take advantage of the only commercially available low-loss solid material—fused silica (Kaiser et al., 1974). The minimum loss reported for a multimode one is 3 dB/km at 1.1 μm.

In the meantime, Corning Glass Works had continued its search for adapting the soot deposition technique to the fabrication of multimode graded-index fibers. Toward the beginning of 1973 the fabrication of a somewhat graded multimode fiber with a NA of 0.14 and the remarkable low loss of 4 dB/km at 0.8, 0.85, and 1.03 μm was anounced (Keck et al., 1973).

Shortly after, people at Bell Labs made low-loss multimode fibers by using a modified chemical vapor deposition (MCVD) process to grow successive layers of silica doped with either germanium or boron on the inside of commercially available silica tubes; these were subsequently collapsed into rods and drawn into fibers (French et al., 1974). The boron-doped fibers had an NA of 0.17 and achieved a new low-loss record: 1.1 dB/km at 1.02 μm.

These fibers were quite attractive for long-distance communication from the point of view of loss, coupling efficiency to sources and handling; however their quasi-step-index profile limited their information carrying capacity to a few tens of megabit kilometers per second. For systems with larger bit-rate length product such as intracity trunk or intercity links the fibers had to be improved to reduce their modal dispersion via the grading of their index profile.

Two types of fibers emerged to satisfy this demand. One called Selfoc®, which is a multimode glass fiber with a quasi-parabolic index achieved by ion exchange (Uchida et al., 1969; Ikeda and Yoshiyagawa, 1976). The other is a silica-based fiber made either by soot disposition or by the MCVB technique. By changing the composition of each deposited layer, a finer control of the index profile is obtained. Using the MCVD technique, fibers with two orders of magnitude larger bit-rate bandwidth products than the step-index fibers with the same NA have been realized (French et al., 1976; Cohen 1976). Theory predicts that an improvement of another order of magnitude is still possible. Thus a fiber with optimum profile and $\Delta = 0.01$ should have a bit-rate bandwidth product of about 13,000 Mbit km/sec. It will not be easy to fabricate such a fiber since index profile control within 1% of theoretical is needed.

The substantial improvement in modal dispersion has left material dispersion as the mechanism limiting the bit-rate length product. In silica fibers using light-emitting diodes at about 0.85 μm, that product is about 140 Mbit km/sec. This limitation can be overcome either by using sources with narrower spectral width or by using sources at about 1.3 μm where the material dispersion of silica-based fibers is much smaller. Operating at these longer wavelengths has the added advantage of low-transmission loss [see, for example, Origuchi and Osanai (1976); Fig. 2.6]. In this phosphor-doped fiber, the losses are approaching the fundamental Rayleigh and infrared-absorption limits; the minimum loss is 0.5 dB/km at 1.2 μm and the only significant contaminant left is a small amount of water responsible for the absorption peaks.

With an insatiable thirst for lower loss and wider bandwidth the designer of intercity optical systems aims toward single mode fibers operating with very narrow spectral-width laser (less than 1 Å wide) emitting in the 1.1- to 1.6-μm wavelength region where a few orders of magnitude larger bit-rate length product than that of ideally profiled multimode fibers is possible and where attenuation smaller than 1 dB/km is likely to be achieved in cables carefully designed to minimize microbending loss.

To facilitate the splicing of these fibers the size of the transmitted single mode beam must be made relatively large without excessively increasing the microbending losses. Typically, core diameters about 10 μm and Δ in the order of a few per thousand provide a reasonable compromise.

An alternative is the W fiber (Fig. 2.1e) where the beam size is given essentially by the core diameter while the single mode propagation is controlled by the value of Δ and Δ_1.

Let us look at the other end of the scale of sophistication where interconnecting fibers within a building provide electrical insulation and avoid electrical interference. These are relatively short fibers, at most a few thousand feet, that require neither extremely low attenuation nor high bit-rate length product. They do require to be inexpensive and to have large cores ($\sim$150 μm) and large NA ($\sim$0.4) for easy splicing and for efficient coupling to light-emitting diodes. The plastic clad fiber meets these needs. It is a step-index guide consisting of a silica or glass core tightly surrounded by a plastic cladding of lower index such as Teflon (Blyler *et al.*, 1975; Kaiser *et al.*, 1975), silicone (Tanaka *et al.*, 1975), etc.

To recap, in less than eight years, fibers have evolved and continue to evolve to satisfy a wide range of requirements. From economical modest loss, modest capacity, high-NA, step-index plastic clad fibers ideally suited for on-premise connections, to multimode low-loss graded-index fibers with intermediate capacity for interoffice trunks, to single mode low-loss fibers capable of very high bit-rate length product for intercity communication, this versatile family of fibers covers an ever-expanding spectrum of capabilities.

REFERENCES

Blyler, L. L., Jr., Hart, A. C., Jr., Jaeger, R. E., Kaiser, P., and Miller, T. J. (1975). Low-loss, polymer-clad silica fibers produced by laser drawing. *Top. Meet. Opt. Fiber Transm., 1st, 1975* p. A5-1–A5-4.

Cohen, L. G. (1976). Pulse transmission measurements for determining near optimal profile gradings in multimode borosilicate optical fibers. *Appl. Opt.* **15,** 1808–1814.

French, W. G., MacChesney, J. B., O'Connor, P. B., and Tasker, G. W. (1974). Optical waveguides with very low losses. *Bell Syst. Tech. J.* **53,** 951–954.

French, W. G., Tasker, G. W., and Simpson, J. R. (1976). Fabrication and optical properties of glass fiber waveguides with graded B_2O_3–SiO_2 cores. *Appl. Opt.* **15,** 1803.

Gambling, W. A., Payne, D. A., and Matsumura, H. (1972). Gigahertz bandwidths in multimode, liquid core, optical fibre waveguide. *Opt. Commun.* **6,** 317–322.

Hondros, D., and Debye, (1910). Elecktromagnetische Wellen an Dielektrischen Drahten. *Ann. Phys. (Leipzig)* [4] **32,** 465–476.

Kaiser, P., and Astle, H. W. (1974). Low-loss single-material fibers made from pure fused Silica. *Bell Syst. Tech. J.* **58,** 1021–1039.

Kaiser, P., Hart, A. C., Jr., and Blyler, L. L. (1975). Low-loss FEP-clad silica fibers. *Appl. Opt.* **14,** 156–162.

Kapany, N. S. (1967). "Fiber Optics Principles and Applications," p. 2. Academic Press, New York.

Kapron, F. P., Keck, D. B., and Maurer, R. D. (1970). Radiation losses in glass optical waveguides. *Appl. Phys. Lett.* **17,** 423–425.

Keck, D. B., Maurer, R. D., and Schultz, P. C. (1973). On the ultimate lower limit of attenuation in glass optical waveguides. *Appl. Phys. Lett.* **22,** 307–309.

Ogilvie, G. J. (1971). Australian Provisional Patent PA 7211/71.

Ogilvie, G. J., Esdaile, R. J., and Kidd, G. P. (1972). Transmission loss of tetrachlorethylene-filled liquid-core-fibre light guide. *Electron. Lett.* **8,** 553–534.

Origuchi, M., and Osanai, H. (1976). Spectral losses of low-OH-content optical fibers. *Electron. Lett.* **12,** 310–312.

Stone, J. (1972). Optical transmission in liquid-core quartz fibers. *Appl. Phys. Lett.* **20,** 239–240.

Tanaka, S., Inada, K., Akimoto, T., and Kojima, M. (1975). Silicone clad fused silica core fiber. *Electron. Lett.* **11,** 153–154.

Uchida, T., Furukawa, M., Kitano, I., Koizumi, K., and Matsumura, H. (1969). A light-focusing fiber guide. *IEEE J. Quantum Electron.* **QE-5,** 331 (abstr.).

Chapter 3

Guiding Properties of Fibers

DIETRICH MARCUSE
DETLEF GLOGE
ENRIQUE A. J. MARCATILI

This chapter contains the theoretical foundations for a description of light propagation in fibers. We begin by describing mode propagation in ideal single and multimode fibers. The discussion of single mode fibers is limited to fibers whose cores have constant refractive indices but, in some instances, compound claddings. Multimode fibers are assumed to have graded-index cores. Single mode fibers are treated mathematically by solving a boundary value problem while the treatment of multimode fibers requires approximate techniques such as ray optics and the Wentzel–Kramers–Brillouin (WKB) method. After the properties of ideal structures have been explained we consider more realistic fibers by allowing the cladding to be of finite extent and study mode losses introduced by finite claddings surrounded by a lossy jacket. Radiation losses caused by constant curvature of the fiber axis are considered as a mechanism that reduces the total number of modes carried by the bent fiber. Radiation effects from the fiber end are of importance as an analytical tool for observing mode excitation and even estimating the refractive index distribution of the fiber core. Closely related to radiation effects from the end of the fiber is the problem of fiber excitation and offset and tilt losses between the junction of two fibers. Coupled mode theory is a powerful tool for treating mode mixing and radiation effects in nonideal fibers with refractive index inhomogeneities and random bends of the fiber axis. Coupled amplitude and coupled power theories will be mentioned. Due to space limitations the theories cannot be derived from first principles; only a brief outline of the methods of derivation will be attempted followed by the statement of results. This chapter is intended primarily as a collection of principles and equations which are needed in other sections.

37

In this text we use the LP-mode notation introduced by Gloge (1971). Its relation with more conventional notations (Marcuse, 1972) is shown in the following table

$$LP_{01}: HE_{11} \qquad\qquad ; \qquad TEM_{00}$$
$$LP_{\nu\mu}: HE_{\nu+1,\mu} \pm EH_{\nu-1,\mu}; \qquad TEM_{\nu,\mu-1}$$

3.1 MODE CONCEPT

An optical fiber is a cylinder made of dielectric materials. A central region, the core, is surrounded by one or more cladding regions and the whole structure is usually protected by a jacket (Miller *et al.*, 1973). Figure 3.1 is a schematic representation of an optical fiber. The optical characteristics of the fiber are determined by its refractive index distribution which is usually circularly symmetric and depends only on the radial coordinate r as indicated by the notation $n(r)$.

The refractive index distributions of optical fibers can assume many shapes. In step-index fibers the refractive index assumes the constant value n_1 in the core and a lower value n_0 in the cladding so that we have

$$n(r) = \begin{cases} n_1 & \text{for } r < a \\ n_0 & \text{for } r > a \end{cases} \tag{3.1}$$

with $n_1 > n_0$ and with a indicating the core radius.

Graded-index fibers have refractive index distributions that vary throughout the core but are usually constant in the cladding region. It is sometimes convenient to express the refractive index in the fiber core as a power law. The index distribution of a graded-index fiber may then be written as (Gloge and Marcatili, 1973)

$$n(r) = \begin{cases} n_1[1 - 2(r/a)^g\Delta]^{1/2} & \text{for } r < a \\ n_0 & \text{for } r > a. \end{cases} \tag{3.2}$$

The constant Δ is the relative-index difference between the value n_1 on axis and the value of the cladding index n_0

$$\Delta = \frac{n_1^2 - n_0^2}{2n_1^2} \approx \frac{n_1 - n_0}{n_1} \tag{3.3}$$

and g is the exponent of the power law. In the special case $g = 2$ we speak of a "parabolic-index" or "square-law" fiber. Values of g very close to two have special significance for multimode fibers of low dispersion (S. E. Miller, 1965).

Electromagnetic light fields traveling in the fiber or being scattered by the fiber can be expressed as superpositions of simpler field configurations—the modes of the fiber. The time dependence of mon-

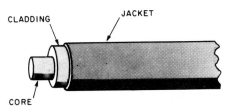

Fig. 3.1 Optical fiber composed of core, cladding, and jacket.

ochromatic light fields of radian frequency ω is given by the factor $e^{i\omega t}$. The fields of fiber modes have, in addition, a simple dependence on the longitudinal z-coordinate that can be expressed in the form $e^{-i\beta z}$. The combined factor $e^{i(\omega t-\beta z)}$ describes a mode traveling in positive z direction. For guided modes the propagation constants β_g cannot assume arbitrary values; instead their values are obtained as solutions of an eigenvalue equation that follows from the requirement that the mode field must satisfy Maxwell's equations and certain boundary conditions (Marcuse, 1974). The finite number of possible propagation constants fall within the range

$$n_0 k < |\beta_g| < n_1 k. \tag{3.4}$$

The subscript g indicates that β belongs to a guided mode. The constant k is the propagation constant of plane waves in vacuum.

In addition to guided modes there are a continuum of unguided or radiation modes that, together with the guided modes, form a complete orthogonal set of modes (Marcuse, 1974) that can be used to express any field configuration as a sum over guided modes plus an integral over the continuum of radiation modes (see Section 3.11). The propagation constants of propagating radiation modes fall within the range

$$-n_0 k < \beta < n_0 k. \tag{3.5}$$

There are also evanescent radiation modes with imaginary propagation constants in the range

$$0 < |\beta| < \infty. \tag{3.6}$$

However, they are of little importance because they do not carry power away from the optical waveguide.

Optical fibers for communication purposes usually satisfy the condition

$$\Delta \ll 1 \tag{3.7}$$

with Δ defined by (3.3). Fibers with the property (3.7) are called weakly guiding fibers (Gloge, 1971); the propagation constants of their guided modes can be approximated by the equation (Snyder, 1969)

$$\beta_g \approx n_0 k. \tag{3.8}$$

It can be shown that the fields of the guided modes of weakly guiding fibers are very nearly linearly polarized (LP modes (Gloge, 1971)) and that all field components can be obtained as derivatives of one dominant transverse component of the electric field vector.

3.2 STEP-INDEX FIBERS

We consider only weakly guiding step-index fibers. We explained in the preceding section that the modes of weakly guiding fibers are very nearly linearly polarized. Certain approximate solutions of the guided mode problem of weakly guiding step-index fibers have been called LP modes (Gloge, 1971) to suggest their linearly polarized character. LP modes arise naturally as approximate mode solutions when the approximate relation (3.8) is used. These mode solutions belong essentially to a scalar problem because it is only necessary to solve the wave equation for the dominant transverse electric field component; all other field components are obtained from this component by differentiation.

It is important, however, to keep in mind that LP modes are not exact modes of the step-index fiber. The exact fiber modes are designated by the symbols HE and EH, and it can be shown that LP modes can locally be expressed as linear superpositions of an HE and EH mode (Marcuse, 1972, 1974). If these two parent modes would have identical propagation constants, their superposition would remain unchanged for all values of the longitudinal coordinate z. Actually, the propagation constants of the two parent modes are slightly different so that the superposition of HE and EH modes changes with z. Therefore LP modes are not true modes, but they are useful because they allow us to derive simple approximate eigenvalue equations for the guided modes. They may also be used to compute coupling coefficients between guided modes that are used to study the interchange of power among the modes of a multimode fiber. For problems of power exchange among guided modes it does not matter that the LP modes are not true modes since the two parent modes, being almost degenerate, exchange their power very quickly so that the concept of LP modes as superpositions of HE and EH modes is sufficient to determine the power exchange between different families of these modes.

The dominant transverse electric field components of LP modes can be expressed as follows (Gloge, 1971; Marcuse, 1974):

$$E_x = \begin{cases} AJ_\nu(\kappa r) \begin{pmatrix} \cos \nu\phi \\ \sin \nu\phi \end{pmatrix} e^{-i\beta_g z} & \text{for} \quad r \le a \\ A \dfrac{J_\nu(\kappa a)}{K_\nu(\zeta a)} K_\nu(\zeta r) \begin{pmatrix} \cos \nu\phi \\ \sin \nu\phi \end{pmatrix} e^{-i\beta_g z} & \text{for} \quad r \ge a \end{cases} \qquad (3.9)$$

The time dependence of the form $e^{i\omega t}$ is omitted from these equations. $J_\nu(x)$

and $K_\nu(x)$ are the Bessel and modified Hankel functions of integral order ν. The field is expressed in terms of a cylindrical coordinate system r, ϕ, and z but its field components are expressed in terms of rectangular Cartesian coordinates x, y, and z; only the x component of the vector $\mathbf{E}$ is given here. The parameter κ is the transverse component of the propagation vector in the fiber core,

$$\kappa = (n_1^2 k^2 - \beta_g^2)^{1/2}, \tag{3.10}$$

and ζ is the transverse decay parameter in the cladding

$$\zeta = (\beta_g^2 - n_0^2 k^2)^{1/2}. \tag{3.11}$$

The sum of the squares of κ and ζ defines a useful quantity,

$$V = (\kappa^2 + \zeta^2)^{1/2} a = (n_1^2 - n_0^2)^{1/2} ka. \tag{3.12}$$

V is a dimensionless number, sometimes called the normalized frequency parameter or simply the V-number. It will be seen later that it determines how many modes the fiber can support.

The amplitude parameter A in (3.9) can be expressed in terms of the power P carried by the mode

$$A = \left\{ \frac{4(\mu_0/\epsilon_0)^{1/2} W^2 P}{e_\nu \pi n_1 a^2 V^2 |J_{\nu-1}(U) J_{\nu+1}(U)|} \right\}^{1/2}. \tag{3.13}$$

A convenient approximation for multimode fibers with large values of V is

$$A \approx \left\{ \frac{2(\mu_0/\epsilon_0)^{1/2} W^2 U^2 P}{e_\nu n_1 a^2 V^2 (U^2 - \nu^2)^{1/2}} \right\}^{1/2}. \tag{3.14}$$

We have used the abbreviations

$$U = \kappa a \quad \text{and} \quad W = \zeta a \quad \text{with} \quad V^2 = U^2 + W^2 \tag{3.15}$$

and

$$e_\nu = \begin{cases} 2 & \text{for} \quad \nu = 0, \\ 1 & \text{for} \quad \nu \neq 0. \end{cases} \tag{3.16}$$

The exact mode fields of step-index fibers have complicated eigenvalue equations. Weakly guiding LP modes have much simpler eigenvalue equations that are obtained from the boundary conditions requiring continuity of the transverse and longitudinal electric field components at the core boundary at $r = a$. The boundary conditions of the magnetic field components are only approximately satisfied in the weak guidance approximation. The eigenvalue equation of LP modes is of the form (Snyder, 1969)

$$U \frac{J_{\nu+1}(U)}{J_\nu(U)} = W \frac{K_{\nu+1}(W)}{K_\nu(W)}. \tag{3.17}$$

Since the eigenvalue equation contains only the variables U and W which, in turn, are connected by the Eq. (3.15) it is clear that U and W are functions of the single variable V. A very good approximation of the solutions of (3.17) is given by the following expression that holds for all integers ν and μ with the exception of $\nu = 0$, $\mu = 1$ ($\mu = 1, 2 \ldots$ labels successive solutions of (3.17) and is called the radial mode number) (Gloge, 1971),

$$U = U_c \exp\{[\arcsin(S/U_c) - \arcsin(S/V)]/S\} \qquad (3.18)$$

with the abbreviation

$$S = (U_c^2 - \nu^2 - 1)^{1/2}. \qquad (3.19)$$

The constant U_c is the cutoff value of U and is defined by

$$J_{\nu-1}(U_c) = 0. \qquad (3.20)$$

Solutions of (3.20) are extensively tabulated and useful approximate solutions are given by the formula (Abramowitz and Stegun, 1964)

$$U_c = A - \frac{B - 1}{8A} - \frac{4(B - 1)(7B - 31)}{3(8A)^3} \qquad (3.21)$$

with (note, this A is not to be confused with the mode amplitude in (3.9))

$$A = [\mu + \tfrac{1}{2}(\nu - 1) - \tfrac{1}{4}]\pi \quad \text{and} \quad B = 4(\nu - 1)^2. \qquad (3.22)$$

For the important LP_{01} mode (in the usual notation this is identical to the HE_{11} mode) with $\nu = 0$ and $\mu = 1$ we have the approximate solution of (3.17),

$$U = (1 + 2^{1/2})V/[1 + (4 + V^4)^{1/4}]. \qquad (3.23)$$

Figure 3.2 shows U and W for the HE_{11} mode as functions of V. The propagation constants of the LP modes can be displayed conveniently in normalized form. We define the normalized propagation constant

$$b = 1 - \frac{U^2}{V^2} = \frac{(\beta/k)^2 - n_0^2}{n_1^2 - n_0^2} \approx \frac{(\beta/k) - n_0}{n_1 - n_0}. \qquad (3.24)$$

Figure 3.3 (Gloge, 1971) shows the normalized propagation constant as a function of V. The figure shows clearly that each mode can only exist for V values that exceed a certain limiting value V_c. This phenomenon is called cutoff. Below $V = V_c$ the mode is cutoff which means that it cannot exist as a properly guided mode. The value of V_c is different for different modes and is given as $V_c = U_c$ by (3.20) or, approximately, by (3.21). For the mode LP_{01} (HE_{11}) the cutoff value is $V_c = 0$. We shall see in Sections 3.3 and 3.4 that wave propagation does not cease abruptly below cutoff. Instead a given mode transforms itself into a leaky wave as V becomes smaller than V_c for the mode in question. Some leaky modes have relatively low losses and may propagate considerable distances along the fiber

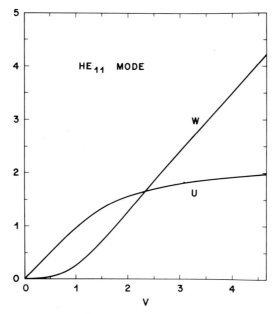

Fig. 3.2 U and W of the HE_{11} mode as functions of V.

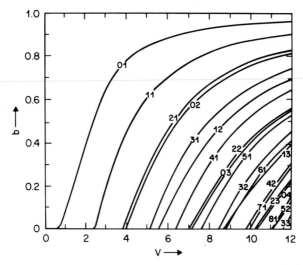

Fig. 3.3 Normalized propagation constants of a number of LP modes as functions of V.

(Snyder and Mitchell, 1974). At cutoff we have $W = 0$ or $\beta = n_0 k$ so that the normalized propagation constant becomes $b = 0$. Very far above cutoff for $V \to \infty$ the propagation constant assumes the value $\beta = n_1 k$ and we have $b = 1$; it is apparent from (3.11) and (3.12) that $W = V$ in that limit.

The number of guided modes is finite. It can be shown that the total number of guided modes is approximately given by the formula (Gloge, 1971)

$$N = V^2/2. \tag{3.25}$$

This number includes modes of both possible polarizations and with both choices of cosine or sine functions indicated in (3.9).

The power carried by the guided modes is not completely contained inside the fiber core. It is apparent from (3.9) that there is an electromagnetic field outside of the core. This field decays exponentially at large distance from the core cladding interface.

Figure 3.4 (Gloge, 1971) shows the ratio P_{clad}/P of the power carried in the cladding to the total power carried by each mode as a function of V for all the guided modes with $V \le 12$. We see that at cutoff, modes with $\nu = 0$ and $\nu = 1$ carry almost all their power in the cladding while all other modes with $\nu \ge 2$ still carry most of their power inside of the fiber core. At cutoff the ratio of power carried in the cladding to total mode power is given by (Gloge, 1971)

$$P_{\text{clad}}/P = 1/\nu \quad \text{for} \quad \nu \ge 2. \tag{3.26}$$

The fiber can be constructed to carry either only one mode or many modes at a given frequency. If only the LP_{01} mode can exist we speak of a

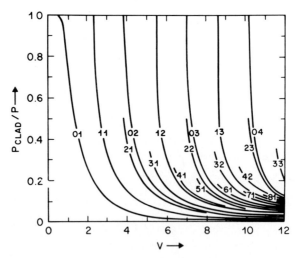

Fig. 3.4 Ratio of power carried in the cladding to total power carried by a number of LP modes as functions of V.

single mode fiber. Actually there are two modes with orthogonal polarization even if $\nu = 0$; the term single mode fiber thus applies to a given polarization of the light power. A fiber operates with a single mode if it satisfies the relation (Snitzer, 1961)

$$V < 2.405 \tag{3.27}$$

because $V_c = 2.405$ is the cutoff frequency of the LP_{11} mode (not to be confused with the HE_{11} mode whose cutoff value is $V_c = 0$).

Up to this point we have considered step-index fibers with the simple refractive index profile defined by (3.1). Another useful step-index fiber is the W fiber (Kawakami and Nishida, 1974) whose refractive index profile is shown in Fig. 3.5. The region $r < a_1$ is the core, $a_1 \le r \le a_2$ is the inner cladding, $a_2 \le r \le b$ is the outer cladding, and the region $r > b$ is a (lossy) jacket. For this fiber we can define two V values, $V_0 < V_2$, via the relations,

$$V_0 = (n_1{}^2 - n_0{}^2)^{1/2} k a_1 \tag{3.28}$$

and

$$V_2 = (n_1{}^2 - n_2{}^2)^{1/2} k a_1. \tag{3.29}$$

The exact solution of the mode problem of the W fiber would require formulating solutions in the three regions $|r| < a_1, a_1 < r < a_2$ and $a_2 < r$ and connecting these solutions by means of boundary conditions. We can get a qualitative understanding of the properties of the W fiber by considering first the modes of the structure with $a_2 \to \infty$ which are identical to the modes of a simple step-index fiber with V value V_2. The values of the propagation constants of these modes lie in the interval $n_2 k < \beta < n_1 k$. However, modes whose β-values are $|\beta| < n_0 k$ lose power by radiation into the outer cladding region; they are leaky modes whose attenuation depends on the width of the inner cladding $a_2 - a_1$. We shall discuss the losses of these leaky modes in Section 3.4. The W fiber makes it possible to design a single mode waveguide with relatively large core dimensions by placing an outer cladding of refractive index n_0 around the inner cladding. There may be a large number of leaky modes in addition to the desired lossless (no power radiation) mode, but these modes do little harm pro-

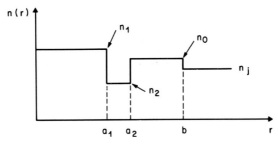

Fig. 3.5   Refractive index profile of the W fiber.

vided their attenuation is sufficiently high. W fibers are useful for easing the splicing problem by providing a single mode fiber with a large core radius. W fibers are also advantageous because they are more tolerant of sharp bends than equivalent simple step-index fibers and because, if properly designed, the dispersion due to waveguide effects tends to compensate the material dispersion (Kawakami and Nishida, 1975).

3.3 GRADED-INDEX FIBERS

The refractive index profiles of step-index and W fibers were composed of piecewise constant sections that are joined together by discontinuous index jumps. Graded-index fibers consist of refractive index distributions that are continuous functions of r. An example of a graded-index profile is provided by (3.2).

It is shown in Chapter 2 how light rays follow continuously curved paths in graded-index fibers. Rays that do not approach the core boundary closely can be regarded as propagating in an idealized infinitely extended optical medium as indicated by the dotted line in Fig. 3.6. This figure shows a graded index profile defining core and cladding regions. The analysis of such fibers is greatly simplified by assuming that the index profile of the core continues indefinitely beyond the core region as indicated by the dotted line in Fig. 3.6. For example, if we consider a parabolic-index fiber [with $g = 2$ in (3.2)] we may express the LP mode solutions of the structure in terms of Laguerre–Gaussian functions (Tien *et al.*, 1965) provided we allow the index profile to continue its parabolic shape beyond $r = a$. The discontinuity of the first derivative of the index profile at the core boundary $r = a$ complicates the analytical description of the mode fields. But most modes have field distributions that decay so rapidly inside the fiber core that they do not interact appreciably with the cladding region and are approximated with sufficient accuracy by assum-

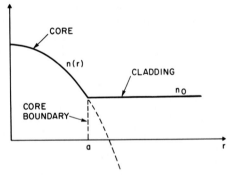

Fig. 3.6 Refractive index profile of a clad graded-index fiber, solid line. The dotted line indicates an infinitely extended parabolic-index profile.

ing that they are solutions of the extended (dotted in Fig. 3.6) profile. Modes with field distributions reaching strongly into the cladding region are near cut off and may often be neglected for approximate calculations by imposing a cutoff condition that separates the modes contained wholly inside the core from the modes that interact with the cladding region.

An extremely useful analytical method for obtaining approximate mode solutions of graded-index fibers with arbitrary profiles is the WKB method named after Wentzel, Kramers, and Brillouin (Morse and Feshbach, 1953). The WKB approach is a geometrical optics approximation that works whenever the refractive index of the fiber varies only slightly over distances on the order of the optical wavelength. The WKB solution expresses the field in the form (Petermann, 1975)

$$E_x = \frac{1}{2}[A_1(r)e^{iS(r)} + A_2(r)e^{-iS(r)}] \begin{pmatrix} \cos \nu\phi \\ \sin \nu\phi \end{pmatrix} e^{-i\beta z}. \tag{3.30}$$

Substitution into the wave equation and neglecting the second derivative of $A_j(r)$ with respect to r yields approximate solutions of the amplitude function $A_j(r)$ and the phase function $S(r)$.

The WKB approximation yields excellent answers in most regions of space with the exception of the so-called turning points of the light rays. We discussed in Chapter 2 that a light ray in a graded-index fiber does not reach every point inside the fiber core but stays between two limiting surfaces—the caustics (Gloge, 1975). If we follow the ray along its trajectory we see it move outward, away from the axis until it reaches the point of greatest departure from the fiber axis—the turning point—at the caustic. The same behavior repeats itself at the inner caustic near the fiber axis. The caustics separate the regions of the oscillatory, propagating behavior of the light field from an evanescent shadow region. The exact solution of the field problem passes smoothly through the regions of the caustics; the approximate WKB solution, on the other hand, has poles at each caustic. Fortunately, these singular regions of the WKB solutions do not destroy its usefulness. It is possible to amend the WKB solution by obtaining approximate solutions of the wave equation in the vicinity of the turning points by replacing the actual refractive index profile by a linear approximation right at the location of the caustic. The solutions at and near the caustic can be used to join two separate solutions of the WKB type together; it is only necessary to connect the undetermined amplitude coefficients of the different solutions by means of continuity requirements of the field and by requiring that the mode field approaches zero as $r \to \infty$ (Morse and Feshbach, 1953).

We state the results of the WKB approximation. The two turning points defining two caustics are designated as r_1 and r_2 with $r_1 < r_2$. In the region $r_1 < r < r_2$ between the caustics the field is oscillatory and we have

$$A_1(r) = A_2(r) = D/\{[(n(r)k)^2 - \beta^2]r^2 - \nu^2\}^{1/4} \qquad (3.31)$$

(where D is an amplitude coefficient) and

$$S(r) = \int_{r_1}^{r} \{[(n(r)k)^2 - \beta^2]r^2 - \nu^2\}^{1/2} \frac{dr}{r} - \frac{\pi}{4}. \qquad (3.32)$$

In the region inside the inner caustic, $r < r_1$, (assuming that r_1 is not too close to $r = 0$) the field solution is monotonically decaying toward the center of the fiber and we have

$$A_2(r) = 0, \qquad (3.33)$$

$$A_1(r) = D/\{\nu^2 - [(n(r)k)^2 - \beta^2]r^2\}^{1/4}, \qquad (3.34)$$

and

$$S(r) = i \int_{r}^{r_1} \{\nu^2 - [(n(r)k)^2 - \beta^2]r^2\}^{1/2} \frac{dr}{r}. \qquad (3.35)$$

Finally, outside the outer caustic, $r > r_2$, the field decays monotonically away from the axis and is described by the equations

$$A_2(r) = 0, \qquad (3.36)$$

$$A_1(r) = De^{i\mu\pi}/\{\nu^2 - [(n(r)k)^2 - \beta^2]r^2\}^{1/4}, \qquad (3.37)$$

and

$$S(r) = i \int_{r_2}^{r} \{\nu^2 - [(n(r)k)^2 - \beta^2]r^2\}^{1/2} \frac{dr}{r}. \qquad (3.38)$$

The solutions that are valid in the vicinity and right at the turning points can be expressed in terms of Hankel functions of order $\frac{1}{3}$ and will not be listed here.

The propagation constant β is a solution of the eigenvalue equation,

$$\int_{r_1}^{r_2} \{[(n(r)k)^2 - \beta^2]r^2 - \nu^2\}^{1/2} \frac{dr}{r} = (2\mu - 1)\frac{\pi}{2} \qquad (3.39)$$

with $\mu = 1, 2, 3, \ldots$.

The eigenvalue equation (3.39) can be solved in closed analytical form only for a few simple refractive index profiles. In most general cases it must be solved numerically or approximately.

The amplitude coefficient D can be expressed in terms of the total power carried by the guided mode. If we restrict ourselves to the power carried between the two turning points r_1 and r_2 we have to a geometric optics approximation,

$$D = \left\{\frac{4(\mu_0/\epsilon_0)^{1/2}P}{n_1\pi a^2 I}\right\}^{1/2} \qquad (3.40)$$

with

$$I = \int_{r_1/a}^{r_2/a} \frac{x\, dx}{\{[(n(ax)k)^2 - \beta^2]a^2x^2 - \nu^2\}^{1/2}}. \tag{3.41}$$

A good understanding of the properties of the WKB solution can be obtained by looking at a graphical representation of the integrand appearing in the function $S(r)$. Figure 3.7 shows plots of the functions (Gloge, 1976)

$$\kappa^2(r) = (n(r)k)^2 - \beta^2 \tag{3.42}$$

and ν^2/r^2. The two curves cross at the turning points $r = r_1$ and $r = r_2$. The field exhibits an oscillatory behavior in regions where $\nu/r < \kappa(r)$ and it decays exponentially as a function of r if $\nu/r > \kappa(r)$. As ν increases, the curve ν^2/r^2 moves higher so that the region between the two turning points becomes narrower. However, even for fixed values of ν the curve $\kappa^2(r)$ shifts up or down depending on the value of the propagation constant β. Modes far from cutoff have large β-values with correspondingly smaller values of $\kappa^2(r)$ and more closely spaced turning points. As the value of β decreases below n_0k, $\kappa^2(r)$ no longer becomes negative for large values of r and Fig. 3.7 changes to Fig. 3.8. At $r = a$ the curve $\kappa^2(r)$ becomes constant and allows the curve ν^2/r^2 to drop below it, a third turning point $r = r_3$ is thus created. The field exhibits an evanescent, exponentially decaying behavior in the region $r_2 < r < r_3$ but for $r > r_3$ the field resumes its oscillatory behavior and carries power away from the fiber core. This picture shows that mode cutoff must occur as soon as $\beta = n_0k$ since the guided mode is no longer perfectly trapped inside of the core but loses power by leakage (tunneling) into the cladding. Modes of this type are called tunneling leaky waves (Snyder and Mitchell, 1974). We shall discuss the properties of leaky modes further in Section 3.4.

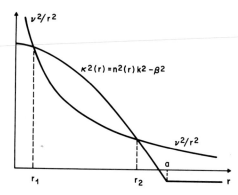

Fig. 3.7   This figure illustrates two functions that are important for the WKB solution and defines the turning points r_1 and r_2 for guided modes of the fiber.

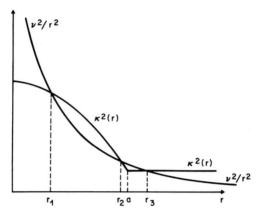

Fig. 3.8 This figure is similar to Fig. 3.7 but applies to leaky mode solutions where a third turning point, r_3, appears.

The guided modes are labeled by ν and μ, the azimuthal and radial mode numbers. Each mode can thus be represented in a mode number plane as shown in Fig. 3.9. Each dot in mode number space represents actually four modes (with the exception of the dots along the horizontal axis $\nu = 0$ each of which represent two modes). For each pair of values ν, μ there are two possible polarizations and two choices of either the sine or cosine function in (3.30). The dotted line in Fig. 3.9 labeled mode boundary separates the guided modes from the leaky and radiation modes. If we define the mode boundary as the function $\mu = F(\nu)$ we can express the total number of guided modes by the formula

$$N = 4 \int_0^{\nu_{max}} F(\nu)\, d\nu \qquad (3.43)$$

because each representation point (representing four modes) occupies an element of unit area in the space ν, μ.

The WKB solution can even be used to describe the behavior of multi-

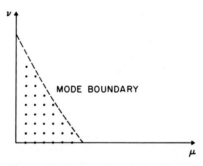

Fig. 3.9 The mode number plane for guided fiber modes.

mode step-index fibers except that we must remember that the turning point r_2 now coincides with the core boundary $r_2 = a$. The method of connecting the different sections of the WKB solutions across the turning point, that is implicit in obtaining the eigenvalue equation (3.39), is now no longer applicable. Instead, we require that the field E_x assumes the value 0 at $r = a$ ignoring the evanescent field outside of the core. This approximate boundary condition is adequate for modes far from cutoff in multimode step-index fibers and leads to the eigenvalue equation

$$\int_{r_1}^{a} \{[n_1{}^2 k^2 - \beta^2]r^2 - \nu^2\}^{1/2} \frac{dr}{r} = \left(2\mu - \frac{1}{2}\right)\frac{\pi}{2}. \qquad (3.44)$$

Integration and rearrangement of terms allows us to write (3.44) in the form

$$(U^2 - \nu^2)^{1/2} - \nu \arccos \frac{\nu}{U} = \left(2\mu - \frac{1}{2}\right)\frac{\pi}{2}. \qquad (3.45)$$

This transcendental equation for U [defined by (3.15)] can easily be solved by an iteration method and provides a less accurate alternate approximate solution of the eigenvalue equation of the step-index fiber.

From (3.30), (3.31), and (3.32) we obtain the following approximate solution for the transverse electric field component of the step-index fiber modes in the region $r_1 < r < a$,

$$E_x = \frac{D \cos\{(\kappa^2 r^2 - \nu^2)^{1/2} - \nu \arccos(\nu/\kappa r) - \pi/4\}}{[\kappa^2 r^2 - \nu^2]^{1/4}} \binom{\cos \nu\phi}{\sin \nu\phi} e^{-i\beta z} \qquad (3.46)$$

with κ defined by (3.10). This approximation is identical to the solution that follows from (3.9) if we replace the Bessel function by its Debye approximation (Abramowitz and Stegun, 1964). The amplitude coefficient D can be expressed with the help of (3.40) and (3.41) as follows

$$D = \{4U^2(\mu_0/\epsilon_0)^{1/2}P/\pi n_1 a^2(U^2 - \nu^2)^{1/2}\}^{1/2}. \qquad (3.47)$$

This expression agrees with (3.14) in the far-from-cutoff limit where the approximation $W = V$ is valid. The additional factor $(2/\pi)^{1/2}$ in (3.47) is explained by the fact that the Debye approximation of $J_\nu(x)$ contains just this factor.

Turning to applications of the WKB solution to graded-index fibers we can derive from (3.43) the following expression for the total number of guided modes for graded-index fibers with the power law profile of (3.2),

$$N = [g/(g + 2)](n_1 ka)^2 \Delta. \qquad (3.48)$$

The step-index fiber can be regarded as a graded-index fiber with the power law exponent $g = \infty$. If we use the condition (3.7) we obtain from (3.12) the approximation

$$V = n_1 ka(2\Delta)^{1/2} \tag{3.49}$$

we see that (3.48) coincides with (3.46) for step-index fibers. For parabolic-index fibers with $g = 2$ we have instead

$$N = \tfrac{1}{2}(n_1 ka)^2 \Delta = \tfrac{1}{4}V^2. \tag{3.50}$$

For the parabolic-index profile the integral in $S(r)$ can be solved resulting in

$$S(r) = \frac{1}{2}(\kappa_p^2 r^2 - G^2 r^4 - \nu^2)^{1/2} + \frac{\kappa_p^2}{4G}\, \arcsin\left[\frac{2G^2 r^2 - \kappa_p^2}{(\kappa_p^4 - 4\nu^2 G^2)^{1/2}}\right]$$
$$- \frac{\nu}{2}\arcsin\left[\frac{\kappa_p^2 r^2 - 2\nu^2}{r^2(\kappa_p^4 - 4\nu^2 G^2)^{1/2}}\right] + \frac{\pi}{8G}(\kappa_p^2 - 2\nu G) - \frac{\pi}{4} \tag{3.51}$$

with

$$\kappa_p^2 = n_1^2 k^2 - \beta^2 \quad\text{and}\quad G^2 = 2n_1^2 k^2 \frac{\Delta}{a^2} = \frac{V_p^2}{a^4} \tag{3.52}$$

The subscript p is a reminder that κ and V belong to the parabolic-index fiber. The eigenvalue equation (3.39) can also be expressed in the form

$$S(r_2) = (2\mu - \tfrac{3}{2})(\pi/2). \tag{3.53}$$

Using (3.51) to get $S(r_2) = \pi(\kappa^2 - 2\nu G)/(4G)$ we obtain from (3.53) the following expression for the propagation constant of the modes of the parabolic-index fiber,

$$\beta = n_1 k \left\{1 - \frac{2}{n_1 ka}(2\Delta)^{1/2}(2p + \nu + 1)\right\}^{1/2} \tag{3.54}$$

with

$$p = \mu - 1 = 0, 1, 2, \ldots . \tag{3.55}$$

It is noteworthy that the WKB value (3.54) of the propagation constant of parabolic-index fiber modes is identical to the exact solution of the wave equation with parabolic-index profile (Marcuse, 1972). However, we must remember that (3.54) is valid only for LP modes and that the effect of the finite core radius has been ignored.

The power normalization of the WKB solution of the parabolic-index fiber modes follows from (3.40) with

$$I = \frac{\pi(\kappa_p a)^2}{2(n_1 ka)^3 (2\Delta)^{3/2}} = \frac{\pi(\kappa_p a)^2}{2V_p^3} \tag{3.56}$$

with $V = V_p$ defined by (3.49).

For the previous derivations, we assumed the index profile of the core to continue beyond the core region and considered the cladding merely by specifying a lower limit for the possible propagation constants. This leads to useful simplifications for most practical multimode fibers, but graded-index fibers designed to transmit only one or a few modes require a more accurate analysis.

Okamoto and Okoshi (1976) have found a variational formulation for the wave propagation in a fiber with the index profile (3.2). Their analysis leads to a modification of the characteristic equation (3.17) of the form

$$\frac{U}{(1 + 2/g)^{1/2}} J_{\nu+1}\left[\frac{U}{(1 + 2/g)^{1/2}}\right]\Big/ J_\nu\left[\frac{U}{(1 + 2/g)^{1/2}}\right]$$
$$= W\frac{K_{\nu+1}(W)}{K_\nu(W)} - \frac{W^2}{g + 2}\left[1 - \frac{K_{\nu-1}(W)K_{\nu+1}(W)}{K_\nu^2(W)}\right]. \tag{3.57}$$

This result is particularly useful in determining the single mode condition of a graded index fiber whose profile obeys (3.2). Since W and hence the right side of (3.57) vanishes at cutoff, one finds the cutoff of LP_{11} at

$$V_0 = 2.405(1 + 2/g)^{1/2} \tag{3.58}$$

and single mode operation for $V < V_0$.

3.4 CLADDING EFFECTS AND LEAKY WAVES

In the preceding sections of this chapter we have discussed a few properties of modes in ideal, lossless fibers. An understanding of the mode properties of perfect structures is essential as the basis for the treatment of real fibers with lossy jackets or lossy claddings. If the materials of fiber core and cladding have identical absorption properties the mode losses are, approximately, equal to the loss coefficients of the fiber materials. However, many fibers are made of core materials with extremely low absorption losses but have claddings whose losses are substantially higher. Furthermore, it is usually necessary to surround a fiber with a lossy jacket to reduce cross talk between fibers and to suppress undesirable waves propagating in the so-called cladding modes (Kuhn, 1975). It is thus necessary to be able to estimate the increase of the mode losses that are caused either by lossy claddings, lossy jackets, or both.

Figure 3.10 shows the cross section through a fiber consisting of a core, cladding, and an outer jacket. For simplicity we assume that the real part of the complex refractive index of the jacket material is identical with the cladding index n_0 and that its imaginary part n_i is much smaller than its real part, $n_i \ll n_0$. These assumptions allow us to assume that the field distribution of the fiber modes with lossy jacket is essentially identical to the modes of a fiber whose lossless cladding extends to infinity, $b \to \infty$. We also assume that the jacket is infinitely thick. Both assumptions tend to overestimate the losses but provide us with an upper bound. The mode

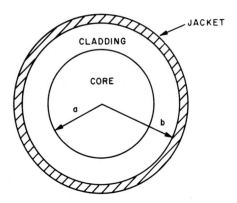

Fig. 3.10   Cross section of an optical fiber.

losses may then be expressed as the ratio of power ΔP that is dissipated per unit length of the fiber divided by the power P carried by the fiber mode,

$$2\alpha = \Delta P/P. \tag{3.59}$$

The power dissipation can be expressed by the formula

$$\Delta P = n_0 \left(\frac{\epsilon_0}{\mu_0}\right)^{1/2} \alpha_j \int_0^{2\pi} d\phi \int_b^\infty r|E|^2 \, dr. \tag{3.60}$$

$|E|$ is the magnitude of the electric vector of the mode field and α_j is the loss coefficient of the jacket material whose relation to the imaginary part of the refractive index of the jacket is given as

$$\alpha_j = n_i k. \tag{3.61}$$

Substitution of approximate expressions for the mode fields into (3.59) and (3.60) results in the following approximate formula for the power losses of the modes of the step-index fiber caused by a lossy jacket

$$\frac{\alpha_{st}}{\alpha_j} = \frac{b^2 U^2}{a^2 V^2} \frac{\nu^2 + W^2}{(\nu^2 + \hat{W}^2)[1 + (\nu^2 + W^2)^{1/2}]}$$
$$\cdot \left\{ \frac{[(\nu^2 + W^2)^{1/2} - \nu][(\nu^2 + \hat{W}^2)^{1/2} + \nu]}{[(\nu^2 + \hat{W}^2)^{1/2} - \nu][(\nu^2 + W^2)^{1/2} + \nu]} \right\}^\nu \tag{3.62}$$
$$\cdot \exp\{-2[(\nu^2 + \hat{W}^2)^{1/2} - (\nu^2 + W^2)^{1/2}]\}.$$

The parameters V, U, and W are defined by (3.12) and (3.15), we define furthermore $\hat{W} = (b/a)W$. Equation (3.62) was derived with the help of the Debye approximation (valid for large ν) of the modified Hankel functions.

For parabolic-index fibers we obtain correspondingly,

$$\frac{\alpha_p}{\alpha_j} = \frac{b^2 V_p^3}{2\pi a^2(\nu^2 + \hat{W}_p^2)U_p^2} \left[\frac{b}{a}\frac{(\nu^2 + W_p^2)^{1/2} - \nu}{(\nu^2 + \hat{W}_p^2)^{1/2} - \nu}\right]^{2\nu}$$

$$\cdot \left(\frac{V_p^2 - 4\nu^2}{V_p^2}\right)^{\nu/2} \left(\frac{V_p + 2\nu}{V_p - 2\nu}\right)^{U_p^2/(4V_p)} \tag{3.63}$$

$$\cdot \exp[-2(\nu^2 + \hat{W}_p^2)^{1/2} + (\nu^2 + W_p^2)^{1/2}]$$

We have used the following abbreviations:

$$V_p = n_1 ka(2\Delta)^{1/2}, \tag{3.64}$$

$$V_c = 2(2p + \nu + 1), \tag{3.65}$$

$$U_p = a(n_1^2 k^2 - \beta^2)^{1/2} = (V_p V_c)^{1/2}, \tag{3.66}$$

$$W_p = a(\beta^2 - n_0^2 k^2)^{1/2} = (V_p^2 - U_p^2)^{1/2}, \tag{3.67}$$

$$\hat{W}_p = (b/a)W. \tag{3.68}$$

The labels ν and p are the azimuthal and radial mode numbers, V_c defined by (3.65) is the cutoff value of the V number of the parabolic-index fiber mode labeled ν,p.

Equations (3.62) and (3.63) do not only approximate the mode losses in the presence of a lossy jacket at radius b but they may also be used to approximate the mode losses in case of a (lossless core and) lossy infinitely extended cladding. In this latter case we set $b = a$ and associate $\alpha_j = \alpha_{cl}$ with the loss coefficient of the cladding material.

Equations (3.62) and (3.63) hold only for small values of the loss coefficient of the jacket material. An upper limit for this value is given at the end of this section.

Figure 3.11 shows mode boundaries in mode number space ν, μ for a step-index fiber with lossy cladding with $V = 32.4$, core radius $a = 25$ μm, $n_0 = 1.458$, and a free-space wavelength of $\lambda_0 = 1$ μm. The solid line is the boundary between the guided and radiation modes defined by the condition $\beta = n_0 k$. The dotted lines indicate boundaries that separate modes with losses that are smaller than a given fixed radio α_{st}/α_{cl} from modes whose values are higher than this preselected value. This graph is useful for assessing the influence of a lossy cladding on the attenuation of the modes in a multimode step-index fiber in the absence of mode coupling. Let us assume that the cladding material has losses of 100 dB/km. The figure shows us that all modes below the line $\alpha_{st}/\alpha_{cl} = 10^{-2}$ have losses less than 1 dB/km while all modes above this line have higher losses. For a fiber with the dimensions and parameters used in the figure more than half of the guided modes would suffer substantial losses due to the lossy cladding. The detrimental influence of a lossy cladding is less serious for larger values of V.

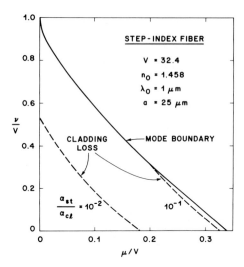

Fig. 3.11 Boundaries in mode number space. The solid line indicates the boundary between guided and radiation modes; the broken lines indicate the boundary between modes whose losses due to a lossy cladding exceed certain values. This figure applies to step-index fibers.

Figure 3.12 shows a similar plot for the parabolic-index fiber. It is immediately apparent that the modes of a parabolic-index fiber are more tolerant of cladding losses than the modes of a step-index fiber because the mode fields of the former are more effectively shielded from interacting with the cladding.

Next we discuss the effects of a lossy jacket surrounding a fiber with lossless core and cladding. Figure 3.13 shows again mode boundaries in mode number space for the step-index fiber for a fixed value of mode loss to jacket loss coefficient $\alpha_{st}/\alpha_j = 10^{-9}$. This very small ratio corresponds to mode losses of 1 dB/km and jacket losses of 1 dB/μm. This example may be somewhat extreme and approaches the limit of applicability of our jacket loss formulas. As the jacket loss increases to infinity we must expect the mode losses to decrease, because the mode field will no longer be able to penetrate into the jacket as deeply as it would for negligibly small jacket losses. In any case, our example serves to set an upper limit on jacket induced mode losses. The solid line in Fig. 3.13 deliniates the boundary between guided and radiation (or cladding) modes. The dotted lines deliniate the boundaries of modes whose losses are below or above the loss ratios $\alpha_{st}/\alpha_j = 10^{-9}$. The parameter values associated with each dotted line give the ratio of jacket radius b to core radius a. It is apparent that b/a must not be too small for a fiber with lossy jacket. For our example one would want to let b/a be larger than 1.5 or even 2.

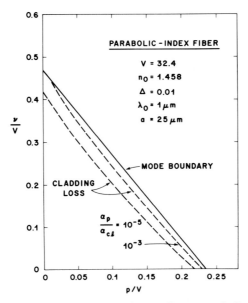

Fig. 3.12 This figure is similar to Fig. 3.11 but applies to parabolic-index fibers.

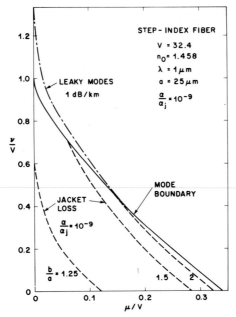

Fig. 3.13 This figure is similar to Fig. 3.11. The broken lines indicate boundaries between modes that exceed certain losses caused by a lossy jacket or by the mechanism of leakage.

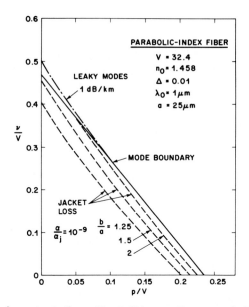

Fig. 3.14 This figure is similar to Fig. 3.13 but applies to parabolic-index fibers.

Figure 3.14 applies to the parabolic-index fiber and is similar to Fig. 3.13. It is again apparent that the parabolic-index fiber is more tolerant of a lossy jacket. Figure 3.14 shows that even a ratio of $b/a = 1.25$ affects far fewer modes of the parabolic-index fiber compared to the step-index case.

The meaning of the dash-dotted lines in Figs. 3.13 and 3.14 will be explained next.

In the preceding section we discussed the mechanism by which a guided mode beyond cutoff loses power by radiative tunneling as indicated in Fig. 3.8. The attenuation coefficients for tunneling leaky waves (Snyder and Mitchell, 1974) can be calculated by computing the power outflow per unit length and dividing it by the power carried in the fiber core. For step-index fibers we obtain the following formula that was obtained by approximating the Hankel function by means of the Debye approximation for large values of ν

$$2\alpha_{\mathrm{st}} = \frac{2(\nu - 1)}{n_0 k a^2}\, e^{2\nu} \left(\frac{U^2 - V^2}{4(\nu^2 - 1)}\right)^{\nu}. \tag{3.69}$$

The value of U may be obtained from (3.18).

For parabolic-index fibers we obtain correspondingly for $V_{\mathrm{p}} < V_{\mathrm{c}}$

$$2\alpha_{\mathrm{p}} = \frac{V_{\mathrm{p}}^2 e^{\psi}}{\pi n_0 k a^2 V_{\mathrm{c}}} \tag{3.70}$$

with

$$\psi = \nu \left[1 - \ln \left(\frac{4\nu^2}{V_p(V_c - V_p)} \right) - \frac{1}{2} \ln \left(\frac{1}{1 - (2\nu/V_c)^2} \right) \right]$$
$$+ \frac{V_c}{4} \ln \left(\frac{V_c + 2\nu}{V_c - 2\nu} \right) \tag{3.71}$$

The parameter V_c is defined by (3.65).

Instead of discussing losses of individual tunneling leaky waves we have plotted in Figs. 3.13 and 3.14 as dash-dotted lines the boundaries in mode number space that separate low-loss modes from high-loss leaky modes. Above the guided mode boundary (solid line) there are, strictly speaking, no guided modes but only the continuum of radiation modes (for infinitely extended cladding). However, the radiation modes can be superimposed to generate transient fields that are confined inside the fiber core for considerable distances along the fiber axis, these are the tunneling leaky modes (Suematsu and Furuya, 1975). Leaky modes close to the mode boundary suffer only very little radiation loss but their losses increase very substantially as we move away from the mode boundary. We can use (3.69) to calculate the U value of the leaky modes of the step-index fiber that have a certain ν value and a preselected amount of loss. This U value can then be used to calculate the radial mode number μ from (3.18) through (3.22). The dash-dotted line shown in Fig. 3.13 was drawn for tunneling leaky modes with a loss of $2\alpha = 1$ dB/km for a step-index fiber with $V = 32.4$, $n_0 = 1.458$, $a = 25$ μm, and $\lambda_0 = 1$ μm. Actually, the value of 2α hardly has any effect on the position of the dotted line. Had we chosen $2\alpha = 10$ dB/km instead of 1 dB/km the resulting dotted line would have been almost indistinguishable from the line shown in the figure.

The area that is occupied in mode number space is proportional to the number of modes. Figure 3.13 thus shows that the total number of effectively guided modes is increased by approximately 5% if we include low-loss tunneling leaky waves in the mode count. The number of low-loss tunneling leaky modes increases with increasing V values. Snyder has shown that the number of low-loss leaky modes becomes equal to the number of truly guided modes as $V \rightarrow \infty$.

The dash-dotted line in Fig. 3.14 is equivalent to that in Fig. 3.13 but applies to the parabolic-index fiber.

The operation of the W fiber, whose refractive index profile is shown in Fig. 3.5, depends critically on leaky mode losses. The W fiber is attractive because it is capable of yielding essentially single mode operation with a much larger core than the conventional step-index fiber, thus easing the splicing problem (Kawakami and Nishida, 1974). If the gap width $a_2 - a_1$ is sufficiently large the modes of the W fiber are primarily determined by the V value $V_2 = (n_1{}^2 - n_2{}^2)^{1/2}ka_1$. The outer cladding of index n_0 has only a slightly perturbing influence on the shape and propagation constants of the modes. However, the outer cladding permits light power to tunnel through the gap region between $a_1 < r < a_2$, so that modes with propagation constant $\beta < n_0k$ suffer radiation losses and do not take part in transmitting signals through the fiber, provided their losses are sufficiently high.

We can calculate the radiation losses of the leaky modes of the W fiber by a perturbation approach. First, we compute the field shape and propagation constants of the modes by assuming that $a_2 \to \infty$. Next we add a "reflected" evanescent wave in the region $a_1 < r < a_2$ to the field expressions and also provide a transmitted propagating wave in the medium with index n_0. The two new amplitude coefficients are determined by satisfying the boundary conditions for E_x and E_z at $r = a_2$. Now the radiation field is known in the region $a_2 < r$ and we can compute the power loss coefficient 2α by calculating the power outflow at infinity per unit length of the fiber and by dividing it through the power carried near the fiber core. After some additional simplifications, involving the use of approximations for cylinder functions of large arguments, we obtain the following result

$$2\alpha = \frac{8\kappa^3\sigma_0\zeta^2 e^{-2\zeta(a_2-a_1)}}{n_0(n_1{}^2 - n_2{}^2)(n_0{}^2 - n_2{}^2)k^5[(\kappa a_1)^2 - \nu^2]^{1/2}} \tag{3.72}$$

The parameters κ and ζ correspond to the step-index fiber of radius a_1 with infinite cladding ($a_2 \to \infty$) and it is $\sigma_0{}^2 = n_0k^2 - \beta^2$. Figure 3.15 is a graphical representation of the leaky mode loss of the LP_{11} mode of the W fiber. The LP_{01} mode is assumed to be able to propagate without radiation loss adjusting the refractive index n_0 continuously as the core radius a_1 changes. LP_{11} is the next higher mode and suffers the least radiation loss of all the leaky modes. We have used the parameters $n_1 = 1.5$, $n_2 = 1.485$, and the wavelength $\lambda = 1$ μm. The radiation power loss coefficient is plotted in decibels per meter as a function of the core radius a_1 and n_0 was chosen so that $n_0k = \beta_{01}$ for all values of a_1. The different curves in Fig. 3.15 apply to W fibers with different relative gap width $(a_2 - a_1)/a_1$. It is apparent that the radiation loss of the leaky LP_{11} mode decreases rapidly with increasing values of the core radius and also with increasing gap width. The fiber in our example supports always only one lossless mode, LP_{01} (or HE_{11} in conventional notation), but the leaky LP_{11} mode can propagate longer distances if its leakage loss is only slight. This consideration puts limits on the permissible core radii and gap widths. Another factor that limits the realizability of a single mode W fiber with large core radius is the fact that the ratio of n_1/n_0 must approach very close to unity if single mode operation is desired. Figure 3.16 shows a plot of $n_1/n_0 - 1$ as a func-

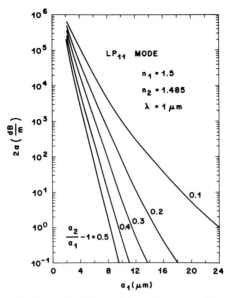

Fig. 3.15 Leaky mode losses of the LP_{11} mode of the W fiber.

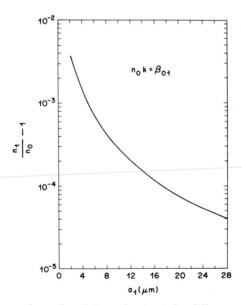

Fig. 3.16 This figure shows the relative refractive index difference between the core and outer cladding of the W fiber as a function of the core radius a_1. It is assumed that the LP_{01} mode is just at the border of becoming a leaky wave.

tion of the core radius a_1 which is obtained from the condition $n_0 k = \beta_{01}$. It is obvious that the refractive index value of the outer cladding must be more critically controlled if we want to let a_1 become large and still maintain single mode operation. If n_0 becomes smaller than the value required by the curve in Fig. 3.16, the LP_{11} mode is no longer a leaky wave, if it becomes larger, LP_{01} becomes a leaky wave and there is no truly guided mode.

Equation (3.72) assumes a maximum as a function of the jacket index n_0. In terms of the notation used for (3.62) this maximum value becomes for small values of ν:

$$2\alpha = (4U^2W/nka^2V^2)e^{-2\zeta(a_2-a_1)}. \tag{3.73}$$

The loss formula (3.62) for step-index fibers with lossy jacket may also be expressed in form of an approximation for small ν values [(3.62) holds for large ν]. If we now consider a jacket whose losses are adjusted to yield the maximum mode loss that can be achieved and compare it with a lossless jacket whose refractive index is designed for maximum leakage loss, we find the following upper value for the loss coefficient of the lossy jacket by equating (the modified equation) (3.62) with (3.73):

$$\alpha_j = 2W^2/nka^2. \tag{3.74}$$

3.5 LOSSES CAUSED BY CONSTANT FIBER CURVATURE

In principle, fibers whose axis is bent into a circle lose power by radiation. For slight bends the losses are so minute as to be unobservable. However, the loss coefficients increase exponentially with decreasing radius of curvature. At a certain critical radius the losses become noticeable, if the fiber is bent just a little bit more the losses become prohibitive and mode guidance is lost for practical purposes. Curvature losses thus act like a switch. Above a certain critical radius of curvature they are negligible; below this critical radius they are fatal. For this reason it is of little interest to describe the curvature loss coefficients exactly. It is more important to know the value of the critical radius of curvature where the losses pass from negligible to prohibitively high values.

To understand the mechanisms of curvature losses we may assume that the field inside the curved fiber is not changed very much from its shape in the straight fiber. Figure 3.17 shows a curved fiber with infinite cladding and indicates the phase fronts of a guided mode traveling in it. The phase velocity of the wave inside the fiber core may be expected to be nearly the same in a straight fiber. Since the phase fronts are pivoted at the center of curvature of the circular fiber there must be a critical distance on the side opposite the center of curvature at which the phase fronts ap-

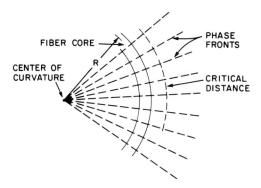

Fig. 3.17 Curved fiber with constant phase fronts and critical radius indicated by dotted lines.

proach the velocity of light in the cladding medium. The light field resists being dragged at a velocity exceeding the velocity of plane waves in the medium and radiates away (Marcatili and Miller, 1969).

An alternate explanation for the occurrence of curvature losses is based on the WKB method. It can be shown that a transformation of coordinates allows us to regard a curved fiber as an equivalent straight fiber with a distorted refractive index distribution of the form (Heiblum and Harris, 1975)

$$n = n(r)[1 + (r/R) \cos \phi]. \tag{3.75}$$

Figure 3.7 thus assumes the form shown in Fig. 3.18. Because of the distortion of the refractive index distribution every mode becomes a tunneling leaky wave according to the explanation given in conjunction with Fig. 3.8.

It can be shown that the curvature loss coefficient can be expressed as

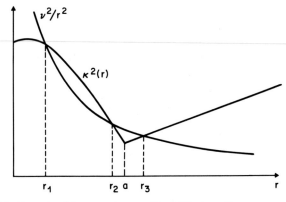

Fig. 3.18 This figure is of the same type as Fig. 3.8 but applies to a curved fiber. It shows that in a curved fiber every mode is a leaky wave.

(Miller and Talanov, 1956; Marcatili, 1969; Lewin, 1974; Arnaud, 1974b; Chang and Kuester, 1975; Marcuse, 1976a)

$$2\alpha = K \exp\left\{-\frac{2}{3} n_0 kR \left(\frac{\zeta^2}{n_0^2 k^2} - \frac{2a}{R}\right)^{3/2}\right\}. \tag{3.76}$$

The parameter ζ is defined as

$$\zeta = (\beta^2 - n_0^2 k^2)^{1/2} \tag{3.77}$$

The coefficient K is different for different types of fibers but its value is of no great importance since it is the exponential function that determines whether the loss is negligibly small or prohibitively high. For any given value of the radius of curvature there exists a value of $\zeta = \zeta_c$ that causes the exponent of the exponential function to become unity. This effective cutoff value of ζ is given by

$$\zeta_c^2 = n_0^2 k^2 \left[\frac{2a}{R} + \left(\frac{3}{2n_0 kR}\right)^{2/3}\right]. \tag{3.78}$$

For $R = \infty$ we obtain $\zeta_c = 0$, the correct cutoff value for the straight fiber.

The total number of modes that can be guided by the curved fiber is smaller than the total number in the straight fiber. Using (3.78) instead of the cutoff condition $\zeta_c = 0$ we obtain from (3.43) the formula for the effective number of modes supported by a curved fiber (Gloge, 1972b).

$$N_{eff} = N_\infty \left\{1 - \frac{g+2}{2g\,\Delta} \left[\frac{2a}{R} + \left(\frac{3}{2n_0 kR}\right)^{2/3}\right]\right\} \tag{3.79}$$

with the total number of modes of the straight fiber

$$N_\infty = [g/(g+2)](n_0 ka)^2 \Delta \tag{3.80}$$

The parameters g and Δ are the exponent of the power law and the relative index difference appearing in (3.2).

Figure 3.19 shows the radius of curvature as a function of fiber radius "a" that causes the mode number to decrease by 50%. The solid lines pertain to step-index fibers ($g = \infty$) while the dotted lines belong to the parabolic-index fibers ($g = 2$). Two values of Δ are used, $\Delta = 0.01$ and $\Delta = 0.001$. The refractive index of the cladding is assumed to be $n_0 = 1.5$ and the vacuum wavelength of the guided light is $\lambda = 1$ μm. The figure shows that the permissible radius of curvature increases with increasing fiber radius; narrower fibers can be bent more sharply. It is also apparent that larger values of Δ permit more abrupt bends. The loss of power is nearly instantaneous. Once a bent fiber section has a local radius of curvature that is equal to the permissible values shown in Fig. 3.19 half the light power is lost. Figure 3.20 applies to the case $(N_\infty - N_{eff})/N_\infty = 0.1$.

The reader is cautioned not to confuse losses caused by constant fiber

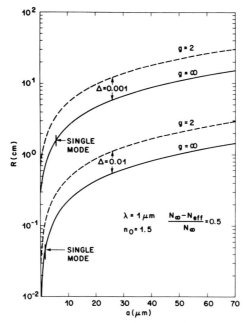

Fig. 3.19 Critical radius of curvature which decreases the number of propagating modes by 50% is shown as a function of the core radius a. In this figure $(N_\infty - N_{eff})/N_\infty = 0.5$.

curvature with losses caused by random or sinusoidal bends. Power loss in constant curvature bends is caused by a leakage process of the guided mode that is analogous to quantum mechanical tunneling through a potential barrier. Curvature losses due to random or sinusoidal bends are caused by coupling between the guided mode and radiation modes—an entirely different process. Losses associated with random bends are discussed in Section 3.13.

3.6 CROSS TALK BETWEEN FIBERS

Optical fiber cables contain many similar fibers whose mutual interaction makes itself felt as cross talk. There are two mechanisms leading to exchange of optical energy between fibers—the directional coupler effect (Marcuse, 1971a) and scattering cross talk (Marcuse, 1971b). Two identical fibers in close proximity may exchange power by the directional coupler effect because the evanescent field tail of one fiber overlaps the core region of its neighbor (Vanclooster and Phariseau, 1970; Arnaud, 1974a). If the fibers are truly identical each of their modes couples to its counterpart in the neighboring fiber delivering the power ΔP to its far end at $z = L$ that

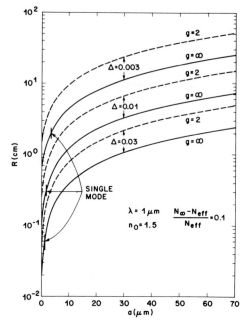

Fig. 3.20 Same as Fig. 3.19 but with $(N_\infty - N_{eff})/N_\infty = 0.1$.

is proportional to $(\Delta P/P \ll 1$, that is L much less than power exchange length is assumed)

$$\Delta P \propto L^2 e^{-2\zeta_{\nu\mu}D}. \tag{3.81}$$

D is the spacing between the core boundaries of the two neighboring fibers and $\zeta_{\nu\mu}$ is the decay parameter of mode ν, μ defined by (3.11). Equation (3.81) shows that cross talk decreases very rapidly with an increase of the spacing D between the fiber cores but it shows also that modes nearer cutoff, whose $\zeta_{\nu\mu}$ values are smaller, couple more strongly than modes far from cutoff.

Light that is scattered out of one fiber by a discontinuity of the refractive index distribution in the fiber core or by a core–cladding interface irregularity cannot be captured by a perfect neighboring fiber. To inject a light ray from the outside into the core of a perfect fiber requires imaginary angles of incidence, that is an evanescent field rather than a radiation field. However, a fiber can capture power that is radiated by its neighbor if it also contains irregularities (Marcuse, 1971b). Because scattering cross talk is caused by the same mechanism that gives rise to scattering loss α_r the amount of power ΔP that is exchanged between two fibers is proportional to the square of the scattering loss coefficient

$$\Delta P \propto (2\alpha_r L)^2. \tag{3.82}$$

Good optical fibers with small scattering losses thus have also very small scattering cross talk.

Cross talk can be reduced by providing each fiber in a cable with a lossy jacket. We have seen in Section 3.4 that lossy jackets increase the losses of the guided fiber modes. Actually, if cross talk between fibers without lossy jackets turns out to be a problem it can be shown that the addition of a lossy jacket increases the losses of the strongly coupled mode beyond acceptable levels (Marcuse, 1971a). It is thus clear that cross talk caused by the directional coupler mechanism must be avoided by sufficiently thick claddings around each fiber that also provide sufficient separation between the fiber cores. However, it appears prudent to provide each fiber in a cable with a protective lossy jacket to make sure that residual cross talk may be suppressed. The cross-talk reduction afforded by a lossy jacket is proportional to the loss that a plane wave would suffer in traveling at right angles through the lossy jacket. A jacket that has a loss of 2 dB/μm and is 10 μm thick would reduce by 20 dB whatever residual cross talk might have existed in its absence.

3.7 EXCITATION OF FIBERS

Two types of sources are used to inject light into optical fibers, light-emitting diodes (LEDs) and diode lasers. In this section we consider the problem of injecting an idealized laser beam into a single mode step-index fiber and of injecting light from an LED into a multimode parabolic-index fiber (Yang and Kingsley, 1975; Marcuse, 1975).

Let us begin by considering the problem of injecting incoherent light of an LED into a multimode fiber with a general profile of the form (3.2). We assume that the LED is in direct contact with the fiber core covering its entire cross section as shown in Fig. 3.21. Each element of the LED of area dA radiates the amount of power

$$\Delta P = B \, dA \, d\Omega \cos \theta \qquad (3.83)$$

into the direction θ and the element of solid angle $d\Omega$; B is the brightness of the source and the cosine function indicates that the LED source is considered to be a Lambert law radiator. Not all the light from the LED is cap-

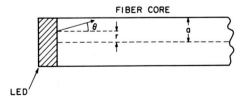

Fig. 3.21 An LED is shown in direct contact with the fiber end.

tured by the fiber core. Rays whose angles are too steep to be trapped by
the fiber escape through the core boundary into the cladding. The trapping angle at each position r in the core is obtained if we consider that the
cutoff value of the propagation constant of each mode is $\beta_c = n_0 k$. The ray
angle associated with a given mode may be defined as

$$\cos \theta = \beta/n(r)k. \tag{3.84}$$

The critical angle at each radius r is thus defined as

$$\theta_c(r) = \arccos\left[n_0/n(r)\right] = \arcsin[1 - (n_0/n(r))^2]^{1/2} \tag{3.85}$$

The total power that the LED injects into the fiber core is thus

$$P_f = B \int_0^a r\, dr \int_0^{2\pi} d\phi \int_0^{2\pi} d\phi' \int_0^{\theta_c(r)} d\theta \sin\theta \cos\theta$$
$$= [2\pi^2 g/(g+2)]Ba^2\Delta \tag{3.86}$$

To obtain this expression we used the definition of the element of solid
angle $d\Omega = \sin\theta\, d\theta\, d\phi'$, $dA = r\, dr\, d\phi$ and (3.2). The injection efficiency I_e
may be defined by the ratio of light power P_f injected into the fiber to the
power P_d that the LED of area πa^2 can maximally radiate into the half-
space solid angle 2π,

$$P_d = \pi^2 a^2 B. \tag{3.87}$$

We thus have

$$I_e = 2g\,\Delta/(g+2) \tag{3.88}$$

For the step-index fiber with $g = \infty$ we have $I_e = 2\Delta$ while the injection efficiency for the parabolic-index fiber with $g = 2$ is $I_e = \Delta$. The injection efficiency increases proportionally to the relative difference between the refractive index value on axis and the cladding value. For a typical value of
$\Delta = 0.01$ only 1% of the LED light is actually trapped in the core of a
parabolic-index fiber.

It is important to know: how critical is the alignment of the LED with
respect to the fiber core? Figure 3.22 shows an LED that is displaced in
longitudinal and transverse direction relative to the core of a parabolic-
index fiber. We present the power P_f that is injected into the fiber core in

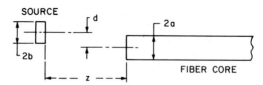

Fig. 3.22 An LED is shown displaced in transverse and longitudinal direction from the
fiber end.

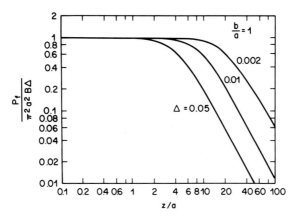

Fig. 3.23   Normalized power injected into a parabolic-index fiber as a function of relative end separation between source and fiber for several values of the relative refractive index difference between core and cladding.

normalized form so that the normalized value of the power is unity for an LED in direct contact with the fiber core. Fig. 3.23 shows the normalized power as a function of (normalized) distance between LED and fiber. The most striking feature of these curves is the horizontal portion for small values of z/a that shows how insensitive the injection process is to longitudinal displacement of the LED from the end of the fiber. For $\Delta = 0.01$ we may displace the LED by approximately five core diameters before the injected power drops to half the maximum value that is achieved when the LED is in direct contact with the fiber core. This tolerance to longitudinal source displacement improves with decreasing values of Δ but of course, the total available power drops also.

The dependence of the injected light power into a parabolic-index fiber on transverse source displacement is shown in Fig. 3.24 for several values of longitudinal source displacement. Fig. 3.24 is drawn for $\Delta = 0.01$ but the shape of the curves is actually independent of Δ. In particular the curve for $z/a = 0$ does not depend on Δ and for the other cases only the limiting value for $d/a = 0$ depends on Δ so that the curves for other Δ values are obtained by vertical displacement. For an LED in direct contact with the fiber the power drops to one-half of its maximum value if the diode (whose radius equals the core radius) is displaced by 0.9 core radii.

We now turn to the problem of exciting a single mode step-index fiber with a laser beam as schematically shown in Fig. 3.25 (Stern *et al.*, 1970; Marcuse, 1970). We assume that the laser beam may be represented as a free-space Gaussian beam mode whose electric and magnetic field vectors are indicated by $\mathbf{E}_g$ and $\mathbf{H}_g$. Lasers whose cavities consist of spherically curved mirrors provide excellent realizations of Gaussian beam modes.

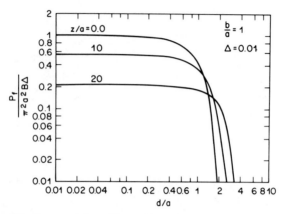

Fig. 3.24 Normalized power injected into a parabolic-index fiber as a function of the relative transverse displacement of the source with respect to the fiber core for several values of the longitudinal separation.

The output from an injection laser is usually not rotationally symmetric around the beam axis and may be represented as a Gaussian beam with different spot sizes in two orthogonal planes perpendicular to its direction of propagation. Such a distorted beam can, in principle, be transformed to the symmetric Gaussian shape by the use of cylindrical lenses. We ignore this complication and address ourselves to the basic problem of the excitation efficiency of the LP_{01} (HE_{11}) mode of a single mode step-index fiber with an ideal Gaussian beam.

The mathematically exact solution of this problem would require to postulate transmitted and reflected waves in free space and in the fiber and determine their expansion coefficients from the boundary conditions at the input plane shown in Fig. 3.25. The exact solution is very difficult, however, and we use an approximate approach that ignores the reflected waves at the input plane and assumes that only transmitted waves exist. With this assumption it is no longer possible to satisfy the boundary conditions exactly. We may define a coupling coefficient I_e by matching the electric field component of the Gaussian input beam to the fiber modes

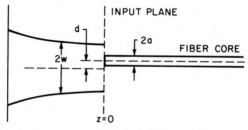

Fig. 3.25 Excitation of a single mode step-index fiber with a Gaussian laser beam.

$$I_e = \frac{1}{2P} \int \int (\mathbf{E}_g \times \mathcal{H}_\nu{}^*)_z \, dx \, dy. \tag{3.89}$$

The script symbol $\mathcal{H}$ indicates the magnetic field vector of a fiber mode labeled ν. The power transmission coefficient of the LP_{01} mode labeled $\nu = 0$ is defined as

$$T = |I_e|^2. \tag{3.90}$$

The dominant transverse field component of the incident Gaussian beam may be represented as

$$E_g = A e^{-(r/w)^2}. \tag{3.91}$$

The beam width parameter w is changing along the beam, but we assume that the narrowest portion of the beam, its waist, is aligned with the fiber end. The transmission coefficient T depends on the relative size of the Gaussian beam with respect to the fiber mode and on its alignment with respect to the fiber axis, both with respect to tilt and offset. If we assume that the Gaussian beam is perfectly aligned with respect to the fiber axis and let its width increase from zero values we find that the transmission coefficient begins to increase at first. It reaches a maximum at the point where the Gaussian beam matches in size with the fiber mode. In this optimum configuration the transmission coefficient can be as high as 99.8%. For a single mode step-index fiber we can express the optimum width of the Gaussian beam by the formula (Marcuse, 1977)

$$w/a = 0.65 + 1.619/V^{3/2} + 2.879/V^6 \tag{3.92}$$

for parabolic-index fibers we obtain instead,

$$w/a = (2/V)^{1/2} + 0.23/V^{3/2} + 1.801/V^6. \tag{3.93}$$

The beam width w of the Gaussian function is defined by (3.91), a is the core radius of the fiber and V is defined by (3.49). For a step-index fiber we have for $V = 2.4$, $w/a = 1.1$.

If the Gaussian input beam is offset with respect to the fiber axis by an amount d the transmission coefficient for passing power from the beam to the single mode fiber is approximately

$$T = \exp(-d^2/w^2), \tag{3.94}$$

where w is given by (3.92) or (3.93). For a tilted input beam we have

$$T = \exp[-(\pi n_0 w \theta/\lambda)^2], \tag{3.95}$$

where θ is the tilt angle.

An offset with $d = w$ reduces T from its maximum value of (nearly) unity to $1/e$. The tilt angle

$$\theta_e = \lambda/\pi n_0 w \qquad (3.96)$$

accomplishes this reduction to $1/e$ for a tilted input beam.

Because the fields of single mode fibers can be approximated by Gaussian field distributions these formulas can also be used to estimate the losses of single mode fiber splices.

3.8 NEAR AND FAR FIELD AT THE FIBER END

The light that reaches the end of an optical fiber escapes into space forming a radiation pattern that depends on the type of fiber and distribution of power among the modes. For single mode fibers with small core area the radiation pattern must be calculated with the help of the Kirchhoff–Huygens diffraction theory. For multimode fibers with large core area the radiation pattern can approximately be obtained with the help of geometrical optics. Consider, for example, a multimode step-index fiber which carries only one mode. The mode field can be decomposed into locally plane waves whose propagation vectors (Marcuse, 1974) (e_r, e_ϕ, and e_z are unit vectors)

$$\mathbf{K} = [n_1^2 k^2 - \beta_{\nu\mu}^2 - (\nu/r)^2]^{1/2} e_r + (\nu/r) e_\phi + \beta_{\nu\mu} e_z. \qquad (3.97)$$

In this expression ν is the azimuthal mode number appearing in (3.9) and $\beta_{\nu\mu}$ is the propagation constant (the z component of the propagation vector). The magnitude of $\mathbf{K}$ is $n_1 k$. The angle θ between the direction of the ray and the z axis in the fiber core is defined by the expression

$$\cos \theta = \beta_{\nu\mu}/n_1 k. \qquad (3.98)$$

On leaving the core the angle is increased to θ_a according to Snell's law,

$$n_a \sin \theta_a = n_1 \sin \theta, \qquad (3.99)$$

with θ_a indicating the radiation angle and n_a the refractive index of the outside medium ($n_a = 1$ in case of air). The mode characterized by the mode labels ν and μ thus illuminates a narrow hollow cone whose cone angle is given by the formula

$$\theta_a = \arcsin\{(n_1/n_a)(1 - \beta_{\nu\mu}^2/n_1^2 k^2)^{1/2}\} \approx \kappa_{\nu\mu}/n_a k. \qquad (3.100)$$

The ring of light that is produced on a screen is not uniformly bright but breaks down into 2ν bright and dark spots (except for $\nu = 0$ which produces a uniformly illuminated ring). Observation of the far-field radiation pattern is thus useful for determining the nature of an individual mode or for determining approximately the number of modes that are excited in a multimode step-index fiber by observing the boundaries of the cone of illumination.

Graded-index fibers do not produce such pronounced, sharp cones of light for a given mode since it is not possible to define a unique mode angle in this case. However, some useful information can still be extracted from observation of the near- and far-field radiation patterns of graded-index fibers.

We consider a graded-index fiber that is excited by an incoherent source. If the source covers the entire core area it excites all fiber modes with equal intensity. We have seen that ray angles can be associated with modes. In graded-index fibers rays do not propagate with constant angles but change their direction continuously inside the fiber core. However, we can still say that rays associated with a given mode cannot exceed a maximum angle that is defined by (3.98). The maximum ray angle θ_m that can be contained inside the fiber core is thus defined by (3.98) (with n_1 replaced by $n(r)$) if we replace the propagation constant of the mode with its cutoff value $n_0 k$,

$$A(r) = n(r) \sin \theta_m = (n^2(r) - n_0^2)^{1/2} = \{2n_1^2[1 - (r/a)^g]\Delta\}^{1/2} \quad (3.101)$$

The right-hand side of this expression is obtained by using (3.2). The function $A(r)$—the local numerical aperture of the graded-index fiber—depends on the radius r at which the light beams are injected into the fiber. If all modes are equally attenuated, the power distribution at the fiber end equals that at its input which is proportional to the solid angle filled with radiation which, in turn, is proportional to the square of the numerical aperture $A(r)$ (Gloge and Marcatili, 1973)

$$P(r) = P_t \frac{g + 2}{\pi a^2 g} \left[1 - \left(\frac{r}{a}\right)^g\right] \propto n^2(r) - n_0^2. \quad (3.102)$$

$P(r)$ is the power density at the fiber end face and P_t the total amount of power integrated over the same end face. Equation (3.102) shows that the near-field power distribution is proportional to the refractive index distribution (3.2) in the fiber core. However, it is important to remember that this result was obtained by ignoring leaky modes. Fibers with very large V values may support a large number of low-loss leaky modes that would distort the power distribution (3.102). Corrections of (3.102) accounting for the contribution of leaky modes have been suggested (Sladen et al., 1975).

The far-field radiation pattern can be obtained just as easily if we consider that each annular ring at the fiber end illuminates a (full) cone whose boundary is defined by the angle θ_m of (3.101). The cone angle, to which the elements on the fiber end face contribute, decrease with increasing distance r from the fiber axis. At any given outside far-field angle θ_a the light contribution stems from and is proportional to an area πr^2 that is capable of illuminating this angle. The radius r is obtained from (3.101) if we use $n(r) \sin \theta_m = n_a \sin \theta_a$,

$$r = a\{1 - (n_0^2 \sin^2 \theta_a / 2n_1^2 \Delta)\}^{1/g} \qquad (3.103)$$

The power density in the radiation far field is thus (Gloge and Marcatili, 1973)

$$p(\theta) = p(0)\{1 - (n_a^2 \sin^2 \theta_a / 2n_1^2 \Delta)\}^{2/g} \qquad (3.104)$$

with $p(0)$ indicating the on-axis light intensity. This far-field power density also ignores the contribution from leaky modes and holds only if all guided modes are equally excited.

3.9 LOSS IN SPLICES

We consider selected phenomena that occur when fibers are joined end to end with flat and perpendicular end faces and have an index matching and lossless material between the ends. Under such conditions, loss may occur as a result of a lateral offset, an angular misalignment, a tilt, or a longitudinal displacement or gap. There may also be discrepancies between the diameters or the index profiles of the two fibers or there may be variations of the cross-sectional symmetry. The following discussion is limited to a theoretical description of some of these phenomena, each considered separately. A statistical evaluation of the expected total loss value for given splicing or connector tolerances is attempted in Chapters 14 and 15.

We assume all differences between fiber or alignment parameters to be small so that expansions in terms of such differences can be curtailed. In the case of lateral offsets, for example, we can write the relation between the radial coordinates r_1 and r_2 of the two fibers (see Fig. 3.26) as

$$r_1 - r_2 \approx s \cos \phi, \qquad (3.105)$$

where ϕ is the azimuthal angle. Let us first consider the case of the single mode fiber. Taylor expansion of the fundamental mode field (3.9) in terms of s yields

$$E_x(r_2, \phi) = E_x(r_1, \phi) - As \kappa \cos \phi \begin{cases} J_1(\kappa r_1) & \text{for } r \le a \\ \dfrac{J_0(\kappa a) K_1(\zeta r_1)}{K_0(\zeta a)} & \text{for } r \ge a \end{cases}. \qquad (3.106)$$

The relation between the amplitude parameter A and the total incident power P is given by (3.13). One obtains the offset loss c_s by squaring the second term of (3.106), integrating it over the fiber cross section, and normalizing it with respect to P. If we assume the cladding to extend to infinity, we find

$$c_s = \tfrac{1}{2}(sWJ_0(U)/aJ_1(U))^2 \qquad (3.107)$$

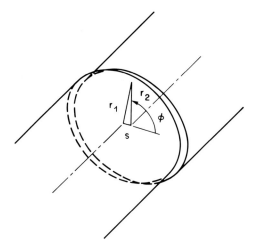

Fig. 3.26 This figure defines the coordinates used to describe the lateral offset of two fibers joined by a splice.

for $s/a \ll 1$. In practice, the result is applicable for s smaller than $a/2$. We can interpret the ratio $2^{1/2} a J_1/W J_0$ in (3.107) as the effective radius of the mode field. It coincides very closely with the effective Gaussian width defined in (3.92). Figure 3.27 shows this effective radius normalized with respect to the core radius and plotted versus V. For $V = 3$, the normalized width is unity and, hence the loss becomes $c_s = (s/a)^2$.

A tilt angle ψ between the axes of two single mode fibers leads to a phase retardation $\beta r \sin \phi \sin \psi$ of the incident field at the tilted front face of the second fiber (see Fig. 3.28). The loss computation for small angles is not as simple as in the previous case because the phase correction indicated above can increase to several radians for large r. However, numerical evaluations (Cook *et al.*, 1973) have shown that the loss c_ψ depends again on

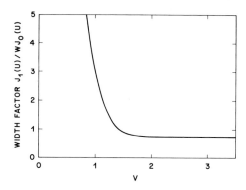

Fig. 3.27 Relative effective mode field radius as a function of the V number.

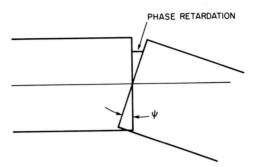

Fig. 3.28 The tilt angle of two fibers joined in a splice is defined in this figure.

the mode width $2^{1/2}aJ_1/WJ_0$. A wide-field distribution is associated with a narrow far-field width, which in turn implies a strong sensitivity to angular misalignment and high tilt loss. If β in the phase correction term is replaced by n_1k, the numerical results can be approximated by the relation

$$c_\psi = (n_1k \sin \psi)^2 [aJ_1(U)/WJ_0(U)]^2. \tag{3.108}$$

At $V = 3$, $c_\psi = 2.25\Delta^{-1} \sin^2 \psi$ or 4.3% for a tilt angle of 0.25° and $\Delta = 0.001$. Since the effective width coincides with the equivalent Gaussian width, we can obtain approximate values of $c = 1 - T$ also from (3.92), (3.94), and (3.95).

Turning now to multimode fibers, we use geometrical ray optics. In the simplest case of uniform power distribution among all modes, splice loss equals the fraction of rays lost in passing the splice (C. M. Miller, 1976). For a step-index fiber under these conditions, every point of the core cross section is passed by the same number and distribution of rays. Hence, a splice offset s results in a loss equal to the fraction of core area which does not overlap with the opposite core. This core area approximately equals a stripe of length $2a$ and width s as shown by the rearrangement of Fig. 3.29. The offset loss therefore becomes $2s/\pi a$.

Here, as in the following discussion, only trapped modes are assumed to arrive at the splice point. The presence of leaky modes leads to a slight correction of the results given here (Gloge, 1976).

The ray distribution in a uniformly excited graded-index fiber is conveniently described by the local numerical aperture $A(r)$ introduced in (3.101). The number of rays passing a point at a distance r from the fiber axis is πA^2. Rays lost in the splice offset are again those rays passing the stripe of Fig. 3.29b. The ratio of rays lost to rays passed is therefore approximately.

$$c_s = \frac{s}{\pi} \left(\int_0^a A^2 \, dr \bigg/ \int_0^a A^2 r \, dr \right). \tag{3.109}$$

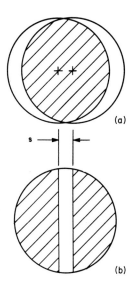

Fig. 3.29   This figure illustrates the core area that is lost due to lack of overlap caused by the transverse displacement of two multimode fibers.

For the index distribution (3.2),

$$c_s = \frac{2s}{\pi a} \frac{g + 2}{g + 1}.$$ (3.110)

The arguments applied to the near field in the case of an offset can be applied to the far-field distribution in the case of an angular misalignment between two multimode fibers. We represent the far field in terms of the angular coordinate ψ introduced by (3.98). If the misalignment or tilt angle ψ between the two fiber axes is small, the angular coordinates θ_1 and θ_2 for the two fibers can be related

$$\sin \theta_1 - \sin \theta_2 = \sin \psi \sin \phi$$ (3.111)

in analogy with (3.105). We assume the incident far-field distribution $p(\theta)$ to be the result of a uniform power distribution among all modes. The analog with (3.109) then leads to

$$c_\psi = \frac{\sin \psi}{\pi} \left[\int_0^{(2\Delta)^{1/2}} p(\theta)\, d\theta \Big/ \int_0^{(2\Delta)^{1/2}} p(\theta)\theta\, d\theta \right]$$ (3.112)

for the tilt loss. In case of the index profile (3.2), we use the far-field power distribution (3.104) with $n_a = n_1$ to obtain

$$c_\psi = \frac{\sin \psi}{(2\pi\Delta)^{1/2}} \frac{\Gamma(2/g + 2)}{\Gamma(2/g + \frac{3}{2})}.$$ (3.113)

For $g = 2$ and $g = \infty$, (3.109) and (3.113) lead to the same relationships between loss and relative misalignment s/a or $\sin \psi/(2\Delta)^{1/2}$. An offset of 10% of the core radius or a tilt angle of 7% of the fiber numerical aperture produces a loss of 6.4% or 0.29 dB in a uniformly excited splice of two step-index fibers and 8.5% or 0.39 dB for square-law fibers.

These results hold for a uniform mode power distribution which rarely exists in practice. More realistic loss values are derived below. In practical fibers, mode coupling and loss deplete the modes near cutoff. Since these modes contribute most of all to offset and tilt loss, the power arriving in these modes at the splice point must be correctly evaluated. Section 12 lists mode power distributions expected under steady state conditions, i.e., after a sufficiently long propagation length when a significant power exchange among the modes has taken place.

According to (3.162), the steady-state power distribution of the step-index fiber follows the Bessel function J_0; because of (3.98), the associated far field distribution is $p(\theta) = J_0[2.405\,\theta/(2\Delta)^{1/2}]$. For a small misalignment angle ψ, we use (3.111) and the first term of the Taylor expansion of $p(\theta)$ at $\theta = (2\Delta)^{1/2}$ [approaching $(2\Delta)^{1/2}$ from angles smaller than $(2\Delta)^{1/2}$]. Properly normalized for unit input power, the tilt loss becomes

$$c_{\psi 0}(g \to \infty) = \frac{1}{8}(2\Delta)^{1/2} \sin^2 \psi \left[\left|\frac{dp}{d\theta}\right|_{\theta \to (2\Delta)^{1/2}}\bigg/ \int_0^{(2\Delta)^{1/2}} p\theta\, d\theta\right] \quad (3.114)$$

or, for the Bessel function distribution mentioned earlier, $c_{\psi 0} = 0.36\Delta^{-1} \sin^2 \psi$.

Equation (3.114) represents the loss incurred at the splice point. Additional splice loss is caused in the fiber immediately following the splice as a result of the fact that the power distribution behind the splice initially deviates from the steady-state distribution. The gradual reconversion to the steady state results in an excess loss which approximately equals the loss in the splice (Gloge, 1976). The total loss in two long step-index fibers spliced together with some angular misalignment ψ is therefore $c_\psi(g \to \infty) = 0.72\Delta^{-1} \sin^2 \psi$. A tilt angle of 7% of the numerical aperture produces a loss of 0.72% as compared to 6.4% in the case of a uniform input distribution.

Within the accuracy of the ray model, the near-field distribution of the step-index fiber is unaffected by the mode power distribution. As a result, the loss caused by a lateral offset is given by (3.109) independent of the incident power distribution. If the steady-state distribution is incident, a lateral offset causes no modification of this distribution behind the splice and hence no additional loss there. Again, these findings are derived from the ray model and hold within the accuracy of that model.

The steady-state distribution in the parabolic-index fiber is given by (3.174). It also follows the Bessel function J_0. The resulting dependence on the near- and far-field coordinates r and θ can be presented in the form $p(\rho) = J_0(2.405\rho)$, where ρ is a new variable

$$\rho = [r^2/a^2 + (\sin^2 \theta)/2\Delta]^{1/2}. \tag{3.115}$$

Similarly, loss as a result of an offset s or a tilt angle ψ can be expressed in terms of a parameter

$$\sigma = [s^2/a^2 + (\sin^2 \psi)/2\Delta]^{1/2}. \tag{3.116}$$

The loss in the splice becomes (Gloge, 1976).

$$c_{\sigma 0}(g = 2) = \frac{\sigma^2}{16} \left(\left| \frac{dp}{d\rho} \right|_{\rho \to 1} \middle/ \int_0^1 p(\rho)\rho^3 \, d\rho \right). \tag{3.117}$$

Additional loss occurs in the fiber following the splice until the steady state is reestablished. This loss approximately equals the loss in the splice so that $c_\sigma = 2c_{\sigma 0}$. For $p(\rho) = J_0(2.405\rho)$, $c_\sigma = 2.3\sigma^2$. An offset of 10% of the core radius or a tilt angle of 7% of the numerical aperture causes a loss of 2.3% or 0.1 dB.

A longitudinal displacement or gap between fibers seems to be more difficult to evaluate. Equation (14.6) presents an estimate for multimode fibers. Note that the end separation can be larger than the offset by a factor equal to the inverse numerical aperture of the fiber before the separation loss approaches the offset loss. Measurements have corroborated such estimates.

Even with perfect alignment, a change in core diameter or in index profile, or a variation of the cross-sectional symmetry can cause splice loss. For multimode fibers, this loss occurs primarily where light passes from a larger to a smaller core diameter or index difference. The result is a loss of modes corresponding to the decrease in mode volume. For the index profiles (3.2), for example, the mode volume N is given by (3.48).

If we assume that the index profile of the second fiber is a replica of the first reduced by the reduction ratio a_2/a_1 of the two core radii, the loss as a result of the mismatch is (Gloge, 1976)

$$c_a = 1 - (a_2/a_1)^2. \tag{3.118}$$

This relation holds independently of the shape of the index profile as long as the profile decreases monotonically from the core axis to its periphery. It is valid only for a uniform mode distribution. A variation of the total index difference under the same condition leads to

$$c_\Delta = 1 - \Delta_2/\Delta_1. \tag{3.119}$$

A 2.5% diameter change or a 5% change in Δ causes a loss of 5%. For the step index fiber, Equation (3.118) is independent of the mode power distribution. A decrease in the index difference Δ of a step-index fiber reduces the acceptable far-field distribution. If the far-field distribution is $p(\theta)$ as a result of a nonuniform mode power distribution, the approach chosen earlier to derive (3.112) leads in this case to

$$c_\Delta = (2\Delta_1)^{1/2}(\Delta_1^{1/2} - \Delta_2^{1/2}) \left[\left(\frac{dp}{d\theta}\right)_{\theta \to (2\Delta_1)^{1/2}} \bigg/ \int_0^{(2\Delta)^{1/2}} p\theta\, d\theta \right].   (3.120)$$

For $p = J_0[2.405\theta/(2\Delta_1)^{1/2}]$, one has

$$c_\Delta = 1.44[1 - (\Delta_2/\Delta_1)^{1/2}]^2   (3.121)$$

or a loss of 0.1% for a 5% change in Δ.

It is interesting to compare the multimode case with splice losses of single mode fibers with no tilt or offset but different core radii. Using the theory of Section 3.7 we obtain the loss expression

$$c_{as} = [(w_1^2 - w_2^2)/(w_1^2 + w_2^2)]^2,   (3.122)$$

where w_1 and w_2 are the width parameters of Gaussian beams that are adjusted to match the fiber mode fields optimally, w_1 is thus the width of the mode of the fiber with radius a_1 and a corresponding relation holds for fiber 2. The width parameters can be obtained from (3.92) for step-index and from (3.93) for parabolic-index fibers. Fig. 3.30 shows the single mode splice loss for step-index fibers as a function of the core radius ratio a_2/a_1. It is assumed that a_1 is held fixed while a_2 changes. The refractive index ratios of both fibers are assumed to be the same. For each curve in Fig. 3.30 the V value of fiber 1 is held constant while V_2 changes proportionally with the core radius a_2 of fiber 2. A 2.5% diameter change causes a single mode splice loss of $c_{as} = 7 \times 10^{-5}$ for $V = 2.4$ or of $c_{as} = 1.5 \times 10^{-4}$ for $V = 3$. A

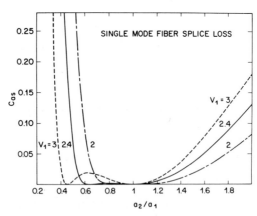

Fig. 3.30 Single mode splice loss for step-index fibers as a function of core radius ratio a_2/a_1.

5% splice loss ($c_{as} = 0.05$) is obtained for a core radius variation of $a_2/a_1 = 1.6$ for $V = 2.4$ or for $a_2/a_1 = 1.47$ for $V = 3$. These numbers show that single mode splices are more tolerant of relative core radius variations than multimode fibers.

In the case of the parabolic profile, the mode power distribution can be expressed in terms of the variable ρ given by (3.115). Because of the equivalence of a and $(2\Delta)^{1/2}$ in that equation, c_a and c_Δ assume a similar form which is analogous to (3.117). We find

$$c_a = \frac{1}{4}\left(1 - \frac{a_2}{a_1}\right)^2\left[\left(\frac{dp}{d\rho}\right)_{\rho\to1}\Big/\int_0^1 p(\rho)\rho^3\,d\rho\right] \tag{3.123}$$

and

$$c_\Delta = \frac{1}{4}\left[1 - \left(\frac{\Delta_2}{\Delta_1}\right)^{1/2}\right]^2\left[\left(\frac{dp}{d\rho}\right)_{\rho\to1}\Big/\int_0^1 p(\rho)\rho^3\,d\rho\right]. \tag{3.124}$$

For $p = J_0(2.405\rho)$,

$$c_a = 4.6[1 - (a_2/a_1)]^2, \tag{3.125}$$

$$c_\Delta = 4.6[1 - (\Delta_2/\Delta_1)^{1/2}]^2. \tag{3.126}$$

A 2.5% diameter change or a 5% change in Δ cause 0.3% of loss.

These results are merely meant to give the reader some help in determining the influence of certain fiber tolerances. In practice, misalignments and dimensional variations are simultaneously present and influence each other. The derivation of actually expected splice loss values requires a statistical approach based on achievable tolerances. These depend on the configuration and the task of specific splices or connections and are discussed in Chapters 14 and 15.

3.10 COUPLED MODE THEORY (MARCUSE, 1974)

We have seen that an optical fiber is capable of supporting a finite number of guided modes. In a perfect fiber these modes propagate independently of each other, arriving at the receiver each with its own group delay and attenuation. Practical fibers are not perfect cylinders so that the guided modes are not really independent of each other. Depending on the nature and magnitude of the existing fiber imperfections power is exchanged among the guided modes and, in addition, guided modes couple to the continuous spectrum of radiation modes. The reader is reminded that the medium outside the fiber core does not extend to infinity and that cladding modes take the place of the continuum of radiation modes. However, for most purposes it does not make an appreciable difference whether we calculate mode coupling to a continuum of radiation modes or actually use the discrete spectrum of cladding modes. If the cladding modes are attenuated either by a lossy jacket or a rough outer cladding boundary the results are practically the same.

Power exchange among guided modes and radiation loss phenomena can be described by coupled mode theory. The central idea of this theory is the observation that an arbitrary field distribution in the fiber can be expressed as a superposition of guided modes and radiation modes. The electric field vector is first decomposed into its transverse and longitudinal (with respect to the fiber axis) components

$$\mathbf{E} = \mathbf{E}_t + \mathbf{E}_z. \tag{3.127}$$

A corresponding decomposition is made for the fields of the guided modes $\mathscr{E}_\nu$ and for the radiation modes $\mathscr{E}_j(\rho)$. The transverse field vector of the total electric field can now be expressed as a superposition of the transverse parts of the mode field vectors,

$$\mathbf{E}_t = \sum_{\nu=1}^{N} a_\nu(z) \mathscr{E}_{t\nu} + \sum_{j} \int_0^\infty a_j(\rho, z) \mathscr{E}_{tj}(\rho) \, d\rho. \tag{3.128}$$

The index j, that is attached to the radiation mode fields and their expansion coefficients, is necessary to distinguish different kinds of radiation modes from each other and the summation over j indicates that the contributions from all types of radiation modes must be included in (3.128). Because this notation is cumbersome we write (3.128) in the simple, symbolic form

$$\mathbf{E}_t = \sum_{\nu} a_\nu \mathscr{E}_{t\nu}. \tag{3.129}$$

The summation symbol is now understood to incorporate summation over radiation modes as well as integration plus summation over radiation modes. The mode field expressions $\mathscr{E}_{t\nu}$ and $\mathscr{E}_{tj}(\rho)$ do not depend on the longitudinal coordinate z; any z dependence is absorbed in the expansion coefficients a_ν.

The longitudinal electric field component is obtained via the relation

$$\mathbf{E}_z = \frac{1}{i \omega \epsilon_0 n^2} \nabla_t \times \mathbf{H}_t. \tag{3.130}$$

The symbol ∇_t indicates the transverse component of the ∇ operator, n is

The symbol ∇_t indicates the transverse component of the ∇ operator, n is the refractive index distribution of the fiber and $\mathbf{H}_t$ is the transverse part of the magnetic field vector. A similar expansion holds, of course for the magnetic field vector.

Substitution of the field expansion into Maxwell's equations allows us to derive a coupled system of differential equations for the expansion coefficients

$$\frac{da_\mu^{(+)}}{dz} = -i\beta_\mu a_\mu^{(+)} + \sum_\nu (K_{\mu\nu}^{(+,+)}a_\nu^{(+)} + K_{\mu\nu}^{(+,-)}a_\nu^{(-)}), \qquad (3.131)$$

$$\frac{da_\mu^{(-)}}{dz} = +i\beta_\mu a_\mu^{(-)} + \sum_\nu (K_{\mu\nu}^{(-,+)}a_\nu^{(+)} + K_{\mu\nu}^{(-,-)}a_\nu^{(-)}). \qquad (3.132)$$

The single expansion coefficient a_ν has been decomposed into the sum

$$a_\nu = a_\nu^{(+)} + a_\nu^{(-)} \qquad (3.133)$$

with $a_\nu^{(+)}$ and $a_\nu^{(-)}$ indicating the expansion coefficients of waves traveling in positive and negative z directions. Since the summation symbol combines summation and integration (3.131) and (3.132) represent actually an infinite system of coupled integrodifferential equations. The coupling coefficients are (approximately) defined as follows;

$$K_{\mu\nu}^{(p,q)} = \frac{\omega\epsilon_0}{4iP} \int\int_{-\infty}^{\infty} (n^2 - n_{\mathrm{id}}^2)[p\mathscr{E}_{\mu t}^* \cdot \mathscr{E}_{\nu t} + q\mathscr{E}_{\mu z}^* \cdot \mathscr{E}_{\nu z}] \, dx \, dy \qquad (3.134)$$

The symbols p and q assume the "values" $+$ and $-$, $n = n(x, y, z)$ is the actual refractive index distribution of the imperfect fiber while $n_{\mathrm{id}} = n_{\mathrm{id}}(x, y)$ (notice the absence of the z dependence) is the refractive index distribution of a hypothetical, ideal fiber that is used to define the modes in the series expansion (3.129). The normalizing factor P is included in the normalization of the mode fields as is apparent from the orthonormality condition

$$\frac{1}{2} \int\int_{-\infty}^{\infty} (\mathscr{E}_\nu \times \mathscr{H}_\mu^*)_z \, dx \, dy = P\delta_{\nu\mu}. \qquad (3.135)$$

The subscript z indicates the z component of the vector inside the expression in parentheses, $\mathscr{H}_\mu^*$ is the complex conjugate magnetic field vector of the mode field. In case that both $\nu = \rho$ and $\mu = \rho'$ label radiation modes the Kronecker delta symbol $\delta_{\nu\mu}$ becomes the Dirac delta function $\delta(\rho - \rho')$.

The coupled wave equations (3.131) and (3.132) are exactly equivalent to Maxwell's equations. An exact solution of this equation system is usually not possible, its value consists in providing perturbation solutions for many important cases of practical interest. For example, we usually neglect coupling between guided waves moving in opposite directions. If only a certain mode $\nu = \mu$ exists at $z = 0$ we can obtain the following approximate expression for the amplitude coefficients of the other modes

$$a_\nu^{(+)}(z) = a_\mu^{(+)}(0)e^{-i\beta_\nu z} \int_0^z K_{\nu\mu}^{(+,+)}(z')e^{i(\beta_\nu - \beta_\mu)z'} \, dz'. \qquad (3.136)$$

This important expression shows that it is the Fourier component of the coupling coefficient at the spatial frequency $\beta_\nu - \beta_\mu$ that is responsible for effective coupling between two modes. A coupling process that is strictly periodic may thus effectively couple only two of the guided fiber modes allowing them to exchange their power periodically.

The power P_N that is carried by all the modes can be expressed as

$$P_N = P \sum_\nu [|a_\nu^{(+)}|^2 + |a_\nu^{(-)}|^2] \tag{3.137}$$

[P is the normalizing coefficient defined by (3.135)]. If we omit the amplitude coefficient of the incident mode $\nu = \mu$ from (3.137) we obtain the expression for the power ΔP that is carried by all the spurious modes which represents the power that is lost from the incident mode $\nu = \mu$. Limiting the sum (integration) in (3.137) to radiation modes we can easily obtain an expression for the power loss coefficient of mode μ for radiation losses,

$$2\alpha_\mu = \frac{\Delta P}{|a_\mu^{(+)}|^2 P L} = \frac{1}{L} \int_{-n_2 k}^{n_2 k} \frac{|\beta|}{\rho} \, d\beta \left| \int_0^L K_{\rho\mu}(z) e^{i(\beta - \beta_\mu)z} \, dz \right|^2. \tag{3.138}$$

We have omitted the superscripts $^+$ and $^-$ from the coupling coefficient because coupling to both forward and backward traveling radiation modes is included in (3.138). The integration over the parameter ρ that should appear in (3.138) was converted to an integration over the propagation constants $\beta = (n_2^2 k^2 - \rho^2)^{1/2}$ of the radiation modes including only propagating radiation modes that carry away power.

Equation (3.138) includes the interaction of any guided mode with the continuum of radiation modes and reduces it to a simple loss coefficient. The interaction among guided modes is far more complicated because guided modes keep on interacting with each other along the entire length of the fiber whereas power coupled from a guided mode to radiation modes radiates away and is effectively lost before it has a chance to interact with the guided mode once more. A description of guided mode interaction in terms of coupled amplitude equations (3.131) is possible only in simple special cases such as two-mode coupling in case coupling to possible other guided modes can be ignored.

It is advantageous to use the coupled amplitude equations (3.131) and (3.132) to derive coupled equations of the average power

$$P_\mu = \langle |a_\mu|^2 \rangle \tag{3.139}$$

carried by each mode. The symbol $\langle \ \rangle$ indicates an ensemble average. Conversion of the coupled wave equations to coupled power equations is possible when certain conditions prevail. For example, the coupling process must be sufficiently weak so that only a small amount of power is exchanged over distances comparable to the correlation length D of the coupling process. In the case of geometrical fiber distortions it is usually possible to decompose the coupling coefficient as follows:

$$K_{\mu\nu} = \hat{K}_{\mu\nu} f(z).$$ (3.140)

$\hat{K}_{\mu\nu}$ is independent of z while $f(z)$ describes the dependence of the coupling process on the length coordinate. The function $f(z)$ is assumed to describe a stationary random process with autocorrelation function

$$R(u) = R(-u) = \langle f(z) f(z + u) \rangle$$ (3.141)

and zero mean

$$\langle f(z) \rangle = 0.$$ (3.142)

The autocorrelation function must decay sufficiently rapidly so that a correlation length D can be defined. It is assumed that $R(u) = 0$ at distances u that are larger but on the order of magnitude of D. If these conditions prevail we can convert the coupled amplitude equations (3.131) to a finite system of coupled power equations

$$\frac{\partial P_\mu}{\partial z} = -2\alpha_\mu P_\mu + \sum_{\nu=1}^{N} h_{\mu\nu}(P_\nu - P_\mu)$$ (3.143)

with the power coupling coefficient defined as

$$h_{\mu\nu} = h_{\nu\mu} = |\hat{K}_{\nu\mu}|^2 \langle |F(\beta_\mu - \beta_\nu)|^2 \rangle.$$ (3.144)

The Fourier transform of $f(z)$ is defined as

$$F(\theta) = \lim_{L\to\infty} \frac{1}{L^{1/2}} \int_0^L f(z) e^{-i\theta z} \, dz$$ (3.145)

It can be shown that the "power spectrum" $\langle |F(\theta)|^2 \rangle$ and the autocorrelation function $R(u)$ are Fourier transforms of each other

$$\langle |F(\theta)|^2 \rangle = \int_{-\infty}^{\infty} R(u) e^{-i\theta u} \, du.$$ (3.146)

The summation in (3.143) extends over all guided modes and the power loss coefficient $2\alpha_\mu$ contains not only dissipative losses but also radiation losses defined by (3.138).

The coupled power equations (3.143) are far simpler than the original coupled amplitude equations because they form a system of only a finite number of equations with constant and symmetric coupling coefficients. Such an equation system can always be solved. This simplification was possible because the coupled power equations contain far less information than the coupled amplitude equations (3.131). The coupled power equations apply only to the average power (in the sense of an ensemble average) of each mode and they do not contain any phase information. However, it is also important to remember that the coupled power equations (3.143) are not equivalent to Maxwell's equations and are not an exact description of the coupling process.

The general solution of (3.143) can be written as follows:

$$P_\mu(z) = \sum_{n=1}^{N} c_n A_\mu^{(n)} e^{-\sigma_n z}. \tag{3.147}$$

$A_\mu^{(n)}$ and σ_n are solutions of an algebraic eigenvalue problem

$$\sum_{\nu=1}^{N} [h_{\nu\mu} + (\sigma_n - 2\alpha_\nu - b_\nu)\delta_{\mu\nu}]A_\nu^{(n)} = 0 \tag{3.148}$$

with

$$b_\mu = \sum_{\nu=1}^{N} h_{\mu\nu}. \tag{3.149}$$

The condition that the homogeneous equation system (3.148) have a solution requires that its system determinant vanishes, this yields N solutions for the eigenvalue σ_n. For each eigenvalue σ_n we have a corresponding eigenvector whose components are $A_\mu^{(n)}$. The expansion coefficients c_n are determined from the initial power distribution at $z = 0$,

$$c_n = \sum_{\mu=1}^{N} A_\mu^{(n)} P_\mu(0). \tag{3.150}$$

The solution (3.147) has a very important general property. Because the eigenvalues of the system (3.148) must all be real and positive we can arrange them in increasing order,

$$\sigma_1 < \sigma_2 < \sigma_3 < \cdots < \sigma_N. \tag{3.151}$$

For large values of z the function $\exp(-\sigma_n z)$ decreases very rapidly so that only the first term in (3.47) remains significant. We thus see that the power distribution (3.147) must settle down to a steady state of the form

$$P_\mu(z) = c_1 A_\mu^{(1)} e^{-\sigma_1 z} \quad \text{for} \quad z \to \infty. \tag{3.152}$$

Clearly, the distribution of power versus mode number μ does not depend on the distribution at $z = 0$ (except for the common amplitude factor c_1) but is determined solely by the first eigenvector $A_\mu^{(1)}$. The attenuation factor of the steady-state power distribution (3.152) is the smallest eigenvalue σ_1. Before the steady-state distribution (3.152) is reached the power distribution (3.147) is in a transient state to which no unique loss coefficient can be assigned. Once steady state is reached the distribution of average power versus mode number settles down to a definite shape and decays with a unique attenuation coefficient that depends only on the statistics of the coupling process and the mode losses.

For a description of pulse propagation we need to introduce the time variable t into the problem. Equation (3.143) can be generalized in a straightforward way to the form

$$\frac{\partial P_\mu}{\partial z} + \frac{1}{v_\mu}\frac{\partial P_\mu}{\partial t} = -2\alpha_\mu P_\mu + \sum_{\nu=1}^{N} h_{\mu\nu}(P_\nu - P_\mu). \tag{3.153}$$

The only new parameter in this equation system is the group velocity v_μ. In most practical cases the group velocities are not too different for the different modes so that (v_{av} is an average value of v_μ)

$$V_\mu = \frac{1}{v_\mu} - \frac{1}{v_{av}} \tag{3.154}$$

can be regarded as a small quantity and (3.153) can be solved by perturbation theory. If the input pulse has Gaussian shape,

$$P_\mu(0, t) = G_\mu e^{-t^2/2\tau^2}, \tag{3.155}$$

the steady-state solution of (3.153) for $z \to \infty$ can be obtained as,

$$P_\mu(z, t) = k_1 \frac{\tau}{T} A_\mu^{(1)} e^{-\sigma_1 z} \exp\left\{-\frac{\left(t - \dfrac{z}{v_{av}}\right)^2}{2T^2}\right\} \tag{3.156}$$

with $A_\mu^{(1)}$ and σ_1 defined as the solutions of (3.148) with the smallest eigenvalue; k_1 is defined as

$$k_1 = \sum_{\mu=1}^{N} A_\mu^{(1)} G_\mu \tag{3.157}$$

and with the length dependent rms width parameter of the Gaussian pulse

$$T = (2\rho z)^{1/2}. \tag{3.158}$$

The parameter ρ is a second-order perturbation of the eigenvalue σ_1,

$$\rho = \sum_{n=2}^{N} \frac{\left[\displaystyle\sum_{\mu=1}^{N} A_\mu^{(1)} V_\mu A_\mu^{(n)}\right]^2}{\sigma_n - \sigma_1} \tag{3.159}$$

The importance of this steady-state pulse solution lies in the dependence of the rms width parameter T on the square root of the distance z along the fiber axis (Personick, 1971). We discuss in Chapter 4 that the width of a

pulse that is carried by many uncoupled modes spreads proportionally to
the distance it has traveled. Mode coupling has the beneficial effect of
shortening the pulse and causing its width to spread only proportionally
to the square root of the traversed distance. However, this advantage has
to be bought at a price. Any coupling mechanism that couples the guided
modes also couples guided and radiation modes. Intentionally introduced
mode coupling not only improves the pulse performance of the fiber but
also increases its radiation losses. It is possible to reduce radiation loss
by careful design of the coupling process but it cannot be avoided completely.

The time-independent coupled power equations (3.143) and the time-
dependent coupled power equations (3.153) can always be solved numerically. If the number of guided modes becomes very large numerical solutions of the eigenvalue problem (3.148) may become impractical. In this
case it is possible to regard the sums in (3.143), (3.148) and (3.159) as integrals over a quasi-continuous variable, the mode number μ. Instead of an
algebraic equation system, integral equations have now to be solved. In
many cases of practical interest it is known that only modes that are
nearest neighbors couple appreciably. If this is true the algebraic eigenvalue problem (3.148) can be converted to a differential equation with the
continuous variable μ. For example, a step-index fiber with randomly
bent axis allows mode coupling only if the azimuthal quantum numbers ν
and ν' of the coupled modes differ by ± 1. Introducing the compound
mode number $M = 2\mu + \nu$ we can approximate (3.148) for the case of
nearest-neighbor coupling by the differential equation (Gloge, 1972a)

$$\frac{2}{M} \frac{\partial}{\partial M} \left(M h_M \frac{\partial A_M}{\partial M} \right) + (\sigma - 2\alpha_M) A_M = 0. \qquad (3.160)$$

The compound mode number is treated as a quasi-continuous variable
and the power coupling coefficient for nearest neighbors is designated as
h_M. We impose the boundary condition

$$A_{N'} = 0, \qquad (3.161)$$

where N' indicates the largest value that M can assume. This boundary
condition implies physically that the modes of highest order do not carry
power because they suffer so much radiation loss that they are depleted
faster than mode coupling can replenish. Equation (3.160) has infinitely
many eigenvalues σ_n and eigenfunctions $A_M^{(n)}$ but only the smallest (positive) eigenvalue has physical meaning as the steady-state power loss while
the corresponding eigenfunction gives the steady-state power distribution.

3.11 MODE MIXING EFFECTS (MARCUSE, 1974)

As an example we consider mode coupling in a step-index fiber with random bends. The form of the power coupling coefficient h_m depends on the "power spectrum" of the distortion function according to (3.144). The factor $\hat{K}_{\mu\nu} = \hat{K}_M$ is proportional to M^4. Assuming that the power spectrum decreases inversely proportional to M^4 we obtain $h_m = h = $ const. Physically we may interpret a coupling process with this property as being caused by random fiber curvature with a constant curvature spectrum. The loss coefficient lumps together absorption losses in the fiber material and scattering losses due to some random imperfection (Rayleigh scattering, say). The mode coupling process due to the randomly bent fibers axis also results in scattering losses by coupling the guided modes to radiation modes. However, this coupling process affects the highest order modes much more than lower order modes and is taken care of by the boundary condition (3.161). Thus we use $\alpha_M = \alpha = $ const. With these assumptions we obtain as solutions of (3.160)

$$A_M^{(n)} = B_n J_0(K_n M) \tag{3.162}$$

with K_n defined as

$$K_n = [(\sigma_n - 2\alpha)/2h]^{1/2} \tag{3.163}$$

The allowed values of K_n follow from the boundary condition (3.161) and from (3.162) as solutions of the equation $J_0(u_n) = 0$ with

$$K_n N' = u_n. \tag{3.164}$$

The maximum value of the compound mode number can be expressed as

$$N' = (2/\pi)V. \tag{3.165}$$

The excess steady-state power loss caused by scattering is thus

$$\sigma_1 - 2\alpha = 2h \frac{u_1^2}{N'^2} = \frac{28.55}{V^2} h \tag{3.166}$$

with $u_1 = 2.405$.

The normalizing factor B_n in (3.162) follows from the normalization condition

$$\sum_{M=1}^{N'} M A_M^2 \approx \int_0^{N'} M A_M^2 \, dM = 1. \tag{3.167}$$

The sum over A_M^2 originally must extend over all radial mode numbers μ and over all azimuthal mode numbers ν. For each value of the compound mode number $M = 2\mu + \nu$ there are M different combinations of μ and

ν's (for a given polarization) so that we obtain the factor M under the summation sign in (3.167). Substitution of (3.162) and use of the boundary condition (3.164) results in

$$B_n = 2^{1/2}/N'J_1(u_n).\tag{3.168}$$

Turning now to the time-dependent problem we can use the results obtained thus far to compute the width T of a short pulse at the end of a fiber of length L. The inverse group velocity differences of the modes of the step-index fiber can be approximated as

$$V_M = \frac{1}{v_m} - \frac{1}{v_{av}} = \frac{\pi^2 M^2}{8n_1 c(ka)^2}\tag{3.169}$$

(n_1 and a are the refractive index and radius of the fiber core, c is the velocity of light in vacuum). Using (3.159), and the results of this section we can calculate the pulse width T (with $z = L$) from (3.158),

$$T = (1.12 \times 10^{-2})\frac{V^3}{n_1 c(ka)^2}(L/h)^{1/2}.\tag{3.170}$$

The pulse width increases with the square root of the fiber length and decreases with the square root of the power coupling coefficient. Stronger coupling thus results in shorter pulses.

More interesting than the absolute value of the pulse width is the ratio R of the width of pulses carried by coupled modes to the pulse width in a fiber without mode coupling. The rms pulse width in a step-index fiber with uncoupled modes is

$$T' = \frac{L}{12^{1/2}c}(n_1 - n_0) = \frac{V^2 L}{2(12)^{1/2}n_1 c(ka)^2}\tag{3.171}$$

The ratio R is called the improvement factor because it measures the reduction in rms pulse width caused by mode coupling,

$$R = T/T' = 0.078V/(hL)^{1/2}\tag{3.172}$$

The product of R^2 times the excess steady-state loss coefficient (3.166) times the fiber length L is independent of the fiber parameters

$$R^2(\sigma_1 - 2\alpha)L = 0.175 = 0.76 \quad \text{dB}.\tag{3.173}$$

Even though (3.173) is independent of the fiber parameters and also independent of the coupling strength its value depends on the assumptions that are made with regard to the power spectrum (3.146). However, it is a general feature of the product $R^2(\sigma_1 - 2\alpha)L$ that it is a universal number independent of the fiber parameters and the coupling strength and dependent only on the type of fiber that is being considered and on the shape of the power spectrum. The actual value of the product does not seem to change dramatically for different types of fibers. To reduce its magnitude it is necessary to shape the power spectrum such that coupling to radiation modes is minimized while coupling among guided mode is maximized. The optimum shape of the power spectrum requires that it remains nearly constant over a given range and drops off abruptly at a desired cutoff "frequency."

The product (3.173) is useful for calculating the loss penalty that must be paid for a certain improvement in pulse width. The ratio R must, of course, be less than unity. If R, as calculated from (3.172), should be larger than unity, the fiber would be too short or the coupling too weak for a steady state distribution to establish itself. If we wish to improve the pulse width by a factor of 3, $R = \frac{1}{3}$, (3.173) tells us that we must pay with an excess scattering loss of approximately 7.6 dB. This figure is independent of fiber length and depends only on the desired pulse width reduction. The loss penalty is thus quite severe for the coupling conditions assumed here (dependence of the power spectrum on the inverse fourth power of the spatial frequency). A power spectrum with a more abrupt cutoff would yield a smaller loss penalty.

A corresponding calculation (Marcuse, 1973; Olshansky, 1975) for the parabolic-index fiber with random bends results in a loss penalty expression, similar to (3.173), of 0.38 dB which is independent of the fiber parameters (as long as it remains a parabolic-index fiber) and independent of the statistics of the random bends. For parabolic index fibers with random bends the loss penalty is truly a universal constant.

The steady-state power distribution for parabolic-index fibers with random bends is also given by a Bessel function of order zero (Marcuse, 1976b), instead of the boundary condition $\partial P_M/\partial M = 0$ at the maximum value of the compound mode number $M = N'$ used in this reference, we must now use $P_{N'} = 0$ in accordance with (3.161). Properly normalized it can be expressed by the formula

$$P_M = AJ_0[2.405(M/N')^{1/2}] \tag{3.174}$$

with

$$A = \frac{1}{0.66N'J_1(2.405)} = \frac{4.13}{nka(\Delta^{1/2})}. \tag{3.175}$$

The compound mode number is defined as $M = 2p + \nu$ with the radial mode number p and azimuthal mode number ν of the Laguerre–Gaussian modes of the parabolic-index fiber. The maximum value of the compound mode number is

$$N' = nka(\Delta/2)^{1/2}. \tag{3.176}$$

3.12 RADIATION LOSS CAUSED BY RANDOM BENDS (MARCUSE 1976c)

In the preceding section we discussed the influence of mode coupling caused by random bends of the fiber axis for a specific statistical model which assumed that the power spectrum of the distortion function of the fiber axis exhibits an inverse fourth power dependence on the spatial frequency. In this section we discuss the power loss of single mode step-index fibers and multimode parabolic-index fibers for more general random bends of the fiber axis.

Random bends of the fiber are a very important loss mechanism that causes loss increases as fibers are wrapped on drums or placed in optical fiber cables. Such bends with small amplitudes have been called micro-bends and microbending losses need to be understood and controlled if low-loss optical cables are to be achieved.

The scattering losses caused by bends of the fiber axis can be calculated for each fiber mode from the general formula (3.138). Such an analysis requires knowledge of the field functions of all guided and radiation modes of the fiber. In case of multimode fibers coupling among the guided modes must be taken into account. The mode losses calculated from (3.138) are substituted into the coupled power equation (3.143) or (3.160) and the steady-state loss that results from the interplay between the individual mode losses and the coupling among the guided modes can be computed. For the modes of the step-index fiber the loss formula appears in terms of Bessel and Hankel functions. A significant simplification of this formula can be achieved if we use a model for the power spectrum of the fiber axis distortion function of the form

$$\langle |F(\theta)|^2 \rangle = \frac{A_m}{(\Delta\theta)^m + \theta^m} \approx \frac{A_m}{\theta^m} \tag{3.177}$$

with

$$A_m = m(\Delta\theta)^{m-1}\sigma^2 \sin(\pi/m). \tag{3.178}$$

The parameter σ represents the rms amplitude of the distortion function $f(z)$ describing the fiber axis, $\Delta\theta$ is a width parameter of the power law distribution, and m is the exponent of the power law. We use (3.177) in its approximate form indicated on its right-hand side. Since most observed power spectra decrease rapidly with increasing spatial frequency (3.177) should be able to furnish a reasonable approximation for most spectra of interest. If we limit the range of V values to $1 < V < 2.4$—the range of practical interest for single mode operation—we may approximate the radiation loss coefficient for the HE_{11} mode for $m \geq 4$ as follows:

$$2\alpha = \frac{n_2 k^4 \langle |F(\Omega)|^2 \rangle}{(m-1)(m-2)} \left(\frac{\Omega}{k}\right)^3 \left[V + 7.58 \times 10^{-5} \frac{m^9 V^m}{e^{1.7m}}\right]. \tag{3.179}$$

The power spectrum (3.177) enters this formula only at the spatial frequency

$$\Omega = \beta_{01} - n_0 k, \tag{3.180}$$

where β_{01} is the propagation constant of the LP_{01} (HE_{11}) mode. For a typical single mode fiber with $V = 2.4$, $n_0 = 1.458$, $n_1/n_0 - 1 = 0.003$, $\lambda_0 = 1$ μm we have $\Omega = 0.014$ μm^{-1}.

If we consider only coupling to the radiation modes but ignore coupling to the LP_{11} mode that begins to propagate at $V = 2.405$ the loss would drop off very sharply at that point. If we now include coupling to the LP_{11} mode and consider the power coupling coefficient between LP_{01} and LP_{11} as the power loss coefficient of the LP_{01} mode, the loss curve continues smoothly through the point $V = 2.405$. In a single mode W fiber designed for high leakage loss for the LP_{11} mode it is justified to regard the power coupling coefficient from the LP_{01} mode to the lossy LP_{11} mode as the loss coefficient of the LP_{01} mode. This loss coefficient can be approximated in the form

$$2\alpha_s = h_{01,11} = \frac{\langle|F(\overline{\Omega})|^2\rangle}{2n_0^2 k^2 a^6} \left\{ 9(V - 1.87) - \frac{0.616}{(V - 2.405)^{1/2}} \right\}. \tag{3.181}$$

The power spectrum enters this formula at the spatial frequency

$$\overline{\Omega} = \beta_{01} - \beta_{11}. \tag{3.182}$$

The loss coefficients (3.179) and (3.181) are plotted in normalized form in Fig. 3.31. For V values in the range $0 < V < 2.405$ only coupling to radiation modes can occur and the normalized loss depends on the exponent of the power law (3.181) (the approximate form is used). For $V > 2.405$ the LP_{11} mode can propagate. It is assumed that this mode has high loss making this curve applicable to a single mode W fiber. If coupling to radiation modes and to the LP_{11} mode were considered simultaneously the curves would join smoothly at $V = 2.405$. It was assumed that $n_1 = 1.515$ and $n_2 = 1.5$.

We have compared our theory based on coupling to radiation modes with a theory that considers coupling to lossy cladding modes instead. The agreement between the two theories is excellent indicating that the assumption of a continuous spectrum of radiation modes is a reasonable approximation.

For a parabolic-index fiber with index profile (3.2) and $g = 2$ the steady-state power loss can be computed with the result,

$$2\alpha_p = 5.8 \frac{\langle|F(\Omega_p)|^2\rangle}{a^4} \Delta = 1.45 \frac{n_1}{N^{1/2}} \left(\frac{\Omega_p}{k}\right)^3 k^4 \langle|F(\Omega_p)|^2\rangle. \tag{3.183}$$

The power spectrum is evaluated at the spatial frequency

$$\Omega_p = (2\Delta)^{1/2}/a \tag{3.184}$$

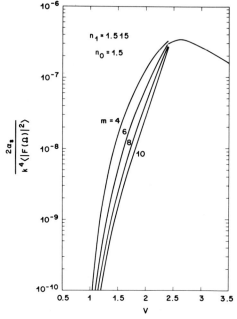

Fig. 3.31 Relative microbending loss as a function of V for several values of the power law coefficient m. Above $V = 2.405$ the LP_{01} mode loses power by coupling to the LP_{11} mode.

and the total number of guided modes is

$$N = (n_1 ka)^2 (\Delta/2). \tag{3.185}$$

It is of interest to compare the losses of a single mode step-index fiber with the losses of a multimode parabolic-index fiber. It is natural to assume the same power spectrum $\langle |F(\theta)|^2 \rangle$ in either case, but both types of fibers sample the power spectrum differently. The guided mode of the step-index fiber is coupled to all the radiation modes requiring a wide band of spatial frequencies, even though the spatial frequency (3.180) is most important since it corresponds to the distance (in β space) of the guided mode from the edge of the radiation mode region. In the parabolic-index fiber the guided modes that are coupled by random bends are very nearly equidistantly spaced so that the power spectrum contributes to mode coupling essentially only at the spatial frequency (3.184). The loss occurs in this case through coupling to the mode group of highest order which, in turn, is tightly coupled to radiation modes. We compare the losses of the step-index and the parabolic-index fibers using the power spectrum (3.177) and normalize the loss coefficient by dividing by the power spectrum at the spatial frequency $\theta = n_2 k \Delta$ with the Δ value of the parabolic-index multimode fiber. This spatial frequency is quite arbitrary

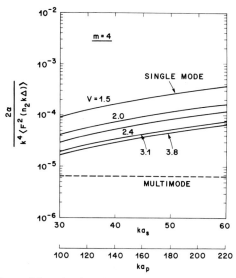

Fig. 3.32 Comparison of the microbending losses of single mode and multimode fibers. In this figure $m = 4$ is used.

and, in particular, has no real significance for the single mode fiber but it is necessary to use a common normalization for both fiber types if loss comparisons are to be made.

Figures 3.32 through 3.35 show the normalized loss coefficient of the single mode step-index fiber as solid curves and the corresponding loss coefficient for the multimode parabolic-index fiber as dashed curves. The only difference between the figures is a different exponent for the power law (3.177); $m = 4$ was chosen for Fig. 3.32 and this value is increased in steps of two until it reaches the value $m = 10$ for Fig. 3.35. The horizontal axes of all figures carry two scales. For both fiber types we use the value of ka but since one is a single mode fiber and the other a multimode fiber the "a" ranges of interest are different. The value for the step-index fiber carries the subscript s, ka_s, while the value for the parabolic-index fiber carries the subscript p, ka_p. For the step-index fiber we have kept the V value constant for a given curve requiring that $n_1 - n_2$ changes along the curve. We allow V to exceed the single mode value $V = 2.405$ making the curves applicable to "single mode" W fibers. For both fiber types we use a fixed value of $n_2 = 1.457$ and let the relative index difference Δ for the parabolic-index fiber be fixed at the value $\Delta = 0.01$. Comparison of the loss curves shows that, in general, the typical parabolic-index multimode fiber has lower loss than the typical single mode step-index fiber. The dependence of the loss of the multimode fiber on ka_p differs for different values of m. For $m = 4$ the normalized loss is actually independent of ka_p

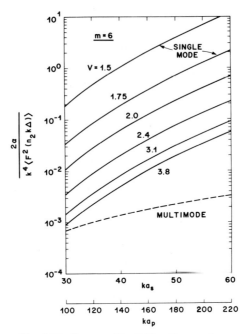

Fig. 3.33 Same as Fig. 3.32 with $m = 6$.

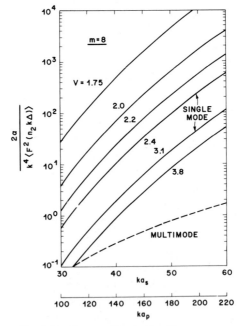

Fig. 3.34 Same as Fig. 3.32 with $m = 8$.

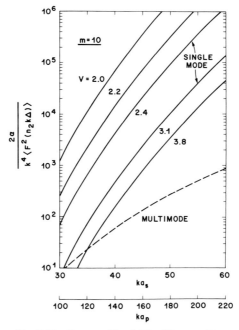

Fig. 3.35 Same as Fig. 3.32 with $m = 10$.

but the dependence becomes steeper for larger values of m. Both fiber types tolerate larger microbends if their ka values are smaller. The step-index fiber performs better for larger values of V. For relatively large V values and large values of m the loss of the step-index fiber can be lower than that of the parabolic-index fiber but for most practical cases the parabolic-index multimode fiber appears to have lower microbending losses.

Calculation of actual numbers for the loss coefficients requires detailed knowledge of the power spectrum of the fiber distortion function. The reader is also cautioned not to lose track of the fact that the loss calculation for the parabolic-index fiber is based on the assumption that the coupling is strong enough so that a steady-state power distribution can establish itself.

Figure 3.36 shows the length $\Lambda = 2\pi/\theta$ of the period associated with the spatial frequency θ for the single mode step-index (Λ_s) and multimode parabolic-index (Λ_p) fibers of our example. We have used $\theta = \Omega$ of (3.180) for the single mode case with $V = 2.4$. The corresponding value for the multimode fiber case is obtained from (3.184). The light wavelength was chosen to be $\lambda = 1 \ \mu$m. The figure shows that the critical spatial period responsible for radiation loss in the single mode step-index fiber is larger than the period of spatial frequencies coupling adjacent modes in the mul-

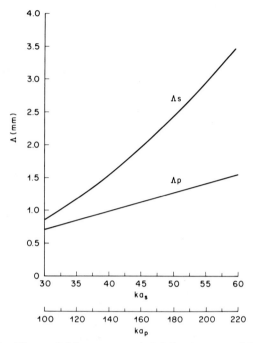

Fig. 3.36 Length of the spatial frequency period Λ that is responsible for mode couping in our example. The subscript s identifies the case of the single mode fiber while the subscript p refers to the multimode parabolic-index fiber.

timode fiber. However, over the ranges of ka values used in Figs. 3.32 through 3.35 the spatial periods differ by no more than a factor of two.

It is not clear whether a comparison of the single and multimode fibers based on the assumption of identical spatial Fourier spectra is meaningful. It is conceivable that the manufacturing process for single mode fibers will introduce Fourier spectra with periods that are quite different from the process for multimode parabolic-index fibers. Our comparison should thus be taken with a grain of salt and not be regarded as a definitive judgment on the matter.

REFERENCES

Abramowitz, M., and Stegun, I. A. (1964). "Handbook of Mathematical Functions," Appl. Math. Ser. No. 55. Nat. Bur. Stand., US Gov. Printing Office, Washington, D.C.

Arnaud, J. A. (1974a) Transverse coupling in fiber optics. I. coupling between trapped modes. *Bell Syst. Tech. J.* **53**, 217–224.

Arnaud, J. A. (1974b). Transverse Coupling in fiber optics. III. bending losses. *Bell Syst. Tech. J.* **53**, 1379–1394.

Chang, D. C., and Kuester, E. F. (1975). General theory of surface-wave propagation on a curved optical waveguide of arbitrary cross section. *IEEE J. Quantum Electron.* **QE-11,** 903–907.

Cook, J. S., Mammel, W. L., and Grow, R. J. (1973). Effect of misalignment of coupling efficiency of single-mode optical fiber butt joints. *Bell. Syst. Tech. J.* **52,** 1439–1448.

Globe, D. (1971). Weakly guiding fibers. *Appl. Opt.* **10,** 2252–2258.

Gloge, D. (1972a). Optical power flow in multimode fibers. *Bell Syst. Tech. J.* **51,** 1767–1783.

Gloge, D. (1972b). Bending loss in multimode fibers with graded and ungraded core index. *Appl. Opt.* **11,** 2506–2512.

Gloge, D., and Marcatili, E. A. J. (1973). Multimode theory of graded-core fibers. *Bell Syst. Tech. J.* **52,** 1563–1578.

Gloge, D. (1975). Propagation effects in optical fibers. *IEEE Trans. Microwave Theory Tech.* **MTT-23,** 106–120.

Gloge, D. (1976). Offset and tilt loss in optical fiber splices. *Bell Syst. Tech. J.* **55,** (1975). 905–915.

Heiblum, M., and Harris, J. H. (1975). Analysis of curved optical waveguides by conformal transformation. *IEEE J. Quantum Electron.* **QE-11,** 75–83.

Kawakami, S., and Nishida, S. (1974). Characteristics of a doubly clad optical fiber with a low-index cladding. *IEEE J. Quantum Electron.* **QE-10,** 879–887.

Kawakami, S., and Nishida, S. (1975). Perturbation theory of a double clad optical fiber with a low-index inner cladding. *IEEE J. Quantum Electron.* **QE-11,** 130–138.

Kuhn, M. H. (1975). Optimum attenuation of cladding modes in homogeneous single mode fibers. *Arch. Elektr. Uebertr.* **29,** 210–204.

Lewin, L. (1974). Radiation from curved dielectric slabs and fibers. *IEEE Trans. Microwave Theory Tech.* **MTT-22,** 718–727.

Marcatili, E. A. J. (1969). Bends in optical dielectric guides. *Bell Syst. Tech. J.* **48,** 2103–2132.

Marcatili, E. A. J., and Miller, S. E. (1969). Improved relations describing directional control in electromagnetic wave guidance. *Bell Syst. Tech. J.* **48,** 2161–2188.

Marcuse, D. (1970). Excitation of the dominant mode of a round fiber by a Gaussian beam. *Bell Syst. Tech. J.* **49,** 1695–1703.

Marcuse, D. (1971a). The coupling of degenerate modes in two parallel dielectric waveguides. *Bell Syst. Tech. J.* **50,** 1791–1816.

Marcuse, D. (1971b). Crosstalk caused by scattering in slab waveguides. *Bell Syst. Tech. J.* **50,** 1817–1831.

Marcuse, D. (1972). "Light Transmission Optics." Van Nostrand-Reinhold, Princeton, New Jersey.

Marcuse, D. (1973). Losses and impulse response of a parabolic-index fiber with random bends. *Bell Syst. Tech. J.* **52,** 1423–1437.

Marcuse, D. (1974). "Theory of Dielectric Optical Waveguides." Academic Press, New York.

Marcuse, D. (1975). Excitation of parabolic-index fibers with incoherent sources. *Bell Syst. Tech. J.* **54,** 1507–1530.

Marcuse, D. (1976a). Curvature loss formula for optical fibers. *J. Opt. Soc. Am.* **66,** 216–220.

Marcuse, D. (1976b). Mode mixing with reduced losses in parabolic-index fibers. *Bell Syst. Tech. J.* **55,** 777–802.

Marcuse, D. (1976c). "Microbending losses of single mode, step-index and multimode, parabolic-index fibers. *Bell Syst. Tech. J.* **55,** 937–955.

Marcuse, D. (1977). Loss analysis of single-mode fiber splices. *Bell Syst. Tech. J.* **56,** 703–718.

Miller, C. M. (1976). Transmission vs. transverse offset for parabolic-profile fiber splices with unequal core diameters. *Bell Syst. Tech. J.* **55,** 929–935.

Miller, M. A., and Talanov, V. I. (1956). Electromagnetic surface waves guided by a boundary with small curvature. *Zh. Tekh. Fiz.* **26,** 2755.

Miller, S. E. (1965). Light propagation in generalized lens-like media. *Bell Syst. Tech. J.* **44**, 2017–2064.

Miller, S. E., Marcatili, E. A. J., and Li, T. (1973). "Research toward optical-fiber transmission systems. I. The transmission Medium; II. Devices and system considerations. *Proc. IEEE* **61**, 1703–1751.

Morse, P. M., and Feshbach, (1953). "Methods of Theoretical Physics," Vol. II. McGraw-Hill, New York.

Okamoto, K., and Okoshi, T. (1976). Analysis of wave propagation in optical fibers having core with α-power refractive-index distribution and uniform cladding. *IEEE Trans. Microwave Theory Tech.* **MTT-24**, 416–421.

Olshansky, R. Mode coupling effects in graded-index optical fibers. *Appl. Opt.* **14**, 935–945.

Personick, S. D. (1971). Time dispersion in dielectric waveguides. *Bell Syst. Tech. J.* **50**, 843–859.

Petermann, K. (1975). "The Mode attenuation in general graded core multimode fibers. *Arch. Elektr. Uebertr.* **29**, 345–348.

Sladen, F. M. E., Payne, D. H., and Adams, M. J. (1975). Determination of optical fiber refractive index profiles by a near-field scanning technique. *Appl. Phys. Lett.* **28**, 255.

Snitzer, E. (1961). Cylindrical dielectric waveguide modes, *J. Opt. Soc. Am.* **51**, 491–498.

Snyder, A. W. (1969). Asymptotic expressions for eigenfunctions and eigenvalues of a dielectric or optical waveguide, *Tans. IEEE Microwave Theory Tech.* **MTT 17**, 1130–1138

Snyder, A. W., and Mitchell, D. J. (1974). Leaky rays on circular fibers. *J. Opt. Soc. Am.* **64**, 599–607.

Stern, J. R., Peace, M., and Dyott, R. B. (1970). Launching into optical-fibre waveguide. *Electron. Lett.* **6**, 160–162.

Suematsu, Y., and Furuya, K. (1975). Quasi-guided modes and related radiation losses in optical dielectric waveguides with external higher index surroundings. *IEEE Trans. Microwave Theory Tech.* **MTT-23**, 170–175.

Tanaka, T., Onoda, S., and Sumi, M. (1967). Frequency response of multimode W-type optical fibers. *Electr. Commun. J.* **J59-C**, 122–130.

Tien, P. K., Gordon, J. P., and Whinnery, J. R. (1965). Focusing of a light beam of Gaussian field distribution in continuous and parabolic lens-like media. *Proc. IEEE* **53**, 129–136.

Vanclooster, R., and Phariseau, P. (1970). "The coupling of two parallel dielectric fibers. I. Basic equations. *Physica (Utrecht)* **47**, 485–500.

Yang, K. H., and Kingsley, J. D. (1975). Calculation of coupling losses between light emitting diodes and low loss optical fibers. *Appl. Opt.* **14**, 288–293.

Chapter 4

Dispersion Properties of Fibers

DETLEF GLOGE
ENRIQUE A. J. MARCATILI
DIETRICH MARCUSE
STEWART D. PERSONICK

4.1 INTRODUCTION

Most applications of fibers are in communications systems which utilize some form of digital envelope modulation of the optical signal. Accordingly, the fiber performance is usually characterized in terms of the degradation of an optical pulse propagating through the fiber. We shall follow this practice for the first five sections of this chapter: alternative descriptions in terms of the baseband frequency characteristic of the fiber will be discussed in Section 4.6.

If transmission by several fiber modes is involved, the pulse energy is divided among the modes. Each mode travels at a different group velocity. The result is a distortion of the total pulse. The individual pulses in each mode are distorted as well, if the input light is not monochromatic. In this case, the wavelength dependence of the group delay leads to a distortion of the pulse portions in each mode.

Both effects combine to cause the complete group delay distortion of the output pulse. To determine the effect, one measures the output pulse shape $h(t)$ as a function of the time t for a very narrow input pulse. This pulse shape is called the impulse response. Usually it is desirable to describe the characteristics of the impulse response $h(t)$ in terms of one single parameter. The rms widths σ has come to be used for this purpose. σ is defined by

101

$$\sigma^2 = \frac{1}{P} \int_{-\infty}^{+\infty} h(t)t^2 \, dt - \tau^2, \tag{4.1}$$

where

$$\tau = \frac{1}{P} \int_{-\infty}^{+\infty} h(t)t \, dt \tag{4.2}$$

is the pulse delay and

$$P = \int_{-\infty}^{+\infty} h(t) \, dt \tag{4.3}$$

is the pulse energy.

In the next section of this chapter, we consider the group delay distortion of a pulse transmitted by a single-mode fiber. We assume the fiber structure to be uniform along its length. In this case, two effects contribute to mode delay distortion: (1) material dispersion and (2) waveguide dispersion. Material dispersion is a result of the fact that the optical wave propagates in glass and that the wavelength dependence of its refractive index causes a wavelength dependence of the group delay. Waveguide dispersion is a result of the fact that the propagating characteristics of the mode are a function of the ratio between the core radius and the wavelength.

Section 4.3 discusses the delay distortion of a pulse transmitted by one individual mode of a multimode fiber. The fiber may have a core index that is graded in a specific way, but the cladding index is assumed to be uniform. The fiber structure is invariant along its length. The important effects are the same as in the single-mode case: material dispersion and waveguide dispersion.

Section 4.4 describes the delay distortion of the output pulse transmitted by a multitude of modes in a lossless fiber with the assumption that all modes are uncoupled. In this case, mode delay contributes to the delay distortion in addition to the two effects mentioned earlier. A suitable design of the refractive index of the core can cause a significant equalization of the mode delays (Miller, 1965; Kawakami and Nishizawa, 1968). This fact has become an important tool of modern fiber design.

The fifth section considers the influence of fiber loss and of mode coupling. Loss affects the pulse shape only if it is different for different modes. Mode coupling usually results in an averaging of the delays and therefore shortens the pulse and reduces group delay distortion.

All of the above effects occur in multimode fibers at the same time, but usually one effect dominates. In most multimode fibers, mode delay is the overriding effect, but if the index profile is well graded, material dispersion may be dominant.

The last section discusses the limits of the fiber characterization presented here and compares time and frequency representations.

4.2 PULSE DISTORTION IN SINGLE-MODE FIBERS

The arrival time of a light pulse transmitted through a fiber of length L is

$$t = L \frac{d\beta}{d\omega} = -L \frac{\lambda}{\omega} \frac{d\beta}{d\lambda}, \qquad (4.4)$$

where β is the propagation constant of the mode and ω the radian light frequency. If the carrier light has a spectral width $\delta\lambda$ which is centered around λ and is broad compared to that of the detected pulse envelope, the pulse spread δt as a result of the change of $d\beta/d\lambda$ can be calculated from (Dyott and Stern, 1971)

$$\frac{\delta t}{\delta\lambda} = \frac{dt}{d\lambda} = -\frac{L}{2\pi c}\left(2\lambda \frac{d\beta}{d\lambda} + \lambda^2 \frac{d^2\beta}{d\lambda^2}\right), \qquad (4.5)$$

where c is the vacuum speed of light.

The propagation constant β is a function of λ not only because the index changes with wavelength (material dispersion), but in addition because β is a function of a/λ where a is the core radius (waveguide dispersion). In single-mode fibers material and waveguide dispersion are interrelated in a complicated way. By computing one in the absence of the other, we show that the material effect usually dominates.

To compute the material dispersion, we assume that the carrier is a plane wave propagating in a dielectric of index $n(\lambda)$. We have $\beta = 2\pi n/\lambda$; by inserting this into (4.4), we obtain the group delay

$$t_m = \frac{L}{c}\left(n - \lambda \frac{dn}{d\lambda}\right). \qquad (4.6)$$

The bracketed expression in (4.6) is usually called the group index (Gloge, 1971)

$$N = n - \lambda(dn/d\lambda). \qquad (4.7)$$

The pulse spread is determined by the derivative of the group delay with respect to wavelength:

$$\frac{dt_m}{d\lambda} = -\frac{L}{c}\lambda \frac{d^2n}{d\lambda^2}\bigg|_{\lambda=\lambda_0}. \qquad (4.8)$$

The solid line in Fig. 4.1 shows $dt_m/Ld\lambda$ for silica (Kapron and Keck, 1971; Payne and Gambling, 1975); the broken line holds for 13% germania and 87% silica.

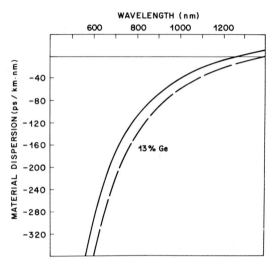

Fig. 4.1 Material dispersion versus wavelength for Si- and Ge-doped Si.

The influence of this material effect on the rms width of the impulse response can be calculated from integrals of the form (4.1) to (4.3) in which $h(t)$ is replaced by the spectral distribution $p(\lambda)$ of the signal, t is replaced by $t_m(\lambda)$ according to (4.6), and the integration is performed over λ instead of t. For an rms spectral width σ_λ, this integration yields approximately

$$\sigma = |dt_m/d\lambda|\sigma_\lambda \qquad (4.9)$$

everywhere except in the vicinity of the wavelength λ_0 at which $dt_m/d\lambda$ vanishes (see Fig. 4.1). In this wavelength range (several σ_λ to both sides of λ_0), σ depends on the shape $p(\lambda)$ of the signal spectrum. For a Gaussian shape centered at λ_0, $\sigma = 2^{-1/2}\sigma_\lambda{}^2|d^2t_m/d\lambda^2|_{\lambda=\lambda_0}$. Typical luminescent diodes made from GaAlAs have a spectral width of about 16 nm and hence produce a pulse broadening of $2\sigma = 3.2$ nsec/km when operated at 800 nm. The effect can be substantially reduced if such sources are operated at longer wavelengths.

Next, let us consider a single-mode step-index fiber made of a dispersionless material. We could use (4.5) to calculate the contribution made by waveguide dispersion. But since β is obtained from (3.17) and usually given in terms of the normalized frequency V defined by (3.12), we use the relation $dV/d\lambda = -V/\lambda$ to write (4.5) in the form

$$\frac{dt_w}{d\lambda} = -\frac{L}{2\pi c}V^2\frac{\partial^2\beta}{\partial V^2} \qquad (4.10)$$

where t_w is the delay resulting from waveguide dispersion (Gloge, 1975). It is convenient to use the definition

$$\frac{dt_w}{d\lambda} = \frac{L}{c\lambda} (n_1 - n_0)D_w(V), \tag{4.11}$$

where $D_w(V)$ is a dimensionless dispersion coefficient pertaining to the fundamental mode of a step-index fiber. Figure 4.2 used with the scale on the right gives a plot of D_w versus V. The left-hand scale indicates $dt_w/Ld\lambda$ for $\lambda = 900$ nm and $\Delta = 0.001$. The upper scale shows the core radius for these parameters. Profile grading caused by dopant diffusion in practical single-mode fibers significantly modifies (flattens) the plot of Fig. 4.2 (Unger 1977).

A comparison of Figs. 4.1 and 4.2 shows that waveguide dispersion can be neglected except in the vicinity of λ_0. In this range, waveguide dispersion essentially shifts $dt_m/d\lambda$ (as given, for example, in Fig. 4.1) by $\Delta\lambda = (dt_w/d\lambda)/(d^2t_m/d\lambda^2)_{\lambda=\lambda_0}$. For a step-index fiber operating at $V = 2.4$, this shift can be as large as 10 nm. An exact determination of σ requires an integration of the form (4.1) to (4.3) over λ with $h(t)$ replaced by $p(\lambda)$ and t from (4.4).

4.3 INDIVIDUAL MODES IN A MULTIMODE FIBER

Delay and distortion of a pulse transmitted by a certain mode in a multimode fiber can be calculated in the same way as in the single-mode case: by forming the first and second derivatives of the propagation constant with respect to the wavelength. However, the possibility of introducing a graded core index profile to achieve special delay effects adds greatly to the variety and complication of these computations. To select and order the possibilities in a useful and manageable way, we introduced in Chapter 3 the class of so-called "power-law profiles" which have the index

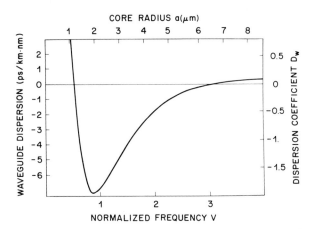

Fig. 4.2 Waveguide dispersion versus V value for single-mode fiber.

distributions (3.2). For this class, the WKB or ray-optics approach yields simple descriptions of the group delay and the pulse distortion for all modes except those very close to cutoff (Gloge and Marcatili, 1973).

Consider the eigenvalue equation (3.39) which is based on the WKB approximation. We introduce the abbreviation

$$f \equiv r^2 n^2(r) k^2 - \beta^2 r^2 - \nu^2 \tag{4.12}$$

so that (3.39) can be written in the form

$$\left(\mu - \frac{1}{2}\right) = \frac{1}{\pi} \int_{r_1}^{r_2} f^{1/2} \frac{dr}{r}. \tag{4.13}$$

Figure 3.7 shows the range of integration. Note that f vanishes at r_1 and r_2. We wish to describe $n(r)$ by (3.2) which represents a power law for $r < a$ and is constant for $r > a$. To simplify matters, we assume the power law to hold also for $r > a$, but we limit the possible propagation constants by specifying a lower limit for β. This approach slightly misrepresents modes close to cutoff. However, many of these modes are strongly attenuated in fibers used for communications purposes which have a limited cladding thickness and a lossy jacket and whose length is of the order of kilometers.

For typical fiber materials, n_1 and n_0 in (3.2) are different functions of the wavelength while g is independent of wavelength (Olshansky and Keck, 1976). This case is considered here. To obtain the group delay, we compute

$$\frac{\partial \mu}{\partial \beta} = -\frac{1}{\pi} \int_{r_1}^{r_2} \beta f^{-1/2} r \, dr \tag{4.14}$$

and

$$\frac{\partial \mu}{\partial k} = \frac{1}{\pi} \int_{r_1}^{r_2} k n_1 N_1 \left[1 - 2\Delta^+ \left(\frac{r}{a}\right)^g\right] f^{-1/2} r \, dr, \tag{4.15}$$

where N_1 is the group index at the core axis according to (4.7) and

$$\Delta^+ = (n_1 N_1 - n_0 N_0)/2 n_1 N_1. \tag{4.16}$$

To solve (4.15), we differentiate (4.12) with respect to r solve the result for $(r/a)^g$ and introduce this into (4.15). Part of the integral in (4.15) is then of the form

$$\int_{r_1}^{r_2} \frac{\partial f}{\partial r} f^{-1/2} \, dr = 2 f^{1/2} \Big|_{r_1}^{r_2} = 0. \tag{4.17}$$

The remaining integral has the same form as the integral in (4.14) and vanishes when one forms the ratio

$$\frac{\partial \beta}{\partial k} = \frac{\partial \mu / \partial k}{\partial \mu / \partial \beta} = n_1 N_1 \left(1 - \frac{2}{2+g} \frac{\Delta^+}{\Delta}\right) \frac{k}{\beta} + \frac{2}{2+g} \frac{N_1 \Delta^+ \beta}{n_1 \Delta k}. \tag{4.18}$$

Using (3.24), (4.4), (4.16), and (4.18), one can therefore write the delay $t = (L/c) d\beta/dk$ in the form

$$t(U) = \frac{LN_1}{c} \left(1 - \frac{4\Delta^+}{2+g} \frac{U^2}{V^2}\right) \bigg/ \left(1 - 2\Delta \frac{U^2}{V^2}\right)^{1/2} \tag{4.19}$$

where U is defined by (3.15) and (3.10). Equation (4.19) describes the modal group delay at a certain wavelength.

To compute the pulse spread of a pulse propagating in a given mode, we must derive (4.18) with respect to λ (Steinberg, 1974) and proceed in a way similar to the derivation of (4.5). If this is done in an approximate way and under the assumption that $\Delta \ll 1$, one finds

$$\frac{dt}{d\lambda} = -\frac{L}{c} \left(\lambda \frac{\partial^2 n(U)}{\partial \lambda^2} - \frac{2N_1^2 \Delta}{\lambda n_1} \left(1 - \frac{4}{2+g} \frac{\Delta^+}{\Delta}\right) \left(1 - \frac{2}{2+g} \frac{\Delta^+}{\Delta}\right) \left(\frac{U^2}{V^2}\right)\right], \tag{4.20}$$

where $n(U)$ varies in the range between n_1 and n_0 as U increases. The second derivative of $n(U)$ essentially determines the contribution of material dispersion. The magnitude of this contribution for typical fiber materials is bracketed by the solid and the broken lines in Fig. 4.1. The remaining part of (4.20) represents the contribution of waveguide dispersion and is plotted in Fig. 4.3 versus g for $\Delta = \Delta^+ = 1\%$ and $n_1 = N_1 = 1.46$ with U as a parameter. A comparison of Figs. 4.1 and 4.3 shows that material and waveguide dispersion in multimode fibers can be of the same magnitude and of opposite sign so that the two effects cancel each other in certain (high-order) modes.

Formula (4.19) for the modal group delay in powerlaw profiles deserves some further discussion. Notice that the group delay does not explicitly depend on the mode numbers μ and ν; all modes with the same propagation constant β have the same group delay. This fact will help us in Section 4.4 to find the temporal distribution of modes at the output. Delay differences among the modes would vanish if the functional dependence on U in the numerator of (4.19) could be made to resemble that in the denominator. This delay equilization is achieved, at least to first order in Δ, when

$$g = g_0 = 4 \frac{\Delta^+}{\Delta} - 2 = 2 - 2 \frac{n_1}{N_1} \frac{\lambda}{\Delta} \frac{d\Delta}{d\lambda}. \tag{4.21}$$

This optimal profile exponent g_0 is a function of the variation of Δ with λ which is a material effect often referred to as profile dispersion (Gloge et al., 1975). Figure 4.4 shows g_0 for a number of dopant materials (Sladen et al., 1978).

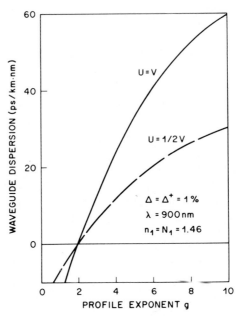

Fig. 4.3 Waveguide dispersion versus profile exponent for high and medium-high mode orders in multimode fibers.

The question of the optimal index profile can be posed in a more general way by starting with an index representation

$$n(r) = n_1[1 - F(r)]^{1/2}, \tag{4.22}$$

where F is the profile function that is to be optimized. Remember that the essential simplication that lead to (4.19) was the transformation of the integral (4.15) into an integral of the form (4.14). A similar transformation can be performed for the general case (4.22) under the condition (Marcatili, 1977)

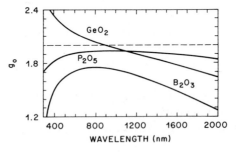

Fig. 4.4 Optimal profile exponent for three dopant materials. (Sladen *et al.*, 1978.)

$$r \frac{\partial F}{\partial r} + D \frac{n_1}{N_1} \lambda \frac{\partial F}{\partial \lambda} - 2(D - 1)F = 0, \qquad (4.23)$$

where D can be a function of λ, but not of r. If F satisfies the partial differential equation (4.23), the group delay $t(U)$ can be written in the form

$$t(U) = \frac{LN_1}{c} \left(1 - \frac{2\Delta}{D} \frac{U^2}{V^2}\right) \Big/ \left(1 - 2\Delta \frac{U^2}{V^2}\right)^{1/2}. \qquad (4.24)$$

Delay equalization to first order in Δ occurs at $D = 2$. If $F(r,\lambda)$ is separable in the form $F_\lambda(\lambda)F_r(r)$, (4.23) has the solution $F_\lambda = 2\Delta$ and $F_r = (r/a)^g$ with g from (4.21) for $D = 2$, as expected. The assumption that $F(r,\lambda)$ is separable is valid for binary core glass systems (Hammond, 1978). For more general cases, (4.23) can be solved by writing $F(r, \lambda)$ as a series of separable terms such that (Olshanky, 1978)

$$F(r, \lambda) = 2 \sum_i \Delta_i (r/a)^{g_i} \qquad (4.25)$$

with

$$\Delta = \sum_i \Delta_i \qquad (4.26)$$

and

$$g_i = 2 - 2 \frac{n_1}{N_1} \frac{\lambda}{\Delta_i} \frac{d\Delta_i}{d\lambda}. \qquad (4.27)$$

This solution has additional degrees of freedom which can be used to achieve desirable propagation characteristics, as, for example, equalization over a broader spectral range than is possible with binary systems (Presby and Kaminow, 1976). To achieve this one can satisfy $dg_i/d\lambda = 0$ in the middle of the spectral range of interest in addition to the equalization condition (4.27).

4.4 PULSE DISTORTION IN IDEAL MULTIMODE FIBERS

We continue to use the power-law description not only for the core as stipulated by (3.2), but also for the cladding region. As mentioned earlier, this requires that the limitations on the possible propagation constants, which are a result of the actual cladding index, be evaluated separately. This approach offers a convenient separation of problems. The first step consists in computing the sequence and density of modes per time increment at a given delay t from (4.19). The time t gives the delay of all modes having the same propagation constant (or the same parameter U). The second step focuses on the limitation of possible propagation constants or U values afforded by the actual cladding.

To find the density of modes per time increment, it is necessary to know the number of modes between U and $U + dU$. This number can be obtained by introducing $\mu(\nu)$ from (3.39) into (3.43). Reversing the order of integration and integrating over ν yields (Gloge and Marcatili, 1973)

$$N_\beta = \int_0^{r_\beta} [k^2 n^2(r) - \beta^2] r \, dr, \qquad (4.28)$$

where r_β denotes the radius at which the integrand vanishes. The integral (4.28) can be solved for the power-law profile (3.2). After introducing U and V from (3.10), (3.12), and (3.15) into the result and differentiating with respect to U, one has

$$dN_\beta/dU = U(U/V)^{4/g}. \qquad (4.29)$$

To find the mode density $dN_\beta/dt = (dN_\beta/dU)(dU/dt)$, (4.19) must be solved for U and the result must then be differentiated with respect to t. Closed-form approximations can be obtained in this way from the Taylor expansion of (4.19). When we write the delay in terms of a normalized delay time

$$\tau = (tc/LN_1) - 1 \qquad (4.30)$$

we obtain

$$h(\tau) = \frac{1}{N} \frac{dN_\beta}{d\tau} = \begin{cases} \dfrac{g+2}{g\Delta} \left| \dfrac{g+2}{g-g_0} \right|^{(2+g)/g} \left| \dfrac{\tau}{\Delta} \right|^{2/g} & \text{for } g \neq g_0 \\[4mm] \dfrac{2}{\Delta^2} & \text{for } g = g_0 \end{cases} \qquad (4.31)$$

with g_0 from (4.21). We also used (3.29). Profiles whose characteristics are very nearly, but not exactly, described by g_0 are not covered by (4.31) and require some further discussion later.

The actual impulse response $h(t)$ depends not only on the density dN_β of modes, but also on the energy these modes carry at the fiber end. This energy depends on mode excitation, the variation of loss with mode number and, in many cases, on the energy exchange among the modes. Since these influences are difficult to evaluate, the assumption of uniform energy distribution in all propagating modes is often used as a first approximation. We therefore devote some thought to this case.

One difficulty associated with this model has to do with the problem of accounting for all propagating modes. Theoretically, the cladding index n_0 limits the propagation constants of propagating modes to values larger than kn_0 ($U < V$). This distinction between propagating or "trapped" modes and radiative modes yields convenient results which we shall list below; but the results of Chapter 3 made it clear that the presence of a lossy jacket and the special features of high aximuthal mode orders modify the boundary for propagating modes to one rather different from the line governed by $U = V$ (see Fig. 3.13). We must consider this fact in later paragraphs.

The introduction of the condition $U = V$ into (4.19) yields the delay $t(V)$ of the highest order trapped modes. The Taylor expansion of this result yields the limits for t in the two cases described by (4.31). We have

$$\tau(V) = \begin{cases} \Delta(g - g_0)/(g + 2) & \text{for } g \neq g_0 \\ \Delta^2/2 & \text{for } g = g_0. \end{cases} \tag{4.32}$$

Figure 4.5 is a plot of the power distributions given by (4.31) and (4.32) for $g = 1, 4, 10$, or ∞ and $g_0 = 2$. If $g > g_0$, the high-order modes arrive later than the fundamental and if $g < g_0$ they arrive earlier. If $g = g_0$, the power distribution is similar to that shown in Fig. 4.5 for $g = \infty$, but its width is

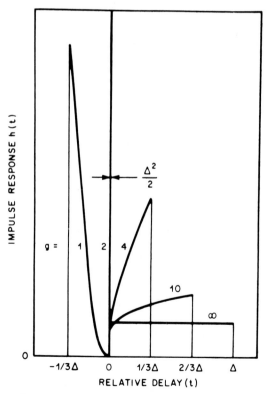

Fig. 4.5 Impulse response for power-law profiles with various exponents g.

$\Delta^2/2$ instead of Δ, i.e., a profile with $g = g_0$ can produce an output pulse that is a factor $\Delta/2$ shorter than that produced by a step-index fiber. Incidentally, inspection of (4.19) shows that a profile exponent $g_0 - 2\Delta$ reduces the total width of the impulse response by another factor of 4 (Gloge and Marcatili, 1973).

Most fiber characteristics (except for delay equalization) do not change much for profiles in the vicinity of the parabolic one. Figure 3.14 gives therefore a satisfactory description of a power profile with an exponent g_0 which, for typical fibers, is in the range between 1.7 and 2.3. Figures 3.13 and 3.14 show the mode boundary $U = V$ for trapped modes and the boundary $\alpha = 1$ dB/km. One can also see that the condition $b = 2a = 50$ μm eliminates some modes of low azimuthal order. The corresponding net change of the power distribution of the output pulse is shown in Fig. 4.6 for various profiles. The pulse tails represent modes of high azimuthal order which, although not trapped, nevertheless propagate with leakage losses of less than 1 dB/km (Adams et al., 1975). In the case of the step-index fiber, the pulse tail is extensive, but represents no more than two or three modes when $V = 32.4$. There are fewer modes with low leakage loss for g in the vicinity of 2.

More important than the total pulse width is the rms width according to

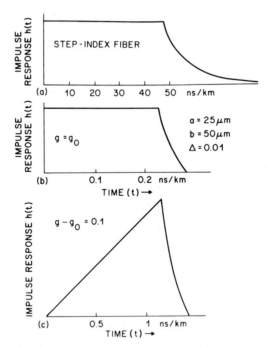

Fig. 4.6 Effect of leaky rays and finite cladding thickness on impulse response.

(4.1). The rms width is $(12)^{-1/2} = 0.288$ of the total width for a rectangular pulse shape and $(18)^{-1/2} = 0.236$ of the total width for a triangular pulse shape. Thus, in the case that all trapped modes are uniformly excited, the rms width is $0.288\, LN_1 \Delta /c$ for the step-index fiber, $0.144\, LN_1 \Delta^2 /c$ for $g = g_0$ and $0.236\, LN_1 \Delta |g - g_0|/c(g + 2)$ for $1 > |g - g_0| \gg \Delta$. The pulse tails indicated in Fig. 4.6 change the numbers given above to $0.388\, LN_1 \Delta /c$ for the step-index fiber, $0.150\, LN_1 \Delta^2 /c$ for $g = g_0$, and $0.246\, LN_1 \Delta |g - g_0|/c(g + 2)$ for $1 > |g - g_0| \gg \Delta$. Figure 4.7 is a plot of the rms pulse width versus g assuming uniform excitation of all trapped modes.

The profile which minimizes the rms width follows very nearly, although not exactly, a power law with $g = g_0 - 2.4\Delta$ (Arnaud, 1975). The associated rms pulse width, which represents the theoretical limit achievable, is

$$\sigma_{\min} = 0.022\, LN_1 \Delta^2 /c. \tag{4.33}$$

The output pulse width of practical multimode fibers is determined by unavoidable deviations from the optimal profile. Fibers fabricated by the

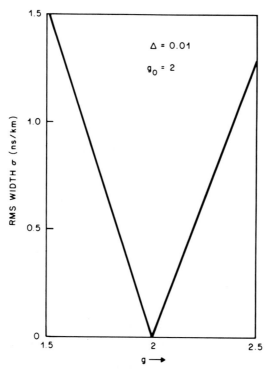

Fig. 4.7 Root-mean-square impulse width versus profile exponent g.

deposition of cylindrical layers usually have circularly symmetric index deviations $\epsilon(r)$. By properly choosing the reference indices n_1 and n_0, one can always find an optimal profile such that the profile error

$$\epsilon(r) = n_{\text{actual}} - n_{\text{optimal}}$$

satisfies   (4.34)

$$\epsilon(0) = \epsilon(a) = 0.$$

We first consider errors periodic in r, that is, concentric index rings that occur, for example, when each individual deposited layer differs slightly from the next or is not quite uniform across its width (Olshansky, 1976). Clearly there is an upper limit to the periodicity of errors that can have an effect on the mode delay, for each mode has a radial periodicity of its own and averages index changes within each of these radial periods. The maximum number m of radial periods of a mode of a near-parabolic fiber is obtained by setting $\beta = n_2 k$ and $\nu = 0$ in (3.54). The result is

$$m = V/4. \tag{4.35}$$

As expected, one finds that errors having a number of radial periods slightly less than $(V/4)$ have the strongest impact on the pulse width. Figure 4.8 shows the rms pulse width obtained for periodic errors of the form

$$\epsilon = 1.46 \times 10^{-4} \sin(2\pi q r/a) \tag{4.36}$$

for $V = 60$ and $\Delta = 0.01$. The pulse width is plotted versus the number of radial periods q. The solid curve is valid for a uniform power distribution in all modes; the dashed curve assumes that the energy in each mode decreases as $1 - U/V$. Note that the maximum number of mode periods is $m = 15$ in this case.

Another class of errors consists of circular index deviations having the form of a singular ring of radius r_ϵ whose width is larger than about $5a/m$. Numerical results obtained by various authors (Miller, 1974; Timmermann, 1974; Olshansky, 1976) indicate that, in this case, a broad estimate of the pulse width can be obtained from the relation

$$\sigma_\epsilon \approx (L/2c)|\epsilon_{\max}|(r_\epsilon/a), \tag{4.37}$$

where $\epsilon_{\max}$ is the peak index deviation in the ring. If the actual index profile follows a power law with exponent g, the maximum index deviation from the optimal profile is

$$\epsilon_{\max} = (\Delta n_1/2e)(g - g_0), \tag{4.38}$$

where e is the base of the natural logarithm; this maximum occurs at $r_\epsilon = e^{-1/g_0}$. Using (4.37), we find $\sigma_\epsilon = 3|g - g_0|$ nsec/km for $\Delta = 0.01$, which is in good agreement with Fig. 4.7.

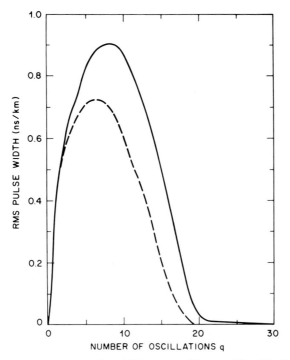

Fig. 4.8 Root-mean-square impulse width computed for a profile with circular symmetric index deviations as defined by (4.36) (Olshansky, 1976).

A frequent error found in multimode fibers fabricated by collapsing a hollow cylindrical structure is an index dip in the core center caused by out-diffusion during the collapsing process. We assume a depression of this kind having a peak deviation ϵ_{max} at $r = 0$, a Gaussian shape, and an rms width Σ. This profile error causes an approximate pulse width (Olshanski, 1976).

$$\sigma_\epsilon \approx (L/c)|\epsilon_{max}|(\Sigma/a). \tag{4.39}$$

If the depression is 5% of the peak profile value and the width Σ is 10% of the core radius, one finds $\sigma_\epsilon = 0.25$ nsec/km.

4.5 INFLUENCE OF EXCITATION, LOSS, AND MODE COUPLING

The condition of uniform power distribution in all modes almost never holds in practical multimode fibers. Most sources do not excite all modes uniformly and, in addition, there are loss phenomena which attenuate different modes by different amounts (Gambling *et al.*, 1973). In either case, the power in each mode at the fiber end is generally a complicated function of both mode numbers μ and ν and of the propagation constant β.

To simplify the following discussion, we consider situations in which the mode power distribution is only a function of the propagation constant or can be approximated by such a function. In this case, we can use the mode parameter U as before to describe the modes and can ignore the principal mode numbers μ and ν.

Consider, for example, a source which illuminates a step-index fiber with a radiation pattern of the form $\exp(-\sin^2\theta/\sin^2\theta_0)$ where the angle θ_0 is significantly smaller than the numerical aperture of the fiber. If we use the relationship (3.100), we find that the power distribution among the modes is proportional to $\exp(-U^2/k^2a^2\sin^2\theta_0)$. This distribution is the same at the input as well as at the output of the fiber, if all modes suffer the same attenuation.

The Taylor expansion of (4.19) and substitution of (4.30) leads to the relationship

$$\tau = \Delta U^2/V^2 \tag{4.40}$$

for the step-index fiber, when higher order terms of U/V are neglected. By introducing (4.40) into the mode power distribution, we find

$$h_g(\tau) = A_g \exp(-2\tau n_1^2/\sin^2\theta_0) \tag{4.41}$$

for $0 < t < \Delta$ and $h_g = 0$ everywhere else. A_g is a normalizing parameter. For $\sin\theta_0 \ll n_1(2\Delta)^{1/2}$, the rms width of (4.41) is approximately

$$\sigma_g = (LN_1/2cn_1^2)\sin^2\theta_0. \tag{4.42}$$

Curve 1 of Fig. 4.9 is a plot of σ_g for $\sin\theta_0 = 0.2$, and a fiber having a radius of 50 μm and a numerical aperture of 0.37.

Numerical apertures of this magnitude can be obtained with plastic-clad silica fibers. In this case, however, a considerable excess loss in the cladding may selectively attenuate certain mode groups so that the mode power distributions at the input and the output of the fiber are different. Assume the excess cladding loss to be α_0 Nepers and approximate (3.62), which describes the loss of a given mode in this case, by

$$\alpha_{st} \approx \alpha_0 U^2/V^2W. \tag{4.43}$$

The power distribution (4.41) at the input of the fiber must now be multiplied by $\exp(-2\alpha_{st}L)$ to find the output distribution. If we introduce (4.40) into the result, we find the impulse response

$$h_\alpha(\tau) = A_\alpha \exp\left(-\frac{2\pi n_1}{\sin^2\theta_0} - \frac{2\alpha_0 L_\tau}{\Delta V(1 - \tau/\Delta)^{1/2}}\right] \tag{4.44}$$

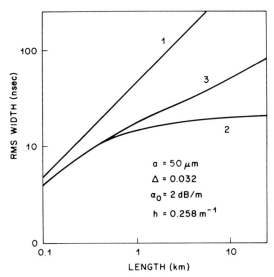

Fig. 4.9   Root-mean-square impulse width versus fiber length for step-index fiber (1) in the lossless case, (2) with lossy jacket, and (3) with lossy jacket and mode coupling.

for $0 < t > \Delta$ and $h_\alpha = 0$ everywhere else. A_α is another normalizing parameter. Curve 2 of Fig. 4.9 is a plot of (4.44) for the same parameter as before and an excess cladding loss of 2000 dB/km. The pulse width approaches a length-independent value

$$\sigma_\alpha(L \to \infty) = N_1 \Delta V / 2c\alpha_0. \tag{4.45}$$

This asymptotic characteristic is a consequence of the relationship (4.43) and specifically of the quadratic increase of α_{st} with U for small U. More generally, if the loss coefficient in the step-index fiber is proportional to U^x, the length dependence of the pulse width for large L will follow the power law $L^{1-x/2}$. This sublinear dependence on L is a consequence of the loss of high-order modes and therefore accompanied by a decrease in the effective numerical aperture with length.

As explained in Section 3, mode coupling replenishes part of the power lost in the high mode orders. To describe its result on the impulse response, we consider only the (usually) dominant power exchange between neighboring modes; in addition, we assume the coupling strength to be equal for all mode pairs so that it can be described by a simple parameter h. The power flow as a result of coupling loss leads to an equilibrium mode power distribution. In the case of the loss function (4.43) and uniform next-neighbor coupling (Gloge, 1972), this equilibrium distribution is of the form $\exp(-U^2/U_\infty^2)$ with

$$U_\infty = \pi^{1/2} V (h/\alpha_0 V)^{1/4}. \tag{4.46}$$

The far-field distribution associated with this equilibrium condition can be obtained by using the relation (3.100) to find the angle θ_∞ at which the far-field radiation has decreased to $1/e$ of the axial density.

   To continue in the description of our earlier fiber model, let us assume a coupling strength $h = 0.258$ m^{-1} so that $\sin\theta_\infty = 0.2$. This case covers the interesting situation in which the input radiation pattern as assumed earlier equals the equilibrium output radiation pattern. Figure 4.10 shows the evolution of the impulse response as a function of the fiber length in this case (Gloge, 1973). The exponential shape stipulated by (4.41) quickly changes and eventually approaches a Gaussian shape. A plot of the corresponding rms width versus fiber length is indicated by curve 3 of Fig. 4.9. This curve approaches the asymptote

$$\sigma_\infty = (N_1 \Delta U_\infty / 2cV^{1/2})(L/\alpha_0)^{1/2} \tag{4.47}$$

for long lengths. We find this asymptotic width to be proportional to $h^{1/4}$ and inversely proportional to $\alpha_0^{3/4}$. In other words, an increase in loss reduces the pulse width—at the expense of reduced output; an increase in coupling increases the pulse width—hence, uniform coupling should be avoided in a step index fiber that has a lossy cladding.

   A comparison of (4.47) with (3.170) shows the fundamental difference between a fiber with a lossy cladding and that with a low-loss cladding. In the latter case, the output pulse also assumes a Gaussian shape and its width also increases with the square root of the fiber length, but this width is inversely proportional to $h^{1/2}$, so that coupling decreases the pulse width.

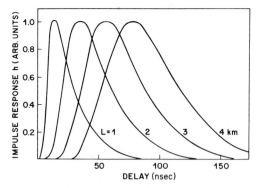

Fig. 4.10 Impulse response for step-index fibers of various lengths in the presence of cladding loss and mode coupling.

4.6 FREQUENCY DOMAIN CHARACTERIZATION OF FIBERS

In the previous discussion, pulse spreading in optical fiber transmission has been described in terms of the fiber impulse response $h(t)$. This is usually referred to as a time domain characterization of fibers. In many applications an alternative known as the frequency domain characterization is required.

The time domain characterization is obtained by injecting narrow pulses of optical energy into the fiber and observing the output response with a broadband detector system. The frequency domain characterization can be obtained in one of two (not necessarily equivalent) ways. One approach is to mathematically Fourier transform the time domain impulse response to obtain the frequency domain transfer function (Gloge and Chinnock, 1972; Chinnock *et al.*, 1973). The other method is to excite the fiber with light which is sinousoidally modulated about a dc level and to measure the reduction in sine wave amplitude (relative to the dc level) and the shift in phase of the sine wave component as a result of transmission (Personick *et al.*, 1974). We can refer to the two methods of measuring the transfer function as the indirect and direct methods, respectively.

There are several reasons for using a frequency domain characterization of fibers. As a practical matter it may be difficult to modulate the optical source in a high peak power mode. Thus sinusoidal modulation about a modest dc level may be necessary. From a performance analysis view, the intuitive time domain characterization may not reveal quantitative information about the effect of fiber pulse spreading on the sensitivity of an optical receiver. Typically, computer programs which calculate such effects require frequency domain data as an input, or obtain these data indirectly from the time domain measurement by Fourier transform.

4.6.1 Comparison of the Direct and Indirect Methods

It is often assumed that the direct and indirect methods of obtaining the transfer function of a fiber are equivalent. In many practical situations this is a reasonable approximation. However, the reader should be aware of the assumptions which must be made before this equivalence can be concluded.

We can always obtain the impulse response of the fiber in the time domain (source modulation constraints permitting) and then Fourier transform the result. Conversely, we can always sinusoidally modulate the source (duty cycle constraints permitting) to obtain the direct measurement. We can call the two measured transfer functions $H_{id}(f)$ and $H_d(f)$, respectively.

In order for these two functions to be the same, it is first of all important that certain relevant properties of the modulated source output are not dependent upon the modulation. Specifically, for directly modulated sources, it is often true that the optical bandwidth of the output light increases as the instantaneous power level increases. Since increased optical bandwith can lead to increased pulse spreading due to dispersion (see Sections 4.2 and 4.3), the transfer function measured with high peak power modulation indirectly in the time domain may have less baseband bandwidth than the transfer function measured with modest peak power directly in the frequency domain. For externally modulated sources it is important that the modulator does not introduce relative phase shifts (time delays) between different portions of the optical spectrum of the source being modulated. Otherwise these delays which will generally be functions of the modulation depth being applied may affect the direct and indirect measurements differently.

Beyond these practical problems associated with variation of the source properties with the modulation techniques there is also the question of baseband linearity (Personick, 1973a).

In order to show the equivalence of $H_{id}(f)$ and $H_d(f)$, we must assume that the output and input powers in an optical fiber satisfy the following linear relationship

$$P_{out}(t) = \int_{-\infty}^{\infty} h_{fiber}(t - \tau)P_{in}(\tau)\, d\tau. \qquad (4.48)$$

From (4.48) it follows that we can excite the fiber with a narrow pulse made up of an infinite weighted superposition of sine waves, and observe the output; or we can excite the fiber with sine waves (superimposed on a dc power level) one at a time and calculate the pulse response as a weighted superposition of the sine wave responses. This superposition of fiber outputs associated with multiple fiber inputs is what is called baseband linearity. One can show that in most practical cases, the fiber is approximately linear in its input–output power relationship (Personick, 1973a). The deviations from exact linearity generally do not affect measurements of impulse response or transfer function. However, it is possible to construct examples where measurements with very coherent sources would lead to results which do not follow from the baseband linearity assumption. Since the subject of baseband linearity involves complicated mathematical arguments, we leave it to the references (Personick, 1973a). For practical purposes we can assume that the baseband linearity approximation is valid, subject to further experimental investigations.

4.6.2 Using the Frequency Domain Characterization

The time domain (impulse response) characterization of fibers is fairly intuitive. If we ignore details such as long low tails on the impulse

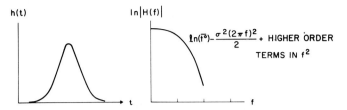

Fig. 4.11   Impulse response and transfer function.

response, then it is clear that the narrower the impulse response the better. Further, it would seem reasonable that pulse spreading will affect the performance of a digital fiber system repeater when the symbol rate increases to the point where adjacent pulses overlap. However, to obtain quantitative information about the effects of pulse spreading, it is generally most convenient to examine the frequency domain (transfer function) characterization.

Figure 4.11 shows a typical fiber impulse response and the corresponding Fourier transform (transfer function) magnitude. The zero frequency magnitude of the transfer function is equal to the area under the impulse response curve. If we divide the curve $H(f)$ by the area of the input power pulse, then the dc value represents the fiber loss. It is easy to show that since the impulse response is a positive function, the transfer function magnitude is always highest at zero frequency.* If we write down the definition of the transfer function

$$H_{\text{fiber}}(f) = \int h_{\text{fiber}}(t) \exp(i2\pi ft) \, dt \tag{4.49}$$

we see that it can be expanded in a power series as follows

$$H_{\text{fiber}}(f) = \int h_{\text{fiber}}(t) \left\{ 1 + i2\pi ft - \frac{(2\pi ft)^2}{2} \cdots \right\} dt$$

$$= \sum_{0}^{\infty} \frac{(i2\pi f)^n}{n!} \overline{t^n}, \tag{4.50}$$

where

$$\overline{t^n} \overset{\Delta}{=} \int t^n h_{\text{fiber}}(t) \, dt \tag{4.51}$$

---

* There are some nonintuitive cases where one must define an impulse response which is not strictly positive. This can occur when different spectral portions of the modulated source output are delayed relative to one another. If the fiber dispersion reverses these delays, then the pulse can actually get narrower in transmission. The resulting impulse response which has positive and negative portions is not physically observable (although it still retains the role described in (4.48)) since such a source cannot emit an impulse by definition. Thus the requirement that power is positive is not violated.

The magnitude of the transfer function $|H_{\text{fiber}}(f)|$ has the following expression:

$$|H_{\text{fiber}}(f)|^2 = (\overline{t^0})^2[1 - (2\pi f \sigma)^2] + \text{higher order terms in } f^2,$$

where

$$\sigma^2 = (\overline{t^2}/\overline{t^0}) - (\overline{t^1}/\overline{t^0})^2. \tag{4.52}$$

Thus we see that as a first-order approximation the transfer function is parabolic in shape and characterized by the area of the impulse response $\overline{t^0}$ and its rms width σ. However, the reader should note that this parabolic approximation begins to fail very quickly for pulses with long tails. Figure 4.12 shows an example of this effect.

Under ideal circumstances, the fiber impulse response is sufficiently narrow so that the transfer function is flat (constant) out to frequencies beyond the digital bit rate. Under such circumstances, the fiber behaves as an optical attenuator. When the fiber transfer function rolls off at frequencies below the bit rate, some sensitivity penalty is incurred at the receiver. The penalty incurred depends on the design of receiver and the exact shape of the transfer function (magnitude and phase). However, roughly speaking, the sensitivity penalty is less than 1 dB if the transfer function magnitude $|H_{\text{fiber}}(f)|$ has not rolled off more than 6 dB (one-half its low-frequency value) for frequencies up to half the bit rate (Personick, 1973b). For fibers with long tails and therefore having a low-frequency bump, this rule can often be relaxed somewhat depending upon the system design.

This rough rule of thumb translates into the time domain as a requirement that the rms pulse width be less than 0.25 time slots (separation between pulses) for a penalty in sensitivity of less than 1 dB.

For the small sensitivity penalties and for impulse responses with reasonably low ratios of tail energy to main pulse energy such rules of thumb based upon the magnitude of the transfer function are adequate. However, it is possible to compute exactly the effect of fiber pulse spreading on

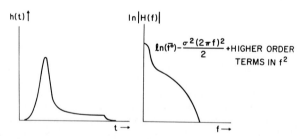

Fig. 4.12 Impulse response and transfer function for an impulse response with a tail.

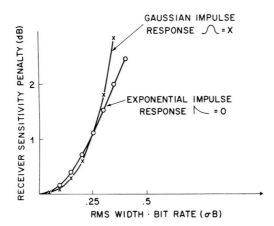

Fig. 4.13 Sensitivity penalties for typical receivers.

the sensitivity of a given receiver by using the complete transfer function (amplitude and phase) (Personick, 1973b). Figure 4.13 shows a typical calculation of the receiver sensitivity penalty as a function of σ for a fiber with a Gaussian impulse response and with an exponential impulse response.

As a final remark, since the impulse response and the transfer function are approximately Fourier transform pairs and since computations done in the frequency domain can be generally transformed to the time domain, it is fair to say that quantitative information about the effect of pulse spreading on the sensitivity of a receiver can be gleaned from the impulse response directly by one who has sufficient experience working with time domain data.

REFERENCES

Adams, M. I., Payne, D. N., and Sladen, F. M. E. (1975). Mode transit times in near-parabolic optical fibers. *Electron. Lett.* **11**, 389–390.

Arnaud, J. A. (1975). Pulse broadening in multimode graded-index fibers. *Electron Lett.* **11**, 167–169.

Chinnock, E. L. *et al.* (1973). The length dependence of pulse spreading in the CGW Bell-10 optical fiber. *Proc. IEEE* **61**, 1499–1500.

Dyott, R. B., and Stern, J. R. (1971). Group delay in glass fiber waveguide. *Electron. Lett.* **7**, 82–84.

Gambling, W. A., Payne, D. N., and Matsumura, H. (1973). The effect of loss on propagation in multimode fibers. *Radio Electron. Eng.* **43**, 683–688.

Gloge, D. (1971). Dispersion in weakly guiding fibers. *Appl. Opt.* **10**, 2442–2445.

Gloge, (1972). Optical power flow in multimode fibers. *Bell Syst. Tech. J.* **51**, 1767–1783.

Gloge, D. (1973). Impulse response of clad optical multimode fibers. *Bell Syst. Tech. J.* **52**, 801–815.

Gloge, D. (1975). Propagation effects in optical fibers. *IEEE Trans. Microwave Theory Tech.* **MTT-23**, 106–120.

Gloge, D., and Chinnock, E. L. (1972). Study of pulse distortion in Selfoc fibers. *Electron. Lett.* **8**, 526–527.

Gloge, D., and Marcatili, E. A. J. (1973). Multimode theory of graded-core fibers. *Bell Syst. Tech. J.* **52**, 1563–1578.

Gloge, D., Kaminow, I. P., and Presby, H. M. (1975). Profile dispersion in multimode fibers: Measurement and analysis. *Electron. Lett.* **11**, 469–470.

Hammond, C. R. (1978). Silica-based binary glass systems: Wavelength dispersive properties and composition in optical fibers. *Opt. Quantum Electron.* **10**, 163–170.

Kapron, F. P., and Keck, D. B. (1971). Pulse transmission through a dielectric optical waveguide. *Appl. Opt.*, **10**, 1519–1523.

Kawakami, S., and Nishizawa, I. (1968). An optical waveguide with the optimum distribution of refractive index with reference to waveform distortion. *IEEE Trans. Microwave Theory Tech.* **MTT-16**, 814–818.

Marcatili, E. A. J. (1977). Modal dispersion in optical fibers with arbitrary numerical apertures and profile dispersion. *Bell. Syst. Tech. J.* **56**, 49–63.

Miller, S. E. (1965). Light propagation in generalized lens-like media. *Bell. Syst. Tech. J.* **44**, 2017–2064.

Miller, S. E. (1974). Delay distortion in generalized lens-like media. *Bell Syst. Tech. J.* **53**, 177–193.

Olshansky, R. (1976). Pulse broadening caused by derivations from the optimal index profile. *Appl. Opt.* **15**, 782–788.

Olshansky, R. (1978). Optical waveguides with low pulse dispersion over an extended spectral range. *Electron. Lett.* **14**, 330–331.

Olshansky, R., and Keck, D. B. (1976). Pulse broadening in graded-index optical fibers. *Appl. Opt.* **15**, 483–491.

Payne, D. N., and Gambling, W. A. (1975). Zero material dispersion in optical fibers. *Electron. Lett.* **11**, 176–177.

Personick, S. D. (1973a). Receiver design for digital fiber optic communication systems, I. *Bell Syst. Tech. J.* **52**, 843–874.

Personick, S. D. (1973b). Baseband linearity and equalization in fiber optic digital communication systems. *Bell Syst. Tech. J.* **52**, 1175–1194.

Personick, S. D., Hubbard, W. M., and Holden, W. S. (1974). Measurement of the baseband frequency response of a 1-km fiber. *Appl. Opt.* **13**, 266–268.

Presby, H. M., and Kaminow, I. P. (1976). Refractive index and profile dispersion measurements in binary silica optical fibers. *Appl. Opt.* **15**, 3029–3036.

Sladen, F. M. E., Payne, D. N., and Adams, M. I. (1978). Profile dispersion measurements for optical fibers over the wavelength range 350 nm to 190 nm. *Proc. Eur. Conf. Opt. Commun., 4th, 1978*, p. 48.

Steinberg, R. A. (1974). "Comment on picosecond mode delays in gradient-refractive-index fibers. *Electron. Lett.* **10**, 375–376.

Timmermann, C. C. (1974). The influence of deviations from the square-law refractive index profile of gradient core fibers on mode dispersion. *Arch. Elektr. Uebertr.* **28**, 344–346.

Unger, H. G. (1977). "Planar Optical Waveguides and Fibers." Clarendon Press, Oxford, p. 506.

Chapter 5

Nonlinear Properties of Optical Fibers

ROGERS H. STOLEN

5.1 INTRODUCTION

Fibers are usually considered to be completely passive or linear media. As the input power is increased, one expects only a proportional increase in output power. This, however, is not strictly true and dramatic power dependent or nonlinear effects can occur which cause strong frequency conversion, optical gain, and many other effects generally associated with strong optical intensities and highly nonlinear materials.

These nonlinear processes depend on the interaction length as well as intensity. In small-core low-loss fibers high intensities can be maintained over lengths of more than a kilometer. If this length is compared with the focal region of a Gaussian beam of comparable spot size, enhancements of 10^5–10^8 are possible using fibers. This enhancement lowers the threshold power for nonlinear processes—in some cases to less than 100 mW. In a communications fiber this can lead to high attenuation, pulse spreading, and physical damage.

While these nonlinear processes could be detrimental in communications fibers the same effects are potentially useful in fiber-optical amplifiers, oscillators, and modulators. The optimal spectral region for such fiber devices is the region of low fiber loss around 1 μm. In the future, many of the optical sources used for measurement of fiber loss, pulse dispersion, etc., may be based on fibers themselves. Fibers also share many of the useful properties of other guided wave structures such as the

125

planar guides used in integrated optics. An important example is the use of the waveguide modes to achieve phase matching for parametric nonlinear processes. In this chapter we concentrate on the basic nonlinear processes which are common to both the communications and device aspects of the problem.

The advantages of fibers for nonlinear optics were pointed out by Ashkin (Ashkin and Ippen, 1970), and the first nonlinear effect using such a geometry was observed by Ippen (1970) in a liquid core fiber. In 3 m of a 12-μm core, CS_2-filled guide the stimulated Raman threshold was less than a watt for long pulses of 514.5 nm light. The implications of such stimulated scattering in optical fibers used for communications were pointed out by Smith (1972). The long interaction length available in glass fibers was used to observe stimulated Raman scattering (Stolen et al., 1972), stimulated Brillouin scattering which is the lowest threshold nonlinear process in glass fibers (Ippen and Stolen, 1972) and optically induced birefringence (Stolen and Ashkin, 1973). This early work has been discussed in a review article by Ippen (1975). Recently the use of long, small-core, quartz hollow fibers has permitted low-power Raman gain in many liquids (Stone, 1975; Görner et al., 1974). The long interaction length in fibers enhances spontaneous Raman scattering as well as stimulated scattering and many weak Raman lines have been observed in both liquid and glass fibers (Walrafen and Stone, 1972, 1975). A similar enhancement occurs for multiphoton effects and two-step absorption has been observed from small numbers of impurities in glass fibers (Stolen and Lin, 1975).

The well-defined geometry of fibers has been used to measure the Raman gain in SiO_2 (Stolen and Ippen, 1973) and to study self-phase modulation in liquid core (Ippen et al., 1974) and glass core (Stolen and Lin, 1978) fibers. Four-photon mixing has been seen in silica fibers by using waveguide modes to phase match (Stolen et al., 1974). Stimulated four-photon mixing occurs for certain favored combinations of modes (Stolen, 1975; Stolen and Leibolt, 1976). Recently, devices using fibers have appeared such as tunable Raman oscillators (Hill et al., 1976b; Jain et al., 1977; Johnson et al., 1977; Stolen et al., 1977; Lin et al., 1977a) and a fiber generated nanosecond continuum useful for time-resolved spectroscopic studies (Lin and Stolen, 1976a).

In this paper we will discuss stimulated Raman scattering (SRS), stimulated Brillouin scattering (SBS), the optical Kerr effect, self-phase modulation (SPM), phase-matched four-photon mixing, and nonlinear absorption. SRS, SBS, and SPM are of particular interest because these processes impose limits on the peak power of fiber transmission systems. Critical powers for these three processes in a single-mode and a multimode fiber are given in a table at the end of this chapter.

5.2 STIMULATED RAMAN SCATTERING (SRS)

5.2.1 SRS for Plane Waves

SRS illustrates particularly well how a long optical fiber reduces the threshold power for nonlinear optical effects. SRS is undesirable in a transmission fiber while on the other hand the Raman interaction has potential as an optical amplifier. We will first deal with spontaneous Raman scattering, Raman gain, Raman absorption, and SRS for plane waves. Next we deal with wave-guide modes as opposed to plane waves and show that the waveguide case can be treated using the concept of an effective core area. Finally we compare some differences between forward and backward SRS. More detailed discussions of SRS are contained in review articles by Bloembergen (1967) and Kaiser and Maier (1972).

Raman scattering can be viewed as modulation of light by molecular vibrations. Upper and lower sidebands appear in the scattered light spectrum, separated from the incident light frequency by the modulation frequency. At high temperatures the molecular modes are highly excited and the sidebands are of equal intensity. At very low temperatures where the modes are not thermally excited this classical picture predicts that both sidebands vanish. Actually, because of quantum effects, it is always possible for an incident photon to scatter, producing a phonon and a "Stokes" photon so that a lower sideband still appears. Figure 5.1 shows the upper and lower bands (anti-Stokes and Stokes) for fused silica at 0 and 300° K. The 0° K Stokes intensity is described by a frequency-dependent cross section $\sigma_0(\Delta\nu)$ which depends on the vibrational density of states and the coupling between light and molecular modes. The Stokes and anti-Stokes

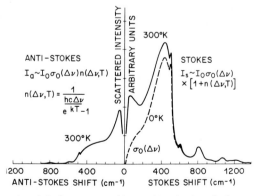

Fig. 5.1 Stokes and anti-Stokes scattered intensities for fused silica at 0° and 300° K. $n(\Delta\nu)$ is the phonon population factor, h and k are the Planck and Boltzman constants, T is temperature in °K, c is the velocity of light and $\Delta\nu$ is the frequency shift in cm^{-1}.

intensities at higher temperatures are related to $\sigma_0(\Delta\nu)$ by the thermal population factor $n(\Delta\nu, T)$ through the relations given in Fig. 5.1. Note that high-frequency phonons (large $\Delta\nu$) are only weakly excited at 300° K.

The Raman bands are much broader in glasses than crystals because the usual wavevector selection rules break down and light scattering occurs from essentially all the normal modes of the material (Shuker and Gammon, 1970). These broad Raman bands were recognized as potentially useful in wide bandwidth optical amplifiers and tunable oscillators (Ippen et al., 1971; Stolen et al., 1972).

A weak Stokes signal injected along with a stronger pump will be amplified while a weak anti-Stokes signal injected with a strong pump will be absorbed (Jones and Stoicheff, 1964). The amplification of Stokes and absorption of anti-Stokes is described by the equations:

$$P_s(L) = P_s(0)e^{GI_0L}, \qquad P_a(L) = P_a(0)e^{-GI_0L}$$

$$G = \frac{\sigma_0(\Delta\nu)\lambda_s^3}{c^2 h \epsilon} \tag{5.1}$$

where I_0 is the pump intensity, c the velocity of light, h is Planck's constant, ϵ the dielectric constant and, λ_s is the Stokes vacuum wavelength. L is the interaction length which for plane waves is the sample length. In a fiber, L is related to the fiber length and the linear absorption as will be discussed in Section 5.2.2. It is clear from Eq. (5.1) that significant amplification could be obtained either with high intensity and a short interaction length or with low intensity and a long length such as would be obtained in a low-loss fiber. The relevant cross section is the low temperature value so that the gain is independent of temperature. The cross section $\sigma_0(\Delta\nu)$ varies as ν^4 so the Raman gain varies linearly with frequency.

Equation (5.1) contains some simplifying assumptions. The first is that the gain is independent of the direction of the two beams. In glasses this is always true because the phonon wavevector is independent of the phonon frequency for the Raman active bands. The present interest in fibers restricts the directions to collinear forward and backward interactions. We also assume that the bandwidth of the pump is much less than the width of the Raman gain curve. The gain coefficient is severely reduced if the opposite is true, for example, in the case of narrow Raman lines in liquids pumped by ultrashort pulses which have large bandwidths. Finally, Eq. (5.1) assumes that the amplification of Stokes does not significantly deplete the pump.

For large I_0 or long L the amplification is so large that spontaneously scattered Raman light can be amplified to intensities comparable to the pump. There is no threshold power but for most purposes it is sufficient to define a critical power as that power for which the Stokes power equals the pump power. This occurs when (Smith, 1972)

$$G_R I_0 L = 16. \tag{5.2}$$

The gain is highest at the maximum of the gain curve which for silica is at 450 cm^{-1} and this is the frequency where SRS will appear. In a long fiber the conversion is almost complete and each Stokes pumps the next higher order Stokes in turn. Figure 5.2 shows the pump and seven orders of Stokes from 100 m of 3.3-μm SiO$_2$ core fiber. The pump was a doubled Q-switched Nd:YAG laser at 532 nm. The unconverted light from the pump and various orders comes from the leading and trailing edges of the pulse where the intensity is small. The smearing of the highest order Stokes is most likely due to self-phase modulation which is discussed in Section 5.4.2. The combination of SRS and SPM has recently been treated theoretically by Lugovoi (1976).

5.2.2 SRS in Fibers

The pump power decreases along a fiber because of linear absorption so the Raman gain is greater near the input end. To take account of this decreasing gain an effective fiber length (L) is used in Eqs. (5.1) and (5.2) which is reduced from the actual fiber length (l) by the linear absorption (α) (Ippen, 1975).

$$L = (1 - e^{-\alpha l})/\alpha. \tag{5.3}$$

The intensity I_0 is then the power at the input end of the fiber divided by an effective core area. This effective area is determined by mode size and the overlap between pump and Stokes modes. Mode overlap, or the lack of overlap, in fibers will reduce the Raman gain from the plane wave expression of (5.1). This can be seen from Fig. 5.3. In Fig. 5.3 the pump propagates in the lowest order mode and the Stokes is injected into a higher order mode such as the LP$_{41}$. The Stokes sees very little of the pump energy and the gain will be significantly reduced. Even with both pump and Stokes in the same mode there is little amplification in the cladding

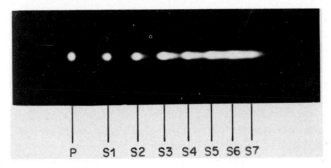

Fig. 5.2   Pump and seven orders of stimulated Stokes from 100 m of 3.3-μm SiO$_2$ core fiber. The output was focused by a 40 × microscope objective and dispersed by a prism.

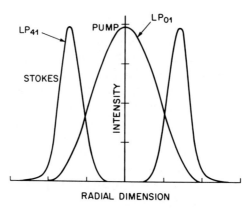

Fig. 5.3 Stokes and pump intensities for a choice of modes with low Raman gain.

where the intensity is low even though the cladding may carry a significant fraction of the power. The plane-wave gain is thus modified by an integral which is a measure of the mode overlap. We describe the electric field in mode j by:

$$|E_j| = |E_j|_{\max} \Psi(r, \theta). \qquad (5.4)$$

The effective intensity in the fiber then becomes

$$I_0 = P_0 \frac{\langle \Psi_1{}^2 \Psi_2{}^2 \rangle}{\langle \Psi_1{}^2 \rangle \langle \Psi_2{}^2 \rangle} = \frac{P_0}{A_{\text{eff}}} \qquad \langle [\ ] \rangle \equiv \int_0^{2\pi} \int_0^{\infty} [\quad] r \, dr \, d\theta. \qquad (5.5)$$

The effective area as defined in Eq. (5.5) is reasonably close to the fiber core area. The effective area is 1.10 times the core area in a step-index fiber of $V = 2.5$. The ratio of effective area to core area is given for several values of V in the Appendix and as expected is larger for low V numbers. If the LP_{01} mode is approximated by a Gaussian,

$$\Psi_1{}^2(r, \theta) = \Psi_2{}^2(r, \theta) = e^{-r^2/w^2}, \qquad (5.6)$$

where r is the radius and w is the $1/e$ radius for the intensity, the effective area becomes $2\pi w^2$. A Gaussian fit to the mode intensity in a step index fiber results in an effective area which is 10–15% too large. This is because there is more energy in the tail of a Gaussian than in the corresponding step-index guide mode. The Gaussian approximation may be better for highly graded index cores.

For the case of an ideal multimode fiber with no mode mixing, the Stokes modes can be treated independently and overlap integrals calculated for the mode of interest with each excited pump mode. In fibers with strong mode coupling the Stokes gain is the average of the individual mode gains (Capasso and DiPorto, 1976). For highly multimode fibers the best procedure is to simply consider the intensity of pump and Stokes to be uniform across the core so the effective area is the core area.

Equation (5.1) assumes that both pump and Stokes have the same linear polarization. For the case of perpendicular polarization the gain in glasses is reduced by more than a factor of 10 with respect to the strong Raman bands. Linear polarization is not maintained in multimode fibers so the gain is reduced from the value given by (5.1) by about a factor of 2. Ideal single-mode guides support the linearly polarized LP_{01} mode, but in practice this occurs only in short lengths or in guides with enough birefringence to lock the polarization (Snitzer and Osterberg, 1961; Stolen et al., 1978). In most long, single-mode fibers, the polarization varies (Kapron et al., 1972) and the best approximation is to divide the gain of (5.1) by a factor of 2 as for the multimode guide.

The expression for the gain in (5.1) is calculated for cw pump and Stokes and is thus valid for pulses only if there is no pulse spreading and if index or waveguide dispersion do not cause separation of the pump and Stokes. Pulse spreading will reduce the peak intensity and thus increase the power and distance required for SRS.

Once physical separation of pump and Stokes occurs there will be no Raman gain. This walk-off distance thus sets a maximum interaction length for use in (5.2) which may be much less than the actual fiber length. Material dispersion usually dominates over waveguide dispersion in a single-mode fiber so the difference in propagation time for two pulses separated by $\delta\nu$ about a frequency ν is

$$\delta t = (L/c)D(\lambda)(\delta\nu/\nu), \tag{5.7}$$

where L is the fiber length, c the velocity of light in vacuum and $D(\lambda) = \lambda^2 d^2 n/d\lambda^2$ is the dispersion (Gloge, 1971). For example, a 1-psec pump pulse at 600 nm separates from a 1-psec Stokes pulse 10 nm away (273 cm^{-1}) in 30 cm of silica-core fiber. It is possible, however, to correct for this walk-off by taking advantage of the group velocity difference between different waveguide modes (Giordmaine and Shapiro, 1968; Lin et al., 1977c).

These walk-off effects cause differences in the apparent gain between long and short fibers and this can show up particularly for the case of pulses whose bandwidth is much larger than the transform limit. For example, a pulse consisting of a train of mode-locked pulses will have a high gain in a short fiber because the stimulated Stokes pulses stay in step with the peaks of the mode-locked pump pulses. In a long fiber, however, the Stokes pulse could pass through many pump pulses and the relevant power would be the average of the mode-locked pulses. Nonmode-locked wide-band pulses should show similar behavior because of strong sharp microstructure in the time behavior of such pulses.

The preceding results are summarized in the Appendix where the critical power is calculated for a 1.0 dB/km single-mode and a 1.0 dB/km multimode silica fiber at 1.0 μm. Notes are included in the Appendix so these values can be translated to different wavelengths and fibers of different core size, length, and loss.

5.2.3 Backward SRS

The gain for backward Raman amplification should be the same as in the forward direction. The two configurations differ, however, in that the maximum forward Stokes power can be no greater than the pump power. The backward amplified Stokes, on the other hand, always sees fresh pump energy. It is in fact possible to extract all the energy of the pump by the leading edge of the Stokes pulse and produce a pulse which is both shorter than the original Stokes pulse and contains all the energy of the pump (Maier *et al.*, 1969).

Figure 5.4 shows backward Stokes amplification in a fiber (Lin and Stolen, 1976b). A 5-nsec pump and a 2-nsec Stokes pulse are produced by two dye lasers. These pulses are coupled into opposite ends of a 1-m-long, 4-μm core fiber. This fiber is slightly longer than the optimum length which would be half the length of the pump pulse. In Fig. 5.4 a weak backward Stokes signal is amplified by a factor of 45. This amplification is consistent with the gain determined in forward scattering measurements. When the Stokes power (300 W) is comparable to the pump (500 W) the total amplification is less than in the small signal case because of pump depletion and there is preferential amplification of the leading edge. In the

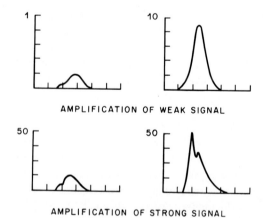

AMPLIFICATION OF WEAK SIGNAL

AMPLIFICATION OF STRONG SIGNAL

Fig. 5.4 Backward Raman amplification. The Stokes signal is measured in the absence and in the presence of the pump pulse. Relative Stokes intensities are shown on the vertical scale; time scale, 2 nsec/div. For a strong Stokes signal most of the pump energy is extracted by the leading edge of the Stokes pulse.

strong signal case over 90% of the pump energy was extracted by the Stokes signal.

There is a limit to the usable fiber or pump pulse length which is set by the conversion of backward amplified Stokes to second Stokes. Because the Stokes pulse is growing and becoming shorter the problem is complicated but some typical backward Stokes pulses have been between 2 and 5% of the pump pulse length in the regime of strong pump depletion (Lin and Stolen, 1976b).

5.3 STIMULATED BRILLOUIN SCATTERING (SBS)

Brillouin scattering is similar to Raman scattering except that acoustic phonons are involved rather than high frequency optical phonons. The frequency of the Brillouin scattered light varies with scattering angle because the frequency of the sound wave varies with acoustic wavelength. This frequency shift is a maximum in the backward direction and is zero in the forward direction so that in fibers only the backward process is of interest. The frequency shift (ν_s) depends on the velocity of longitudinal sound waves (V_s) and the pump vacuum wavelength λ. If we assume the pump and shifted wavelengths are approximately equal

$$\nu_s = 2nV_s/\lambda, \tag{5.8}$$

where n is the refractive index.

The gain for SBS in glasses is more than two orders of magnitude larger than that for SRS. Despite this large Brillouin gain, however, SRS is usually the dominant process because the pump linewidth is much larger than that of the Brillouin line. The pumping efficiency is reduced approximately by the ratio of the Brillouin and pump linewidths (Denariez and Bret, 1968). Brillouin linewidths are typically around 100 MHz as compared to Raman bands in glasses which are very broad (400 cm^{-1} or 10^{13} Hz).

The Brillouin gain coefficient for the case where the pump linewidth ($\Delta\nu_p$) is much less than the Brillouin linewidth ($\Delta\nu_B$) is:

$$G_{B0} = \frac{2\pi n^7 p_{12}^2}{c\lambda^2 \rho V_s \Delta\nu_B} \text{ cm/W}. \tag{5.9}$$

where ρ is the density, p_{12} is the elasto-optic coefficient, and c is the velocity of light in vacuum (Tang, 1966). For plane waves in fused silica at a wavelength of 535.5 nm the calculated frequency shift is 32.25 GHz and $G_{B0} = 4.5 \times 10^{-9}$ cm/W. The Brillouin linewidth at this acoustic frequency is 134 MHz (Pelous and Vacher, 1975) and $p_{12} = 0.271$ (Primak and Post, 1959). The gain, G_{B0} is almost independent of frequency. This is because the Brillouin frequency shift (ν_s) varies linearly with pump frequency (Eq. 5.8) and the linewidth ($\Delta\nu_B$) varies as the square of the phonon frequency (Pelous and Vacher, 1975) which cancels λ^2 in Eq. (5.9). If $\Delta\nu_p \gg \Delta\nu_B$ then

$$G_B \approx G_{B0}(\Delta\nu_B/\Delta\nu_p). \tag{5.10}$$

Now the gain decreases at long wavelengths by the factor $1/\lambda^2$.

Because of the large Brillouin gain only SBS will occur with a narrow line pump. In an experiment to observe SBS in fibers, a pulsed Xenon laser was operated with an internal etalon so its output was restricted to a single longitudinal mode (Ippen and Stolen, 1972). Figure 5.5 shows the oscilloscope traces of the input and transmitted signals and the stimulated backward scattering as observed with a beam splitter at the input. The critical power for backward stimulated scattering is related to the gain by (Smith, 1972):

$$P_{crit} = 21A_{eff}/GL_{eff}. \tag{5.11}$$

Note the difference between the thresholds for forward and backward stimulated scattering as given by Eqs. (5.2) and (5.11). The overlap integral for SBS is the same as for SRS in Eq. (5.4). The measured gain coefficient, reduced to the plane wave case, was 4.3×10^{-9} cm/W which is in excellent agreement with the calculated value.

The periodic nature of the scattered signal is explained in terms of oscillations due to finite sample length. The buildup of the backward Brillouin wave results in a depletion of the pump wave near the input of the fiber. This in turn reduces the gain in the fiber until the depletion region has passed out of the fiber. The temporal period of the resulting oscillation is $2nl/c$ which is the round-trip transit time of the fiber.

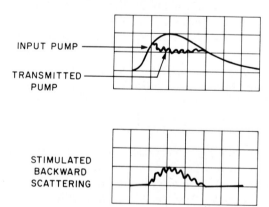

Fig. 5.5 Oscilloscope traces of input and transmitted pump and stimulated backward Brillouin scattering.

In the Appendix the SBS threshold power is calculated for cw excitation of a single-mode and a multimode fiber. As in the Raman case the Brillouin gain is severely reduced for perpendicular polarizations so the gain is reduced by a factor of 2 in multimode fibers. The calculated cw critical power would, of course, not represent the limiting peak power in such a fiber used for the transmission of information. First, because of the backward nature of the amplification the Brillouin wave sees the average of a train of pulses rather than the peak power. Second, the pulse bandwidth would be significantly larger than the Brillouin linewidth.

CW-Stimulated Brillouin scattering has been seen by employing an optical fiber ring resonator (Hill *et al.*, 1976a). In single-mode fibers a single Brillouin shifted line can contain 50% of the pump light and has a frequency width of 20 MHz.

5.4 INTENSITY-DEPENDENT REFRACTIVE INDEX

5.4.1 Optical Kerr Effect

The refractive index usually increases slightly in regions of high optical intensity which leads to self-focusing (Chiao *et al.*, 1964). In fibers, any additional confinement caused by self-focusing is negligible but the small phase shifts caused by changes in refractive index will add up over the length of a fiber and have important consequences. One such case is the optical Kerr effect, i.e., optically induced birefringence, which has been used as a fast optical shutter for picosecond optical pulses (Duguay and Hansen, 1969) and has been observed in silica-core fibers (Stolen and Ashkin, 1973).

In the optical Kerr effect the birefringence induced by a strong linearly polarized pump pulse is probed by a weaker signal at a different wavelength. A typical arrangement is illustrated in Fig. 5.6. The probe is introduced with its polarization 45° to that of the pump and the polarizer is set

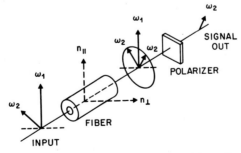

Fig. 5.6 Optical Kerr effect. The birefringence induced by a linearly polarized pump pulse at ω_1 is probed by a weaker signal at ω_2.

to eliminate probe light in the absence of a pump. The induced birefringence is then related to the transmission of the polarizer. Linear pump
polarization must be maintained so that a single-mode birefringent fiber
is required (Stolen et al., 1978). The pump polarization must be along a
principal axis of the fiber and a phase retardation plate (not shown in the
figure) is inserted ahead of the polarizer to correct for the effect of the intrinsic birefringence on the probe light.

The power transmission of the system is related to the induced phase
difference of the components of the probe by $\sin^2(\Delta\phi/2)$ where $\Delta\phi$ is the
phase difference in radians and is proportional to the difference between
the intensity dependent indices parallel and perpendicular to the pump
polarization (Duguay and Hansen, 1969).

$$\Delta\phi = (2\pi L/\lambda)(\delta n_\parallel - \delta n_\perp), \qquad \delta n_\parallel - \delta n_\perp = \tfrac{1}{2}n_{2B}E^2,$$

$$E^2 = \frac{8\pi P}{ncA_{\text{eff}}} \times 10^7. \tag{5.12}$$

The Kerr coefficient is n_{2B} and is expressed in esu, L is the effective guide
length which was defined in Eq. (5.3), λ is the probe vacuum wavelength,
E is the peak field amplitude, and P is the pump power in watts. The effective area is the same as for SRS and SBS and is defined in Eq. (5.5) (Bjorkholm, 1975). The Kerr coefficient is often expressed in terms of a "B_0" coef-
ficient or as parallel and perpendicular susceptibilities with $\chi_\perp \approx \tfrac{1}{3}\chi_\parallel$
(Maker and Terhune, 1965; Wang, 1966; Owyoung, 1971).

$$n_{2B} = B_0\lambda = (24\pi/n)(\chi_\parallel - \chi_\perp). \tag{5.13}$$

Experimentally, a probe transmission of 5% of maximum has been
achieved in a 3-m silica core fiber with a pump power of 7 W (Stolen and
Ashkin, 1973). The value of n_{2B} was found to be $1.4 \pm 0.3 \times 10^{-13}$ esu as
compared with previous results which fall in the range $2 \pm 1 \times 10^{-13}$ esu
(Duguay and Hansen, 1969). The probe transmission increases quadrati-
cally with pump power as expected from (5.12).

The response time of the Kerr effect in glass is expected to be considerably faster than a picosecond (Owyoung et al., 1972). Group velocity dispersion, as discussed for Raman gain in Section 5.2.2, requires that for
short pulses either the pump and probe must be very close in frequency or
that they propagate in different waveguide modes in order to match group
velocities.

5.4.2 Self-Phase Modulation (SPM)

The intensity dependent refractive index also leads to a phase modulation within a single optical pulse. Because of the index change, the phase
is retarded at the peak with respect to the leading and trailing edges. This

phase shift adds up in a long fiber and results in a sizable phase modulation. The frequency broadening associated with this phase modulation will, in combination with the group velocity dispersion, cause additional pulse spreading. SPM may be the most important power-limiting nonlinear process in high-capacity single-mode fibers.

Self-phase modulation was first observed as a modulated spectrum extending both above and below the laser frequency after self-focusing had occurred in a liquid-filled cell and was explained as phase modulation due to the intensity dependent refractive index by Shimizu (1967). Because this process took place in self-focused filaments the intensity was high and there were problems with competing nonlinear effects and uncertainty concerning the filament size. These problems were overcome by the observation of SPM in optical fibers (Ippen et al., 1974; Stolen and Lin, 1978).

The approximate frequency shift at any time (t) is given by the time derivative of the phase perturbation which in turn is proportional to the power (Shimizu, 1967)

$$\delta\omega(t) = -\frac{d\,\delta\phi}{dt} = -\frac{2\pi L}{\lambda}\frac{d\,\delta n}{dt}. \qquad (5.14)$$

In glasses the refractive index increases with intensity so that the leading edge of the pulse is downshifted in frequency and the trailing edge is upshifted. For normal dispersion the group velocity is less for higher frequencies so the effect of SPM plus dispersion is to lengthen the pulse. In regions of anomalous dispersion a pulse tends to shrink and Hasagawa and Tappert (1973a) have shown that there exist stable solutions which propagate without distortion. In fibers with normal dispersion a stationary "dark" pulse should exist (Hasagawa and Tappert, 1973b).

The frequency spectrum after SPM is given by the Fourier transform of the pulse amplitude, $P(t)^{1/2}$.

$$F(\omega) = \frac{1}{2\pi} \int_{-\infty}^{\infty} P(t)^{1/2} e^{i\Delta\phi(t)} e^{-i(\omega-\omega_0)t}\, dt,$$

$$\Delta\phi(t) = (2\pi L/\lambda)\delta n, \qquad \delta n = n_2(E^2/2), \qquad (5.15)$$

where δn is the intensity dependent refractive index, n_2 is the coefficient for self focusing, λ is the vacuum wavelength, and the field is the same as in Eq. (5.12). For a Gaussian pulse the broadening $\delta\omega$ is proportional to the initial linewidth $\Delta\omega$ and the maximum phase shift in radians by

$$\delta\omega = 0.86\ \Delta\omega\Delta\phi_{max}. \qquad (5.16)$$

Calculated frequency spectra are shown in Fig. 5.7 for Gaussian pulses with different peak powers. From Eq. (5.14) the frequency shift is proportional to the derivative of the instantaneous power and there will generally

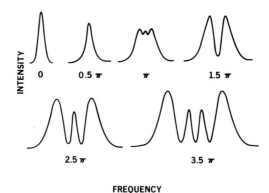

Fig. 5.7 Calculated frequency broadening by SPM for a Gaussian pulse $P(t) = e^{-t^2/\tau^2}$. The powers were chosen such that $\Delta\phi(t)_{max}$ of Eq. (5.15) is a multiple of $\pi/2$. (From Stolen and Lin, 1978.)

be two times when the frequency shift is the same. The structure in the spectrum has been interpreted as the interference of these different pairs of frequencies (Shimizu, 1967). In Eq. (5.15) power and length are equivalent; this would not be true in a very long fiber where group velocity dispersion was important. For long fibers where both SPM and dispersion are large, calculations are performed by splitting the fiber up into short sections and alternately calculating the effects of SPM and dispersion (Hasagawa and Tappert, 1973a,b; Fisher and Bischel, 1975).

The combination of frequency broadening from SPM and group velocity dispersion will increase the pulse dispersion. A critical length can be defined where the high-frequency part of the pulse is retarded by one pulse length with respect to the low-frequency side. The expression for this critical length (Fisher and Bischel, 1975) is

$$z_c = (D(\lambda)\delta n_{max})^{-1/2} cT_p \tag{5.17}$$

where $D(\lambda) = \lambda^2 d^2 n/d\lambda^2$ is the dispersion, cT_p is the pulse length, and δn_{max} is the maximum index change.

As an example, we choose a 10-psec pulse at 1.0 μm with a 1-W peak power in a 10-μm core single-mode silica fiber. The value of n_2 for silica is 1.2×10^{-13} esu (Owyoung et al., 1972). The material dispersion dominates in a single-mode guide and at 1.0 μm in silica $D(\lambda) = 0.011$ (Gloge, 1971). The critical length from (5.17), (5.15), and (5.12) is then 1.42 km. A comparison with the SRS threshold power of 3.1 W in a much longer fiber (see Appendix) shows that SPM is a problem well below the SRS threshold.

A critical power for SPM can be defined as the power at which the initial pulse doubles in frequency width. If the capacity of a single-mode fiber is limited by pulse dispersion then any additional frequency broadening is

undesirable. Note that because of linear loss most of the frequency broadening takes place near the beginning of a long fiber while the pulse
spreading takes place along the entire fiber. From Fig. 5.7 the frequency
width doubles at $\Delta\phi_{max} \approx 2.0$ so if we use this to define a critical power
then from Eqs. (5.15) and (5.12):

$$P_c = (nc\lambda A/4\pi^2 n_2 L) \times 10^{-7}\text{W}, \qquad (5.18)$$

where A and L are an effective area and an effective length as for the
Raman and Brillouin cases. P_c is calculated for a single mode and a multimode fiber in the Appendix.

For a linear polarization n_2 is related to $\chi_\parallel$ of Eq. (5.13)

$$n_2 = (12\pi/n)\chi_\parallel. \qquad (5.19)$$

For the case of circular polarization $\chi_\parallel$ of Eq. (5.19) is replaced by $2\chi_\perp$
(Wang, 1966). In a multimode fiber or a long single-mode fiber that does
not maintain linear polarization we must average the effect of polariza-
tions ranging from linear to circular. In silica $\chi_\perp \approx \frac{1}{3}\chi_\parallel$ so that if we simply
average the two extreme cases:

$$n_{2av} = (12\pi/n)(\chi_\parallel + 2\chi_\perp)/2 \approx \tfrac{5}{6}n_2. \qquad (5.20)$$

No satisfactory treatment of the effect on SPM of modal dispersion in a
multimode fiber exists, although some qualitative idea can be obtained
from experimental results on CS_2-filled glass fibers (Ippen et al., 1974).
Here modal dispersion doubled the pulse length in the fiber used and the
measured frequency broadening was half that calculated in the absence of
pulse spreading.

5.4.3 Continuum Generation

The series of Stokes lines produced by sequential SRS becomes progressively broader to higher orders. The broadening is quite similar in
liquid-filled fibers which have narrow Raman lines (Ippen, 1970) and in
glass fibers where the Raman gain curve is broad. This frequency
smearing is poorly understood but in liquid cells the broadening is larger
for a multiline pump than for a single-frequency pump (Stoicheff, 1963)
and is believed to be due to self-phase modulation.

If a pump with a bandwidth comparable to the Raman shift is used to
generate SRS the result is a continuum which can extend from the pump
frequency into the infrared (Lin and Stolen, 1976a). It is not possible to
separate the individual contributions of SRS and SPM but by analogy with
the narrow-band excitation case one assumes that sequential SRS produces the wide frequency range which is filled in and smoothed by SPM.

A white-light pulse of 10-nsec duration has been produced by passing a broadband blue dye laser pulse through a glass fiber. The spectrum of the white-light pulse extends from the blue to the deep red. This pulse is potentially useful for nanosecond flash photolysis and for studies of optically induced absorption in fibers (Lin and Stolen, 1976a).

5.5 PHASE MATCHED PARAMETRIC INTERACTIONS

5.5.1 Four-Photon Mixing

Anti-Stokes lines (frequencies higher than the pump) as well as Stokes are frequently produced in SRS experiments in bulk samples. This anti-Stokes light cannot come from the Raman effect because as discussed in Section 5.2.1 anti-Stokes is absorbed rather than amplified. The source of this anti-Stokes is mixing between pump and Stokes by means of the $\chi_3 E^3$ term in the polarization expansion. This same term also gives rise to frequency tripling, the optical Kerr effect, and SPM. By way of comparison $\chi_2 E^2$, which gives frequency doubling and three-photon parametric interactions, is zero in glass because of inversion symmetry. Because of index dispersion, the wavelength of the polarization produced by the mixing is slightly different than that of the free running wave. The correction for this mismatch is called phase matching and is usually accomplished by the use of birefringent crystals or by mixing the beams at an angle. An important advantage of fibers is that the differing phase velocities of the waveguide modes of a multimode fiber can be used to achieve phase matching.

A four-photon mixing experiment is illustrated in Fig. 5.8. Here two green pump photons at frequency ω_p are mixed with a red signal photon at ω_s to produce a blue idler photon at ω_a. Energy conservation requires that $2\omega_p - \omega_s = \omega_a$. In a thin sample there is no problem with phase matching. In a thicker sample, however, there will be some point where the electric field at ω_a is out of phase with the polarization produced by mixing

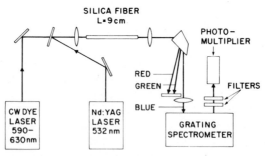

Fig. 5.8 Four-photon mixing experiment. (From Stolen *et al.*, 1974.)

$2\omega_p - \omega_s$ so that ω_a will be reabsorbed. This distance is the coherence length and is defined as:

$$\delta k L_c = 2\pi, \tag{5.21}$$

where δk is the wavevector difference between the free running and polarization idler waves. In silica $L_c \approx 2$ mm at a frequency shift of 1000 cm^{-1} from a 514.5 nm pump. $L_p \sim (\omega_p - \omega_s)^{-2}$ so that at a frequency difference of 1 cm^{-1}, which is the Brillouin shift in silica at this pump wavelength, L_c is about 2 km. Such four-photon mixing has been observed in a cw Brillouin oscillator (Hill *et al.*, 1976c). For larger frequency shifts, however, L_c becomes very small and phase matching is necessary if one is to take advantage of the long interaction length available in a fiber.

The use of waveguide modes to phase match is illustrated in Fig. 5.9a. Because of index dispersion $k_s + k_a > 2k_p$ while phase matching requires that $k_s + k_a = k_p$. In the guide, however, the propagation constant of a higher order mode is less than that of a lower order mode so that phase matching is then possible if either the signal or the idler is in the appropriate higher order mode. In the example of Fig. 5.9a the pump and signal are in the LP$_{01}$ mode while the idler is in the LP$_{02}$ mode. Note that because of the symmetry of the modes it will not work to mix the LP$_{01}$ and LP$_{11}$ modes but that LP$_{11}$ will mix with LP$_{21}$. The phase mismatch which has to be made up with the fiber modes is rather small and the phase velocity differences between the modes for typical waveguides are more than sufficient. Figure 5.8 illustrates an actual experiment (Stolen *et al.*, 1974) for

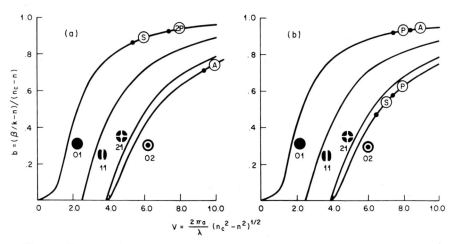

Fig. 5.9 Pump, Stokes, and anti-Stokes modes for phase matched four-photon mixing. (a) and (b) are representative of processes with short and long effective coherence lengths. The black dots indicate the change in effective index with a decrease in fiber diameter. (From Stolen, 1975.)

which phase matching occurred at a signal of 626 nm and an idler of 463 nm for a pump of 532 nm. The fiber had a core radius of 5.0 μm and $V = 8.0$ at 532 nm. For smaller core diameters and larger index differences the signal and idler will be even further apart. In the actual experiment many different modes were excited both by signal and pump so that as the cw dye laser signal was tuned a blue idler appeared whenever a particular combination of modes was phase matched.

It should be possible to observe phase matched up conversion or second harmonic generation in fibers using the magnetic dipole or electric quadrupole terms in the polarization expansion. However, in contrast to four-photon mixing, the phase velocities of modes in typical fibers do not differ enough to easily permit phase matching and fibers with large index differences and small cores would be necessary.

At the phase matching frequency, the useful fiber length is ideally limited only by the linear absorption. In actual fact the idler peaks approached the ideal $\sin^2 x / x^2$ form only for fiber lengths around 9 cm while for longer fibers the corresponding peaks were broader in frequency and relatively weak. The problem comes about because in real fibers the core diameter and index difference vary along the length of the fiber. This changes the V number and the original frequencies are not phase-matched. For ideal phase matching the frequency bandwidth decreases as $1/L$ and the peak power increases as L^2. This means that for short fiber lengths any fluctuation in the phase matching frequency is usually less than the bandwidth but for long fibers the phase-matching frequency will wander outside the bandwidth. This leads to the concept of an effective coherence length where the wave vector difference δk in Eq. (5.21) is replaced by a wave vector difference determined by fiber imperfections. The effective coherence length is still several orders of magnitude larger than the coherence length in the bulk material.

5.5.2 Stimulated Four-Photon Mixing

An alternative way of viewing the four-photon mixing process is to consider the modulation of the dielectric constant at $2\omega_p$ by the pump at ω_p. A signal at ω_s introduced into a medium with such a varying dielectric constant will be amplified (Akhmanov and Khokhlav, 1963). Conservation of energy requires generation of the idler. In the limit of negligible amplification this is just the mixing experiment previously discussed. Four-photon mixing is symmetric with respect to signal and idler and the signal could be on either the Stokes or anti-Stokes side.

If fiber imperfections always limited the effective coherence length to a few centimeters there would be little incentive to further explore the potential of the four-photon mixing process as an optical amplifier. In practice stimulated four-photon mixing is commonly observed in small-core multimode fibers (Stolen, 1975) and the effective coherence length found to be several meters. These long coherence lengths apply only to certain favored combinations of modes for which, at least to first order, δk is insensitive to changes in V.

Both ordinary and long coherence length phase matching is illustrated in Fig. 5.9. The pump energy in Fig. 5.9b is divided between two modes. The black dots indicate the changes in pump, signal, and idler effective indices for a change in core diameter which in turn changes V. The change in the wavevector difference is

$$\delta k = (\delta k_{P1} - \delta k_s) + \delta(k_{P2} - \delta k_a). \tag{5.22}$$

If the wave vector is expanded as a power series in $\Delta\nu$ about the pump frequency, L_c will be controlled by terms in $(dk/dV)\delta V$ except for the divided-pump processes for which these terms cancel and L_c is determined by terms of the order of $(d^2k/dV^2)\Delta\nu\delta V$. The difference between the coherence lengths in the two cases amounts to more than two orders of magnitude (Stolen, 1975).

Figure 5.10 shows output spectra from two silica core fibers. These pictures show both SRS at 460 cm^{-1} and various Stokes and anti-Stokes pairs from stimulated four-photon mixing. Stokes and anti-Stokes from the four-photon mixing process appear in different modes and the Stokes mode is always the higher order of the two; the pump is divided between these same two modes. Different processes can be favored by adjusting the input coupling. The frequency shifts for the long-coherence-length divided-pump processes are comparable to the Raman shift so that the presence of Raman gain increases the amplification of the Stokes and absorbs some of the anti-Stokes. The four-photon process by itself would produce equal intensity Stokes and anti-Stokes. In the photographs, neutral density filters are used to attenuate the Stokes patterns.

The gain coefficient for the mixing process is about a factor of five larger than for the Raman process. The photographs clearly show that SRS is the weaker of the two effects. The net four-photon gain, however, is reduced somewhat by the overlap integral. For example we approximate the LP_{01} and LP_{02} modes of Fig. 5.10 by Gaussians

$$\Psi_a{}^2 \approx \Psi_{p1}^2 = \Psi_1{}^2 = e^{-r^2/w^2},$$

$$\Psi_s{}^2 \approx \Psi_{p2}^2 = \Psi_2{}^2 = \left(1 - \frac{r^2}{w^2}\right)^2 e^{-r^2/w^2}. \tag{5.23}$$

The overlap integral and effective area take the same form as Eq. (5.5) with Ψ_2 and Ψ_1 referring to the Stokes and anti-Stokes mode fields. The effective area is then $4\pi w^2$ as compared to $2\pi w^2$ for Raman gain with both pump and Stokes in the LP_{01} mode.

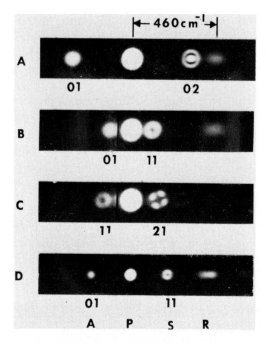

Fig. 5.10 Stimulated four-photon mixing. The fiber output is dispersed by a prism and ND filters are used to attenuate the pump and Stokes with respect to the weaker anti-Stokes. A, B, C are all from a 10-μm core fiber. D is the same combination of modes as B but in a 7-μm core fiber. P, S, A, and R refer to the pump, the stimulated Stokes, and anti-Stokes from the four-photon mixing process, and the stimulated Raman output. (From Stolen, 1975.)

Stimulated four-photon mixing shows up in continuum generation in small core multimode fibers (Lin and Stolen, 1976a). There is about 500 cm^{-1} of frequency generation on the anti-Stokes side of the pump and on the Stokes side the output appears in one of the highest order modes of the fiber.

5.6 DAMAGE

5.6.1 Multiphoton Absorption

Fibers often become lossy after prolonged operation at high-peak or cw power (Stolen *et al.*, 1972). This is particularly serious in doped-core fibers but such loss has recently been observed in pure SiO_2 core fibers as well (Hill *et al.*, 1976b). This problem has not yet received much attention but the most likely source of the absorption is color centers created by multiphoton absorption.

Preliminary studies of multiphoton effects in glass fibers have revealed a variety of phenomena (Stolen and Lin, 1975). The experimental approach is to measure the absorption induced in a weak probe by a stronger pump at a different wavelength. The long interaction lengths provide great sensitivity to weak effects so small impurity concentrations can cause significant absorption. If the induced absorption shows the same time dependence as the pump pulse then the absorption probably arises from simultaneous absorption of two photons. If the absorption persists after the passage of the pump pulse the absorption is a two-step process in which an impurity is excited by a pump photon and raised to a higher level by a probe photon. In general, two-step absorption is the dominant process. Both short- and long-term bleaching and saturation effects have also been seen.

5.6.2 End Damage

The input ends of fibers are often damaged if the full pump power is applied before good coupling to the fibers is obtained. In these cases the intensity at the focus is close to the damage threshold for fused silica which is around 10^{10} W/cm² (Smith *et al.*, 1975). The damage threshold should be higher in the high quality core and cladding of a CVD fiber than in the lower quality substrate. In all cases a pump beam of good spatial quality is desirable for efficient coupling and minimum damage.

Somewhat different input damage behavior took place during studies of SBS (Ippen and Stolen, 1972; Ippen, 1975). This damage occurred at much lower power than the usual end damage and only when good coupling was achieved. It is believed that the cause was the acoustic shock generated by the mixing of the pump and backward amplified Brillouin wave.

5.7 FUTURE DIRECTIONS

The study of nonlinear processes in fibers has so far concentrated on step-index silica-core fibers. This is partly because such fibers are most readily available and partly because silica is one of the most transparent and damage-resistant glasses. Silica also has one of the lowest peak Raman cross sections and one of the smallest nonlinear susceptibilities of any material. The observation of nonlinear effects at low powers despite the small magnitude of these coefficients emphasizes the utility of the fiber geometry for nonlinear optics and the necessity of considering nonlinear processes in the design of any system employing fibers.

For fiber devices it is desirable that pump powers be as low as possible and glasses with larger coefficients would be useful. An example of such a glass is Schott LaSF-7 which has a peak Raman cross section 8 times that of silica and a nonlinear susceptibility 5.5 times that of silica (Hellwarth *et al.*, 1975). Another potentially useful glass is vitreous germania which has a peak Raman cross section 10 times that of silica (Hass, 1970; Lin *et al.*, 1977b). These glasses have high refractive indices so a small core can be used and the intensity will be higher than in an equivalent silica fiber. It remains to be seen whether such glasses will approach silica in transparency and damage resistance when pulled into fibers.

Different fiber geometries are also of interest. For example, the single material fiber can be used to make fiber guides out of glasses which are incompatible with other glasses in conventional core–cladding fibers (Kaiser *et al.*, 1973). There is also the possibility of nonlinear interactions in such structures as fibers with two adjacent cores and concentric core and ring guides.

5.8 CONCLUSION

Work on nonlinear properties of fibers has been of an exploratory nature with an eye toward the effects of nonlinear processes on optical fibers used for communications. The present discussion of these basic nonlinear processes, although by no means complete, also points to various device applications such as tunable Raman oscillators, four-photon parametric amplifiers, and Kerr modulators. Indeed, tunable fiber Raman oscillators have now appeared and promise to be particularly useful as sources of both cw and pulsed tunable radiation in the near-infrared spectral region.

When one balances the potential deleterious effects of nonlinear processes on single-mode communications fibers with the advantages of these same processes in fiber devices one finds that Nature has been kind. The critical powers collected in the Appendix show that while there are circumstances where nonlinear effects could cause difficulties, the nonlinear susceptibilities are small enough so that these limits are not usually serious. On the other hand, the susceptibilities are large enough so that devices such as tunable Raman oscillators work very well.

Of course, larger susceptibilities are always desirable for device applications and in this respect exploration of new materials and geometries is particularly useful. There is a great deal to be done and every reason to expect growth in interest and applications of nonlinear processes in fibers.

APPENDIX: CRITICAL POWERS (P_c) FOR STIMULATED RAMAN SCATTERING (SRS), STIMULATED BRILLOUIN SCATTERING (SBS), AND SELF-PHASE MODULATION (SPM)

As examples we choose two long silica core fibers with a loss of 1.0 dB/km. The first fiber is a single-mode fiber with a 10.0-μm core diameter and $V = 2.50$; the second is a 50.0-μm core diameter multimode fiber. The wavelength is 1.0 μm. Notes are include so these results can be translated to different wavelengths, core sizes, and losses.

		10-μm single-mode	50-μm multimode
SRS	$P_c = \dfrac{16A}{G_R L}$ W	3.3 W	150 W
SBS	$P_c = \dfrac{21A}{G_{B0} L}$ W	9.8 mW	440 mW
SPM	$P_c = \dfrac{nc\lambda A}{4\pi^2 n_2 L} \times 10^{-7}$ W	185 mW	5.0 W

(a) The critical power for SRS and SBS is that power for which the Stokes power builds up from noise to equal the pump power. The critical power for SPM is defined as the power for which the frequency bandwidth doubles. All powers are measured at the input end of the fiber.

(b) The power for SRS and SBS are either cw powers or the average power for many closely spaced pulses. For single pulses P is the peak power and the interaction length is the walk-off distance between pump and Stokes as determined by group velocity dispersion. P_c for SPM refers to the peak power of a single transform limited pulse.

(c) L is an effective length: $L = (1 - e^{-\alpha l})/\alpha$ where l is the fiber length and α is the linear absorption coefficient. P_c was calculated in the limit of large l for which $L = 1/\alpha = 4.34$ km.

(d) The Raman gain G_R varies linearly with pump frequency. G_{B0} and n_2 are approximately independent of pump frequency.

(e) P_c for SBS calculated assuming the bandwidth of the pump $(\Delta\nu_p)$ is much less than the Brillouin linewidth $(\Delta\nu_B)$. For $\Delta\nu_p > \Delta\nu_B$, G_B, is reduced by the factor $\Delta\nu_B/\Delta\nu_p$. $\Delta\nu_B$ is estimated to be 38.4 MHz for a 1.0-μm pump and increases as ν_p^2.

(f) The effective area (A) for the multimode fiber is the core area. For a single-mode fiber A varies with the V number. The ratios of A to the core area for V = 1.5, 2.0, 2.5, and 3.0 are 2.43, 1.47, 1.10, and 0.93.

(g) P_c for SRS, SBS, and SPM in the single-mode fiber was calculated assuming linear polarization. P_c in the multimode fiber was calculated assuming complete polarization scrambling. In a single-mode fiber that does not maintain linear polarization G_R and G_{B0} are multiplied by $\frac{1}{2}$ and n_2 by $\frac{3}{8}$.

REFERENCES

Akhmanov, S. A., and Khokhlav, R. V. (1963). Concerning one possibility of amplification of light waves. *Sov. Phys.—JETP (Engl. Transl.)* **16**, 253.

Ashkin, A., and Ippen, E. P. (1970). Optical stimulated emission devices employing optical guiding. U.S. Patent 3,399,012.

Bjorkholm, J. E. (1975). Has pointed out that the overlap integral for the Kerr effect should be the same as for the Raman effect and that Stolen and Ashkin (1973) are in error in this respect.

Bloembergen, N. (1967). The stimulated Raman effect. *Am. J. Phys.* **35**, 989.

Capasso, F., and DiPorto, P. (1976). Coupled-mode theory of Raman amplification in lossless optical fibers. *J. Appl. Phys.* **47**, 1472.

Chiao, R. Y., Garmire, E., and Townes, C. H. (1964). Self-trapping of optical beams. *Phys. Rev. Lett.* **13**, 479.

Denariez, M., and Bret, G. (1968). Investigation of Rayleigh wings and Brillouin-stimulated scattering in liquids. *Phys. Rev.* **171**, 160.

Duguay, M. A., and Hansen, J. W. (1969). An ultrafast light gate. *Appl. Phys. Lett.* **15**, 192.

Fisher, R. A., and Bischel. W. K. (1975). Numerical studies of the interplay between self-phase modulation and dispersion for intense plane-wave laser pulses. *J. Appl. Phys.* **46,** 4921.

Giordmaine, J. A., and Shapiro, S. L. (1968). Highly efficient Raman emission device. U. S. Patent 3,571,607.

Gloge, D. (1971). Dispersion in weakly guiding fibers. *Appl. Opt.* **10,** 2442 (1971). Note that $\lambda^2 d^2 n/d\lambda^2 = kd^2(kn)/dk^2$.

Görner, H., Maier, M., and Kaiser, W. (1974). Raman gain in liquid-core fibers. *J. Raman Spectrosc.* **2,** 363.

Hasegawa, A., and Tappert, F. (1973a). Transmission of stationary nonlinear optical pulses in dispersive dielectric fibers. I. Anomalous dispersion. *Appl. Phys. Lett.* **23,** 142.

Hasegawa, A., and Tappert, F. (1973b). Transmission of stationary nonlinear optical pulses in dispersive dielectric fibers. II. Normal dispersion. *Appl. Phys. Lett.* **23,** 171.

Hass, M. (1970). Raman spectra of vitreous silica, germania and sodium silicate glasses. *J. Phys. Chem. Solids* **31,** 415.

Hellwarth, R. W., Cherlow, J., and Yang, T. (1975). Origin and frequency dependence of nonlinear optical susceptibilities of glasses. *Phys. Rev. B* **11,** 964.

Hill, K. O., Kawasaki, B. S., and Johnson, D. C. (1976a). cw Brillouin laser. *Appl. Phys. Lett.* **28,** 608.

Hill, K. O., Kawasaki, B. S., and Johnson, D. C. (1976b). Low-threshold cw Raman laser. *Appl. Phys. Lett.* **29,** 181.

Hill, K. O., Johnson, D. C., and Kawasaki, B. S. (1976c). cw generation of multiple Stokes and anti-Stokes Brillouin shifted frequencies. *Appl. Phys. Lett.* **29,** 185.

Ippen, E. P. (1970). Low power Quasi-cw Raman oscillator. *Appl. Phys. Lett.* **16,** 303.

Ippen, E. P. (1975). Nonlinear effects in optical fibers. *In* "Laser Applications to Optics and Spectroscopy" (S. F. Jacobs, M. O. Scully, and M. Sargent, eds.), p. 213. Addison-Wesley, Reading, Massachusetts.

Ippen, E. P., and Stolen, R. H. (1972). Stimulated Brillouin scattering in optical fibers. *Appl. Phys. Lett.* **21,** 539.

Ippen, E. P., Patel, C.K.N., and Stolen, R. H. (1971). Broadband tunable Raman-effect devices in optical fibers. U.S. Patent 3,705,992.

Ippen, E. P., Shank, C. V., and Gustafson, T. K. (1974). Self-phase modulation of picosecond pulses in optical fibers. *Appl. Phys. Lett.* **24,** 190.

Jain, R. K., Lin, C., Stolen, R. H., Pleibel, W., and Kaiser, P. (1977). A high-efficiency tunable cw Raman oscillator. *Appl. Phys. Lett.* **30,** 162.

Johnson, D. C., Hill, K. O., Kawasaki, B. S., and Kato, D. (1977). Tunable Raman fiber-optic laser. *Electron. Lett.* **13,** 53.

Jones, W. J., and Stoicheff, B. P. (1964). Inverse Raman spectra: Induced absorption at optical frequencies. *Phys. Rev. Lett.* **13,** 657.

Kaiser, P., Marcatili, E. A. J., and Miller, S. E. (1973). A new optical fiber. *Bell Syst. Tech. J.* **52,** 265.

Kaiser, W., and Maier, M. (1972). Stimulated Rayleigh, Brillouin and Raman spectroscopy. *In* "Laser Handbook" (F. T. Arecchi and E. O. Schulz-DuBois, eds.), p. 1077. North-Holland Publ., Amsterdam.

Kapron, F. P., Borrelli, N. F., and Keck, D. B. (1972). Birefringence in dielectric optical waveguides. *IEEE J. Quantum Electron.* **QE-8,** 222.

Lin, C., and Stolen, R. H. (1976a). New nanosecond continuum for excited-state spectroscopy. *Appl. Phys. Lett.* **28,** 216.

Lin, C., and Stolen, R. H. (1976b). Backward Raman amplification and pulse steepening in silica fibers. *Appl. Phys. Lett.* **29,** 428.

Lin, C., Stolen, R. H., and Cohen, L. G. (1977a). A tunable 1.1 μm fiber Raman oscillator. *Appl. Phys. Lett.* **31,** 97.

Lin, C., Cohen, L. G., Stolen, R. H., Tasker, G. W., and French, W. G. (1977b). Near-infrared sources in the 1-1.3 μm region by efficient stimulated Raman emission in glass fibers. *Opt. Commun.* **20**, 426.

Lin, C., Stolen, R. H., and Jain, R. K. (1977c). Group velocity matching in optical fibers. *Opt. Lett.* **1**, 205.

Lugovoi, V. N. (1976). On stimulated combinational emission and frequency scanning in an optical wave guide. *Sov. Phys.—JETP (Engl. Transl.)* **71**, 1307.

Maier, M., Kaiser, W., and Giordmaine, J. A. (1969). Backward stimulated Raman scattering. *Phys. Rev.* **177**, 580.

Maker, P. D., and Terhune, R. W. (1965). Study of optical effects due to an induced polarization third order in the electric field strength. *Phys. Rev.* **137**, A801.

Owyoung, A. (1971). The origins of the nonlinear refractive indeces of liquids and glasses. Ph.D. Thesis, California Institute of Technology, Pasadina (Clearinghouse for Federal Scientific and Technical Information Report No. AFOSR-TR-71-3132).

Owyoung, A., Hellwarth, R. W., and George, N. (1972). Intensity-induced changes in optical polarizations in glasses. *Phys. Rev. B* **5**, 628.

Pelous, J., and Vacher, R. (1975). Thermal Brillouin scattering measurements of the attenuation of longitudinal hypersound in fused quartz from 77 to 300 K. *Solid State Commun.* **16**, 279.

Primak, W., and Post, D. (1959). Photoelastic constants of vitreous silica and its elastic coefficient of refractive index. *J. Appl. Phys.* **30**, 779.

Shimizu, F. (1967). Frequency broadening in liquids by a short light pulse. *Phys. Rev. Lett.* **19**, 1097.

Shuker, R., and Gammon, R. W. (1970). Raman-scattering selection-rule breaking and the density of states in amorphous materials. *Phys. Rev. Lett.* **25**, 222.

Smith, R. G. (1972). Optical power handling capacity of low loss optical fibers as determined by stimulated Raman and Brillouin scattering. *Appl. Opt.* **11**, 2489.

Smith, W. L., Bechtel, J. H., and Bloembergen, N. (1975). Dielectric-breakdown threshold and nonlinear-refractive-index measurements with picosecond laser pulses. *Phys. Rev. B* **12**, 706.

Snitzer, E., and Osterberg, H. (1961). Observed dielectric waveguide modes in the visible spectrum. *J. Opt. Soc. Am.* **51**, 499.

Stoicheff, B. P. (1963). Characteristics of stimulated Raman radiation generated by coherent light. *Phys. Lett.* **7**, 186.

Stolen, R. H. (1975). Phase-matched stimulated four-photon mixing in silica-fiber waveguides. *IEEE J. Quantum Electron.* **QE-11**, 100.

Stolen, R. H., and Ashkin, A. (1973). Optical Kerr effect in glass waveguide. *Appl. Phys. Lett.* **22**, 294.

Stolen, R. H., and Ippen, E. P. (1973). Raman gain in glass optical waveguides. *Appl. Phys. Lett.* **22**, 276.

Stolen, R. H., and Leibolt, W. N. (1976). Optical fiber modes using stimulated four photon mixing. *Appl. Opt.* **15**, 239.

Stolen, R. H., and Lin, C. (1975). Two-photon and two-step absorption in glass optical waveguide. *In* "Optical Properties of Highly Transparent Solids" (S. S. Mitra and B. Bendow, eds.), p. 307. Plenum, New York.

Stolen, R. H., and Lin, C. (1978). Self-phase modulation in silica optical fibers. *Phys. Rev. A* **17**, 1448.

Stolen, R. H., Ippen, E. P., and Tynes, A. R. (1972). Raman oscillation in glass optical waveguide. *Appl. Phys. Lett.* **20**, 62.

Stolen, R. H., Bjorkholm, J. E., and Ashkin, A. (1974). Phase-matched three-wave mixing in silica fiber optical waveguides. *Appl. Phys. Lett.* **24**, 308.

Stolen, R. H., Lin, C., and Jain, R. K. (1977). A time-dispersion-tuned fiber Raman oscillator. *Appl. Phys. Lett.* **30**, 340.

Stolen, R. H., Ramaswamy, V., and Kaiser, P. (1978). Linear polarization in elliptically-clad birefringent, single-mode fibers. *IEEE/OSA Top. Meet. Integr. Guided Wave Opt., 1978* Paper PD-1.

Stone, J. (1975). cw Raman fiber amplifier. *Appl. Phys. Lett.* **26**, 163.

Tang, C. L. (1966). Saturation and spectral characteristics of the Stokes emission in the stimulated Brillouin processes. *J. Appl. Phys.* **37**, 2945.

Walrafen, G. E., and Stone, J. (1972). Intensification of spontaneous Raman spectra by use of liquid core optical fibers. *Appl. Spectrosc.* **26**, 585.

Walrafen, G. E., and Stone, J. (1975). Raman characterization of pure and doped fused silica optical fibers. *Appl. Spectrosc.* **29**, 337.

Wang, C. C. (1966). Nonlinear susceptibility constants and self-focusing of optical beams in liquids. *Phys. Rev.* **152**, 149.

Chapter 6

Fiber Design Considerations

DETLEF GLOGE

WILLIAM B. GARDNER

6.1 INTRODUCTION

This chapter introduces the reader to the considerations and compromises required in choosing the fiber parameters for a given optical system. Cross-sectional dimensions, jacket and cladding materials, and the core-index profile all influence the various loss and dispersion phenomena and must be chosen to achieve a satisfactory compromise for a given application. In some cases, this compromise can be described in mathematical form derived from relationships given in other chapters, but often qualitative arguments pointing to manufacturing problems or cost penalties, etc., are the only help we can offer in making a design decision.

We begin with some general thoughts concerning the fiber dimensions. A discussion of the index difference and its influence on the fiber-to-source coupling efficiency and the overall transmission loss follows. Sensitivity to microbending and the associated loss receives special attention. The final sections are devoted to considerations concerning the bandwidth of the fiber medium.

In this latter discussion, the bandwidth f and the bit rate B of a fiber are defined as

$$B = 2f = \tfrac{1}{4}\sigma, \tag{6.1}$$

where σ is the rms width of the impulse response $h(t)$ as indicated by (4.1). Figure 4.13 shows that digital transmission at a bit rate B causes approximately a sensitivity penalty of 1 dB independent of the shape of $h(t)$. The bandwidth f as defined by (6.1) coincides approximately with the frequency at which the received optical power decreases to 0.707 of its low-frequency value (that is, where the electrical power behind a wide-

151

band receiver decreases by 3 dB). Another bandwidth often quoted is that at which the received optical power reaches 0.5 of its low-frequency value. Depending on the pulse shape, this latter bandwidth is in the range between 1.5f and 2.2f.

6.2 FIBER DIAMETER

The cost of raw materials is a significant part of the manufacturing cost of low-loss fibers. Minimizing these costs is a strong incentive in the direction of small fiber diameters.

Another equally important impetus in this direction is the dual requirement of high strength and good fiber flexibility. A fiber fracture usually originates at the surface where tensile strain is a maximum when the fiber is bent. For a fiber of radius b bent to a curvature radius R, the surface strain is

$$\epsilon = b/R. \tag{6.2}$$

Although suitably prepared silica-based fibers have been found to withstand strains of several percent (see Chapter 12), a fraction of 1% is a more typical strain limit necessary to guarantee survival of the fiber under all conditions in a cable environment. Thus, if a typical minimum curvature radius of 15 mm and a strain of $\frac{1}{2}$% should be admissible, the fiber radius must not exceed about 75 μm. Thicker fibers have been made and are viable if restricted by a stiff plastic jacket from normal flexure.

In general, the incentive to maximize the core diameter is equally strong. Splicing tolerances for lateral offset typically increase with increasing core dimensions (see Section 3.9 or Fig. 14.25). In the case of incoherent transmission, the desire to maximize the fiber-to-source coupling efficiency may be another motive (see Section 6.5).

In most applications, therefore, the fiber diameter is limited to 150 μm and the cladding is chosen as thin as possible. What is possible in this respect depends on the requirement that the light propagating in the core is well isolated from the lossy outside fiber surface.

6.3 CLADDING THICKNESS

Beginning with the simpler case of a uniform cladding index, we discuss the cladding requirements for single mode and then for multimode operation. Subsequently, we consider the effect of a two-step cladding of the W type. In this latter case, a low-index barrier layer is surrounded by a second layer of higher index. Both single mode and multimode fibers of this type have been considered for long-distance transmission purposes.

If there is just one uniform cladding layer surrounded by a plastic jacket with the absorption coefficient α_j, the loss suffered by a certain propagating mode can be calculated from (3.62) or (3.63). Unfortunately, α_j is rarely known with accuracy and may also vary significantly along the transmission path. In this case, it is convenient to use the worst case condition (3.74) which leads to

$$\alpha = (8.7U^2W/nka^2V^2)\, e^{-2W(b/a-1)} \tag{6.3}$$

in decibels for the loss suffered by a low-order propagating mode, where a and b are core and fiber radius, respectively, and U, V, and W define the mode. Figure 6.1 shows the ratio b/a necessary to limit the jacket loss of a single mode fiber to 1 dB/km in straight sections. Under typical operating conditions ($V = 2.4, \ldots , 2.8$), b must exceed 6 core radii to guarantee negligible jacket loss. Since fiber curvature increases the penetration of the evanescent field into the cladding, the required b of Fig. 6.1 must be considered as a lower limit. A ratio $b/a = 8$ constitutes a reasonable design objective. For this ratio, the next higher mode suffers losses of the order of several decibels per meter in the region $2.4 < V < 3$ so that single mode operation can be extended into this regime even though this higher mode is theoretically not a "cutoff" mode.

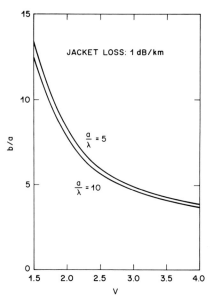

Fig. 6.1 Ratio of cladding radius b and core radius a plotted versus the V value for 1 dB/km jacket loss in single mode fibers.

For multimode operation, it is usually sufficient to protect some, but not all, of the propagating modes from the lossy outside fiber surface. Since the extent of the evanescent cladding field varies rapidly with the mode order, a given cladding thickness divides the total of the propagating modes into those that are practically extinguished and those that are essentially not affected by the fiber surface. Figures 3.13 and 3.14 show dividing lines between these groups of modes for the step profile based on (3.62) and for the square-law graded profile based on (3.63). We can assume a worst-case jacket loss α_j of about 1 dB/μm according to (3.74) so that $\alpha = 1$ dB/km at the dividing line. For $b/a = 2$, only a few percent of the propagating modes suffer a loss larger than 1 dB/km in the fiber jacket. Note that this fraction of lossy modes includes most of the so-called "leaky modes" which are expected to propagate in fibers with very thick claddings.

Figures 3.13 and 3.14 are valid for straight multimode fibers without diameter variations along the length of the fiber. Gradual variations of the fiber diameter do not lead to power exchange between the modes, but they increase the jacket loss where the fiber core narrows down (Olshansky and Nolan, 1976). Computations of the kind illustrated by Figs. 3.13 and 3.14 should therefore be based on the minimum fiber diameter, not the average one. Fiber curvature increases the evanescent fields of all modes and hence causes additional jacket loss in the high-order modes; estimates of this effect are not available.

The results of Chapter 3 are in agreement with independent theoretical studies (Kuhn, 1975; Cherin and Murphy, 1975) and with experimental results (Miller, 1976; Kashima and Uchida, 1977). A cladding thickness equal to the core radius ($b/a = 2$) is considered adequate for multimode fibers. This choice not only conserves a sufficient number of propagating modes, but it also eliminates those high-order modes that propagate significantly faster than all other modes and would otherwise contribute strongly to signal distortion (Olshansky, 1977). This choice of fiber dimensions makes the fiber characteristics essentially independent of the optical properties of the surrounding jacket. Where several fiber channels are to be placed in close proximity, the jacket should be highly lossy to prevent cross talk between adjacent fibers.

The loss threshold between desirable and undesirable modes can be substantially increased with the help of a two-step cladding of the W type (Fig. 3.5). Consider a single mode fiber designed with this concept. Figure 3.15 shows how the loss in decibels per meter increases for the first unwanted mode as the width $a_2 - a_1$ of the low-index barrier layer is reduced. At the same time, the fundamental mode loss can be kept arbitrarily small by making the outer cladding radius b large.

Alternatively, the barrier layer can be employed to minimize the overall cladding thickness required to insulate the fundamental from the lossy jacket. A 30% reduction of the fiber diameter (corresponding to a 50% reduction in material) seems possible compared to single mode fibers without a barrier layer. A suitable design would have the dimensions $a_1 = 5$ μm, $a_2 = 10$ μm, $b = 30$ μm, $1 - n_2/n_1 = 0.5\%$, and $1 - n_0/n_1 = 0.13\%$.

In multimode fibers, a barrier layer in the cladding can serve to eliminate certain unwanted high-order modes. For example, a graded-index fiber with a very thick uniform cladding supports modes (close to cutoff) that are significantly faster than all other modes and cause undesirable signal distortion (Olshansky, 1977). A suitably designed barrier layer eliminates these modes. Figure 6.2 is a plot of the leakage loss as a function of the mode parameter V for a step-index fiber with $a_1 = 30$ μm, $\lambda = 0.835$ μm, $1 - n_0/n_1 = 0.7\%$, and $1 - n_2/n_1 = 0.2\%$. This representation shows how the loss threshold can be adjusted with the help of the barrier width $a_2 - a_1$.

6.4 COMPOSITION SCATTERING AND INDEX DIFFERENCE

Most low-loss fibers are made from silica to which one or several constituents are added to create a refractive index variation. Typical constituents are germania, boron oxide, or phosphorus pentoxide. Germania can be added in concentrations up to 25% and increases the index of silica by more than 3% in this concentration (O'Connor et al., 1976). It is widely used as a core dopant for the preparation of low-loss fibers. Unfortunately, germania belongs to those glasses which, when added to silica cause addi-

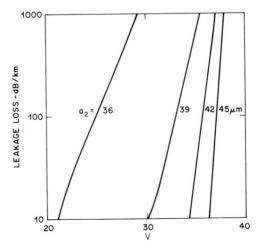

Fig. 6.2 Leakage loss versus mode parameter V for $a_1 = 30$ μm, $\lambda = 0.835$ μm, $1 - n_0/n_1 = 0.7\%$, $1 - n_2/n_1 = 0.2\%$. [From Tanaka et al. (1977).]

tional Rayleigh scattering over and above the scattering exhibited by silica alone. As a rule of thumb, an increase of the index of silica by $\Delta = 1\%$ as a result of the admixture of germania doubles the Rayleigh scattering component of silica. Thus

$$\alpha_s = 0.8\ \lambda^{-4}(1\ +\ 100\Delta) \qquad \text{for}\quad \text{Ge–Si Glass,} \qquad (6.4)$$

where λ is in microns and α_s in decibels per kilometer. This relationship has been predicted by theory (Ostermeyer and Pinnow, 1974) and found in Ge–Si samples and fibers of various kinds (Kaiser, 1977). On the other hand, since the admixture of third materials can improve or worsen the degree of miscibility of a compound glass it is not clear at this time, if the Ge–Si composition scattering described above must be considered as unavoidable. Some of the following arguments which assume (6.4) as given are merely meant to reflect the state of the art. Other admixtures to silica show other variations of the Rayleigh scattering component. A few have been shown to decrease the scattering of silica (Schroeder *et al.*, 1973). The following discussion is restricted to Ge as a representative and important example. A large index difference is particularly useful in LED systems since it increases the amount of light collected by the fiber. In addition, it makes the fiber more resistant to microbending loss (see Section 6.6).

6.5 INJECTION LOSS IN LED SYSTEMS

The injection efficiency of a typical LED-fiber arrangement was considered in Section 3.7. It was assumed that the active area of the LED completely covered the end face of the fiber core and that the fiber was a multimode fiber with an index profile of the type (3.2). The power injected into the fiber was in this case

$$P_f = \frac{2g\,\pi^2}{2\ +\ g}\ Ba^2\Delta, \qquad (6.5)$$

where B is the brightness of the LED, a is the core radius and Δ is the relative index difference between core center and cladding. To maximize, P_f, g, a, and Δ should be chosen as large as possible. We recommend a step-index fiber ($g = \infty$) unless bandwidth requirements make mode velocity equilization necessary (see Section 6.7). As explained in Section 6.2 the radius b of a typical glass-clad multimode fiber is limited to 75 μm so that typically $a < b/2 = 37\ \mu$m. The relative index difference can be increased to 3.2% with the admixture of germania to the core glass (O'Connor *et al.*, 1976). However, the scattering loss of this fiber increases with Δ according to (6-4). Therefore, if the desired transmission distance is L, the power available at the fiber end is proportional to

$$\Delta \exp(-18.4\Delta L)$$

with L in kilometers. This power has a maximum when

$$\Delta = 1/18.4L \tag{6.6}$$

and for $\Delta = 3.2\%$, (6.6) yields $L = 1.7$ km. Thus, the maximum index difference of 3.2% should be chosen when the desired transmission distance is less than 1.7 km. For longer links, the index difference should be computed from (6.6). The power available at the end of the link is then (assuming $g = \infty$)

$$P_e = \frac{2\pi^2}{e} Ba^2\Delta \exp(-\alpha_{si}L), \tag{6.7}$$

where e is the base of the natural logarithm and α_{si} the loss caused by all attenuation effects excepting Raleigh scattering. Naturally, if the index difference so computed is relatively small, other dopants should be considered which may offer the same index difference with less increase in scattering.

A third possibility for short optical links is the plastic-clad silica fiber. Because it is the core radius which determines the flexibility of this fiber in the way explained in Section 6.2, the radius of its core can be about twice that of the glass-clad fiber. As a consequence, the light power collected from a large-area LED is four times that collected by a glass-clad fiber of equal NA. This advantage is not fully realized if a small-area LED is used whose active area may be limited to a diameter of 50 μm or less.

The NA achievable in plastic-clad fibers is limited to 0.4 by the refractive index of available low-loss silicone resins (Tanaka et al., 1975). The loss spectrum of the best of these resins has a minimum of 1000 dB/km in the vicinity of 0.85 μm. The cladding loss affects mostly the modes of high order as is evident from (3.62) and (4.43). Integration of (4.43) over all modes yields the power P_e available at the end of the link:

$$P_e = 2\pi^2Ba^2\Delta(L_0/4.4L)^{1/2} \exp(-\alpha_{si}L) \tag{6.8}$$

where L_0 is defined by (6.13) and α_{si} is again the loss caused by all attenuation effects except Rayleigh scattering. A comparison of (6.8) with (6.7) shows that for short links (700 m or less) plastic-clad fibers may have an advantage over germania fibers with respect to overall transmission efficiency.

There may be other considerations which may favor all-glass fibers, as, for example, the risk of loss increases in the plastic cladding or in the plastic-glass interface with age. Also, the uniform silica core of the plastic-clad fiber produces mode velocity differences of several percent and therefore offers a narrower transmission bandwidth than a graded-index germania fiber. As discussed in the following section, however, these bandwidth differences may not be of great importance in short optical links. Most importantly the production cost of plastic-clad fiber tends to be lower than that of all-glass fiber.

6.6 MICROBENDING LOSS

In addition to composition scattering and injection efficiency, there is a third transmission characteristic that is strongly influenced by the index difference between core and cladding. We refer to random bends and the associated loss as described in detail in Sections 3.11 and 3.12. Although fibers produced in a vibration-free and well-controlled pulling process are almost perfectly straight, nonuniformities of the fiber sheath or a nonuniform lateral pressure applied to it can cause microscopic deviations of the fiber axis from the straight condition. These microscopic random deviations $f(z)$ are commonly called "microbends." After analyzing $f(z)$ in terms of its Fourier spectrum, one can identify "spectral" components with specific spatial frequencies Ω which are detrimental to the guidance properties of the fiber and cause a microbending loss α. In the case of a parabolic or near-parabolic index profile, for example, a multimode fiber with core radius a and index difference Δ is affected solely by one spectral component, namely that having the spatial frequency $\Omega_p = (2\Delta)^{1/2}/a$ as indicated by (3.184). For $\Delta = 1\%$ and $a = 25$ μm, $\Omega_p = 5.66$ mm^{-1}.

Because of the high Young's modulus E_f of glass, fibers exhibit a surprisingly strong resistance to deformations by outside pressures in spite of their minute thickness. This property is of course more effective against deformations with high spatial frequencies (short wavelengths). More specifically, a fiber with a radius b embedded in a soft jacket of modulus E_j (see Fig. 6.3) will resist deformations with frequencies higher than

$$\Omega_r = \frac{1}{b}\left(\frac{4}{\pi}\frac{E_j}{E_f}\right)^{1/4} \tag{6.9}$$

and stay relatively straight inside the jacket while the latter is being de-

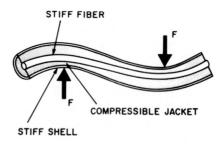

STIFF FIBER

F

COMPRESSIBLE JACKET

STIFF SHELL

Fig. 6.3 Stiff fiber embedded in compressible jacket.

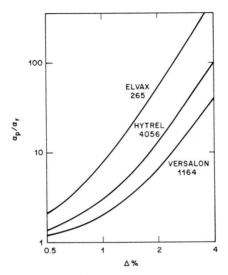

Fig. 6.4 Microbending loss reduction factor α_p/α_r expected for a near-parabolic fiber with index difference Δ when the fiber is embedded in Elvax 265 ($E_j = 2.1$ kg/mm²), Hytrel 4056 ($E_j = 5.9$ kg/mm²), or Versalon 1164 ($E_j = 10.6$ kg/mm²). A ratio $b/a = 2$ was assumed.

formed. As a result, the deformation at Ω_p of a fiber so jacketed is a function of the ratio Ω_p/Ω_r. In the case of a near-parabolic profile, the microbending loss α_r of the jacketed fiber is reduced by a factor

$$\alpha_r/\alpha_p = [1 + (\Omega_p/\Omega_r)^4]^{-2} = [1 + \pi\Delta^2(b/a)^4(E_f/E_j)]^{-2} \qquad (6.10)$$

compared to the microbending loss α_p suffered without the jacket (Gloge, 1975).

Figure 6.4 illustrates the reduction achievable with the three common jacket materials Elvax®* 265 ($E_j = 2.1$ kg/mm²), Hytrel®* 4056 ($E_j = 5.9$ kg/mm²), and Versalon®† 1164 ($E_j = 10.6$ kg/mm²) for a diameter ratio between cladding and core of 2. The loss reduction gained with a soft jacket and a moderately high index difference is significant. For $\Delta = 1\%$ and Elvax 265, an improvement of 10 is possible.

Notice also the strong dependence of the improvement on the index difference Δ. This improvement is independent of, and occurs in addition to, a strong dependence of the power spectrum $F(\Omega)$ on Δ (see Section 3.10 for a physical interpretation of F). According to experiments (Gardner, 1975), a good empirical model is based on F increasing as Ω^6. The introduction of this relationship into (3.183) and (3.184) leads to a microbending loss α_p (without jacket improvement) that decreases as Δ^2. Because of (6.10), the

* Dupont.
† General Mills.

microbending loss α_r of a suitable jacketed fiber therefore decreases at least as Δ^2, but possibly as Δ^6 depending on the magnitude of the second term in (6.10) compared to unity.

The improvement (6.10) is a consequence of high flexural rigidity (stiffness) together with lateral compressibility of the jacketed fiber. The concept of this design can be further improved by introducing a thin outer rigid tube as indicated in Fig. 6.3 (Gloge, 1975). This tube adds resistance to lateral deformations as a consequence of the high moment of inertia of the tube. Moderately hard plastic materials like Nylon 6-12 with a Young's modulus in the vicinity of 100 kg/mm² can serve as the tube material. An added advantage of the outer hard material is the improved abrasion resistance of the fiber unit which is usually only fair or poor for a soft material like Elvax or Hytrel.

The concept of combining a hard and a soft plastic jacket was used by C. M. Miller (1976) in the preparation of laminated fiber ribbons. Residing in a soft polyethylene layer, the fibers are shielded from outside forces, while two polyester cover films provide rigidity, protection, and dimensional stability. The insert in the upper right of Fig. 6.5 shows the structure of the ribbon. The points in Fig. 6.5 indicate microbending losses measured in fibers of various kinds after they had been incorporated into such ribbons (Miller, 1976). The loss is plotted as a function of the parameter $\Delta^{1/2} (b/a)$ which is proportional to the ratio Ω_p/Ω_r used in (6.10). The dashed line indicates a possible interpretation of the measured functional dependence in terms of a loss reduction as described by (6.10) with $\Omega_r = 0.07/b$. In this case, Ω_r can be interpreted as a characteristic parameter of the compound ribbon structure.

Another approach of gaining flexural stiffness with simultaneous compressibility involves oriented polymerization of a suitable jacket material (Jackson et al., 1977). This is done by extruding a very loose jacket which is subsequently stretched to orient the cross-linked molecules in axial direction. During the stretching process additional fiber is fed through the loose tube. While the concept works, it suffers from the difficulty of achieving good contact between the jacket and the fiber after stretching. If this contact is not achieved, thermally induced shrink-back of the jacket can cause kinking of the fiber and thus introduce a worse microbending problem than it is designed to prevent.

The reduction of microbending loss with the help of the jacket works equally well in the case of single mode fibers. In this case, Ω_p in (6.10) must be replaced by $\Omega_s = 2\pi/\Lambda_s$ from Fig. 3.36. Similarly, α_p must be replaced by the microbending loss α_s applicable to single mode fibers. For the case for which F is proportional to Ω^6, for example, α_s can be found in Fig. 3.33.

To compare the microbending loss of single mode fibers with that of typical parabolic-index fibers, we consider the following example. A

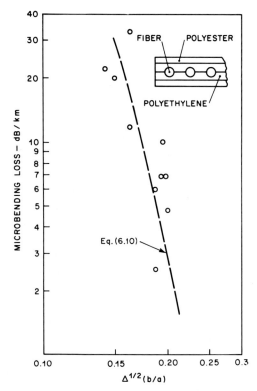

Fig. 6.5 Microbending loss as a function of the fiber parameters Δ, a, and b for fibers laminated into a flat ribbon structure.

parabolic-index fiber with $\Delta = 1\%$, $a_p = 25$ μm, $b_p = 55$ μm, and $\lambda = 0.825$ μm so that $ka_p = 190$. For the single mode fiber, $\Delta = 0.25\%$, $a_s = 5$ μm and $b_s = 40$ μm so that $V = 3$ and $ka_s = 38$. We find $\Lambda_s = \Lambda_p$ from Fig. 3.36 and $\alpha_s/\alpha_p = 0.25$ from Fig. 3.33. Because the outside radius b_s of the single mode fiber is smaller than that of the graded-index fiber, the same Elvax jacket provides about four times less microbending protection to the single mode fiber than to the graded-index fiber. As a result, the overall sensitivity to microbending is about the same for both fibers, if jacketed with the same material.

6.7 LOW-BIT-RATE SYSTEMS

The dispersion characteristics of plastic-clad fibers are discussed in Section 4.5 and illustrated in Fig. 4.9. In the absence of mode coupling, the rms impulse width increases as

$$\sigma = (12)^{-1/2}LN_1\Delta/c \tag{6.11}$$

for short fiber lengths L and approaches the asymptote

$$\sigma = N_1 \Delta V / 2c\,\alpha_0 \qquad (6.12)$$

for large L where N_1 is the group index of the core and α_0 the cladding loss in Nepers/km. It is convenient to define a characteristic length

$$L_0 = \tfrac{1}{2}(12)^{1/2}(V/\alpha_0) = 1.73(V/\alpha_0) \qquad (6.13)$$

so that (6.11) holds for $L < L_0$ and (6.12) for $L > L_0$. For $a = 50\ \mu m$, $\lambda = 1\ \mu m$, $\Delta = 3.2\%$, and $\alpha_0 = 460$ Neper/km, $L_0 = 436$ m. The asymptotic rms width according to (6.12) is 20 nsec for the above parameters. A fiber of this kind has a bandwidth of 12.5 MHz and can transmit 25 Mbit/sec. Such bit rates are commensurate with inexpensive integrated-circuit speeds and hence adequate for internal and peripheral computer bus connections. If somewhat higher speeds are desirable, a trade-off between transmission efficiency and bandwidth is possible by a modification of the V values. For LED systems, a reduction of the V value increases the bandwidth, but reduces the power available at the receiver.

Higher bit rates can in principle be transmitted with graded-index fibers. In the case of a typical LED source, the limit is set by material dispersion. If we assume an rms source spectral bandwidth of 16 nm centered at 850 nm, we find a length-bandwidth product of 68 MHz km from Fig. 4.1 if the fiber contains 13% germania; this corresponds to a relative index difference $\Delta = 1.6\%$ (NA = 0.26). If the fiber contains less or no germania, the length-bit rate product may go up to 92 MHz km. Note that these numbers hold for a wavelength of operation of 0.85 μm and that higher bit rates can be obtained by operating at longer wavelengths where material dispersion is smaller (see Section 6.9).

6.8 HIGH-BIT-RATE SYSTEMS

The emission spectrum of a single-mode semiconductor laser is in principle not wider than a fraction of 1 nm. However, such lasers are available only as laboratory models; lasers for system use have an rms spectral width that is more typically 1 nm. To estimate the material dispersion effect produced by this source, we refer again to Fig. 4.1. If no other sources of dispersion are present, the length-bandwidth product is between 1086 and 1470 MHz km.

To maintain other dispersion effects at comparable or smaller levels, either a multimode fiber with an accurately shaped index profile or a single mode fiber must be used. Equations (4.36) to (4.39) can be used to compute the tolerances to which the index profile must be controlled if mode delay effects are to be smaller than the material effects mentioned above. In order for σ to be smaller than 0.125 nsec/km, for example, concentric index variations (profile ripples) must be controlled to 0.5% of the total index difference or 8.10^{-5} if $\Delta = 1.6\%$. These are estimates based on a complete and uniform mode power distribution. It has been found that the more realistic nonuniform distributions present in practical fibers tend to reduce mode delay effects. Hence, the above tolerances may be considered as conservative. Whether such tolerances are too costly to maintain in large scale production remains to be seen.

The index profile of a single mode fiber needs hardly any control at all. If it is graded for technological convenience, its V value can be computed from (3.58) where g is the exponent of the closest power-law profile. The V-values so computed should be in the range between 2 and 3 to obtain single-mode propagation for typical cladding thicknesses (see Section 6.3).

Minimum microbending loss occurs in a narrower range of V; hence if microbending is critical, V must be controlled in a narrower tolerance range (see Fig. 3.31). Waveguide dispersion is usually negligible in the range of $V = 2$ to 3 (see Fig. 4.2).

A critical challenge in the application of single mode fibers is their small core size. The core diameter of a single mode fiber is typically between 8 and 12 wavelengths ($g = 2$, $V = 3.4$) for practical index differences between 0.1 to 0.2% of the axial index. Concentricity of the core and its alignment in a splice are in this case technological challenges indeed. A 3-μm core offset (2.5% of the outside diameter) leads to 0.35 dB splice loss (see Section 3.9), which is probably more than one can tolerate in the presence of other splice imperfections. An axial misalignment of 0.35° increases the total splice loss to 0.5 dB. It is certainly not impossible to achieve such precision in preparing and splicing fibers, but the cost involved must be weighed against the cost of precise control of the index profile of a multimode fiber.

A modest increase of the V value suited for single mode operation can be obtained by the introduction of an index barrier in the cladding (see Section 3.2 for the W structure) or an index depression in the core center (Marcuse and Mammel, 1973). The barrier region between cladding and core which is the characteristic of the W structure may be required anyway for reasons which have to do with the chemical vapor deposition process. In this case, no extra cost is involved in fabricating the W structure; an increase of V to 3.5 or 4 is possible.

The bandwidth of single mode fibers is limited by material dispersion and is in the range between 1 and 2 GHz km for available GaAlAs lasers as mentioned earlier. Technological advances leading to better contol of semiconductor laser operation can be expected and should lead to narrower emission spectra. A reduction of the 1-nm width mentioned earlier by one order of magnitude is theoretically possible. Alternately, the bandwidth of single mode systems can be increased by shifting the wavelength of operation into a range where the material dispersion of silica is smaller.

6.9 WAVELENGTH OF OPERATION

The variation of group delay with wavelength that is plotted in Fig. 4.1 diminishes in its magnitude with increasing wavelength and vanishes at a certain wavelength in the range between 1.2 and 1.6 μm depending on the fiber composition. At this wavelength, fiber delay distortion assumes a minimum value of about 0.025 psec/km nm^2 (Kapron, 1977). Even though light-emitting diodes that emit in this wavelength range have an rms spectral width in excess of 40 nm, the delay distortion associated with this source spectral width permits a bandwidth-distance product close to 1 GHz km when the source is operated at the dispersion minimum. The corresponding product for a 1-nm laser spectral width is 2000 GHz km. In practice it may be difficult to operate the source exactly at the dispersion minimum.

An additional incentive to operate fiber systems at longer wavelength comes from the fact that the dominant sources of loss in highly purified low-loss fibers are Rayleigh scattering which decreases as the fourth power of the wavelength and infrared absorption which rises sharply with wavelength in a region beyond 1 μm (Osanai et al., 1976). As a result, there is typically a loss minimum between 1.0 and 1.6 μm. Since both Rayleigh scattering and infrared absorption are a function of the amount and species of glass constituents present, the loss minimum and its position depends on the index-forming dopants in the fiber core. For boron-doped fibers, for example, the minimum appears close to 1 μm while Ge-doped fibers have a minimum at 1.5 μm. The loss minimum of P-doped fibers is in the vicinity of 1.2 μm.

The OH ion resonance has its first overtone at 1.4 μm. Depending on the OH contamination present, it may affect the loss spectrum in the region between 1.3 and 1.5 μm or beyond. Only 30 parts per billion of OH ions are sufficient to cause a loss of 1 dB/km at 1.4 μm. Thus, if OH removal proves difficult or if Boron is desirable as an index-forming or otherwise modifying agent, operation between 1.0- and 1.2-μm wavelength may prove adequate for certain applications. The bandwidth-distance product of such a fiber is 10 GHz km for a 1-nm source spectral width centered at 1.1 μm.

Other considerations relevant for the wavelength choice in an overall systems design are discussed in Chapter 20.

6.10 TOLERANCES ON FIBER PARAMETERS

Once their values are chosen, the fiber parameters discussed in this chapter must be held within a narrow range if excessive loss is to be avoided in splices. A quantitative summary of the implications of this "intrinsic" splice loss for fiber tolerances is given at the end of Section 3.9.

In addition to splice considerations, a rather tight tolerance on the profile shape parameter g is necessary if a large fiber bandwidth is desired. This is discussed quantitatively in Section 4.4.

REFERENCES

Cherin, A. H., and Murphy, E. J., An analysis of the effect of lossy coatings on the transmission energy in a multimode optical fiber. *Bell Syst. Tech. J.* **54**, 1531–1546.

Gardner, W. B. (1975). Microbending loss in optical fibers. *Bell Syst. Tech. J.* **54**, 457–465.

Gloge, D. (1975). Optical-fiber packaging and its influence on fiber straightness and loss. *Bell Syst. Tech. J.* **54**, 245–262.

Jackson, L. A. Reeve, M. H., and Dunn, A. G. (1977). Optical fiber packaging in a loose fitting tube of oriented polymer. *Opt. Quantum Electron.* **9**, 493–498.

Kaiser, P. (1977). Numerical aperture dependent spectral loss measurements of optical fibers *Proc. Int. Conf. Integr. Opt. Opt. Fiber Commun. 1977* Paper B6-2.

Kapron, F. P. (1977). Maximum information capacity of fiber-optic waveguides. *Electron. Lett.* **13**, 96–97.

Kashima, N., and Uchida, N. (1977). Excess loss caused by an outer layer in multimode step-index fibers: Experiment. *Appl. Opt.* **16**, 1320–1322.

Kuhn, M. H. (1975). Lossy jacket design for multimode cladded core fibers. *Arch. Elektr. Uebertr.* **29**, 353–355.

Marcuse, D., and Mammel, W. L. (1973). Tube waveguide for optical transmission. *Bell Syst. Tech. J.* **52**, 423–435.

Miller, C. M. (1976). Laminated fiber ribbon for optical communication cables. *Bell Syst. Tech. J.* **55**, 929–935.

O'Connor, P. B., MacChesney, J. B., and Di Marcello, F. V. (1976). Large-numerical-aperture, germanium-doped fibers for LED applications. *Proc. Eur. Conf. Opt. Fiber Commun. 2nd, 1976* pp. 55–58.

Olshansky, R. (1977). Effect of the cladding on pulse broadening in graded-index optical waveguides. *Appl. Opt.* **16**, 2171–2174.

Olshansky, R., and Nolan, D. A. (1976). Mode-dependent attenuation of optical fibers: Excess loss. *Appl. Opt.* **15**, 1045–1047.

Osanai, H., Shioda, T., Moriyama, T., Araki, S., Horiguchi, M., Izawa, T., and Takata, H. (1976). Effect of dopants on transmission loss of low-OH-content optical fibers. *Electron Lett.* **12**, 549–550.

Ostermeyer, F. W., and Pinnow, D. A. (1974). Concentration fluctuation scattering applied to optical fiber waveguides. *Bell Syst. Tech. J.* **53**, 1359–1402.

Schroeder, J., Mohr, R., Macedo, P. B., and Montrose, C. J. (1973). Rayleigh and Brillouin Scattering in K_2O–SiO_2 Glasses. *J. Amer. Ceram. Soc.* **56**, 510–514.

Tanaka, S., Inada, K., Akimoto, T., and Kozima, M. (1975). Silicone clad fused silica core fiber. *Electron. Lett.* **11**, 153–154.

Tanaka, T. P., Yamada, S., Sume, M., and Mikoshiba, K. (1977). Microbending losses of doubly clad (W type) optical fibers. *Appl. Opt.* **16**, 2391–2394.

Chapter 7

Materials, Properties, and Choices

BRIAN G. BAGLEY
CHARLES R. KURKJIAN
JAMES W. MITCHELL
GEORGE E. PETERSON
ARTHUR R. TYNES

7.1 INTRODUCTION

Although a number of different waveguides have been considered for the transmission medium suitable in an optical communications system (for review, see Miller, 1970), attention now centers (Chynoweth, 1976) on a fiber lightguide consisting of a glass core surrounded by a lower refractive index cladding. The current interest, and emphasis, on this system for the transmission medium is due to recent, extraordinary, technological advances in the preparation of an acceptable lightguide, coupled with several practical advantages (e.g., easy installation).

The design requirement for the fiber lightguide is stated simply that it guide light efficiently; be mechanically strong enough for installation; be optically, chemically, and mechanically stable with time; and be economically viable. This simple statement, however, translates in materials requirements and understanding which are at the forefront of our present knowledge. Indeed, extraordinary progress has been made in the fabrication of acceptable lightguides without the benefit of a large background of pertinent scientific knowledge, even in the most fundamental aspects. Our scientific understanding has, in general, followed the empirical successes, and the background is only now developing. In addition, the high purity and low optical loss required for fiber lightguides have put stringent demands on the chemical and optical characterizations of materials.

Those aspects which govern the choice of the materials appropriate for use in a glass lightguide, together with their chemical and optical characterizations, are reviewed in this chapter.

7.2 MATERIALS ASPECTS—BASIC CONSIDERATIONS

The optical properties of a material are of primary interest if the material is to be used in a fiber lightguide. At the present, material dispersion is not a limitation in optical communications (Di Domenico, 1972) although it may be an important consideration in future high bandwidth applications. To guide light efficiently, however, a glass must have low optical loss (attenuation). Thus, an understanding of the contributions to the optical loss in a material is of considerable importance.

In addition to a material's optical properties, there are also "engineering" properties such as the compositional dependences of the refractive index, the glass transition temperature, and thermal expansion, and the temperature dependence of the viscosity; all of these must be considered in order to produce an appropriately clad, intact, fiber lightguide.

These aspects, together with a description of the glass systems actively being considered, are discussed in more detail in the balance of this section.

7.2.1 Optical Loss

In this chapter, we treat the optical losses which are intrinsic to the glass and its composition and thus can be viewed as bulk, or material, losses. Those losses related to the fiber, its configuration, and its manufacture are treated in Chapters 2, 3, 6, and 11.

It is convenient to divide the total optical loss into those contributions due to absorption and those due to scattering.

7.2.1.1 Absorption.   The search for ultralow loss materials for lightguides is faced with the problem of understanding the nature of the bandgaps of amorphous materials (Weaire, 1971) and, most important, the roll-off characteristics into the forbidden region. At first sight this seems rather formidable. As will be indicated in Section 7.2.2, silicate glasses are commonly used now and are likely to be used for sometime. This is because the engineering properties of other glasses, e.g., phosphate and borate glasses, are generally unsatisfactory. (For example, with these latter materials, problems are encountered with working behavior and durability.) The materials science of lightguides is still rather young, however, and we may be in for some surprises. Materials thought at this juncture to be unsuitable may well turn out to have important applications. In order to illustrate the principles and because of the current interest in silicates we shall limit our discussion to them.

In a qualitative sense, the maximum passband of a lightguide will be determined by the bandgap on the high frequency end and vibronic states on the low frequency end. Figure 7.1 shows the reflection spectrum (Philipp, 1966) and one possible energy level scheme (Ibach and Rowe, 1974) for fused SiO_2 in the uv. The spectrum is dominated by four strong bands. The transitions at 10.2, 14.0, and 17.3 eV are (according to Ibach and Rowe (1974)) excitonic while the transition at 11.7 eV is to the conduction band. The bandgap is 8.9 eV giving a rough upper limit for transmission for this material at about 1400 Å.

Other energy level schemes have been put forward (Koma and Ludeke, 1975; Pantelides and Harrison, 1976; Stephenson and Binkowski, 1976; Schneider and Fowler, 1976). The theory is still in a state of flux and a recent review of the situation has been given by Griscom (1977). For our purposes the data presented in Figure 7.1 will suffice.

As is well known (Dow, 1976), both the phonon absorption edge and the low-energy electronic edge (Urbach, 1953) roll off into the forbidden gap as approximately exponential functions of photon energy. Thus, in the case of the electronic edge the following expression (Dow, 1971) for the absorption coefficient, $\alpha(\omega)$, can be written

$$\alpha(\omega) = A \, \exp[\sigma(\hbar\omega - \hbar\omega_0)/kT]. \qquad (7.1)$$

Here k is Boltzman's constant, T is the absolute temperature and σ, A, and ω_0 are parameters characteristic of the material in question.

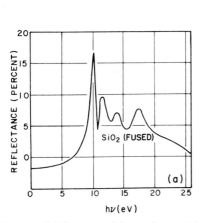
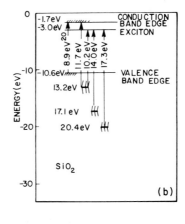

Fig. 7.1 (a) The spectral dependence of the reflectance of fused quartz (Philipp, 1966), (b) Part of the experimentally observed electron energy levels and transitions for amorphous SiO_2 (Ibach and Rowe, 1974).

Convincing proof (Bagley *et al.*, 1976) has been given for the validity of Eq. (7.1) for silicate glasses. In fact, it has been shown that Eq. (7.1) holds accurately over a range of five decades for the above material. The rate of falloff is in general so rapid that even though the absorption in the uv is quite large the absorption in the red or near-infrared is negligible.

The physics behind this roll-off is important because it provides insight into what we might expect from glasses in general. The random structure of glasses gives rise to varying local electric fields on a microscopic scale. The dominant source of the microfields will be expected to vary to some extent from glass to glass but should be associated with optical phonons, impurities, and dangling bonds. What matters is the way in which these random fields affect the optical absorption.

The absorption by excitons in a uniform electric field E (Dow and Redfield, 1972) can be shown to have an exponential dependence. Thus

$$\alpha(\omega, E) = \exp C(\hbar\omega - \hbar\omega_0)/E. \tag{7.2}$$

If the probability density function for the random electric microfield distribution in the glass is $p(E)$ one can write:

$$\alpha(\omega) = \int \alpha(\omega, E)p(E) \, dE. \tag{7.3}$$

The crucial question is whether the exponential edge of Eq. (7.2) will survive the averaging represented by Eq. (7.3). Fortunately it does, and even more important, the exponential shape turns out to be rather insensitive to the exact details of the microfield distribution. Consequently, the form of Eq. (7.1) is justified and there are strong reasons to believe that most glasses will obey it. This implies that almost any glass with a reasonable bandgap might be optically suitable for lightguides, given that other engineering parameters, such as strength, are satisfactory. For example, the addition of network modifying cations such as alkali ions to silica causes only small shifting of the uv edge (Sigel, 1971). Likewise the addition of network formers such as GeO_2, B_2O_3, or P_2O_5 has only a small effect (Hensler and Lell, 1969; Horiguchi and Osanai 1976). Thus there is every reason to believe that these glasses would be just about as transparent in the visible as pure fused silica. However, as technology improves and the losses in lightguides get lower and lower, small effects may become important. Obviously, a study of the vacuum uv spectrum of any potential new lightguide material class would be useful.

Absorption edge tail. It has been suggested that there are additional states in the bandgap associated with the disorder inherent in amorphous materials (Anderson, 1958; Mott and Davis 1971). This could lead to a tail on the Urbach edge which would cause excessive loss in optical fibers (Tauc, 1974, 1975; Pinnow *et al.*, 1973). Theoretical arguments have been given for the existence of such states (Mott and Davis 1971; Cohen *et al.*, 1970) and while they affect some properties, they do not affect optical properties. It now seems quite clear that the "apparent" uv edges are the result of impurities, most probably Fe^{3+} (Douglas, 1967; Sigel and Ginther, 1968; Kurkjian and Peterson 1974). As techniques for making "pure" sodium silicate glass improved the "apparent" edge moved. Figure 7.2 gives a historical account of this phenomenon (Kurkjian and Peterson, 1974). The absorptions in this figure are given for 1 ppm of transition metal ion, both in terms of optical density (or absorbance) per centimeter and decibels per kilometer (absorbance/cm = dB/km $\times$ 10^{-6}). Curve 1 is from the work of Starkie and Turner (1928), curve 2 is from Stanworth (1950), and curves 3 and 4 are from Ginther and Kirk (1971). Curve 5 is the absorption due to 1 ppm Fe^{3+}. The large movement of the edge is startling but in retrospect not

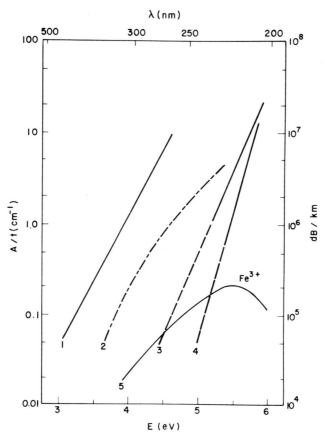

Fig. 7.2 Apparent uv edge for fused silica (see text). (C. R. Kurkjian and G. E. Peterson, unpublished, 1974).

too surprising. It is quite clear that an Fe^{3+} level of the order of 0.1 to 10 ppm can account for the "in the gap" absorption in these glasses. It is interesting to note that each of the above workers emphasized the controlling importance of iron absorption in determining the "apparent" uv edge. If indeed states do exist in the bandgap they do not appear to have been observed optically.

From a practical point of view, both the Urbach edge and the impurity tail can usually be fitted by exponential functions, the steeper of the two being the Urbach edge. The optical loss due to the Urbach edge can in SiO_2 be ignored. Regardless of the exact nature of the impurity tail, it can be extrapolated to longer wavelengths to ascertain its effect on waveguide loss. For example, for reasonably pure silica it has been shown (Keck et al., 1972) that at photon energies of less than 2 eV ($\lambda > 620$ nm) the impurity tail causes a loss of less than 1 dB/km. Thus, it too can sometimes be ignored. There is no reason to believe that in other glasses there will be any appreciable difference in these impurity tail phenomena.

Drawing induced coloration. Kaiser (1974) reported a drawing-induced coloration in vitreous silica fibers. Peak losses of up to 500 dB/km were found at a band centered at 6300 Å. Although this band appears in fibers made from many different grades of silica, it was particularly pronounced in those drawn from low OH material. He tentatively identified this loss band with drawing-induced network defects associated with ruptured Si–O–Si bands. Previously (Stroud, 1962), absorption bands with maxima centered around 6200 and 4400 Å had been found in various silicate, phosphate and borate glasses after irradiation with X and γ rays. Those bands had been associated with holes trapped at network-forming SiO_4 tetrahedra due to one or more nonbridging oxygens.

Recently Yoshida et al. (1977) have shown that a loss at 6200 Å depends upon the drawing temperature. In particular, silica fibers drawn at 2100°C and higher showed drawing-induced coloration. By studying the birefringence of clad fibers, Yoshida et al. found that the core of high-loss fibers contained extensive (tensile) strain. Kaiser reported that when fibers are aged after drawing the loss peak disappears. This was confirmed by Yoshida for some of his fibers. In addition he found that the extensive strain of the cores became less with time. He believes that when fibers contain extensive strain the Si–O–Si bonds rupture. This is of course the mechanism of Kaiser. This very interesting phenomenon deserves further study.

Fundamental absorption edge infrared. It is tempting to associate the ir absorption in glass with vibrations of small fundamental groups. For example, in the case of fused silica the spectral bands have been classified in terms of normal modes of vibration of the SiO_4 tetrahedra (Adams and Douglas, 1959). This is a fairly low order of approximation as no individual tetrahedra exist. Clearly the vibrations should be associated with much lrger groupings. Fortunately, more realistic treatments are possible (Bell and Dean 1970; Dean and Bell, 1970). These analyses, which are based on random network theory, are essential for a full understanding of infrared absorption in glass. It is useful however, to continue to use the labels assigned to the various bands given by the more elementary treatments, that is, by assuming discrete fundamental groups. However, the modes of vibration should not be taken too seriously.

Figure 7.3 shows the absorption spectrum for SiO_2 in the range from 0 to 1400 cm^{-1} (Gaskell and Johnson, 1976). There are two striking peaks, one at 460 cm^{-1} and a second at 1076 cm^{-1}. To be consistent with the earlier work (Adams and Douglas, 1959) we label these ν_4 and ν_3, respectively. Similarly we have labeled the band near 804 cm^{-1} as ν_1 and the unresolved band near 380 cm^{-1} as ν_2.

Bell and Dean (1970) have given theoretical normal mode assignments for the main spectral bands in SiO_2. Modes in the 1000–1200 cm^{-1} region are associated with Si–O–Si stretching vibrations, in which the O atoms move out of phase with their Si neighbors and parallel to the Si–Si lines. Bands throughout the 400–850 cm^{-1} range are dominated by Si–O–Si bending motion, in which the O atoms move parallel to the bisectors of the Si–O–Si angles (although in the neighborhood of 600 cm^{-1} a significant proportion of Si–O–Si stretching motion is also involved); again the vibration of neighboring atoms tends to be out of phase. The infrared and

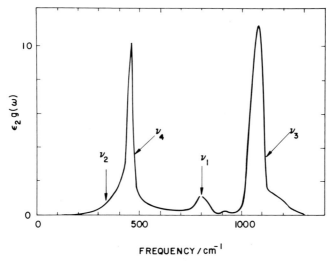

Fig. 7.3 Experimental values for the imaginary part of the dielectric constant for vitreous silica (Gaskell and Johnson, 1976).

Raman inactive modes around 350 cm^{-1} are associated with Si–O–Si rocking motion, in which the O atoms vibrate perpendicularly to the Si–O–Si planes. Modes below 350 cm^{-1} have been assigned to overall network translation or deformation motions.

If only the fundamental vibrations were important, the ir edge for fused silica would be at about 1200 cm^{-1} (8 μm). However, combination and overtone bands exist. In terms of the labeling previously given the stronger ones are Adams and Douglas, 1959):

$$\nu_1 + 2\nu_3 \rightarrow 3.2 \ \mu\text{m}, \qquad \nu_2 + 2\nu_3 \rightarrow 3.8 \ \mu\text{m}, \qquad 2\nu_3 \rightarrow 4.4 \ \mu\text{m}.$$

Thus, for dry silica the ir edge is at roughly 3 μm. A transmission curve of commercial fused silica in the 1- to 6-μm range is shown in Fig. 7.4. This sample is fairly dry, but a water band at 2.73 μm is evident.

The addition of alkali ions to silica causes some distinct changes in the infrared absorption in the 7–12 μm range (Jellyman and Proctor, 1955). Likewise doping with GeO_2 and B_2O_3 alters the ir absorption (Wong and Angell, 1976). Even though the change in absorption near 1 μm due to the above dopings is small ($\sim$1 dB/km), Osanai et al. (1976) suggest that for very low-loss lightguides such as changes could be important. Thus except for very low-loss lightguides a sodium borosilicate would be expected to perform just about as well as pure fused silica.

Impurity absorption. Impurity absorption plays a crucial role in determining fiber loss. This absorption is due mainly to the presence of iron group (3d) transition metal ions. Two types of electronic transitions are possible in these ions, resulting from transitions within the d-shell (ligand

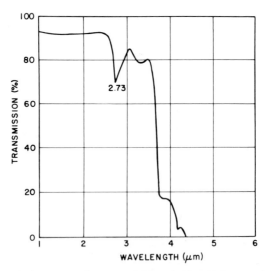

Fig. 7.4 Transmission curve of commercial fused silica (Adams and Douglas, 1959).

field transitions) or from the transition metal ion to a ligand ion (charge transfer transitions). Ligand field transitions occur at fairly low energies (<4 eV or 30,000 cm^{-1}), and because they are "parity forbidden," have low intensities (generally having extinction coefficients, ~ 20–30, but may be as high as 200). The charge transfer absorptions occur at higher energies (>4 eV) and because they are "allowed" transitions, have generally much greater intensities. An illustration in a typical silicate glass of the behavior of some important transition metal ions is given in Fig. 7.5. As in Fig. 7.2 the absorptions in this figure are also given both in terms of optical density (or absorbance) per centimeter and decibels per kilometer (absorbance/cm = dB/km $\times 10^{-6}$). The curves for Fe^{3+}, Cr^{6+}, and the straight-line, high-energy portion of the Fe^{2+} curve, represent charge transfer absorptions. The lower energy portion for Fe^{2+} and the curve for Co^{2+} represent ligand field absorptions. In addition, absorption levels are indicated for 1 ppm of ions having various extinction coefficients from 5 to 25. Three points

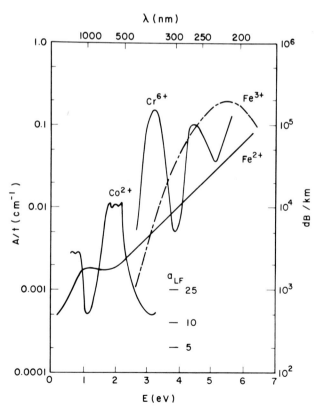

Fig. 7.5 Absorption of some ligand field and charge transfer transitions in glass (Kurkjian and Peterson, 1974).

become clear from a consideration of this figure: (1) both ligand field and charge transfer processes must be considered in assessing the importance of absorption from a given ion; (2) in the wavelength region of most interest at present (800–1000 nm), total concentrations of only 1-10 ppb of transition metal ions can be tolerated; and (3) in the range of 500 nm, almost all of the transition metal ions, regardless of valence or the transition being observed, have absorptions of the order of 10^3 dB/km/ppm. This latter information is valuable since it allows an estimate to be made of total impurity levels, from a knowledge of the absorption at 500 nm. That is, an "impurity equivalent" of 1 ppm results in 10^3 dB/km at this wavelength.

Without doubt impurity absorption is the most important single factor determining the loss of a fiber lightguide. Any durable silicate which can be prepared largely free from transition metals ought to be an excellent candidate for lightguides. In fact, extremely pure silicates would be expected to have applications in the uv where previously only pure silica was thought to be useful.

Water bands. Water in fused silica can cause excessive loss in fiber lightguides (Kaiser, 1973; Keck et al., 1972). The two fundamental water vibrations occur at 3663 cm^{-1} (2.73 μm) and 1600^{-1} (6.25 μm). These correspond to stretching and bending motions, respectively. The usual notation for the stretch is ν_3^1 and for the bend, ν_2^1. At first sight neither of the vibrations would be expected to be troublesome. However, overtones and combination vibrations strongly influence the loss in the near infrared and visible. Figure 7.6 shows some of these vibrations. Table 7.1 (Kaiser, 1973) identifies the overtone or combination frequency and gives its loss in de-

TABLE 7.1

OH Overtones and Combinational Vibrations In Suprasil 1 Vitreous Silica and Their Peak Intensities.[a]

Wavelength (μm)	Frequency	Loss (dB/km)
0.60	$5\nu_3^1$	6
0.64	$2\nu_1 + 4\nu_3^1$	1
0.68	$\nu_1 + 4\nu_3^1$	4
0.72	$4\nu_3^1$	70
0.82	$2\nu_1 + 3\nu_3^1$	4
0.88	$\nu_1 + 3\nu_3^1$	90
0.945	$3\nu_3^1$	1,000
1.13	$2\nu_1 + 2\nu_3^1$	110
1.24	$\nu_1 + 2\nu_3^1$	2,800
1.38	$2\nu_3^1$	65,000
1.90	$2\nu_1 + \nu_3^1$	10,300
2.22	$\nu_1 + \nu_3^1$	260,000
2.72	ν_3^1	10,000,000

[a] The OH level is about 1200 ppm. From Kaiser (1973).

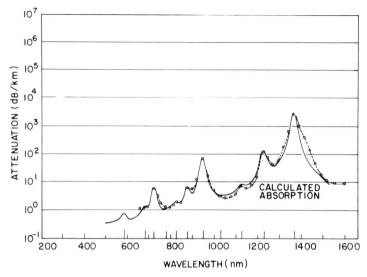

Fig. 7.6 Calculated absorptive attenuations as a function of wavelength (Keck *et al.*, 1973).

cibels per kilometer. It is quite obvious that very large losses indeed can be expected from OH vibrations. A very interesting observation is that *all* the absorptions shown in Fig. 7.6 can be accounted for by a combination of ν_1 and ν_3^1.

It is interesting to inquire as to the reason for this peculiar coupling of ν_1 and ν_3^1. We can understand this qualitatively from a study of the vibrations of the Bell and Dean (1970) model of fused silica. We recall that the ν_1 vibration is a Si–O–Si bending motion. Near a chain end in the Si–O network this bending causes a stretching of the terminal oxygen where presumably the hydrogen is attached. Thus we have a very good potential for the OH stretch and ν_1 bend to couple. Figure 7.7 shows this motion and the arrows indicate the direction of movement. Clearly great care should be exercised to keep water out of optical waveguide glass.

7.2.1.2 Scattering. In the pure glasses appropriate for lightguides, light is scattered principally by the interaction with phonons through nonlinear optical effects and by spatial variations in the dielectric constant (due to the presence of an inhomogeneous microstructure). The scattering by phonons is a Raman or Brillouin stimulated emission. While this scattering would be intrinsic, it does not make a significant contribution to the loss if the optical power density is below a threshold level. As the presently anticipated power density levels for optical communications are below this threshold, this is not expected to be an important contribution to the loss. This is discussed in detail in Chapter 5.

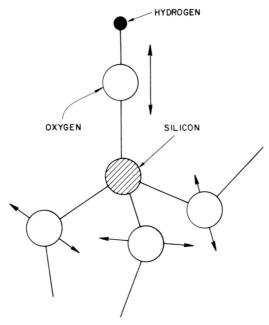

Fig. 7.7 Vibrations near a chain end for a model of fused silica (Bell, unpublished, 1974).

The contribution to the optical loss due to scattering from variations in the dielectric constant is a significant one. Indeed, in the very pure, dry, low-loss lightguide glasses which have been prepared, at the energies of interest for optical transmission it is this scattering which dominates the total loss (Osanai *et al.*, 1976). It is convenient to separate the various sources of inhomogeneities (which lead to variations in the dielectric constant) into two groups; those which are extraneous (extrinsic) to the glass and are the result of the preparation procedure, and those which are intrinsic to the glass and its composition.

Extrinsic inhomogeneity sources. Those inhomogeneities which result from the preparation procedure include bubbles (bulk and interface), unreacted starting materials, motes (nonsoluble particulate inclusions), and particulates resulting from the reaction crucible. Platinum, an otherwise attractive crucible material for the preparation of bulk compound glasses, is particularly troublesome. In glass, at concentrations above a few parts per million, it exists as submicron-sized metallic crystals (Ginther, 1971; Shibata and Takahashi, 1977), the presence of which contributes significantly to the total loss. Methods of preform manufacture have evolved (see Chapter 8) which, in addition to producing high-purity low-absorption-loss material, are also particularly effective in eliminating these extraneous sources of inhomogeneity and thereby reducing the extrinsic scattering loss.

Intrinsic inhomogeneity sources. We discuss now the sources of inhomogeneities which are intrinsic to the glass and its composition and their effect on the scattering loss. As shown in Fig. 7.8, current technology is providing fibers pure enough such that the intrinsic scatter loss is the predominant contribution to the loss in the wavelength region of interest for optical communication. The intrinsic inhomogeneities include density fluctuations, compositional inhomogeneities (static compositional fluctuations frozen in at a higher temperature, phase separation, and crystallites with a composition different than the matrix), and structural inhomogeneities (crystals in a glassy matrix). Two of these sources are intimately connected, and therefore intrinsic, to the glass composition; they are partial crystallization and phase separation.

The *crystallization* of an amorphous material proceeds by a two-step process consisting of the formation of crystalline nuclei and their subsequent growth. The overall crystallization rate is suppressed by limiting either (or both) of these steps. The thermodynamics and kinetics of crystallization have been recently reviewed (Bagley, 1974; Turnbull and

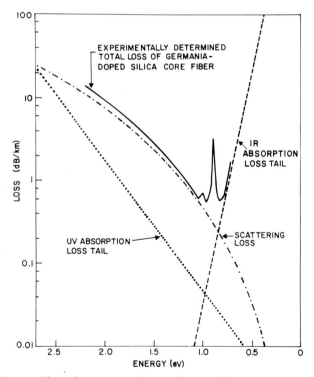

Fig. 7.8 Measured loss of a germania-doped silica core fiber, in the energy region of the loss minimum, resolved into contributions from scattering and absorption. Peaks at 1.00 and 0.89 eV are due to residual water. After Osanai *et al.* (1976).

Bagley, 1975). Crystallization, while it cannot be thermodynamically precluded, can be avoided. Several empirical rules can be used to narrow the search for glass compositions which are resistant to forming microcrystalline inhomogeneities. We are interested particularly in glasses with (1) a high viscosity at the melting point (T_M); (2) a high value for the ratio T_g/T_M where T_g is the glass transition temperature; and (3) compositions at, or near, a eutectic in the equilibrium phase diagram. The possibility also exists that a material with a very high viscosity at its melting point may permit fiber drawing above T_M; this coupled with the very fast quench obtained in the fiber would certainly bypass crystallization. Stability against crystallization is very important because during fiber drawing the glass spends time at a temperature at which the crystallization kinetics are fastest.

Phase separation is the unmixing of an initially homogeneous multicomponent material into two or more amorphous phases with differing compositions. The thermodynamics and kinetics of phase separation have been reviewed by Bagley (1974), and the experimental evidence in oxide systems reviewed by Levin (1970).

The immiscibility region may be either thermodynamically stable with respect to crystallization, in which case its occurrence will be reflected in a phase diagram, or it may be thermodynamically metastable with respect to crystallization (subliquidus), in which case its occurrence may not be reported in published phase diagrams.

The driving potential for the unmixing process is a reduction in the total-system free energy. Phase separation can occur whenever the polyphase system free energy is lower than that for the homogeneous single phase. The process is analogous to precipitation in crystals. There are two important differences however. In a liquid immiscibility gap there is no elastic energy contribution to the total free energy and, because of an easy atomic accommodation at the interface, liquid–liquid interfacial energies are generally small.

These two effects (small strain and interfacial energies) have two important consequences. First, it is kinetically difficult to prevent the unmixing process as the kinetic barriers are small or nonexistent; the separation kinetics are limited only by diffusion. Increasing the quenching rate simply decreases the dispersion size. Second, once unmixed, the second phase can exist as a very fine dispersion (often tens of angstrom units) as the driving force for coarsening is negligible. Thus, in multicomponent systems phase separation should always be considered a possibility unless evidence to the contrary is obtained.

There are *density (and in a polycomponent system, compositional) fluctuations* which are in thermodynamic equilibrium (see Münster, 1969, for review) in the dense liquid state; we are not concerned here with critical point phenomena. The density of the total sample is the superposition of local fluctuations in a dynamic time-dependent equilibrium. If these fluctuations are considered random with no correlation, then the mean squared value for the density fluctuation ($\Delta\rho$) is given by the equation

$$\langle(\Delta\rho)^2\rangle = \frac{-kT\rho^2}{(\Delta V)^2}\left(\frac{\partial\Delta V}{\partial P}\right)_T,$$

where ρ is the density, V is volume, and P is the pressure. ΔV is a fixed (but arbitrary) volume element under consideration, but it must be large enough to give meaning to the thermodynamic quantities used and to make a connection to experiment through macroscopic parameters such as the isothermal compressibility. As the liquid in thermodynamic equilibrium is cooled through the glass transition, these fluctuations, dynamic in time, become static in space. This type of compositional fluctuation differs from phase separation in that, with time, the driving force favors homogenization whereas with phase separation it favors an increasing inhomogeneity.

Light scattering phenomenology. The inhomogeneities and fluctuations just described will lead to spatial variations in the dielectric constant. Light traversing such a medium will be scattered (see reviews by Fabelinskii, 1968; Kerker, 1969). For isolated (noninterfering) scattering regions smaller than $\sim\lambda/10$, the scattering (termed Rayleigh) has an attenuation coefficient α (base e) given by (see, for example, Kerker, 1969, Chapter 9);

$$\alpha = \frac{8\pi^3}{3}\frac{1}{\lambda^4}\frac{\overline{(\delta\epsilon)^2}}{\epsilon^2}\,\delta V. \tag{7.4}$$

Here λ is the wavelength of light, δV is the scattering volume element, and ϵ is the dielectric constant, which is averaged over the volume element δV. Allowing for density and compositional fluctuations, equilibrium thermodynamics gives

$$\overline{(\delta\epsilon)^2} = (\partial\epsilon/\partial\rho)^2_{T,m}\overline{(\delta\rho)^2} + (\partial\epsilon/\partial m)^2_{T,\rho}\overline{(\delta m)^2},$$

where ρ is the density, m is the molality, and T is the temperature. Thus the expression for α consists of two terms, one of them arising from the density fluctuations and the other from compositional fluctuations.

If the glass system and composition are chosen so as to preclude phase separation and avoid partial crystallization (i.e., producing ostensibly homogeneous materials) there still exist density and, in a polycomponent system, compositional fluctuations. The Rayleigh scattering from these fluctuations is important, for it represents the lowest loss obtainable in the material. From the interpretation of X-ray scattering data, Weinberg (1963a,b) concluded that the density fluctuations in vitreous silica were frozen in at the glass transition. Pinnow *et al.* (1973) reached the same conclusion in interpreting their optical scattering data. Thus, in both these works, the thermodynamic equilibrium formalism was applied, except that T_g (or the fictive temperature) was applied instead of the actual temperature T. In a quenching process the situation that one obtains is as follows: as the temperature of the liquid is lowered toward that of the glass transition, a temperature will be reached at which the liquid cannot maintain thermodynamic equilibrium on the time scale for the cooling conditions. At that point, with decreasing temperature, the density and compositional fluctuations become irreversibly time-dependent. There are two hypothetical limiting cases. In one, an infinitely slow cooling rate maintains thermodynamic equilibrium to the ideal glass; the equilibrium formalism is applicable and the scatter loss is the lowest possible for that glass. In the other limiting case, the liquid in equilibrium at its fictive temperature is quenched infinitely fast to a temperature low enough so that no molecular transport occurs. In this case, what were dynamic fluctuations in time become static fluctuations in space and the most elementary treatment of this glass is then as a thermodynamic system with one additional parameter, the fictive temperature. Such a material would have a higher compressibility and a higher scattering loss than the glass in case one. In actually quenching a material, however, neither of these two limiting cases occurs. The glass relaxes irreversibly and the state of the system is dependent upon its entire thermal history. Obviously, the scattering loss will fall somewhere between that represented by our two limiting quenching rates. The use of a higher temperature (T_g or the fictive temperature) in the equilibrium formalism is an approximation for obtaining this loss.

Daglish (1970) has determined the scattering loss at 900 nm in a number of glasses and obtains values from 0.5 dB/km in vitreous silica and a soda-lime-silica to 7.1 dB/km in a lead flint glass. Rich and Pinnow (1972) determined the scattering loss in vitreous silica to be 0.64 ± 0.04 dB/km at 1.06 μm. A number of authors have observed scattering losses in polycomponent silicate glasses which are lower than that observed in vitreous SiO_2. Thus Pearson (1974) observed a scattering loss in a soda-lime-silica glass slightly less than that in silica. Schroeder *et al.* (1973) observed a scatter loss $\frac{2}{3}$ that of pure silica in a K_2O-SiO_2 glass, and Pinnow *et al.* (1975) observed a scatter loss less than $\frac{1}{4}$ that of pure fused silica in a soda-aluminosilicate glass. Obviously, in these glasses, any possible contribution to the scatter loss due to compositional fluctuations caused by the introduction of a second or third component is far outweighed by the decrease in the density fluctuations in vitreous silica. Any attempt to assess the relative contributions from these two sources would be speculative. The lowest loss obtainable in those systems in which heavy atoms are used to increase the NA will be higher because of the increased scatter loss as observed by Daglish (1970).

It is seen from Eq. (7.4) that Rayleigh scattering has a λ^{-4} dependence. The experimental verification of this λ^{-4} dependence for the scattering in glass is described in Section 7.3.2.4. One important consequence of this λ^{-4} dependence is that, in pure dry glasses, the total loss in the visible and near-ir is dominated by the intrinsic scattering loss which is high in the visible and decreases with increasing wavelength until the ir absorption edge is encountered whereupon the total loss increases (Fig. 7.8). The most attractive wavelength region for optical communications lies in this region of highest transparency.

If the scattering inhomogeneity size is larger than $\sim \lambda/10$ the scattered intensity has an angular dependence, termed Mie scattering (Mie, 1908), which can be very intense if the size approaches λ. Figure 7.9 shows the angular dependence of radiation scattered by a phase-separated sodium silicate glass (Andreev *et al.*, 1970).

7.2.1.3 Summary—Contributions to the Total Loss. In a dielectric, there is a transparent "window" between the region of high absorption at high energies due to electronic transitions and the region of high absorption at low energies due to vibrational states. Absorption in the fundamental edge of the electronic transition region varies exponentially with energy (Urbach behavior) and likewise, the high-energy edge of the absorption due to vibrational states is exponential with energy (sometimes with structure) due to multiphonon processes. There are low absorption tails on both these edges. The tail on the high-energy edge is due principally to extrinsic effects, such as impurities, although there may be an intrinsic lower limit. The tail on the low-energy edge is due to vibrational states introduced by impurities and/or dopants. In general, this tail is produced by elements lighter than those which produce the main absorption. Water in fused silica is an appropriate example; water is the principle contribution to the tail on the low-energy edge in a pure fused silica.

In an "impure" material the region of minimum loss will be where the tails from the high- and low-energy edges cross. If the purity of the material is increased, the contribution from the absorption edge tails will decrease and a situation will occur where the scattering contribution can dominate the total loss (Fig. 7.8). In this case, because the scattering loss varies as λ^{-4}, the region of minimum loss will be in the region where the scattering loss curve meets the tail on the low-energy absorption edge (Fig. 7.8). For oxide glasses this occurs, fortuitously, in the region of zero material dispersion (1.27 μm). Clearly, improving upon the present state

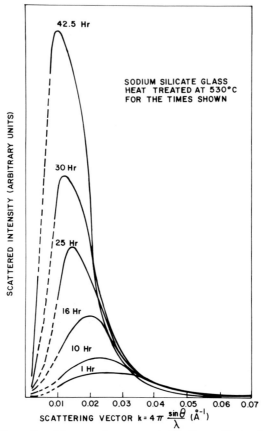

Fig. 7.9 Intensity scattered from a phase separated sodium silicate glass. $k < 0.004$ Å^{-1} are light scattering results and $k > 0.01$ are X-ray scattering results. After Andreev *et al.* (1970).

of the art will require reducing the scattering loss (without shifting the tail to higher energies) and/or going to systems whose low-energy absorption edge and tail are at lower energies.

A practical limit on the lowest loss in an oxide glass may already have been reached (~ 0.5 dB/km) due to the kinetic difficulty encountered in reducing further the scattering loss. However, the ultimate intrinsic level in a very pure, very homogeneous material will be the combined contributions from Raman and Brillouin scattering—about 0.05 dB/km at 1 eV and 300°K.

7.2.2 Engineering Considerations

Glasses suitable for use in an optical communications system must obviously meet all of the requirements of any normal commercial glass, perhaps modified by the specific manufacturing process used. Some of these requirements involve the following parameters:

(a) melting characteristics—melting point, vapor pressures, corrosiveness, ease of homogenization, etc.;

(b) refining behavior;

(c) forming characteristics—viscosity, devitrification, phase separation;

(d) chemical durability;

(e) mechanical properties, in particular, strength.

In the specific case of high-silica glasses, (a) and (b) have been partially avoided by making use of vapor-deposition techniques.

Characteristics (c), (d), and (e) are particularly important in dealing with fibers because of their geometry and high surface area to volume ratio. In addition, optical fibers for long haul use have extremely stringent optical requirements. A further consideration in optical fibers is the result of using a glass-clad, glass configuration. In this case the clad fiber is essentially a composite structure since the achievement of the required Δn will normally lead to differences in other physical properties as well. Probably the two most important of these properties from an engineering standpoint are the coefficient of thermal expansion (α) and the viscosity (η).

The difference in the coefficient of thermal expansion between the core and cladding is of importance because it results in the development of the well-known thermal expansion stresses

$$S \sim (\Delta\alpha)(\Delta T)E, \qquad E = \text{Young's modulus,}$$

which for a fiber are tangential, radial, and axial. These stresses may be both harmful and beneficial. While Brugger (1971) has shown that the refractive index change due to these stresses will probably be negligible, the polarizations produced may have important effects (Cohen, 1971; Kapron et al., 1972; Papp and Harms, 1975). In addition to their effect on optical properties, it is obvious that these stresses may influence the strength of the fiber (Krohn and Cooper, 1969). In the case of thermal expansion stresses it is desirable to have the glass with the smaller expansion as the cladding glass. This will lead to axial compressive stresses in the cladding and thus to a strengthening of the fiber. In addition to being present in the clad fiber, these stresses will appear in vapor-deposited preforms as soon as they are allowed to cool. Most additions to silica result in an increased thermal expansion (see Section 7.2.3). Thus if a doped silica glass is deposited *inside* of a silica tube, axial compressive stresses will be generated on the outside surface of the silica tube and axial tensile stresses will be generated on the inside surface of the tube. In some cases it may be necessary to maintain the tube at an elevated temperature until it is collapsed in order to keep these tensile stresses from developing. In general, even though the stresses in the collapsed tube preform and the resultant fiber are similar, they tend to have more serious consequences in the case of the preform. One reason for this is that the preform will not normally be protected from mechanical damage and therefore its practical strength will be of the order of 50 ksi or less.

It has recently been shown (Paek and Kurkjian, 1975; Kurkjian and Paek, 1978) that axial stresses which are proportional to the force used in drawing the fiber can be generated in a clad fiber if the viscosities of the core and cladding glasses are different. Since residual compression is produced in the softer (more fluid) glass, it is desirable that this glass be the cladding. It has been suggested (Rongved, 1978) that the large (≥ 100 ksi) compressive stresses which may be generated in this way can be useful in strengthening fibers.

A more obvious effect of differing viscosities is directly related to the drawing process itself. Since the viscosity of the more viscous glass will tend to dictate the temperature at which the fiber is drawn, the softer glass may have to be drawn in a very fluid condition. While this may tend to distort both a CVD preform and its index gradient, L. Rongved (personal communication, 1978) has recently shown that if the temperature gradient across the preform and fiber are small, plane flow will be maintained and the index profile will not be affected. Mixing in very fluid cores may still be a problem in the preform collapse, however.

Another property of importance to the overall fiber process is the diffusion coefficient of various glass constituents as well as foreign or impurity ions. In general, it is considered that the ions are effectively immobile throughout the processing so that gradients and interfaces are maintained throughout. One exception (the Selfoc® process) is noted in Section 7.2.3.

A final important property is materials dispersion. The refractive index of glass varies with wavelength and thus the group velocity of light changes. This causes a spreading in a pulse of light propagating through the glass. This spreading is proportional to $d^2n/d\lambda^2$ and to the linewidth of the source. For many silicate glasses the materials dispersion goes to zero at about 1.3 μm (Payne and Gambling, 1975). It is possible to shift the zero cross over in silicate glasses a small amount by adding modifiers. For a large bandwidth lightguide it would be desirable to operate as close to the zero crossover as possible (Fleming, 1978).

7.2.3 Systems

At the present time, two basically different approaches, resulting in different glass compositions and different processes, are used in optical waveguide fiber production. The first is the upgrading of the transmission of "classical" glass compositions, e.g., multicomponent silicate glasses. This was a natural evolution and makes use of optical glass experience and prior optical fiber developments. The second approach—the use of a silica–high-silica glass pair—was the result of the recognition that low optical loss is more easily achieved in systems of this sort. This realization was primarily due to the early work of Jones and Kao (1969) who measured a minimum attenuation of about 5 dB/km at 850 nm in a bulk sample of commercial "Infrasil."*

After the illustration that very low attenuation could be achieved in a simple high-purity silica glass, the challenge was to develop a suitable glass which was physically and chemically compatible with silica but which had a slightly different refractive index. Infrasil is made by bulk melting of natural quartz crystal and it is very difficult to melt these crystals with a minor (~10 mole%) addition and obtain a homogeneous, high-purity melt. Thus, the waveguide announced by the Corning Glass Works in 1970 (Kapron *et al.*, 1970) as the first really low-loss glass fiber waveguide with a loss of 20 dB/km, made use of a vapor deposition technique (Hyde, 1942) and a TiO_2–SiO_2 composition developed much earlier by Nordberg (1943). Since the TiO_2–SiO_2 glass used had a refractive index higher than that of silica it had to be used as the core with pure silica as the cladding. Although both of the glasses were basically of high purity, this pair was not entirely satisfactory because of high losses resulting from the reduction of Ti^{4+} to Ti^{3+} during the fiber pulling process (Carson and Maurer, 1973). These losses could be reduced by a subsequent anneal; however, the resultant decrease in mechanical strength would be troublesome (Proctor *et al.*, 1967).

Because of the low absorption obtainable with undoped silica it would seem desirable to use it as the core material. This is difficult, however, because oxide additions to SiO_2 generally tend to increase its index. Although pure B_2O_3 and pure SiO_2 have approximately the same indices, it was found by Van Uitert *et al.* (1973) that the high cooling rates obtained upon drawing a borosilicate glass into a fiber, produced a somewhat lower refractive index than that of silica. It is thus possible to produce a borosilicate-clad silica fiber. Glasses in this system tend to have limited Δn and somewhat questionable thermal and temporal stability.

Recently Rau and co-workers (1977) at Heraeus have succeeded in doping fused silica with up to 3 wt% fluorine by means of a plasma torch arrangement. This amount of doping produces a decrease in refractive index from 1.458 to 1.445 and a numerical aperture of 0.2 with a pure silica core. The optical loss is satisfactory and certain other measured properties such as thermal expansion and viscosity are compatible with those of sil-

---

* Infrasil is a tradename of Thermal Syndicate. It is electrically fused crystalline quartz.

ica. An assessment of the usefulness of this glass pair must await further studies, however.

The bulk of the work on glass composition has concentrated on systems in which silica is the cladding glass and a doped silica is the core glass. Although alkali and alkaline earths have been proposed as dopants, the most satisfactory modifiers have been found to be those oxides which are more similar to silica itself. In fact, the successes to date have been those glasses based on silica and modified by the addition of other glass-forming oxides, e.g., GeO_2 (MacChesney *et al.*, 1973; Black *et al.*, 1974), B_2O_3, and P_2O_5 (Payne and Gambling, 1974). Other quadrivalent oxides (TiO_2, ZrO_2, SnO_2) have also been tried with some success. Since all of these oxides (except B_2O_3) either singly or in combination will increase the index of silica, silica is, of course, used as the cladding. Except from the point of view of optical attenuation, this is desirable. Compared with the doped silicas, pure silica has better chemical and mechanical stability. In addition its lower thermal expansion results in an axially compressive thermal expansion stress.

Very little detailed information is available concerning the pertinent properties of these doped silicas. Because of the difficulties involved in making satisfactory bulk samples, there is some disagreement in the data from different investigators. A large body of useful information regarding these compositions is available from the thin film literature (see, for instance, Hass and Thun, 1967). Some of the available data on properties of mixed glass former compositions are shown in Figs. 7.10A,B,C.

As indicated, most oxide additions (1) increase the refractive index, (2)

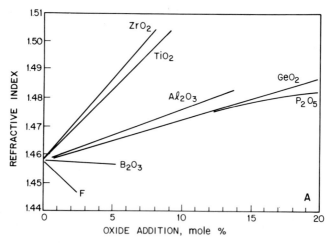

Fig. 7.10A   Refractive index (n_D) of doped silica glasses. ZrO_2, TiO_2, Al_2O_3 from Maurer and Schultz (1971). GeO_2 from Schultz (1977a). P_2O_5 from Schultz (1977b). B_2O_3 from Van Uitert *et al.* (1973). F from Rau *et al.* (1977).

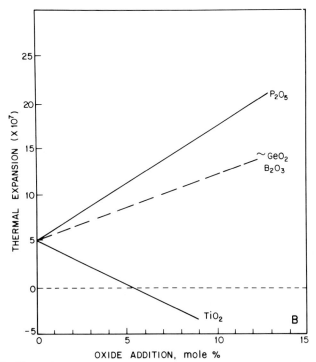

Fig. 7.10B Thermal expansion (α) of doped silica glasses. P_2O_5 from Schultz (1977b), GeO_2 from Riebling (1968). B_2O_3 from Brückner and Navarro (1966). TiO_2 from Nordberg (1943).

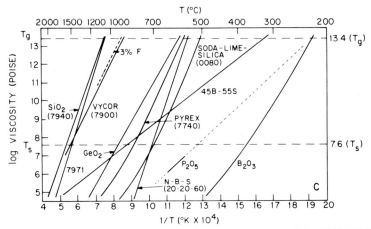

Fig. 7.10C Viscosity of some waveguide glasses and dopants. SiO_2 (7940) CGW Tech. Bull. LEM 1969. Vycor, Pyrex (7740) and soda-lime-silica (0080) from CGW Bull. B-83, 1949. 3% F from Rau *et al.* (1977). N.B.S. (20-20-60) from English and Turner (1923). 45B-55B from Brückner and Navarro (1966). P_2O_5 from Cormia *et al.* (1963). B_2O_3 from Napolitano *et al.* (1965). GeO_2 from Kurkjian and Douglas (1960).

increase the coefficient of thermal expansion, and (3) decrease the viscosity of silica. Exceptions have been noted above and are seen in the figures. Although these glass-forming mixtures are not ideal solutions, useful qualitative estimates of property changes can be made by simply considering them as ideal.

Multicomponent glasses. An alternative to the use of high-silica glasses is the modification of some type of commercial, multicomponent glass composition. Early attempts were made in this direction by starting with either simple soda-lime-silica (Pearson, 1974) or lead crown glasses (Faulstich *et al.*, 1975). While both of these could be purified to the extent of showing losses in the neighborhood of 25–50 dB/km, more recent work has tended to concentrate on alkali borosilicate compositions (Newns *et al.*, 1974). The reason for this is the somewhat lower melting temperatures required and the overall lower attenuation per unit impurity presumably due to favorable redox conditions and lower extinction coefficients. To date, minimum losses of the order of 5 dB/km have been achieved (Beales *et al*, 1975).

Interesting variations in glass processing have been employed for both multicomponent and high-silica glasses. The first is "Selfoc," marketed by Nippon Sheet Glass Co. of Japan. This fiber has a continuous, almost parabolic index profile and it is produced by ion-exchanging potassium in the core glass for thallium. This is accomplished by causing ion exchange to occur while drawing from a double crucible or by ion-exchanging into the preform rod from a molten salt bath. This obviously requires reasonably large diffusion coefficients for the exchanging ions as indicated in the previous section. Typically, losses of the order of 20 dB/km are realized (Koizumi *et al.*, 1974).

A bridge between the very different processes and compositions of multicomponent silicate glasses on the one hand and pure silica on the other, was achieved by the "Vycor"* process of Hood and Nordberg (1934). In this process a low-melting sodium borosilicate is melted in the normal way. Reheating of this composition causes it to separate into a "96% silica" skeleton and a sodium borate phase which can be leached out with hydrochloric acid. The skeleton is then sintered to full density. This is an attractive technique for producing-high silica glasses. In the case of high-purity waveguide glasses (Macedo and Litovitz, 1976) this "Phasil" technique provides an additional benefit. Presumably because the high-silica phase tends to be covalently bonded with respect to the borate phase the distribution coefficient or transition metal ion impurities is quite favorable and impurity concentration reductions of as much as 100 to 1 have been achieved. The required index gradient is produced by an ion-stuffing technique of the type practiced in the production of filters,

---

* Vycor is a brand name of Corning Glass Works.

etc. The desired ion is adsorbed from solution onto the surface of the pores of the skeleton previously produced by leaching and the normal sintering is carried out. Losses as low as 5 dB/km have been claimed for this process (Macedo *et al.*, 1976). It has been suggested that both partial leaching and suitable ion stuffing may lead to desirable compressive strengthening stresses in fibers produced by this process (Mohr *et al.*, 1977; Drexhage and Gupta, 1977).

All of the glasses so far studied for optical waveguides have been based on silica as the major glass-forming oxide. The great majority of commercial glasses for any purpose are silicates as well. There are good reasons for this. Other inorganic glass formers that might be considered are GeO_2, B_2O_3, P_2O_5, BeF_2, and $ZnCl_2$. While some of these may possess advantages in terms of optical properties and processing simplicity, none of these glass formers used by themselves, in combination with others, or in combination with another cation in a simple two-component glass, is known to be as chemically stable as are silicate glasses. Thus, although multicomponent glasses based on one of these glass formers might be attractive for one reason or another, it is probable that a silica-based glass will be found most satisfactory overall.

7.3 MATERIALS ASPECTS—CHARACTERIZATION

Synthesizing soda-lime-silicate glass with ≤ 10 dB/km loss was an early objective in some optical waveguide research programs. Attaining this goal was projected to require starting materials of sodium and calcium carbonates and silica sufficiently pure to yield a glass melt with concentrations of Co, Cr, Cu, Fe, Mn, Ni, and V less than 2, 20, 50, 20, 100, 20, and 100 ng/g, respectively (Pearson and French, 1972). Carbonates of this extreme purity had not been prepared previously although commercial fused silica of high purity was available. Evaluating such materials for use in glass fabrication requires extremely sensitive and reliable quantitative analytical techniques for screening commercially available materials, for detecting contaminants, and for identifying modes of contamination operative during the glass-melting process. Furthermore, analytical methods for these applications were needed that could distinguish accurately between small changes in submicrogram amounts of various impurities present in raw materials and in the product glass.

A wide variety of techniques are available with sufficient sensitivity for detecting trace (1–100 μg) or ultratrace (<1 μg) elements. Extreme sensitivity is provided by methods listed in Table 7.2. Not all extremely sensitive methods are suitable for providing the required analytical information. Thus the selection of the method for a specific characterization problem must be based on the nature of the material to be char-

TABLE 7.2
Techniques for Trace and Ultratrace Analysis

Atomic absorption spectroscopy
Atomic fluorescence spectroscopy
Coulometric titration
Electron-capture gas–liquid chromatography
Electron probe microanalysis
Emission spectroscopy
Fluorimetry
Ion-specific electrodes
Kinetic measurements
Mass spectroscopy
Neutron activation
Nuclear track counting
Polarography
Radioisotope dilution
RF induction coupled atomic emission
Stable isotope dilution
Spectrophotometry
Titrimetric methods
X-ray fluorescence

acterized, the analytical information desired, constraints of accuracy, speed of the determination, and cost of the analysis. Where a reliable analysis of the material is paramount to the success of an entire research program, economic considerations become less important than the reliability of data on which future materials research decisions are based. To obtain the analytical information listed previously, thermal neutron activation analysis (NAA) was considered the best technique for quantitative determinations of ultratrace elements (concentrations ≤ 1 $\mu g/g$) in waveguide materials. The primary advantages of the method include (1) very high sensitivity for the detection and practical measurement of most of the detrimental elements, (2) a substantial degree of freedom from blank (contamination) problems, and (3) utility for reliable single-element determination or multielement survey analyses. However, for adequate, broad spectrum capabilities for characterizing ultrapure materials a clean room facility, ultrapure analytical reagents, and standards are needed and special procedures have to be developed to eliminate environmental contamination as the causative factor limiting the accuracy of trace element determinations (Mitchell, 1973).

7.3.1 Chemical

A review of optical waveguide materials characterization by activation analysis, X-ray fluorescence, atomic absorption, radioisotope techniques, and infrared is reported in this section.

7.3.1.1 Materials for Synthesis of Glass by Melting

7.3.1.1.1 Sodium and Calcium Carbonates. Survey analyses by γ-ray spectrometry can provide quick screening of commercial samples of sodium and calcium carbonate to detect Fe, Cr, and Co impurities. Several samples can be irradiated simultaneously under identical conditions and equal weights subsequently counted at a constant geometry for the same length of time (Mitchell *et al.*, 1973a). Visual inspection of recorded γ-ray spectra of several samples of sodium carbonate analyzed nondestructively are shown in Fig. 7.11. Sample A is distinguishable clearly as the most pure product.

Spectra of $CaCO_3$ samples distributed by suppliers of lightguide grade materials are shown in Fig. 7.12. Characteristic photopeaks, for ^{60}Co, ^{51}Cr, and ^{46}Sc are present in Sample B. Additional impurities in Sample C include full energy peaks for ^{182}Ta, ^{183}Ta, ^{192}Ir, ^{191}Pt, and ^{195m}Pt. The detection of ^{191}Pt ($T_{1/2} = 3.00d$) after a 13d decay period showed considerable contamination by this rare trace element; this was subsequently related to the use of platinum containers for thermal decomposition of a high-purity precursor compound to obtain calcium carbonate. Iridium, an additive of platinum to increase hardness, is detected with very high sensitivity. Because of its large cross section and long half-life, detection of this isotope serves as a sensitive probe to determine contamination from platinum containers.

Quantitative determinations of impurities in carbonates by activation analysis require chemical separations to eliminate radioisotopic interferences. For example, Cu and Mn in sodium carbonate have been measured simultaneously by irradiating and dissolving samples and then passing the carrier-doped solution through hydrated antimony pentoxide (HAP) columns to remove ^{24}Na. Interferences from other matrix isotopes are eliminated by simultaneously extracting ^{64}Cu and ^{56}Mn at pH 6 to 8 into $CHCl_3$ containing 1% pyrrolidine dithiocarbamic acid and 0.01% dithizone (Mitchell *et al.*, 1973b). Cu and Mn in samples of calcium carbonate have been determined in a similar manner but in this case, it was not necessary to remove ^{24}Na with the HAP column since the organic extracts of dissolved calcium carbonates contained only low background activity from any matrix isotopes.

The reliability of these procedures has been assessed by performing replicate analyses (six determinations) of synthetic standard solutions prepared by dissolving extremely pure sodium carbonate ($[Cu] = 18.9 \pm 7$ ng/g, $[Mn] = 3.3 \pm 0.7$ ng/g) and doping with 1.000 μg/ml of Cu and 0.1107 μg/ml of Mn. The results, 0.992 $\pm$ 0.055, and 0.1050 $\pm$ 0.0135 μg/ml for Cu and Mn, respectively, indicate errors of 0.8 and 5.3%. Measurable detection limits of 0.3 and 0.03 ng were obtained for Cu and Mn, respectively, under the conditions of 15-min irradiations at 2×10^{13}

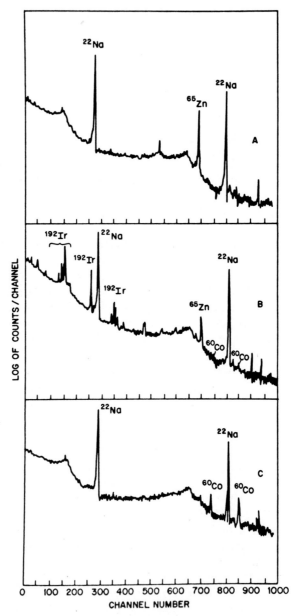

Fig. 7.11 Computer generated γ-ray spectra of impurities in sodium carbonates.

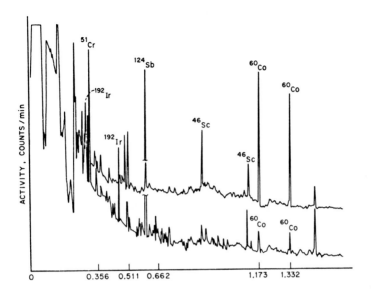

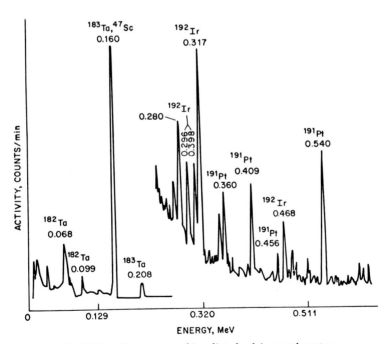

Fig. 7.12 γ-Ray spectra of irradiated calcium carbonates.

n/cm^2 sec, ≤ 1 hr decay, and 30 min counting intervals on a 9% efficient sodium iodide well-type detector.

For determinations of manganese in calcium carbonates a new substoichiometric method was developed. (Mitchell and Ganges, 1974). With this method highly sensitive, precise analyses were obtained as indicated by the mean of five determinations, 0.151 ± 0.009 μg. Comparison of the determined value with that calculated from the recommended SRM value, 0.17 ± 0.013 μg, indicated an absolute error of 5.9 to 11.7% for the determination of Mn at the 0.2-μg level. Additional data on calcium carbonate purity with respect to Cu and Mn are given in Table 7.3.

Other trace elements in sodium carbonate with long lived isotopes, ^{60}Co, ^{51}Cr, and ^{59}Fe, for example, can be measured instrumentally following 100-hr irradiations at a neutron flux of 2×10^{13} n/cm^2 sec and 10- to 20-day decay periods. Detection limits of 120, 2, and 0.5 ng were indicated by measuring respective activities from appropriate photopeaks of ^{59}Fe, ^{51}Cr, and ^{60}Co induced in synthetic standard solution.

The determination of these traces in calcium carbonate is complicated by the decay of the matrix isotope ^{47}Ca. Although the matrix isotope could be allowed to decay before measuring low levels of activity from ^{59}Fe, ^{51}Cr, and ^{60}Co, separating matrix isotopes chemically permits analytical results to be obtained more quickly. Sc and Ca have been removed by extraction with TOPO-cyclohexane and HTTA-TOPO-cyclohexane, respectively. Carbonates of sodium and calcium analyzed via these methods have usually shown levels of Cr, Co, and Fe below detection limits (Mitchell *et al.*, 1973b).

7.3.1.1.2 *Silica.* Instrumental γ-ray spectroscopy has proved extremely useful for screening high-purity silica because of the favorable nuclear decay of the matrix. Characteristic photopeaks can often be assayed immediately after irradiating the samples. The spectra of the sample in Fig. 7.13

TABLE 7.3
NAA Determination of Mn and Cu in $CaCO_3$

Sample	Conc. found (ng/g)	
	Cu	Mn
1[a]	ND[b]	43.8 ± 7.8
2	26 ± 6	29.8 ± 1.0
3	ND	340 ± 9.0
4	38 ± 5	523 ± 12
5	ND	658 ± 13

[a] Four determinations.
[b] Not detected.

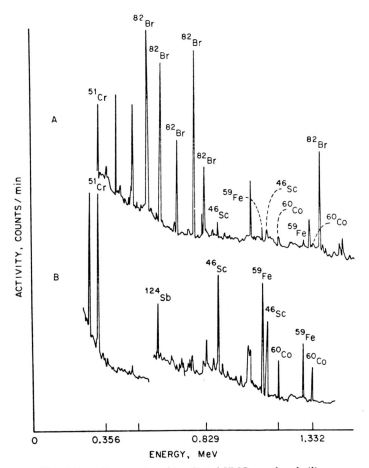

Fig. 7.13   γ-Ray spectra of irradiated KMC powdered silica.

were recorded 24 hr after the end of a 100-hr irradiation. In spectrum A, ^{82}Br, ^{24}Na, and ^{51}Cr are visible. After decay of the more abundant shorter lived radionuclides, spectrum B was obtained by counting for 23 min. Characteristic photopeaks of ^{59}Fe, ^{60}Co, and ^{51}Cr are more distinct than observed in spectrum A.

Series of different samples of silica from various suppliers can be irradiated simultaneously and compared with synthetic standards. The semiquantitative survey reported in Table 7.4 indicated that Fe contamination in all of the commercially available powdered silica exceeded tolerance limits. Although synthetic fused silica rods (Suprasil) were the purest form of silica available commercially and appeared to meet specifications for most elements, powdered starting materials were required for the production of bulk glass by melting techniques. Cu and Mn in silica can be

TABLE 7.4
Detection of Trace Elements in Silica[a]

Supplier	Concentration (μg/g)		
	Fe	Co	Cr
(1)	1.24	0.030	1.90
(2)	0.82	0.005	2.16
(3)	1.60	0.041	0.17
(4)	1.00	0.008	0.19
(5)	0.68	0.009	0.009
Suprasil	ND	ND	ND
	<0.1	<0.003	<0.005

[a] Semiquantitative γ-ray spectrometric surveys using synthetic comparison standards.

determined quantitatively after 15- or 30-min irradiations and volatilization of ^{31}Si with HF. After 100-hr irradiations and 2-day decay periods, Cr and Co are assayed nondestructively, while up to 60 days irradiations are necessary to increase detection limits for Fe.

The reliability of instrumental methods should be assessed by analyzing standard reference materials. Results for the analysis of SRM 112 (a silicon carbide) and SRM 85B (an aluminum alloy) are reported in Table 7.5. Reasonable precision and acceptable accuracy are shown for the determination of relatively large amounts of Fe and Co. The precision percentages for iron determinations in SRMs 112 and 85B are ±5 and ±2.1%, respectively, as measured by relative standard deviations. The average absolute error of the value determined for SRM 112 was 2.5% compared to the recommended NBS value. Based on the certified value for SRM 85B an absolute error of 4.6% occurred. Accuracy of the results for cobalt could not be assessed since no certified data were available. However, the accuracy of these measurements should be similar to that obtained for iron in view of the comparable precision for cobalt determinations at much lower levels than iron.

TABLE 7.5
Fe and Co Content of Silicon Carbide (SRM 112) and Aluminum Alloy (SRM 85B)[a]

Sample	SRM 112		SRM 85B	
	Fe	Co	Fe	Co
Mean of 6 determinations	4570 ± 230	2.50 ± 0.13	2320 ± 50	1.48 ± 0.03

[a] Certified Fe values, SRM 112, 4500 μg/g; SRM 85B, 2400 μg/g.

Using the instrumental methods described above, measurable limits for trace elements in fused silica were 1.0, 0.1, 5.0, and 85 ng for Cu, Mn, Co, Cr, and Fe, respectively. During most analyses very low levels of Cu and Mn were found. One example of results obtained by neutron activation for the determination of various elements in silica prepared by special procedures is listed subsequently: Cr (0.008 $\mu g/g$), Co (ND = <0.001 $\mu g/g$), Cu (0.009 ± 0.004 $\mu g/g$), Fe (<0.100 $\mu g/g$), and Mn (0.0089 ± 0.0004 $\mu g/g$). Other results obtained after concentration of trace impurities and measurement by X-ray fluorescence indicated 0.008, 0.001, 0.03–0.09, and <0.01 $\mu g/g$ of Cr, Co, Cu, and Mn, respectively. The supplier's emission spectrographic analysis of this sample after preconcentration of trace elements showed Cr, <0.01; Co, 0.01; Cu, 0.01; Fe, 0.01; Mn, 0.007; Ni, <0.01; and V, <0.01 $\mu g/g$.

Extremely pure raw materials have been prepared successfully (see Section 7.3.1.2) and analyzed by the methods described previously. Additional data for purity characterizations are reported in Tables 7.6 and 7.7. The results for these materials are reported in chronological order with respect to the sequence of their preparation in the laboratory. With few exceptions, the results for silica indicate that specifications were being met for Cu, Mn, Co, and Cr. Although no clear trend of increasing purity with time of preparation is seen, silica of extreme purity was obtained routinely. In contrast to the high purity of silica, samples of the carbonates (Table 7.7) have Cu (usually above specifications) and periodic Mn contamination. Sodium carbonates (1 and 2) and calcium carbonates (1–4) were prepared during the period in which the purification processes were

TABLE 7.6
Transition Metal Impurities in Silica (ng/g)

Sample[a]	Cu	Mn	Co	Cr	Fe
1A	ND	13	1.0	86	ND
B	18	2.0	1.1	17	ND
C	34	5.8	2.2	37	10^3
2A	50	2.8	1.5	ND	ND
B	41	1.7	1.4	ND	ND
3A	79	2.7	1.0	5.6	ND
B	43	3.4	0.5	ND	ND
C	28	6.3	0.5	ND	ND
D	36	0.5	0.2	2.2	150
4A	35	0.9	ND	1.1	ND
B	12	0.4	ND	1.5	ND
C	15	1.9	ND	1.4	ND
D	45	0.9	ND	ND	ND

[a] A, B, C designate separate analyses of various large batches of silica obtained from the same supplier.

TABLE 7.7
Transition Metal Impurities In Waveguide Raw Materials

Sample	Cu	Mn	Co	Cr	Fe
$NaCO_3$					
1.	38 ± 8	46 ± 8	—	—	—
2.	18.9 ± 7	3.3 ± 1.7	—	—	—
3.	303	ND[a]	1.0	43.0	ND
4.	17.7 × 10³	ND	ND	0.6	ND
5.	104 ± 4	2.5 ± 0.3	ND	21 ± 2.0	ND
$CaCO_3$					
1.	—	—	ND	12 ± 4	ND
2.	—	—	ND	17 ± 3	ND
3.	ND	35.7 ± 4.2	—	—	—
4.	26 ± 6	29.7 ± 1.1	—	—	—
5.	65	ND	<1	ND	ND
6.	64 ± 3	3.8 ± 0.2	ND	147	ND

[a] Not detected <1.0, <0.1, <0.1, <1.0, and <200 ng for Cu, Mn, Co, Cr, and Fe, respectively. Measurements in nanograms per gram.

carefully monitored by analysis at various steps. Preliminary samples were analyzed also prior to the delivery of the large sample. Very low levels of impurities were found in these samples, whereas the remaining samples (Table 7.7) of sodium carbonate were highly contaminated with Cu. Most of these samples were delivered without any prior analyses.

7.3.1.2 Ultrapurification Methods.   Where readily available commercial raw materials are not sufficiently pure to meet waveguide tolerance limits for trace impurities materials of sufficient purity must be prepared specifically. This project can be best attacked by a special task force consisting of purification specialists and analytical chemists with expertise in quantitative, ultratrace, and inorganic analyses. Appropriate purification schemes must be devised and evaluated for producing the desired reagent. Radioisotope techniques and activation analyses become powerful tools for establishing the optimum conditions for executing the procedure and in pinpointing specific deficiencies of each step of the purification process.

7.3.1.2.1 *Prepurifying Reagent Grade Material by Solvent Extraction.*   Solvent extraction is an excellent method for preliminary purification of reagent grade solutions of sodium or calcium. The efficiency of step A (Fig. 7.14) was evaluated by measuring distribution ratios for Cu, Fe, Cr, Mn, Co, and V using appropriate radioisotopes. The measured distribution ratios and the corresponding data showing percentage extraction results indicated excellent removal of trace quantities of Mn, Co, Fe, and Cu from sodium feed solutions, but the distribution ratios for Cr^{3+} and VO_2^+ were low. No Cr^{3+} could be extracted with a 0.2% solution of pyrrolidine-

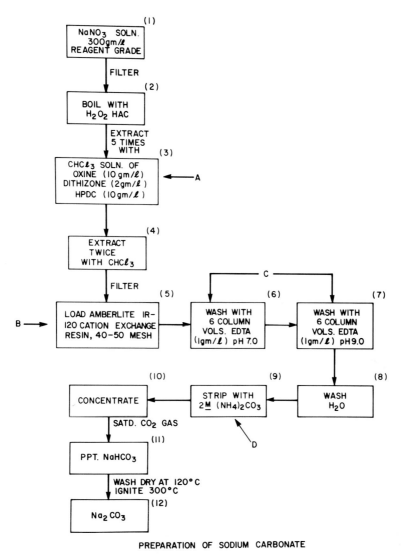

PREPARATION OF SODIUM CARBONATE

Fig. 7.14 Schematic outline of method for purifying Na_2CO_3 by ion exchange.

dithiocarbamic acid in $CHCl_3$ (in the range, 6 N HCl to pH 3.04), with dithizone, or with 0.10–0.01 M 8-hydroxyquinoline in $CHCl_3$ at any pH.

Other investigations with radiotracers aided in evaluating solvent extraction methods for preparing ultrapure calcium carbonate. Multiple extraction of 5.0 M calcium nitrate solutions with $CHCl_3$ containing 1% dithiocarbamic salt showed effective removal of nanogram per milliliter amounts of Fe^{3+} and Mn^{2+}. Distribution ratios at pH 7.4 were ~14 and 900, respectively. This extraction system was used effectively to decontaminate calcium solution containing Mn impurities. Samples of an original and extracted calcium solution were irradiated. ^{56}Mn and ^{64}Cu were separated simultaneously by extraction into $CHCl_3$ solutions of 8% HPDC–0.12% dithizone. Results by activation analysis showed 109 ± 10 ng and <0.1 ng of Mn in the original and purified solutions. Copper was below detection limits in both samples.

7.3.1.2.2 Ion Exchange Procedures.   Ion exchange procedures are very effective for removal of ultratrace impurities. A flow diagram of a procedure for preparing sodium carbonate is shown in Fig. 7.14. Procedures of this kind require many handling steps and chemical manipulations. Therefore extreme care is necessary to minimize contamination from containers and to ensure that the most pure chemical reagents are used. Other major sources of impurities, for example, the reagent grade $NaNO_3$ feed solution and exchangeable cationic impurities on the Amberlite IR 120 resin, must be recognized and minimized. Steps A and C were designed to purify the feed solution and to remove impurities from the loaded ion exchange resin. Ion exchange resins were also purified by elution with EDTA prior to loading with feed solutions.

The effectiveness of the entire procedure for removing submicrogram per milliliter amounts of Fe, Mn, Cu, Cr, Co, and V was determined by radioisotope techniques. One portion of a prepurified sodium nitrate feed solution was doped with carrier-free or high specific activity ^{59}Fe, ^{54}Mn, ^{51}Cr, and ^{60}Co and the second aliquot with ^{64}Cu and ^{48}VO$_2$$^+$. The effectiveness of each step of the procedure from the purification of the feed solution by solvent extraction (A) to stripping of the loaded sodium resin with 2 *M* ammonium carbonate (D) have been examined and detailed experimental procedures and important results of this investigation are reported elsewhere (Mitchell and Riley, 1976).

7.3.1.2.3 Mercury Cathode Electrolysis.   Methods based on mercury cathode electrolysis have been perfected as final processes for decontaminating sodium and calcium solutions (Zief and Hovath, 1974). The development of the successful procedure for preparing large quantities of the material depended ultimately on effective analytical support. Small 100-gm test samples of $CaCO_3$ (samples 1 and 2, Table 7.3) were obtained for evaluation of purity prior to the beginning of large-scale preparations. Neutron activation results for Mn and Cu showed test samples to pass specifications for these traces. When kilogram quantities of the reagent were supplied (samples 3, 4, and 5) the level of contamination by manganese made the material unsuitable. Activation analyses were performed to monitor the decrease in the concentration of Mn in samples periodically removed during the electrolysis process. Accurate control of the pH and time of electrolysis resulted in large-scale preparation of calcium carbonate meeting the specification for manganese. An additional essential factor in ultrapurification was adequate control of particulate contamination. X-ray fluorescence techniques disclosed the significant contribution of particulate contamination to the total transition element content of the final product (Mitchell *et al.*, 1973a).

7.3.1.3 Bulk Glass and Fibers

7.3.1.3.1 Detecting Contamination During Glass Melting.   In order to perfect techniques for glass-making, material scientists performed exploratory glass-melting experiments using impure reagents. As anticipated, glasses with absorbance losses in the 200–600 dB/km range were fabricated. The γ-ray spectrum of a sodium-calcium-silicate glass synthesized from raw materials from a single supplier is given in Fig. 7.15. The presence of silver was unexpected and both Cr and Co were well above tolerance limits. One of the better optical quality samples (still well above the desired 10 dB/km), prepared from materials from another source, gave the γ-ray fingerprint shown in Fig. 7.16. Characteristic photopeaks of ^{51}Cr, ^{192}Ir,

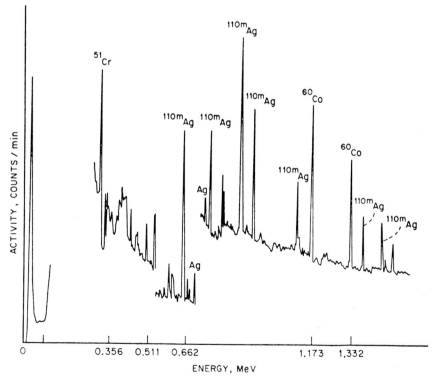

Fig. 7.15   γ-Ray spectrum of glass produced with BTP chemicals.

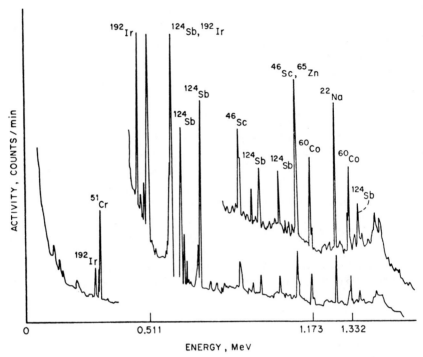

Fig. 7.16 γ-Ray spectrum of glass synthesized from KMC chemicals.

^{124}Sb, ^{46}Sc, ^{65}Zn, and ^{60}Co are present in the glass, but no ^{192}Ir was detected in any of the spectra of the raw materials used in preparing this sample. Samples of platinum from crucibles used in the melting operation were irradiated and substantial activity ($>10^6$ counts/min per 100 mg of Pt) from ^{192}Ir was detected. Gamma-ray spectrometric analysis of glasses melted in an oxygen atmosphere also contained photopeaks of iridium. However, in nitrogen atmospheres less iridium was detected in the glass. Thus, preparation of glasses by melting in a nitrogen atmosphere reduced contamination from platinum vessels.

7.3.1.3.2 Detecting Contamination During Fiber Production. Modified chemical vapor deposition (MCVD) methods are now being used routinely for the production of preforms from which optical waveguide fibers are drawn (MacChesney *et al.*, 1974). In this method suitable volatile compounds of silicon, and/or phosphorus, germanium, and boron are thermally decomposed to form oxide composites inside fused silica tubes. Initially during these investigations fibers failed to meet optical quality specifications and causative factors were sought by performing neutron activation analyses. The most probable suspects were contaminants present initially in the MCVD starting materials and surface contaminants in the silica tube. The source of optical loss was investigated as follows. Two tubes of approximately equal size (0.63 cm × 8.16 cm), one with a thin film of the CVD deposit, TiO_2–SiO_2, and the other without, were cleaned, prepared, and sealed under the condition existing during the preparation of fiber preforms. These tubes and an aliquot of a standard solution containing known amounts of Na, Mn, Cl, and Cu were irradiated simultaneously for 20 min. After the irradiation, sample tubes were immersed in 1:1, HNO_3 and rinsed with water to remove any contaminants from the exterior surface. One end of each sealed tube was then carefully opened and each capsule was filled with the same volume of 1:1 HF and HNO_3. The TiO_2–SiO_2 thin film was completely dissolved during a 5-min equilibration at room temperature. Gamma-ray spectra of these solutions are reproduced in Figs. 7.17(a), (b). The solution containing the dissolved film was contaminated with 1.3 μg of Na, 2.5 μg of Mn, and 0.08 μg of Cu. Quantities of these impurities in the solution from the blank tubes were Na = 0.93 μg, Mn = 3.2 ng, and Cu = 0.15 μg. Thus, excessive amounts of contamination are present already in the tube prior to deposition of the thin film. A qualitative comparison of the photopeaks in spectrum (c) shows the concentration of impurities in the silica tube to be negligible in comparison to the amounts found in the solutions obtained by rinsing the blank and sample tubes. The results indicated that more thorough cleaning of tubes prior to MCVD was needed. Effective procedures for cleaning fused silica tubes serving as containers for samples during long neutron irradiations have proved to induce negligible amounts of impurities into contained samples (Zief and Mitchell, 1976). Similar procedures can be adopted for cleaning fused silica tubes used in MCVD production of waveguides.

In addition to the purity of raw materials and freedom from contamination during the fabrication of bulk glass, it is also necessary to prevent contamination when drawing fibers. On one occasion low-loss SiO_2 fibers (~2 dB/km) were drawn from commercially available pure fused silica and coated with Teflon (FEP). In a subsequent drawing using identical procedures, but with a new furnace, a fiber with a loss over 300 dB/km was produced. Samples of the acceptable and rejected fiber (15.24-cm lengths × 0.016 cm/diameter, sample wt ~40 mg) were activated to identify the impurities responsible for high absorbance losses in the bad fiber. Results (ng/g) for good and bad fibers were Cu, 89 and 311; Mn, 2.2 and 60; and Co <0.1 and 2.2, respectively. The Cr and Fe levels in each fiber were below detection limits of 10 and 200 ng, respectively. No W was found in the low-loss fiber, but 900 ng/g was found in the bad fiber. After stripping the FEP coating from a sample of the high-loss fiber and irradiating it, some of the W could be removed by rinsing the fiber in warm

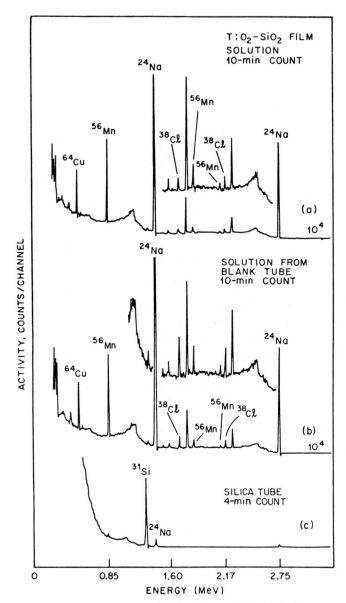

Fig. 7.17 γ-Ray spectra of CVD deposited SiO_2–TiO_2 films.

1:1 HNO_3. This suggested that the impurity was probably located at the surface of the fiber and was most likely deposited during drawing. The walls and heating elements of the furnace were contaminated with these impurities.

7.3.1.3.3 Analytical Methods for Elemental Analysis. Other elements potentially capable of producing absorbance losses in glass and silica include titanium, rhodium, palladium, platinum, and the rare earths. Spectrophotometric methods have been applied for the determination of some of these impurities present in fused silica (Sugarwara and Su, 1975). A general overview of the analysis of glass has been reported (Campbell and Adams, 1969). These investigators preferred optical emission and colorimetry for the determination of elements. Other analytical chemists using these methods met with many difficulties due to the chemical preparation processes essential to spectrophotometry, colorimetry, and atomic absorption spectrophotometry. Nevertheless, a method for the determination of platinum in glass by X-ray fluorescence (Fuller *et al.*, 1971) and the determination of trace metals in glass and raw materials by flameless atomic absorption (Fuller and Whitehead, 1974) have been reported.

Spark source mass spectrometry, a method requiring less sample pretreatment than previously mentioned methods, has been used as an alternative procedure for analyzing glass (Ikeda *et al.*, 1972). Routine determinations of trace elements in glass, ranging from 100 ng/g to several tens of ng/g have been reported. Despite successful applications of various techniques for determining trace impurities in glass and glass-making materials, neutron activation methods in many cases are superior in accuracy and reproducibility since blank problems are reduced substantially. Applications of separations in activation analyses offers an added advantage of specificity. Substoichiometric neutron activation results have been found to be particularly reliable (Kudo *et al.*, 1975), and trace amounts of platinum have been detected after neutron activation and chemical separation (Becker, 1972).

7.3.1.4 Chemical Vapor Deposition (CVD) Reagents

7.3.1.4.1 Silicon Tetrachloride. The effect of the initial purity of commercially available silicon tetrachloride, the primary starting material for preparing optical waveguides by MCVD, on the optical quality of the waveguide was unknown during the early development of optical waveguide MCVD technology. Material scientists therefore used the precaution of experimenting with the most pure $SiCl_4$ available from commercial sources. This starting material is available in several stages of purity ranging from technical to semiconductor grade. Potential metal containing impurities include elements forming reasonably volatile halides, Ti, Sb,

As, Ga, W, and Mo, for example. Other elements forming halide complexes Co, Ni, Cu, In, Sn, Tl, Fe, and Sc may contaminate the reagent during improper storage. Since the volatilities of these species are low compared to that of $SiCl_4$, the reagent can usually be decontaminated during a simple distillation.

Determining trace impurities in $SiCl_4$ by neutron activation requires conversion to the oxide prior to analysis because the physical and chemical properties of the reagent, a fuming, hygroscopic, and highly volatile liquid, preclude a direct irradiation. Successful irradiations were accomplished following the sampling procedure described below (J. W. Mitchell and W. R. Northover, unpublished results, 1972). Ten-milliliter portions of silicon tetrachloride (0.087 moles) were transferred to Teflon beakers and chilled with liquid nitrogen. The samples were than hydrolyzed by dropwise addition from a 100-ml FEP separatory funnel into 20 ml of deionized quartz-distilled water, which had previously been cooled to nearly 0°C. While the mixture was spontaneously reacting, it was stirred for 1 hr and brought to room temperature. After the supernatant liquid was concentrated by evaporation, the gel-like silica precipitate was dispersed and the mixture was frozen on the walls of the Teflon beaker by rotating the container in a dry ice–acetone bath. The beaker containing the product was placed inside a specially fabricated Teflon round bottom flask (500 ml), which was capable of being separated by unscrewing the top and bottom halves. The flask was connected via its standard taper (24/40) joint to a vacuum apparatus and excess water was removed from the precipitate by freeze-drying. The resulting powdered sample was spread on a sheet of Teflon, covered with a Pyrex dome and dried under a heat lamp in a laminar flow hood. Weighed samples of prepared SiO_2 and standards were encapsulated in precleaned quartz vials and irradiated for 100 hr with thermal neutrons. After a 14-day decay period the samples were nondestructively analyzed with a lithium-drifted germanium detector. An examination of γ-ray spectra showed no characteristic photopeaks for ^{59}Fe, ^{60}Co, and ^{51}Cr, indicating concentrations below 500, 3, and 5 ppb, respectively. Determinations of ^{64}Cu and ^{56}Mn after 15-min irradiations were not successful due to interferences resulting from intense radioactivity from decay of ^{38}Cl and ^{31}Si.

X-ray fluorescence analyses were then performed to circumvent difficulties encountered during activation analyses. The solid hydrolyzed samples now had to be chemically dissolved in order that trace impurities could be coprecipitated for X-ray analyses. A special Teflon chamber in which silica matrices could be destroyed by gaseous phase reactions with HF vapors was constructed (Mitchell and Nash, 1974). With this technique contamination was minimized and blank values were precisely controlled at low levels (Zief and Mitchell, 1976).

A survey of materials from various suppliers using the X-ray method gave the results reported in Table 7.8. The results substantiated that storage of the reagent is critically important to purity. Small quantities of reagent (≤ 464 g) stored in glass bottles were purer than materials shipped in steel cylinders, the most widely used storage vessel.

Silicon tetrachloride spontaneously evaporates at room temperature while most of its suspected metal halide impurities have low vapor pressures. Thus, evaporation would appear to offer an attractive way of preconcentrating impurities prior to their analytical determination. However, the extent of simultaneous loss of metal halides via azeotropic-like distillation has not been determined quantitatively. Consequently, initial analyses of this reagent at Bell Laboratories were based on completely hydrolyzing liquid nitrogen cooled samples by mixing with deionized water. Hydrolyzing the reagent, however, provides no preconcentration of impurities and only the most prevalent elements can be determined.

Alternative procedures permitting preconcentration of impurities in $SiCl_4$ prior to analyses are now being developed by applying highly sensitive flameless atomic absorption spectrometry (Kometani, 1976). Evaluations of a variety of procedures for treatment of the sample prior to analysis have been made. Investigations of volatilization losses of iron-doped reagents have shown that a straightforward evaporation of the sample provides a rapid preconcentration step. Extremely low detection limits have been measured for various impurities.

Most experimental data now available suggest that the initial purity of $SiCl_4$ is not a critical factor affecting the optical quality of fibers produced by MCVD (K. Nassau, A. W. Warner, and J. W. Shiever, unpublished results). The desired negligible dependence of the optical quality of the fiber on reagent purity most likely results from the intrinsic prepurification of the reagent due to the vaporization step in MCVD. Since impurities do concentrate in the bubbler, determining whether or not there are maxi-

TABLE 7.8
Coprex X-Ray Fluorescence Analysis of Silicon Tetrachloride[a]

Supplier	Cu	Co	Ni	Fe	Cr
1. Air Products	ND[b]	ND	0.114	1.73	0.528
2. Alpha Inorganics	ND	ND	ND	0.307	0.085
3. Apache Chemicals	ND	ND	0.133	1.76	0.523
4. Atomergic Chem Metals	ND	ND	ND	0.307	0.159
5. Stauffer Chemical	ND	ND	0.432	4.05	1.03
6. Texas Instruments	ND	ND	ND	0.312	0.044

[a] Performed by J. E. Kessler, BTL. Measurements in micrograms per milliliter.
[b] ND = Concentration less than 0.01 μg/ml.

mum impurity levels at which the reagent quality affects the optical quality of the fiber is important for commercial production of fibers by
MCVD. Investigations by atomic absorption and radioisotope techniques
are now in progress to determine the concentration limits at which
impurities are simultaneously volatilized with $SiCl_4$.

The susceptibility of $SiCl_4$ to contamination by molecular species is due
to (1) its exceptionally good solvent properties for many organic materials,
and (2) the formation of hydrolysis and decomposition products when exposed to the atmosphere. Molecular contaminants of $SiCl_4$, particularly
oxygen and hydrogen containing species can be detected conveniently by
ir spectrophotometry (W. G. French, J. P. Luongo, and J. R. Simpson,
1975). Figure 7.18A shows the spectrum of a reasonably pure sample of
$SiCl_4$, and for comparison, Fig. 7.18B shows the spectrum of a commercially available $SiCl_4$ sample identified as a low grade of $SiCl_4$ containing a
relatively high concentration of contaminants. The (B) spectrum in Fig.
7.18 illustrates the additional absorption bands which are attributable to
contaminants. Relative intensities of bands are proportional to the concentration of the impurities identified below.

In Fig. 7.18 the sharp, medium-intensity band at 3680 cm^{-1} results from
the stretching mode of the free (nonbonded) OH group due to $Si(OH)Cl_3$,
a hydrolysis product of the $SiCl_4$ (Rand, 1963). The band at 3360 cm^{-1} is

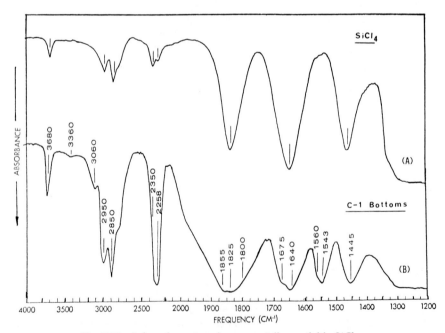

Fig. 7.18 Infrared spectra of commercially available $SiCl_4$.

also assigned to OH groups, but most likely in a polymeric type structure where both the oxygen and hydrogen atoms are involved in bonding (West, 1956). A haloform (CH–Cl) containing contaminant is associated with the band at 3060 cm^{-1} (Robinson *et al.*, 1962). The strongly absorbing bands at 2950/2850 cm^{-1} are due to HCl (Herzberg, 1950). The band in the 2300 cm^{-1} region is composed of a CO_2 band at 2353 cm^{-1} and a Si–H band characteristic of Si(H)Cl$_3$ at 2258 cm^{-1} (Rand, 1963). The broadband near 1850 cm^{-1} is actually composed of three bands. The band head (maximum absorption) at 1855 cm^{-1} is due to Si_2OCl_6; at 1825 cm^{-1} is due to the $SiCl_4$ band and at 1800 cm^{-1} is a band ascribed to $COCl_2$. The band envelope at 1650 cm^{-1} also consists of two bands. The maximum absorption at 1640 cm^{-1} is characteristic of $SiCl_4$ and the shoulder at 1675 cm^{-1} is additional evidence of $COCl_2$.

The band near 1550 cm^{-1} is a combination of two bands due to contaminants. The shoulder at 1560 cm^{-1} is associated with CCl_4 and at 1543 cm^{-1} there is another band due to Si_2OCl_6. The 1445 cm^{-1} band is $SiCl_4$ absorption.

Primary standards containing known amounts of Si(OH)Cl$_3$, SiHCl$_3$, the different haloforms, and the various decomposition and hydrolysis products are not available. Synthetic standards could be prepared but high-purity specimens of the contaminants are not yet available. Consequently, quantitative determinations of molecular impurities have not yet been made by ir. However, the relative intensities of contaminant bands can be used to compare the purity of various samples.

Results for several samples are shown in Table 7.9.

7.3.1.4.2 Other Reagents. Other important CVD reagents for waveguide fabrication include germanium tetrachloride (GeCl$_4$), phosphorus

TABLE 7.9
Relative Amounts of Impurities in SiCl$_4$

Sample[a]	Absorbance			
	Si(OH)Cl$_3$	Haloforms	HCl	SiHCl$_3$[b]
Purified	0.11	0.0	0.24	0.15
Dow high purity	0.27	0.0	0.27	0.60
Dow low purity	0.26	0.70	0.12	4.15
Texas instruments				
epitaxial grade	1.03	0.67	0.14	5.31
Low grade	0.18	0.28	1.14	4.25

[a] Analysis performed on freshly opened sample.
[b] Due to the strong intensity of the SiHCl$_3$ band, the band areas were used to measure the relative concentration of the SiHCl$_3$.

oxychloride ($POCl_3$), phosphorus trichloride (PCl_3), boron trichloride (BCl_3), arsenic trichloride ($AsCl_3$), aluminum trichloride ($AlCl_3$), and di-borane (B_2H_6). Characterizing these reagents by quantitative methods for trace metal impurities can be based on procedures similar to those described previously for $SiCl_4$. As demonstrated earlier for $SiCl_4$, Fe is also the most prevalent impurity in $POCl_3$. Determining ultratraces of iron in $POCl_3$ is complicated by the presence of phosphate. Thus separation techniques are necessary to assay $POCl_3$. An ion exchange separation of iron has been developed for this purpose (Mitchell and Gibbs, 1977).

The analytical chemistry of CVD reagents is complicated by the high reactivity, volatility, and hygroscopicity of the compounds. These physi-cal characteristics cause decomposition and other dynamic changes in the reagent composition due to side reactions. Since standards for the various impurity species as well as pure samples of the by-products are unavaila-ble, preparing synthetic standards and maintaining their stability is not yet accomplished routinely. In spite of these obstacles gas chromato-graphy, mass spectrometry, Fourier transform ir, and laser absorption spectroscopy are powerful tools for identifying molecular impurity species in various CVD reagents. On the other hand, flameless atomic absorption spectrophotometry and Coprex X-ray fluorescence are the most useful tools for the quantitative measurements of trace elemental impurities in volatile CVD halides.

7.3.1.5 Summary.   Gamma-ray spectrometric survey analyses are effec-tive for screening raw materials for fabricating optical waveguides. Examinations of γ-ray spectra and semiquantitative survey analyses indi-cate that most commercially available samples are too insufficiently pure for waveguide applications. However, highly pure samples can be pre-pared successfully under carefully controlled conditions when purification processes are adequately monitored by analyses. Radioisotope techniques and neutron activation analyses have been vital in the development of successful procedures for purifying various reagents from which glass could be subsequently fabricated by melting techniques. These methods are also useful for identifying contamination sources during fiber produc-tion.

7.3.2 Optical Characterization of Glass

Techniques used to measure both absorption losses and scattering losses in bulk glasses are reported in this section.

7.3.2.1 Importance of Bulk Loss Measurements.   As discussed in Chapters 8 and 9, present-day low-loss optical fibers are made by first fa-bricating a preform about 1 or 2 cm in diameter, starting with vapor-phase constituents. No large uniform bulk glass samples are involved. It has proven most effective to evaluate the very low losses which have become feasible by measurement on the fibers drawn from the preforms, and these techniques are described in Chapter 11. This state of the art was only achieved after valuable insight was obtained from more classical optical measuring techniques, which are still very useful. The older fiber fabrication techniques, drawing from a double crucible or from a preform assembled using a rod and tube separately prepared, have been characterized. Although fibers with optical losses approaching those of bulk materials have been made, in some cases the scattering and absorption loss of the fibers are different from those of the bulk materials, as shown by Tynes *et al.* (1971).

There are several reasons why the losses of the fibers could be different from those of the bulk glass from which they were drawn. The most obvious of these is the entrapment of bubbles or foreign particulate matter at the core–cladding interface where their presence can be catastrophic. The effect of such fiber interface defects on scattering loss and the technique used to distinguish them from ordinary Rayleigh scattering has been discussed by Tynes *et al.* (1971).

Fiber scattering loss can also be changed by processes that occur during the drawing process itself. Thus, devitrification or phase separation can occur during the heating of the preform to the drawing temperature. Time exposure photography, as discussed in Section 7.3.2.4, produces a visual comparison of the nature of the scattering centers in both the fiber and the bulk glass. If scattering loss measurements are made in both the fiber and in the parent bulk glass they can be compared quantitatively as described by Tynes (1970, 1972) and Tynes *et al.* (1971). Thus, if the scattering loss in the fiber is different from that in the bulk glass, we have available both qualitative and quantitative techniques that expose the differences.

Absorption losses in fibers can be different from those in bulk glasses for two reasons. First, Kaiser (1973) has observed a drawing-induced coloration in low-water-content silica fibers. See Section 7.2.1 for more details. This absorption is attributed to the formation of color centers that are formed during the fiber drawing process. A similar loss in bulk glass has not yet been reported. Second, the absorption spectrum due to transition metal ions present as impurities depends not only on wavelength and the nature of the glass host itself, but also on the chemical valence of the ion. Whether the oxidation or reduction of impurity ions decreases or increases the absorption in the near-infrared depends on which impurities are present.

Absorption loss measurements are also important as an aid in diagnosing the effect the host glass has on the absorption due to ionic impurities of the transition series elements. Schultz (1974) discusses this problem and lists other pertinent references. One of the important conclusions of such studies is that for a fixed quantity of a given impurity, the absorption is generally greater in a compound silicate glass (e.g., soda-lime-silicate glass) than in a more-or-less pure vitreous silica.

Scattering loss measurements are important because there are glass compositions that vastly reduce scattering loss. Schroeder et al. (1973) have reported the measurement of scattering loss in binary alkali silicate glasses that is lower than that in pure vitreous silica. This reduction in scattering loss might not yet represent the lowest loss attainable. If lower fiber losses are to be obtained it could very well be through simultaneously minimizing the net loss due to scattering and absorption in bulk glass as discussed above. Thus, loss measurements in bulk glass can be a very useful guide for choosing glass compositions that could reduce fiber losses even further.

7.3.2.2 Absorption Loss Measuring Techniques.   Many techniques exist for measuring absorption loss in bulk glass. When light is absorbed in glass it is converted into heat. Therefore, in addition to the temperature rise itself, any physical property of the glass that changes with temperature might be used as a measure of the absorption loss. There are non-thermal loss measuring techniques as well. The most obvious of these is the direct measurement of the attenuation of a beam of light on passing through the glass sample. Another nonthermal measuring technique is to measure the Q, or finesse, of an optical cavity in which the sample is placed. Because the cavity Q depends on the optical losses of the cavity, a measurement of the Q with and without the sample present yields a measure of the sample loss.

The direct attenuation measurement schemes are by far the most widely used, the most versatile, and exist in the greatest variety of forms. To illustrate versatility, we list below some of the measurements that can be performed. They are (1) bulk insertion loss measurement on liquids as well as solids, (2) Rayleigh scattering in pressurized gases, (3) losses in optical fibers, (4) interferometry of glass plates in which an interferometer is inserted in one arm of a two-beam measuring set as described by Tynes and Bisbee (1967), (5) losses in thin films, (6) very small changes in the state of polarization of a light beam, (7) changing reflectivity of freshly deposited optical coatings, (8) reflectivity of dielectric surfaces near the Brewster angle as a means of producing small known attenuations, (9) reflectivity at a gas–glass interface as a function of gas pressure as a means of producing very small known attenuations, and (10) light beam position sensing as described by Tynes (1968).

Direct attenuation measurement schemes are widely varied in form. Thus, there are single- and double-beam measuring sets. Some use one detector, others two; some use ac detection, others dc. Some use broad-band incoherent sources, others narrow-band coherent laser sources. Some use differential recording, others use ratio recording. For all of these variations, however, most of the problems of any one are common to all.

Kao and Davis (1968) describe a single-beam technique that uses a tungsten–halogen lamp as a light source, a monochromator to select the wavelength and bandwidth, and dc detection by means of a multiplier phototube and a digital voltmeter. The technique requires two samples of different thickness and has an accuracy of $\pm 5 \times 10^{-5}$ cm^{-1} when the thicknesses of the samples differ by 20 cm. Two samples are required to offset the surface reflections of about 4% per surface which at normal incidence are large compared to the internal losses of the samples.

In a continuation of Kao's work described above, Jones and Kao (1969) describe a double-beam measuring set which again uses two glass samples whose lengths typically differ by about 20 cm. One glass sample is placed in each arm of the set so that the light is incident on each sample normal to its surfaces. A light chopper and beam splitter combination permit the use of ac detection of the output of a single detector. A tungsten–halogen lamp is used as a light source and filters select the wavelength. The instrument has a balance stability of $\pm 1 \times 10^{-5}$ and has made measurements of attenuation coefficients of 10^{-4} cm^{-1}. An accuracy of $\pm 10^{-5}$ cm^{-1} for samples whose lengths differ by 20 cm is quoted. The useful spectral range is 0.5–1.0 μm. Spectral losses of some low-loss vitreous silica samples are included. In a further continuation of this study, Wright and Kao (1969) studied the variations in surface reflectance of polished surfaces by means of an ellipsometer and concluded that spurious variations in surface reflectance are not a serious problem in evaluating sample losses.

Sell (1970) described a spectrophotometer for measuring both relative and absolute optical reflectance and transmittance (loss). A prismlike rotating transparent refracting chopper was used to produce both the sample beam and the reference beam. This permitted the use of ac detection of the output of a single multiplier phototube used as a detector. A monochromator allowed the wavelength to be varied from 0.2 to 2.0 μm which is the useful range of the instrument. Changes in reflectance and transmittance as small as a few parts in 10^5 can be measured.

Jacobsen et al. (1971) described a single-beam measuring set that uses a conventional light source and a monochromator to select the wavelength. Two glass samples of different length are alternately placed in the light beam and the difference in transmitted intensity is a measure of the differential loss. Instead of allowing the light to fall directly on the detector, a lens converges the light so that it all passes through the entrance port of an Ulbricht sphere that has a second port to permit entry of a photodetector.

Although it is not described, the schematic diagram of their apparatus implies direct dc detection. The accuracy of the measurements is stated to be ±30 dB/km over the wavelength range 0.4 to 2.0 μm.

A balanced two-beam measuring set has been described by Tynes (1972) and some of its uses and characteristics have also been described (Tynes and Bisbee, 1967; Tynes, 1968; Tynes *et al.*, 1971). A schematic diagram of the apparatus is shown in Fig. 7.19. Light from a He–Ne or YAG laser (not shown) is incident on the beam splitter shown on the right side of Fig. 7.19 and is divided into two beams which then pass through the rotating light chopper. The light-chopping sequence is such that the light is directed alternately to the two detectors so that the light is incident on one or the other, but not on both simultaneously. The beam that passes through the glass sample first passes through a polarizer that establishes the proper plane of polarization so that the light incident on the surface of the sample satisfies the Brewster condition. After the light passes through the glass sample it falls on a solar cell detector (operated as a photovoltaic detector) mounted on a x–y micromanipulator so that the light can be made to fall on the same area of the detector as it did before the sample was inserted. To accomplish this a small aperture is placed over the detector and the light beam is scanned in order to locate the beam center pre-

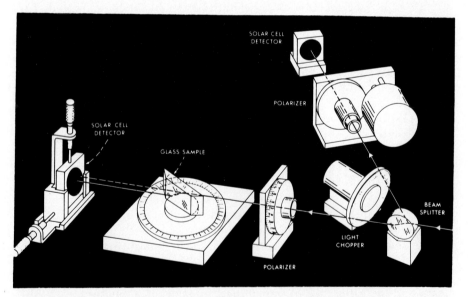

Fig. 7.19 Two-beam optical loss measuring set. Not shown are the laser light source and the first polarizer to the right of the beam splitter and the 0.25-mm-diameter aperture that fits over the solar cell detector in the sample arm of the measuring set so that the beam position can be determined accurately before and after the sample is inserted.

cisely both before and after the sample is inserted. The other beam from the beam splitter passes through a second polarizer used as an attenuator to equalize the intensity of the two beams and then passes on to a fixed solar cell reference detector.

Because of the displacement of the light beam caused by the insertion of the glass sample and because the detectors are the most critical elements in the entire measuring set, the detector must be relocated very precisely in the displaced light beam to assure that the light falls on the same area on the surface of the detector. To minimize any error caused by inaccurate re-location of the detector, only those detectors which have regions on their surfaces where the sensitivity is independent of position of the light beam are selected for use. These flat-response regions are located by scanning the detector across one of the two light beams in the balanced-beam loss measuring set. Changes in sensitivity of 1×10^{-5} can thus be detected. Figure 7.20 shows the results of such a scan (in the x-direction only) and shows that the region of flat response is large compared to the position re-setability error.

The two preselected and matched silicon solar cell detectors are con-nected electrically in parallel. The manner in which the detectors are made to have equal sensitivities has been described by Tynes (1970). Narrow-band phase-sensitive ac detection is used. The minimum sample loss that can be measured is about 4×10^{-5} which for a 5-cm-thick sample amounts to a loss of about 4 dB/km at the two discrete laser wavelengths used (0.633 μm He–Ne and 1.06 μm YAG).

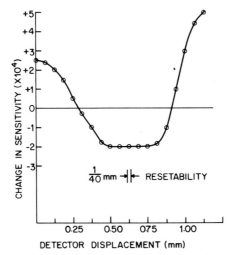

Fig. 7.20 Variation of sensitivity over surface of detector. In the region of flat response, the detector can be reset so as to produce an error of no more than plus or minus one part in 10^5. (Tynes and Bisbee, 1967.)

In the calorimetric measuring technique one measures either the temperature rise of the sample caused by the partial absorption of the light passing through it or measures the distortion of the transmitted beam due to the temperature gradient as was done by Gordon *et al.* (1965) and Solimini (1966). In either case, if lasers are used, there is a distinct advantage to be gained by placing the sample inside the laser cavity where the optical power is much higher. Thus, Dowley and Hodges (1968) measured the losses in ADP and KDP crystals by placing them in an argon ion laser cavity and measuring the temperature rise. Similarly, Kushida and Geusic (1968) measured a refrigeration effect in single crystal YAG while subject to intense 1.064-μm laser irradiation. Rich and Pinnow (1972) later applied the temperature-rise technique to glasses by placing the samples in a YAG laser cavity. Rich reports the measurement of an extinction coefficient measured at 1.06 μm of 0.53 $\pm$ 0.12 $\times$ 10^{-5}/cm (2.3 dB/km) in a commercially available vitreous silica.

In the Q-measuring technique, the sample is placed inside an oscillating-mirror Fabry–Perot interferometer. Because the cavity Q can be defined as the ratio 2π (energy stored)/(energy lost/cycle) it is seen that the sample loss directly affects the cavity Q. Rack and Biazzo (1964) measured the cavity finesse and related this directly to the sample insertion loss which is made up of bulk absorption and scattering as well as a small surface loss. The technique requires extreme mechanical stability, very efficient optical decoupling from the laser cavity, and the narrow spectral linewidth usually available only from laser sources.

7.3.2.3 Calibration of Loss Measuring Sets.   Ideally, various types of loss measuring sets would be calibrated by the use of a set of standard samples whose losses had already been determined. Unfortunately this has not been done and in practice would be difficult to accomplish. This is due partly to the lack of low-loss samples whose losses are accurately known and due partly to the wide range of sample sizes required for the different loss measuring techniques. Thus, the calorimetric loss measuring set requires samples about 1 mm in diameter by 20 mm in length whereas some of the conventional loss measuring sets require samples about 2 cm in diameter by 20 cm in length. Clearly, this latter size is larger than the vessels in which most batches of experimental low-loss glasses are melted.

There are calibration techniques that rather closely simulate the presence of the sample in the loss measuring sets. The techniques that are now described were used by A. R. Tynes and D. L. Bisbee (1965–1972) over a period of several years time and mostly represent the results of unpublished work.

In the calorimetric loss-measuring technique the presence of the sample can be simulated by attaching the temperature-sensing thermocouple to a piece of capillary tubing of the same length and outside diameter as that of the sample and whose inside diameter is the same as that of the laser beam in the sample when the sample is inside the laser cavity. If the capillary is filled with a liquid or wire of known resistivity, then the conducting medium can be heated electrically and simulate very closely the heating effect produced by the absorption of the laser beam in the sample. By this means, the rate of transport of heat across the surface of the sample and across the junction between the sample and the thermocouple can be monitored very carefully. If the thermal coupling of the thermocouple to the sample is a variable of the attachment process, then such variability would be readily apparent.

In the case of the single- and two-beam loss measuring sets, the entire electronic detection system (solar cell detector, narrow-band preamplifier, phase-sensitive lock-in amplifier, and chart recorder) can be calibrated using several techniques, the three most useful of which will now be described.

The multiple-beam calibration has been described by Tynes and Bisbee (1967). In this calibration shown in Fig. 7.21a, three auxiliary light beams of predetermined low intensity are superimposed on the sample beam of the measuring set in such a manner that any of them individually or in combination can be added to one arm of the balanced set. The intensities are adjusted from about 10^{-3} to 50% of that of either of the two balanced beams. The adjustment of these intensities is made using auxiliary electronics independent of the measuring set (e.g., an oscilloscope to observe the pulsed detector output). This calibration shows the measuring set response to be linear and reproducible to within ± 2 parts in 10^5 of the balanced-beam intensity. Figure 7.21b shows the results of such a calibration.

The second calibration uses the very small changes in transmitted intensity that can be produced by passing the light beam through a thick glass plate oriented near the Brewster angle. Very small rotations produce intensity changes of a few parts in 10^5 with ease. No nonlinearities or errors in measuring set performance are indicated.

The third calibration uses a 5-cm-long pressure cell to produce very small changes in intensity of the light beam transmitted through it. As the gas pressure increases, so also does its refractive index. As the refractive index of the gas increases, the reflectivity at the gas-to-glass interfaces of the entrance and exit windows of the pressure cell decreases. The effect can be enhanced by placing one or more thick glass plates inside the pressure cell. The use of materials in the vapor phase (i.e., those whose critical temperatures are above the ambient room temperature) should be avoided because after repeating the pressurizing cycle many times, the slope of the

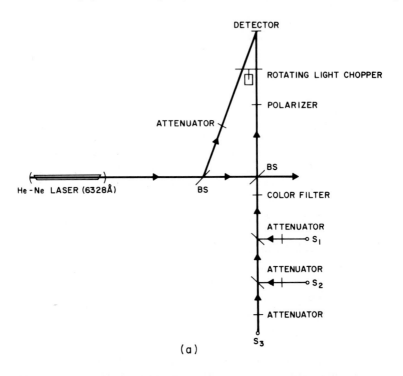

(a)

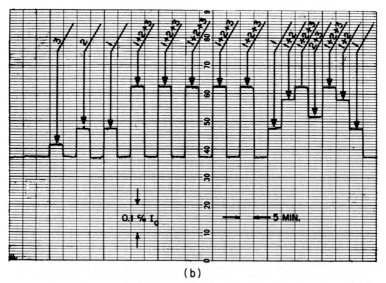

(b)

Fig. 7.21 (a) Schematic diagram of multiple-beam calibration set. S_1, S_2, and S_3 represent three auxiliary incandescent light sources. (b) Results of multiple-beam intensity calibration. (Tynes and Bisbee, 1967.)

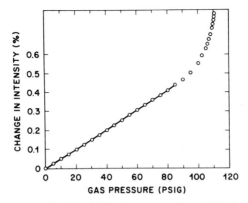

Fig. 7.22 Short pressure chamber calibration curve. One thick plate is placed inside the pressure chamber so there are four gas–glass interfaces for the light beam to pass through. Note the nonlinear behavior at high pressures when the chamber is pressurized with propane. The solid line through the experimentally determined points (circles) is the predicted behavior.

intensity vs pressure curve increases and finally behaves erratically. This appears to be due to a build-up of an adsorbed surface layer on the glass surfaces. The use of gases (critical temperatures below ambient room temperature) eliminates the problem. Intensity changes of the order of 10^{-5} can be made. Again, the calibration shows no nonlinear effects or errors in the measuring set performance. Figure 7.22 shows a calibration curve and demonstrates the sensitivity and usefulness of this calibration technique.

The measuring capabilities of the various loss-measuring techniques are summarized in Table 7.10 for comparison. Unfortunately, not all of the quoted numbers express the same quantity. Thus, some authors quote a minimum measured loss followed by a plus-or-minus quantity which implies a measured accuracy, some quote only the spread in the probable loss which presumably applies to all measured losses large and small, and

TABLE 7.10
Measuring Set Sensitivities

Measuring apparatus	Wavelength range	Sensitivity	Reference
Single-beam	400–1000 nm	5×10^{-5} cm^{-1}	Kao and Davis (1968)
Two-beam	500–1000 nm	10^{-5} cm^{-1}	Jones and Kao (1969)
Two-beam	200–2000 nm	$\delta I \sim 10^{-5}$	Sell (1970)
Single-beam	400–2000 nm	$\pm 7 \times 10^{-5}$ cm^{-1}	Jacobsen *et al.* (1971)
Two-beam	$\lambda = \dfrac{0.633\ \mu m}{1.06\ \mu m}$	$\delta I \sim 4 \times 10^{-5}$	Tynes (1972)
Calorimetric	$\lambda = 1.06\ \mu m$	$\pm 0.12 \times 10^{-5}$ cm^{-1}	Rich and Pinnow (1972)

some quote only the minimum reliable sample loss (independent of thickness) that can be measured. The reader should take note of these differences.

7.3.2.4 Scattering Loss Measuring Techniques. The origins and characteristics of stimulated and ordinary light scattering are discussed in this book in Chapters 5 and 7, respectively, and update the subject matter to the present time. Fabelinskii (1968) has written an extremely valuable book which covers not only the theoretical aspects of both ordinary and stimulated scattering processes, but also covers experimental techniques as well. The many references cited throughout the book greatly magnifies its usefulness. This discussion of measuring techniques will not attempt to be comprehensive, but rather will touch only those points that are important for the measurement of the Rayleigh scattering in bulk glass.

There are three characteristics of Rayleigh scattering that are important from an experimental point of view. They are wavelength dependence of the scattered intensity, angular scattering dependence, and depolarization ratio. For pure Rayleigh scattering, theory predicts that the scattering intensity varies as λ^{-4}. Once this dependence is confirmed for typical glasses, one is justified in extrapolating measurements made in the visible to the infrared region of the spectrum where measurements are difficult. The scattering loss in the infrared region is important for communications purposes.

If the angular dependence of the scattering follows theory, then measurements of the 90° scattering alone can be used to calculate extinction coefficients. If the depolarization ratios are found to be small, then one need not be concerned about the presence of optical elements through which the scattered light passes that have polarization dependent reflective or transmissive properties. Finally, the glass sample must be highly strain-free so that the state of polarization of the light inside the sample volume under observation is precisely known.

Daglish (1970) describes scattering loss measurements on a series of optical glasses including vitreous silica. He uses a commercially available Fica light-scatter photometer for his measurements. He finds that the scattering loss is somewhat steeper than predicted by λ^{-4} but that the angular dependence of the scattering is very close to that predicted by theory. The depolarization ratio was found to range from about 0.05 to 0.33 for the glasses studied. These values are not insignificant and great care should be exercised in interpreting the results of scattering measurements especially for the larger depolarization ratios.

Laybourn et al. (1970) describe the construction and calibration of a light scattering photometer they used to measure scattering in optical glasses. In a continuation of this study, Dakin and Gambling (1972) report on measurements on Schott SSK2 and F7 glasses. They find a λ^{-4} dependence for the scattering and that the angular scattering dependence agrees with theory. The measured depolarization ratio of 0.1 does not create serious problems.

Rank and Yoder (1970) describe a light-scattering photometer and use it to measure polarization properties of the light scattered from optical glasses. Geindre *et al.* (1973) describe a light-scattering photometer used to measure scattering in gases and use a low power ($\sim$3 mW) He–Ne laser as a light source at 0.633 μm and use photon-counting techniques to measure the intensity of the scattered light.

Tynes (1972) used a metallurgical microscope with an apertured photomultiplier placed at the plane where the photographic film is usually located. The aperture when projected backward through the microscope optics measures 10×20 μm and this is the area under observation within the sample. This is depicted in Fig. 7.23. The focused laser beam within the sample typically has a diameter of about 4 μm. If the microscope objective through which the incident laser light passes is translated about 20 μm, the image of the scattered light will fall on the opaque part of the aperture on the front face of the photomultiplier and a reliable background correction will be obtained. If the incident laser light is chopped, one can use conventional narrow-band phase-sensitive detection techniques. The

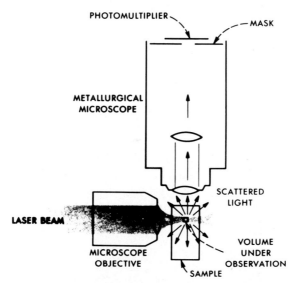

Fig. 7.23 Commercial metallurgical microscope used as a light-scattering photometer. The microscope objective through which the incident laser light passes and the sample are mounted on independent X, Y, Z micromanipulators. (*Bell Laboratories, Record,* November 1972.)

scattered light is sufficiently intense that an integration time in the dc output of the phase-sensitive detector of less than 3 sec is adequate. This permits the background correction to be made in a few seconds. An incident power of 1 mW is sufficient for measurements of the scattered intensity. Scattered light has been measured with as little as 1 μW of incident power although the signal-to-noise ratio is low. The instrument can be calibrated for absolute scattering measurements or one can use a comparison specimen for relative Rayleigh scattering measurements. In either case an entire measurement can be made in a few minutes time. Because of the very small volume under observation (typically 2×10^{-9} cm^3), the sample can be scanned through the focused laser beam and local variations in scattering intensity can be quickly measured or observed visually. Most homogeneous glasses show no local variations. Scattering loss measurements have been made on rods as small as 2 mm in diameter by 2 mm long. Index-matching immersion eliminates the effect of the curved surface and reduces the background caused by the light multiply reflected within the sample.

A time exposure photograph of the laser beam focused inside a glass sample taken solely from the scattered light is shown in Fig. 12 of Tynes *et al.* (1971) and is reproduced here in Fig. 7.24. The discrete scattering centers are clearly visible. For comparison, Fig. 11 of the same reference is shown here in Fig. 7.25 and shows the nature of the scattering centers in the light-guiding region of a fiber with laser light propagating in its core. The scattering centers in the bulk glass and in the fiber core appear to be identical. The section of the fiber shown in Fig. 7.25 is totally devoid of any scattering centers due to defects at the core–cladding interface. Therefore, it seems reasonable to attribute those scattering centers that are visi-

Fig. 7.24 Laser beam focused inside a bulk glass sample photographed by means of scattered light. Original magnification was 200×. (Tynes *et al.*, 1971.)

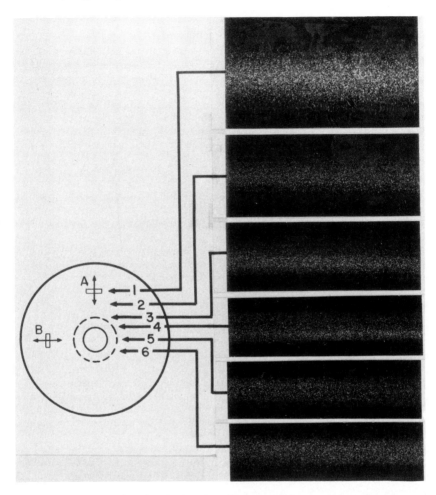

Fig. 7.25 Sequence of uniformly distributed microscopic scattering centers. These scattering centers are believed to be those that produce Rayleigh scattering in glass. Ten-hour exposure at 500× (original magnification) except at level 1 which is 57 hr to emphasize scattering in this region. (Tynes *et al.*, 1971.)

ble to pure Rayleigh scattering. That the scattering centers are not due to the coherence properties of the laser (i.e., laser speckle) is attested to by the fact that whereas the focused laser beam in Fig. 7.24 is highly coherent, the laser light propagating in the multimode fiber shown in Fig. 7.25 is incoherent.

Techniques have been shown to exist which permit both qualitative and quantitative comparisons of the scattering in bulk glass and in light-guiding fibers. Any differences in the characteristics of the scattering can be made readily apparent.

REFERENCES

Adams, R. V., and Douglas, R. W. (1959). Infrared studies on various samples of fused silica with special reference to the bands due to water. *J. Soc. Glass Technol.* **43**, 147T.

Andreev, N. S., Boiko, G. G., and Bokov, N. A. (1970). Small-angle scattering and scattering of visible light by sodium-silicate glasses at phase separation. *J. Non-Cryst. Solids* **5**, 41.

Bagley, B. G. (1974). The nature of the amorphous state. *In* "Amorphous and Liquid Semi-conductors" (J. Tauc, ed.), p. 1. Plenum, New York.

Bagley, B. G., Vogel, E. M., French, W. G., Pasteur, G. A., Gan, J. N., and Tauc, J. (1976). The optical properties of a soda-lime-silica glass in the region from .006 to 22 eV. *J. Non-Cryst. Solids* **22**, 423.

Beales, K. J., Duncan, W. J., and Newns, G. R. (1975). Sodium borosilicate glass for optical fibers. *Proc. Eur. Conf. Opt. Fibre Commun., 1st, 1975* p. 27.

Becker, D. A. (1972). Trace analysis for platinum in glass by neutron activation. *Anal. Chim. Acta* **61**, 1.

Bell, R. J., and Dean, P. (1970). Atomic vibrations in vitreous silica. *Discuss. Faraday Soc.* **50**, 55.

Black, P. W., Irven, J., Byron, K., Few, I. S., and Worthington, R. (1974). Measurements on waveguide properties of GeO_2–SiO_2-cored optical fibers. *Electron. Lett.* **10**, 239.

Bruckner, R., and Navarro, F. (1966). Physikalisch-chemische Untersuchungen in System B_2O_3-SiO_2. *Glastech. Ber.* **39**, 283.

Brugger, K. (1971). Effect of thermal stress on refractive index in clad fibers. *Appl. Opt.* **10**, 437.

Campbell, D. E., and Adams, P. B. (1969). An evaluation of the problem of the chemical anal-ysis of trace coloring oxides in optical glasses. *Glass Technol.* **10**, 29.

Carson, D. S., and Maurer, R. D. (1973). Optical attenuation in titania silica glasses. *J. Non-Cryst. Solids* **11**, 368.

Chynoweth, A. G. (1976). The fiber light guide. *Phys. Today* **29**, No. 5, 28.

Cohen, L. G. (1971). Measured attenuation and depolarization of light-transmitted along glass fibers. *Bell Syst. Tech. J.* **50**, 23.

Cohen, M. H., Fritzsche, H., and Ovshinsky, S. R. (1970). Simple band model for amorphous semiconducting alloys. *Phys. Rev. Lett.* **22**, 1065.

Cormia, R. L., Mackenzie, J. D., and Turnbull, D. (1963). Viscous flow and melt allotropy of phosphorous. *J. Appl. Phys.* **34**, 2245.

Corning Glass Works (1949). "Properties of Selected Commercial Glasses," Bull. B-83. Corning Glass Works.

Corning Glass Works (1969). "Low Expansion Materials," Tech. Bull. LEM. Corning Glass Works.

Daglish, H. N. (1970). Light scattering in selected optical glasses. *Glass Technol.* **11**, 30.

Dakin, J. P., and Gambling, W. A. (1972). Angular distribution of light scattering in bulk glass and fiber waveguides. *Opt. Commun.* **6**, 235.

Dean, P., and Bell, R. J. (1970). A model approach to glasses. *New Sci.* **15**, 104.

Di Domenico, M. (1972). Material dispersion in optical fiber waveguides. *Appl. Opt.* **11**, 652.

Douglas, R. W. (1967). The spectra of Glasses. *Interact. Radiat. Solids, Proc. Cairo Solid State Conf., 1st., 1966* p. 563.

Dow, J. D. (1975). Urbach's rule. *In* "Optical Properties of Highly Transparent Solids" (S. S. Mitra and B. Bendow, eds.), p. 131, Plenum, New York.

Dow, J. D., and Redfield, D. (1971). Theory of exponential absorption edges in ionic and co-valent solids. *Phys. Rev. Lett.* **26**, 762.

Dow, J. D., and Redfield, D. (1972). Toward a unified theory of Urbach's rule and exponential absorption edges. *Phys. Rev. B* **5**, 594.

Dowley, M. W., and Hodges, E. B. (1968). B-1-Studies of high-power CW and quasi-CW parameteric uv generation by ADP and KDP in an argon-ion laser cavity. *IEEE J. Quantum Electron.*, **QE-4**, 552.

Drexhage, M. G., and Gupta, P. K. (1977). A new technique for strengthening glass. *Am. Ceram. Soc., Bull.* **56**, 804 (1977).

English, S., and Turner, W. E. S. (1923). The physical properties of boric oxide containing glasses and their bearing on the general problem of the constitution of glasses. *J. Soc. Glass Technol.* **7**, 155.

Fabelinskii, I. L. (1968). "Molecular Scattering of Light." Plenum, New York.

Faulstich, M., Krause, D., Neuroth, N., and Reitmayer, F. (1975). Highly transparent glasses for producing optical fibers for telecommunication. *Tech. Dig., Top. Meet. Opt. Fiber Transm., 1st, 1975* No. TuB 5-1.

Fleming, J. W. (1978). Material dispersion in lightguide glasses. *Electron. Letts.* **14**, 326–328.

Fuller, C. W., and Whitehead, J. (1974). The determination of trace metals in high purity glasses by flameless atomic absorption spectrometry. *Anal. Chim. Acta* **68**, 407.

Fuller, C. W., Himsworth, G., and Whitehead, J. (1971). The determination of traces of platinum in glass. *Analyst,* **96**, 177.

Gaskell, P. H., and Johnson, D. W. (1976). The optical constants of quartz, vitreous silica and neutron-irradiated vitreous silica. *J. Non-Cryst. Solids* **20**, 171.

Geindre, J. P., Gauthier, J. C., and Delpech, J. F. (1973). Rayleigh light scattering with a low-power He–Ne laser. *Phys. Lett. A* **44**, 149.

Ginther, R. J. (1971). The contamination of glass by platinum. *J. Non-Cryst. Solids* **6**, 294.

Ginther, R. J., and Kirk, R. D. (1971). Luminescence due to impurity traces in silicate glasses. *J. Non-Cryst. Solids* **6**, 89.

Gordon, J. P., Leite, C. C., Moore, R. S., Porto, S. P. S., and Whinnery, J. R. (1965). Long-transient effects in lasers with inserted liquid samples. *J. Appl. Phys.* **36**, 3.

Griscom, D. L. (1977). The electronic structure of SiO_2: A review of recent spectroscopic and theoretical advances. *J. Non-Cryst. Solids* **24**, 155.

Hass, G., and Thun, R. E., eds. (1967). "Physics of Thin Glass Films," Vol. 4. Academic Press, New York.

Hensler, R. J., and Lell, E. (1969). U. V. absorption in silicate glasses. *Proc. Annu. Meet. Int. Comm. Glass* p. 51.

Herzberg, G. (1950). "Spectra of Diatomic Molecules," 2nd ed., p. 53. Van Nostrand-Reinhold, Princeton, New Jersey.

Hood, H. P., and Nordberg, M. E. (1934). U.S. Patent 2,106,744.

Horiguchi, M., and Osanai, H. (1976). Spectral losses of low-OH-content optical fibers. *Electron. Lett.* **12**, 310.

Hyde, J. F. (1942). Method of making transparent silica. U.S. Patent 2,272,342.

Ibach, H., and Rowe, J. E. (1974). Electron orbital energies of oxygen adsorbed on silicon surfaces and of silicon dioxide. *Phys. Rev. B* **10**, 710.

Ikeda, Y., Unayahara, A., Kubota, E., Aoyoma, T., and Watorabe, E. (1972). Trace element analysis of glass by high resolution spark source mass spectrometry. *20th Annu. Conf. Mass., Spectrom. Allied Top., 1972.*

Jacobsen, A., Neuroth, N., and Reitmayer, F. (1971). Absorption and scattering losses in glasses and fibers for light guidance. *J. Am. Ceram. Soc.* **54**, 186.

Jellyman, P. E., and Procter, J. P. (1955). Infrared reflection spectra of glasses. *J. Soc. Glass Technol.* **39**, 173T.

Jones, M. W., and Kao, K. C. (1969). Spectrophotometric studies of ultra low loss optical glasses. II. Double beam method. *J. Sci. Instum.* **2**, 331.

Kaiser, P. (1973). Spectral losses of unclad fibers made from high-grade vitreous silica. *Appl. Phys. Lett.* **23**, 45.

Kaiser, P. (1974). Drawing-induced coloration in vitreous silica fibers. *J. Opt. Soc. Am.* **64,** 475.

Kaiser, P., Tynes, A. R., Astle, H. W., Pearson, A. D., French, W. G., Jaeger, R. E., and Cherin, A. H. (1973). Spectral Losses of unclad vitreous silica and soda-lime-silicate fibers. *J. Opt. Soc. Am.* **63,** 1141.

Kao, K. C., and Davis, T. W. (1968). Spectrophotometric studies of ultra low loss optical glasses. I. Single beam method. *J. Sci. Instrum.* **1,** 1063.

Kapron, F. P., Keck, D. B. and Maurer, R. D. (1970). Radiation losses in glass optical waveguides. *Appl. Phys. Letts.* **17,** 423.

Kapron, F. P., Borrelli, N. F., and Keck, D. B. (1972). Birefringence in dielectric optical waveguides. *IEEE J. Quantum Electron.* **qe-8,** 222.

Keck, D. B., Schultz, P. C., and Zimar, F. (1972). Attenuation of multimode glass optical waveguides. *Appl. Phys. Lett.* **21,** 215.

Keck, D. B., Maurer, R. D., and Schultz, P. C. (1973). On the ultimate lower limit of attenuation in glass optical waveguides. *Appl. Phys. Lett.* **22,** 307.

Kerker, M. (1969). "The Scattering of Light and Other Electromagnetic Radiation." Academic Press, New York.

Koizumi, K., Ikeda, Y., Kitano, I., Furukama, M., and Sumimoto, T. (1974). New light-focusing fibers made by a continuous process. *Appl. Opt.* **13,** 225.

Kolesova, V. A., and Sher, E. S. (1973). Two component glasses of the system GeO_2–SiO_2. *Izv. Akad. Nauk SSSR, Neorg. Mater.* **9,** 909.

Koma, A., and Ludeke, R. (1975). Core-electron excitation spectra of Si, SiO and SiO_2. *Phys. Rev. Lett.* **14,** 107.

Kometani, T. Y. (1976). Analysis of silicon tetrachloride by flameless atomic absorption spectrometry. *Pap., Fed. Anal. Chem. Spectrosc. Soc. Meet.,* 1976.

Krohn, D. A., and Cooper, A. R. (1969). Strengthening of glass fibers: Cladding. I. *J. Am. Ceram. Soc.* **52,** 661.

Kudo, K., Shigematsu, T., Kobayashi, K., and Iso, H. (1975). Substoichiometric determination of Co, Cu and Mn in glass and glass-making materials by radioactivation analysis. *J. Radioanal. Chem.* **24,** 261.

Kurkjian, C. R., and Douglas, R. W. (1960). The viscosity of glasses in the system Na_2O–GeO_2. *Phys. Chem. glasses* **1,** 19.

Kurkjian, C. R., and Paek, U. C. (1978). Effect of drawing tension on residual stresses in clad glass fibers. *J. Am. Ceram. Soc.* **61,** 176.

Kurkjian, C. R., and Peterson, G. E. (1974). Some materials problems in the design of glass fiber optical waveguides. *Proc. Cairo Solid State Conf., 2nd, 1973.* Vol. 2, p. 61.

Kushida, T., and Geusic, J. E. (1968). Optical refrigeration in Nd-doped yttrium aluminum garnet. *Phys. Rev. Lett.* **21,** 1172.

Laybourn, J. R., Dakin, J. P., and Gambling, W. A. (1970). A photometer to measure light scattering in optical glasses. *Opto-electronics* **2,** 36.

Levin, E. M. (1970). Liquid immiscibility in oxide systems. *Phase Diagrams* **3,** 144.

MacChesney, J. B., O'Connor, P. B., Simpson, J. R., and DiMarcello, F. V. (1973). Multimode optical waveguides having a vapor deposited core of germania doped borosilicate glass. *Am. Ceram. Soc., Bull.* **52,** 704 (abstr.).

MacChesney, J. B., O'Connor, P. B., DiMarcello, F. V., Simpson, J. R., and Lazay, P. D. (1974). Preparation of low loss optical fibers using simultaneous vapor phase deposition and fusion. *Proc. Int. Congr. Glass, 10th, 1974* pp. 6–40.

Macedo, P. B., and Litovitz, T. A. (1976). Method of producing optical waveguide fibers. U.S. Patent 3,938,974.

Macedo, P. B., Simmons, J. H., Olson, T., Mohr, R. K., Samanta, M., Gupta, P. K., and Litovitz, T. A. (1976). Molecular stuffing of phasil glass for graded index optical fibers. *Proc. Eur. Colloq. Transm. Fiber Opt., 2nd 1976* p. C1.4.

Maurer, R. D., and Schultz, P. C. (1971). Optical waveguide. Japanese Patent S46(1971)-6423.

Mie, G. (1908). Beiträge zur Optik trüber Medien, speziell koloidaler Metallösungen. *Ann. Phys. (Leipzig)* [4] **25**, 377.

Miller, S. E. (1970). Optical communications research progress. *Science* **170**, 685.

Mitchell, J. W. (1973). Ultrapurity in trace analysis. *Anal. Chem.* **45**, 492A.

Mitchell, J. W., and Ganges, R. (1974). Determination of manganese by substoichiometric neutron activation. *Talanta* **21**, 735.

Mitchell, J. W., and Gibbs, V. (1977). Quantitative ion exchange separation of submicrogram amounts of iron from phosphate medium. *Talanta* **24**, 741.

Mitchell, J. W., and Nash, D. L. (1974). Teflon apparatus for vapor phase destruction of silicate materials. *Anal. Chem.* **46**, 326.

Mitchell, J. W., and Riley, J. E. (1976). Practical application of neutron activation and radio-isotope techniques in optical waveguide research and development. *Proc. Int. Conf. Mod. Trends Act. Anal., 1976.* Paper B324.

Mitchell, J. W., Luke, C. L., and Northover, W. R. (1973a). Techniques for monitoring the quality of ultrapure reagents—Neutron activation and X-ray fluorescence. *Anal. Chem.* **45**, 1503.

Mitchell, J. W., Northover, W. R., and Riley, J. E. (1973b). Analysis of ultrapure sodium and calcium carbonate by neutron activation. *J. Radioanal. Chem.* **18**, 133.

Mohr, R. K., Mukherjee, S. P., and Gupta, P. K. (1977). Strength measurements on abraded fibers with surface compression. *Am. Ceram. Soc., Bull.* **56**, 807.

Mott, N. F., and Davis, E. (1971). "Electronic Processes in Non-Crystalline Materials." Oxford Univ. Press (Clarendon), London and New York.

Münster, A. (1969). "Statistical Thermodynamics," Vol. 1. Springer-Verlag, Berlin and New York.

Napolitano, A., Macedo, P. B., and Hawkins, E. G. (1965). Viscosity determination of boron trioxide. *J. Am. Ceram. Soc.* **48**, 613.

Newns, G. R., Beales, K. J., and Duncan, W. J. (1974). Low loss glass for optical transmission. *Electron. Lett.* **10**, 201.

Nordberg, M. E. (1943). Glass having an expansion coefficient lower than that of silica. U.S. Patent 2,326,059.

Osanai, H., Shioda, T., Moriyama, T., Araki, S., Horiguchi, M., Izawa, T., and Takata, H. (1976). Effect of dopants on transmission loss of low-OH-content optical fibers. *Electron. Lett.* **12**, 549.

Paek, U. C., and Kurkjian, C. R. (1975). Calculation of cooling rate and induced stresses in drawing of optical fibers. *J. Am. Ceram. Soc.* **58**, 330.

Pantelides, S. T., and Harrison, W. A. (1976). Electronic structure, spectra and properties of $4:2$-coordinated materials. I. Crystalline and amorphous SiO_2 and GeO_2. *Phys. Rev. B* **13**, 2667 (1976).

Papp, A., and Harms, H. (1975). Polarization optics of index-gradient optical waveguide fibers. *Appl. Opt.* **14**, 2406.

Payne, D. N., and Gambling, W. A. (1974). New silica-based low-loss optical fiber. *Electron Lett.* **10**, 289.

Payne, D. N., and Gambling, W. A. (1975). Zero material dispersion in optical fibres. *Electron. Letts.* **11**, 176–178.

Pearson, A. D. (1974). Progress in compound glass optical waveguides. *Proc. Int. Congr. Glass, 10th, 1974* p. 6–31.

Pearson, A. D., and French, W. G. (1972). Low loss glass fibers for optical transmission. *Bell Lab. Rec.* **50**, 102–109.

Philipp, H. R. (1966). Optical transitions in crystalline and fused quartz. *Solid State Commun.* **4**, 73.

Pinnow, D. A., Rich, T. A., Ostermayer, F. W., Jr., and DiDomenico, M., Jr. (1973). Fundamental optical attenuation limits in the liquid and glassy state with application to fiber optical waveguide materials. *Appl. Phys. Lett.* **22,** 527.

Pinnow, D. A., Van Uitert, L. G., Rich, T. C., Ostermayer, F. W., and Grodkiewicz, W. H. (1975). Investigation of the soda aluminosilicate glass system for application to fiber optical waveguides. *Mater. Res. Bull.* **10,** 133.

Proctor, B. A., Whitney, I., and Johnson, J. W. (1967). The strength of fused silica. *Proc. R. Soc. London, Ser. A* **297,** 534.

Rack, A. J., and Biazzo, M. R. (1964). A technique for measuring small optical loss using an oscillating spherical mirror interferometer. *Bell Syst. Tech. J.* **43,** 1563.

Rand, M. J. (1963). Purity clamination of silicon and germanium halides by long-path infrared spectrophotometry. *Anal. Chem.* **35,** 2126–2131.

Rank, D. H., and Yoder, P., Jr. (1970). Polarization of light scattered from optical glass. *Mater. Res. Bull.* **5,** 335.

Rau, K., Muhlich, A., and Treber, N. (1977). Progress in silica fibers with fluorine dopant. *Top. Meet. Opt. Fiber Transm., 2nd, 1977* Paper TuC$_4$.

Rich, T. C., and Pinnow, D. A. (1972). Total optical attenuation in bulk fused silica. *Appl. Phys. Lett.* **20,** 264.

Riebling, E. F. (1968). Nonideal mixing in binary GeO_2–SiO_2 glasses. *J. Am. Ceram. Soc.* **51,** 406.

Robinson, C. C., Tare, S. A., and Thompson, H. W. (1962). Intensities of CH vibrational bands in haloforms and methylene halides. *Proc. R. Soc. London Ser. A* **269,** 492–499.

Rongved, L. (1978). Stress in glass fibers induced by the draw force. ASME Paper No. 78-WA/APM-21, to be published in *J. Appl. Mech.*

Rongved, L. (1978). Evaluation of the draw force induced frozen stress in glass fibers. (To be published.)

Schneider, P. M., and Fowler, W. B. (1976). Band structure and optical properties of silicon dioxide. *Phys. Lett.* **36,** 425.

Schroeder, J., Mohr, R., Macedo, P. B., and Montrose, C. J. (1973). Rayleigh and Brillouin scattering in K_2O–SiO_2 glasses. *J. Am. Ceram. Soc.* **56,** 510.

Schultz, P. C. (1974). Optical absorption of the transition elements in vitreous silica. *J. Am. Ceram. Soc.* **57,** 309.

Schultz, P. C. (1977a). Ultraviolet absorption of titanium and germanium in fused silica. *Proc. Int. Congr. Glass, 11th, 1977* Vol. 3, p. 155.

Schultz, P. C. (1977b). Fused P_2O_5 type glasses. U.S. Patent 4,042,404.

Sell, D. D. (1970). A sensitive spectrophotometer for optical reflectance and transmittance measurements. *Appl. Opt.* **9,** 1926.

Shibata, S., and Takahashi, S. (1977). Effect of some manufacturing conditions on the optical loss of compound glass fibers. *J. Non-Cryst. Solids* **23,** 111.

Sigel, G. H. (1971). Vacuum ultraviolet absorption in alkali doped fused silica and silicate glasses. *J. Phys. Chem. Solids* **32,** 2373.

Sigel, G. H., and Ginther, R. J. (1968). The effect of iron on the ultraviolet absorption of high purity soda-silica glass. *Glass Technol.* **9,** 66.

Solimini, D. (1966). Accuracy and sensitivity of the thermal lens method for measuring absorption. *Appl. Opt.* **5,** 1931.

Stanworth, J. E. (1950). Development of glasses transmitting bactericial radiation. *J. Soc. Glass Technol.* **34,** 153.

Starkie, D., and Turner, W. E. S. (1928). The influence of ferric oxide content on the light transmission of soda-lime-silica glass, with special reference to the ultraviolet. *J. Soc. Glass Technol.* **12,** 324.

Stephenson, D. A., and Binkowski, N. J. (1976). X-ray photoelectron spectroscopy of silica in theory and experiment. *J. Non-Cryst. Solids* **22,** 399.

Stroud, J. S. (1962). Color centers in a cerium-containing silicate glass. *J. Chem. Phys.* **37,** 836.

Sugawara, K. F., and Su, Y.-S. (1975). Spectrophotometric determination of ultra-trace amounts of Ti, V, Fe, and Al in fused silica. *Anal. Chim. Acta* **80,** 143.

Tauc, J. (1974). Optical properties of amorphous semiconductors. *In* "Amorphous and Liquid Semiconductors" (J. Tauc, ed.), p. 159. Plenum, New York.

Tauc, J. (1975). Highly transparent glasses. *In* "Optical Properties of Highly Transparent Solids" (S. S. Mitra and B. Bendow, eds.), p. 245. Plenum, New York.

Turnbull, D., and Bagley, B. G. (1975). Transitions in viscous liquids and glasses. *Treatise Solid State Chem.* **5,** 513.

Tynes, A. R. (1968). A novel method for the measurement of small displacements of a light beam and for optical ranging. *Appl. Opt.* **7,** 145.

Tynes, A. R. (1970). Integrating cube scattering detector. *Appl. Opt.* **9,** 2706.

Tynes, A. R. (1972). Measuring loss in optical fibers. *Bell Lab. Rec.* **50,** 302.

Tynes, A. R., and Bisbee, D. L. (1967). Precise interferometry of glass plates. *IEEE J. Quantum Electron.* **QE-3,** 459.

Tynes, A. R., Pearson, A. D., and Bisbee, D. L. (1971). Loss mechanisms and measurements in clad glass fibers and bulk glass. *J. Opt. Soc. Am.* **61,** 143.

Urbach, F. (1953). The long-wavelength edge of photographic sensitivity and of the electronic absorption of solids. *Phys. Rev.* **92,** 1324.

Van Uitert, L. G., Pinnow, D. A., Williams, J. C., Rich, T. C., Jaeger, R. E., and Grodkiewicz, W. H. (1973). Borosilicate glasses for fiber optical waveguides. *Mater. Res. Bull.* **8,** 469.

Weaire, D. (1971). Existence of a gap in the electronic density of states of a tetrahedrally bonded solid of arbitrary structure. *Phys. Rev. Lett.* **26,** 1541.

Weinberg, D. L. (1963a). Absolute intensity measurements in small-angle X-ray scattering. *Rev. Sci. Instrum.* **34,** 691.

Weinberg, D. L. (1963b). X-ray scattering measurements of long range thermal density fluctuations in liquids. *Phys. Lett.* **7,** 324.

West, A., ed. (1956). "Chemical Applications of Spectroscopy," Vol. IX, p. 417. Wiley (Interscience), New York.

Wong, J., and Angell, C. A. (1976). "Glass Structure by Spectroscopy." Dekker, New York.

Wright, C. R., and Kao, K. C. (1969). Spectrophotometric studies of ultra low loss optical glasses. III. Ellipsometric determination of surface reflectances. *J. Sci. Instrum.* **2,** 587.

Yoshida, K., Sentsui, S., Shii, H., and Kuroha, T. (1977). Optical fiber drawing and its influence on fiber loss. *Int. Conf. Integr. Opt. and Opt. Fiber Commun.,* 1977.

Zief, M., and Horvath. (1974). Ultrapure calcium salts. *Laboratory Practice* **175.**

Zief, M., and Mitchell, J. W. (1976). "Contamination Control in Trace Element Analysis," Chapter 2. Wiley (Interscience), New York.

Chapter 8

Fiber Preform Preparation

WILLIAM G. FRENCH
RAYMOND E. JAEGER
JOHN B. MACCHESNEY
SUZANNE R. NAGEL
KURT NASSAU
A. DAVID PEARSON

8.1 INTRODUCTION

We are concerned here with the preparation of preforms for several types of optical fibers (Fig. 8.1). Only one type of bulk glass is required for the single material, plastic clad, and liquid-filled fiber preforms. Two different bulk glasses are required for step index preforms and as feed material for the two-crucible technique. The majority of graded-index preforms are made by vapor deposition techniques, where a bulk glass is used as the initial deposition surface. In some variants, this bulk glass is subsequently removed; in others, it remains as part of the preform as in Fig. 8.1F.

It is convenient to separate bulk glass preparation techniques into two groups. In Section 8.2 are considered relevant multicomponent glasses which can be melted in the range 800–1200°C. The high-silica glasses are not amenable to conventional glass processing techniques because of their refractory nature and are discussed in Section 8.3. Finally, Section 8.4 deals with the various vapor deposition techniques for preparing preforms.

233

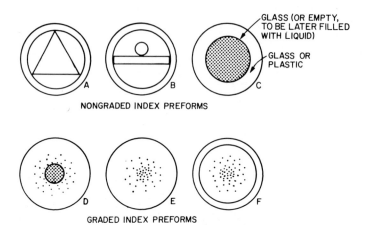

Fig. 8.1 Types of preforms for optical fibers (stippled areas represent a higher refractive index).

8.2 PREPARATION OF MULTICOMPONENT GLASSES AND FIBERS

8.2.1 Introduction

In contrast to the high-silica compositions, multicomponent glasses are usually prepared by fairly conventional melting techniques. The compositions consist of one or more of the glass formers such as silica, boron oxide, or phosphorus pentoxide, modified by the incorporation of such nonglass formers as sodium oxide, calcium oxide, aluminum oxide, etc. These latter oxides, when fused together with silica, for example, change the structure of the silica by causing the rupture of Si–O bonds. This results in large decreases in softening temperature. Thus, these glasses can be melted in crucibles in resistance heated furnaces.

The starting materials for multicomponent glass preparation are either the oxides themselves or compounds which decompose to the oxides during glass melting. For instance, sodium carbonate or sodium nitrate can be used as a source of sodium oxide. Since no methods are presently available for the purification of glass after melting, it is necessary to use very high purity starting materials for glass melting and to carry out all the melting operations without introducing deleterious contaminants which would cause unacceptable absorption loss in the finished glass.

The preparation of the high purity starting materials is in itself a difficult task. Transition metal ion impurities must be reduced to a level of a few tens of parts per billion in most cases. The methods used vary depending upon the material to be prepared, and have included electrolysis, distillation, solvent extraction, and ion exchange. Chemical analysis of the starting materials is also very important as a guide to the purification chemist. At the low impurity levels encountered it is also very difficult. The techniques employed include neutron activation analysis, atomic absorption spectrometry, spark source mass spectrometry, and chemical co-precipitation followed by X-ray fluorescence analysis. These subjects are discussed in detail in Chapter 7.

In addition to maintaining the high purity of the material, the melting techniques must also be capable of producing a glass of good optical quality, free from chemical and optical inhomogeneities, bubbles, inclusions, and unreacted starting material.

The chief sources of contamination are the crucible in which the glass is melted and the furnace refractories which can give rise to airborne dust particles that may become incorporated into the melt. Thus, the choice of crucible material and the type of furnace design are important factors in producing good quality low-loss glass.

The melting techniques which have been employed for making high-purity glasses are fairly straightforward but must be carried out with great care.

After melting has been completed, the glass must be formed so as to be useful for fiber drawing. If the rod and tube technique (see Chapter 9) is to be employed, the glass is allowed to cool to a solid block in the crucible. The rods and tubes of appropriate compositions are then cut and polished from these blocks. For the double-crucible fiber drawing method (see Chapter 9) it is more convenient to have the finished glass in the form of long rods of 5–10 mm diameter. These rods are usually known as "cane." To form them, the crucible of pure glass is placed in a pot furnace at a suitable temperature to hold the glass in a viscous condition. A rod of glass of similar composition to the melt is then lowered mechanically until its end just touches the melt surface and fuses to it. The rod is then slowly raised, and under the proper conditions of temperature and rate, a cane will be drawn directly from the melt surface.

8.2.2 Soda-Lime Silicate Glasses

Methods for melting soda-lime silicate glasses have been described by a number of workers. The technique described by Pearson and French (1972) is typical. The method employed as starting materials powdered silica, sodium carbonate, and calcium carbonate, all of high purity. These dry powders were premixed and then melted in a platinum crucible using R.F. power with the crucible acting as the susceptor. The R.F. melting was carried out in a class 100 clean hood. The crucible containing the crudely melted glass was placed in a high-temperature vertical muffle resistance heated furnace. First, the glass was homogenized by heating to 1500°C and stirring rapidly with a platinum paddle. Large gas bubbles were removed by soaking at 1500°C without stirring, whereupon the bubbles floated to the surface and burst. A slow stirring followed to ensure homogeneity. Microscopic bubbles were removed by lowering the temperature to 1300°C and soaking again. Under these conditions the gas in the bubbles is reabsorbed into the melt.

The crucible and its contents were then removed from the furnace and allowed to cool over a period of a few hours in an insulated enclosure. After cooling, the platinum was cut with a carborundum wheel and peeled away from the glass. The glass block was then annealed on a 24-hour cycle. Bulk glass losses as low as 20–30 dB/km over the range 700–800 nm have been measured on glasses made by this technique (Pearson, 1974). These authors have also used other crucible materials such as alumina and graphite for glass preparation, but no low-loss results were obtained. However, Scott and Rawson (1973a) noted a correspondence between crucible purity and glass loss after melting in both platinum and alumina. Newns *et al.* (1973) also prepared soda-lime glasses in platinum crucibles. The losses observed were not low but a correspondence did exist between the values measured and those calculated from the known impurity contents of the starting materials. They also reported that a gradual increase in impurity content occurs when these glasses are held at fiber drawing temperatures in the range 1180–1380°C in platinum crucibles for long times. Lead-containing multicomponent silicate glasses have also been studied (Faulstich *et al.*, 1975). Total losses of 15 dB/km at 850 nm were reported.

It is evident from a study of the literature cited above that in most cases, the glasses show a higher absorption loss and hence contamination than would be expected from the impurities contained in the starting materials. It usually seemed most likely that the source of this extra contamination was the crucible in which the glass was melted. Therefore, many workers have employed high-purity fused silica crucibles for glass melting. In some cases R.F. power has been employed to melt the glass (Scott and Rawson, 1973a, b). In the molten state, the electrical conductivity of multi-component glasses is high enough so that R.F. power in the megacycle range will couple directly to the melt. The technique employs a water- or gas-cooled silica crucible, so that the layer of premixed starting material next to the crucible walls never melts. Thus the molten batch is contained in a "skin" of its own starting material, and never comes into contact with the crucible. It should therefore be possible to avoid crucible contamination entirely. However, it is not so easy to obtain good optical homogeneity by this method. Since the center of the melt may be at 1500°C and the outside skin at about room temperature, severe temperature gradients

exist which give rise to convection currents in the melt which makes homogenization and bubble removal difficult.

Various workers have drawn clad glass fibers from soda-lime silicate
glasses, and have reported total optical losses of 20 to 60 dB/km (Pearson,
1974; Stewart *et al.*, 1973; Stewart and Black, 1974). In one case (Pearson,
1974) the total optical loss of the fiber was compared to that of the core
glass from which it was drawn (Fig. 8.2). The two curves are almost identical, indicating that a near perfect optical interface between core and cladding has been obtained. Furthermore, the glass had not been measurably
contaminated during the fiber drawing operation, which was done by
the double-crucible method.

The only really low-loss soda-lime silicate fibers which have been reported also contained germanium, lithium, and magnesium oxides (Takahashi and Kawashima, 1977). The starting materials were prepared and
precipitated together from solution in a closed system, and the glasses
were melted in silica crucibles at 1400°C in a dry N_2-O_2 atmosphere. Minimum fiber loss was 4.23 dB/km at 850 nm, and excess loss due to OH was

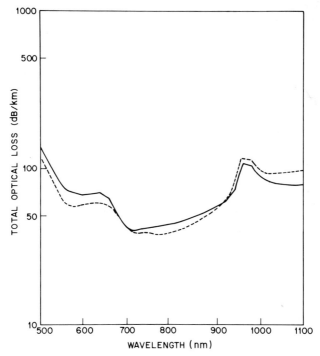

Fig. 8.2 Spectral loss of a clad soda-lime silicate glass fiber (——) compared to that of the
core glass (———) from which it was drawn.

6.5 dB/km at 970 nm. The core diameter and overall diameter of the fiber were 93 and 150 μm, respectively, and the numerical aperture was 0.23. The fibers were drawn from a platinum double crucible at 20 m/min.

8.2.3 Borosilicate Glasses

Over the past few years more work has been done on borosilicate glasses than on the soda-lime compositions. This is because of two factors. The first is that the "optical loss coefficients" (Ikeda *et al.*, 1974) of transition metal ion residual impurities in the borosilicates are generally lower than in the soda-lime glasses. This is particularly true for iron, one of the most deleterious impurities. Thus, in general, a given concentration of iron will cause less loss in a borosilicate than the same concentration would cause in a soda-lime system. The second factor is that the borosilicate glasses are generally lower melting than the soda-limes. At these lower temperatures, both during preparation and fiber drawing, less leaching of impurities from the crucible walls into the glass melt would be expected.

One of the major programs for the study of borosilicate glasses for optical waveguide use has resulted in the development of the Selfoc® fibers at the Nippon Sheet Glass Company. The core glass of these fibers is a sodium borosilicate composition which is doped with thallium oxide (Ikeda *et al.*, 1974; Ikeda and Yoshiyagawa, 1976). The cladding is an undoped sodium borosilicate. The fibers are drawn by the double-crucible technique (see Chapter 9). The nozzle of the core glass crucible was located so that the molten core glass came into contact with the molten cladding glass some distance above the cladding crucible orifice (Fig. 9.4). As the two concentric columns of glass flow toward the cladding orifice, diffusion of thallium and sodium ions takes place across the interface between them. This produces a core with a graded refractive index. The clad fiber was drawn directly from the cladding crucible orifice by a capstan device.

The refractive index profile of the fiber core is governed by the diffusion constants of the diffusing ions, the nozzle length over which the diffusion takes place, the radius of the fiber core and the drawing speed. Graded-index fibers with 60-μm cores and numerical apertures of 0.2 have been made using 30-cm-long nozzles at drawing speeds of 17 m/min. Using a GaAlAs laser input pulse of 0.3-nsec rms. pulse width, broadening was found to be less than 0.5 nsec/km. This is equivalent to a baseband bandwidth of more than 1 GHz/km (Yamazaki and Yoshiyagawa, 1977).

The loss minimum for these fibers is in the wavelength range 800–850 nm and has been gradually reduced from 17 to 20 dB/km (Koizumi *et al.*, 1974; Ikeda *et al.*, 1974) to about 6.7 ± 0.3 dB/km (Yamakazi and Yoshiyagawa, 1977) using incoherent excitation.

Newns *et al.* (1974) and Beales *et al.*, (1975, 1976) have studied the losses of fibers prepared from undoped sodium borosilicates. The core–cladding refractive index difference was achieved in this case by using different compositions within the ternary system for core and cladding. The lowest losses obtained with these fibers were 5 to 7 dB/km at 800 nm (Newns *et al.*, 1977). The glasses were prepared in silica crucibles using bubble stirring. The oxygen partial pressure in the bubbling gas was carefully controlled in order to achieve minimum loss by obtaining the best ratio of oxidized to reduced impurity ions.

In a similar series of experiments (Pearson and Northover, 1978) sodium borosilicate glass has been prepared with a total optical loss of 7 dB/km at 820 nm. These glasses were made in fused silica crucibles in a silica-lined furnace using controlled atmosphere bubble stirring. The glasses were evaluated in the form of 150-μm-diameter fibers clad with low-loss silicone resin.

Sodium aluminoborosilicate glasses have also been melted in platinum crucibles (Imagawa and Ogino, 1977). This was done by preparing starting materials with very high chemical reactivities so that the glass melting reactions could take place at comparatively low temperatures and therefore reduce contamination from the crucible. Double-crucible fibers drawn from these glasses have minimum losses of about 5 dB/km at 840 nm and numerical apertures of 0.3.

Figure 8.3 shows the spectral loss curves of all the "low-loss" multicomponent glass fibers reported to date. The loss minima all occur at about 840 nm and range from 7.2 to 4.2 dB/km. All the fibers show a strong absorption at about 970 nm, which is the second overtone of the OH stretching vibration. Four of the fibers show an absorption at about 730 nm, the third OH overtone. The strong absorption centered at 640 nm in the Takahashi and Kawashima (1977) fiber has not been explained. The shoulder at 900 nm in the Pearson and Northover (1978) fiber probably results from a C–H stretching vibration in the silicone cladding. The remarkably low OH overtone at 970 nm in the Takahashi and Kawashima (1977) fiber indicates a low "water" content. This is attributed to the absence of boron oxide in the glass composition. Fibers made under the same conditions from a similar glass substituted with 10% B_2O_3 showed an excess absorption loss of 41 dB/km due to the OH overtone absorption at 970 nm. The authors explain the difference by noting that B_2O_3 has a greater affinity for water than GeO_2 does.

8.2.4 Alkali Oxide–Silica Glasses

In principle, low-loss fibers can be fabricated with metal oxide-doped cores provided that the vapor pressure of the metal is high compared to

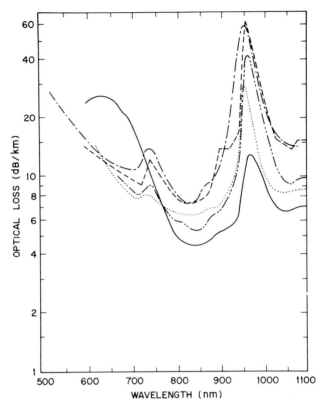

Fig. 8.3 Spectral loss curves for various multicomponent glass fiber waveguides:
– – –Ikeda and Yoshiyagawa (1976); – – – –Imagawa and Ogino (1977); ...... Newns *et al.*
(1977); ——— Pearson and Northover (1978); ——— Takahashi and Kawashima (1977).

that of deleterious contaminants such as iron or copper. In practice, the alkali metals have been examined using a method described by Nagel *et al.*
(1976). It involves the evaporation of the metal into a fused silica tube
under vacuum and subsequent sealing and heating of the tube to form the
metal silicide. The tube is then opened at both ends and heated in a glass
working lathe with a steam of oxygen flowing through it. This forms the
metal oxide which reacts with the silica tube surface to form a binary metal
oxide-silica glass. The tube is then collapsed to a solid rod and drawn into
fiber. Using potassium metal, losses of 10 and 11 dB/km were obtained at
650 and 800 nm, respectively (Nagel *et al.*, 1976).

8.2.5 The Phasil Process

In this technique (Macedo *et al.*, 1976), which is based in part on the
Vycor® process, an alkali borosilicate glass was first melted from reagent

grade starting materials and then pulled into rods. The rods were then heat treated so that phase separation occurred forming two interconnected phases. One phase was silica rich (92–98% SiO_2) while the other contained the alkali oxide and most of the boron oxide. The silica-rich phase was found to be resistant to attack by dilute acids (except HF) whereas the other phase was susceptible to such attack. In addition, transition metal impurities were preferentially soluble in the alkali-rich phase. The rods were therefore leached in dilute acid until all the soluble phase, together with most of the impurities, was dissolved. The silica-rich phase remained in the form of a porous rod.

The rod was soaked in a solution of a suitable dopant until all the pores had been filled. Controlled reverse diffusion was then used to create the cladding layer and/or a graded refractive index profile.

The rods were subsequently heated under vacuum to remove the solvent and dopant compound decomposition products. Further heating to higher temperatures in oxygen caused the porous structure to collapse to a dense glass and diffuse to give a homogeneous structure on the molecular scale.

Fibers drawn from such rods have shown losses as low as about 20 dB/km at 800 nm and a numerical aperture of 0.28 has been reported (Macedo *et al.*, 1976).

8.3 HIGH SILICA BULK GLASSES

8.3.1 Introduction

SiO_2 is unique among glass-forming compositions. Part of this uniqueness derives from its exceptionally high melting temperature, thus requiring preparation techniques distinct from those used for other glasses, and part of it lies in its unusual thermal and mechanical properties. These appear to originate from inefficient packing of SiO_4 tetrahedra and from the nature of the Si–O–Si Bond which has a very low force constant for bending.

Nevertheless, techniques exist to prepare vitreous silica of ultrahigh purity, thus also providing the advantage of the low losses to be expected from a single-component optical medium. Its refractive index can be either increased or decreased by appropriate additions with minimal degradation of its properties. Bulk vitreous silica and high-silica compositions have been used in optical fiber preforms of the various configurations of Figs. 8.1A–D. In addition, fused silica of lower quality is usually used in tube form for the vapor deposition type preforms of Fig. 8.1F.

8.3.2 Vitreous Silica Preparation Techniques

Four major types of vitreous silica have fairly distinct characteristics. Their preparation and some commercially available examples are given in Fig. 8.4. The starting material for all vitreous silica is either natural quartz, or a highly purified volatile silicon compound such as $SiCl_4$ or one of the silanes.

Natural crystal quartz is found widely distributed in Brazil and elsewhere. For "fused quartz" production, it is hand-sorted, crushed, acid-washed, and then fused in one of two ways. Type I vitreous silica is made by fusion of crushed quartz in an electrically heated furnace, usually under vacuum or an inert atmosphere. Some details are given by Dumbaugh and Schultz (1968); most processes of this type are covered by patents. The product is typically very low in water content but has significant aluminum and iron content from the quartz.

Type 2 vitreous silica is made by passing the crushed quartz through a vertical downward-pointing oxygen hydrogen-flame in the Verneuil torch configuration (Verneuil, 1904). A slight reduction of the metallic impurity content results from volatilization in the flame, accompanied however, by a rise in OH content.

Type 3 vitreous silica is made from volatile silicon compounds, such as

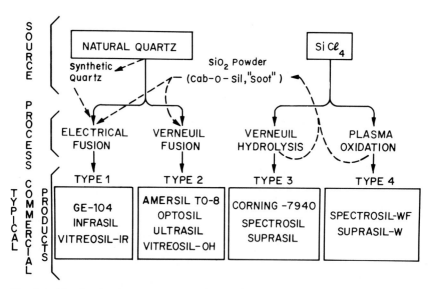

Fig. 8.4 The four major processes used in making bulk vitreous silica. (Cab-O-Sil is made by Cabot Corp; GE-104 by General Electric; Infrasil, Optosil, Ultrasil, and Suprasil by Heraeus; Vitreosil and Spectrosil by Thermal Syndicate; Amersil by Amersil, Inc.; and Corning-7940 by Corning Glass Works.)

$SiCl_4$ or one of the silanes, with hydrolysis in the Verneuil flame:

$$SiCl_4 + 2H_2O \rightarrow SiO_2 + 4HCl.$$

Due to the intimate exposure of the SiO_2 to water vapor in the flame, very high water content vitreous silica results. Since the volatile silica compounds can be purified by fractional distillation and are used in the vapor phase, an extremely low metallic impurity content can be achieved. The same is true in the Type 4 material, where a plasma torch is used to eliminate the water content as described in the next section.

Typical impurity contents of the four types of materials are given in Table 8.1. Similar values have been reported for Soviet domestic vitreous silica (Leko *et al.*, 1974).

Modifications of these four techniques may be used. Thus, a certain purification occurs when synthetic quartz is made from natural quartz; the synthetic quartz can subsequently be vitrified by electrical fusion (R. A. Laudise and E. D. Kolb, unpublished observations, 1975). Particulate amorphous silica can be made from $SiCl_4$ by either the Verneuil or plasma processes, and this also can be fused.

At times, a granularity can be seen in vitreous silica (and also in compound glasses) made from particulate feed. This can be eliminated in some instances by "working" the glass, a process that can be used to improve the optical quality of most fused silica.

8.3.3 Plasma Torch Preparation of Vitreous Silica

The plasma torch (Reed, 1961a, b) in its simplest form consists of a closed fused silica tube with a tangential gas feed surrounded by a radio-frequency coil, as shown in Fig. 8.5A. Starting with a flow of argon (which is easily ionized), a spark-initiated coupling of R.F. energy from the coil (typically 4MHz, 10 to 30kW) produces a plasma, a swirling ball of ionized gas with a temperature near 10,000°C. Oxygen can now be added to the argon.

In the Tafa torch (Humphreys Corp., Tafa Division, Bow, New Hamp-

TABLE 8.1
Typical Impurity Levels of Four Types of Transparent Vitreous Silica[a]

Type of silica	Metals	Cl	OH
1 (Melt—Natl. quartz)	20–200[b]	0	<10
2 (Verneuil—Natl. quartz)	10–100[b]	0	100–500
3 (Verneuil—$SiCl_4$)	<5	100	1000–2000
4 (Plasma—$SiCl_4$)	<5	200	<5

[a] In ppm.
[b] Major impurity is typically 50 ppm Al; Fe is about 1 ppm.

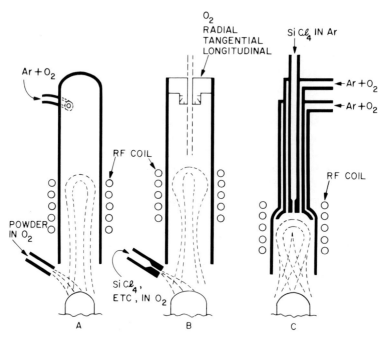

Fig. 8.5 The plasma torch as used for the preparation of vitreous silica or high-silica glasses: in its simplest form (A); as used by Nassau and Shiever (1975) (B); and as used by Kikuchi *et al.* (1975) (C).

shire) three separate gas flows (radial, tangential, and longitudinal) are used to provide easy ignition and maintenance of a stable plasma by controlling the swirling gas flow in a reproducible manner. This torch may be operated with pure oxygen and operating parameters for Type 4 vitreous silica have been described by Nassau and Shiever (1975) in the arrangement shown in Fig. 8.5B.

Feed material may be introduced into the center of a plasma torch from above, or from the side into the hot exit stream. With center feed, there is a tendency for product to deposit on the sides of the torch, and Kikuchi *et al.* (1975) have used the arrangement shown in Fig. 8.5C with a flowing gas curtain near the walls to reduce this tendency.

Semiconductor grade $SiCl_4$, redistilled into a fused silica bubbling flask, is vaporized at constant temperature into a dried oxygen stream which is fed to the plasma torch. The oxidation reaction with the hot ($\sim 2000°C$ at the torch exit) oxygen stream

$$SiCl_4 + O_2 \rightarrow SiO_2 + 2Cl_2$$

is highly exothermic with an equilibrium constant (Audsley and Bayliss, 1969)

$$K_p = \frac{p^2_{Cl_2}}{p_{SiCl_4} \cdot p_{O_2}} \approx 100,000.$$

The vitreous silica deposits as a "boule" on a rotating pedestal which is lowered slowly. The uncorrected optical pyrometer temperature of the boule surface is 1700 to 1800°C.

Detailed studies on bubble formation, boule shape, and water incorporation have been presented for side injection (Nassau and Shiever, 1975). Under optimum conditions both this side injection as well as top injection (Kikuchi *et al.*, 1975) can produce vitreous silica at a 30 to 40% yield containing 5 ppm OH or less. A minimum optical loss of 1.2 dB/km at 1.06 μm was observed (Nassau *et al.*, 1974) and the transition metal impurity content was found by long-time (up to 60 days) neutron activation analysis to be in parts per billion: Cr < 10, Fe $\leq$ 10, Cu $\sim$ 3, Mn $\sim$ 0.4, Co $\sim$ 0.2 (Mitchell and Riley, 1976).

High-silica glasses can be made in the plasma torch in two ways. The other ingredients may be vaporized into the feed gas stream if a sufficiently volatile compound is available and if the resulting oxide does not vaporize at the deposition temperature (as it does in the case of GeO_2). Silica-alumina compositions, as an example, have been prepared in this way (Nassau *et al.*, 1975).

Alternatively, presintered powder has been vitrified in boule form by injection into the exit stream from a plasma torch (R. E. Jaeger, T. J. Miller, and J. W. Fleming, unpublished observations, 1974). The aim was to permit the preparation of bulk high-silica glasses, as above, without significant loss of volatile constituents. These glasses, even in relatively impure form, were required for the characterization of the various compositional candidates used in the CVD processes in terms of their physical and optical properties. The glass feed powder was prepared by "liquid drying" doped aqueous silica slurries (injecting them into a swirling bath of acetone) (Jaeger and Miller, 1974), or by the hydrolysis of miscible organometallic liquids or of the appropriate chlorides (Miller and Sigety, 1975). In this manner homogeneous, reactive, hydrous oxide powders were produced. Loss of volatile components was avoided in the final plasma fusion by prereaction of the powders at relatively low temperatures to form glassy particulate feed.

A number of high-silica glasses have been prepared by these techniques containing varying amounts of GeO_2, B_2O_3, P_2O_5, Na_2O, and Al_2O_3. The properties of some of these compositions are given in Section 7.2.3. The minimum total optical loss of an early unclad fiber prepared from a bulk germanium silicate composition was measured to be $\sim$100 dB/km. This loss level is a combined result of the impurity content of the starting material and the granularity present in the fused glass as mentioned in Section 8.3.2.

8.3.4 Properties of Vitreous Silica

The general properties of pure vitreous silica have been well-described in reviews by Brückner (1970) and Dumbaugh and Schultz (1968). The point has been made that the designation "vitreous silica" does not indicate a specific material with unique properties; the exact nature depends on both the technique of preparation as well as on the thermal history.

The differing impurity contents of the four types of vitreous silica of Table 8.1 lead directly to the spectral loss properties of Fig. 8.6. [The effects of the transition elements on the absorption in vitreous silica have been studied by Schultz (1974).] The relatively high overall loss of Types 1 and 2 originate in the metallic impurities present in the natural quartz starting materials. Types 3 and 4 have much lower overall losses, particularly at longer wavelength in Type 4 with very low water. However, it will be

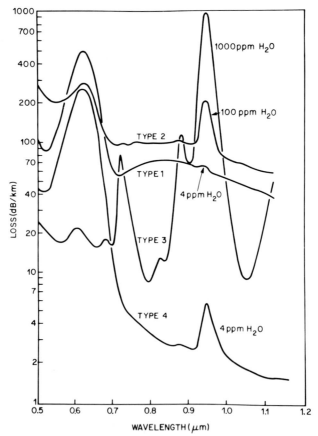

Fig. 8.6 Spectral losses of representative samples of the four types of vitreous silica (Type 1: Infrasil; Type 2: Ultrasil; Type 3; Suprasil 1; Type 4: Spectrosil WF).

noted that the "drawing-induced absorption" near 0.64 μm is most prominent in low water materials. The chlorine present in Types 3 and 4 does not produce losses in the transmission region of vitreous silica. Details of the spectral loss features have been discussed in Section 7.2.1.

The refractive index of vitreous silica is lowered by the addition of B_2O_3 and raised by other additives. Details were given in Section 7.2.3.

Anisotropies in the optical absorption, paramagnetic resonance, and viscosity were observed in vitreous silica made by several processes (Nassau and Shiever, 1976). Oxygen annealing decreased the absorption anisotropy, due to Fe^{2+}, and increased the resonance anisotropy, due to Fe^{3+}, both apparently in a nonisotropic environment similar to a crystal field (presumably originating in the residual stress field). The viscosity anisotropy was observed in deformation experiments and was also present in annealed or worked material.

8.3.5 Uses of Bulk Vitreous Silica

Vitreous silica of relatively low quality (usually Type 2) is used in vapor deposition processes. In the "soot" process (see Section 8.4.2) it is used as the mandrel on which soot is deposited; the mandrel is later removed. In the CVD and MCVD process, vitreous silica is used as the cladding tube onto the inside of which layers of material are deposited from the vapor as described in Section 8.4, leading to a fiber of the type shown in Fig. 8.1F.

Complete waveguides have been made from vitreous silica as "single-material fibers," Figs. 8.1A, B, using Type 3 or 4 for the light guiding members and Type 1 or 2 for the outer tube (Kaiser and Astle, 1974).

Type 4 vitreous silica has also been used as the core rod onto which borosilicate claddings are deposited for subsequent fiber pulling (Dabby et al., 1974) leading to a fiber of the type shown in Fig. 8.1D; alternatively a tube lined with a vapor deposited borosilicate layer has been collapsed onto such a rod either concentrically (MacChesney et al., 1973a) or eccentrically (Miyashita et al., 1974). Finally, vitreous silica rod may be pulled and covered with polymer and fused silica tubing has been pulled and later filled with liquid.

8.4 HIGH-SILICA FIBERS PRODUCED BY VAPOR DEPOSITION METHODS

8.4.1 Introduction

The most successful fiber preparation methods to date have involved deposition from the vapor phase. These methods do not suffer from the contamination problems which plague bulk glass methods based on multicomponent glasses, nor are the refractive index profiles limited to step-index or diffusion-controlled-index profiles of fibers made from bulk glasses.

The vapor deposition processes, described below, share one important feature; the core glass in the waveguide is derived from vapor phase reagents. It is this simple fact which can lead to high-purity, low-loss fibers. Because the glass is made from constituents transported as vapor, contamination results only from gaseous transition metal compounds or particles borne by the gas stream. There are no crucibles or furnaces to contribute impurities. The vapor pressures of the transition metal compounds are typically very low compared to glass-forming reagents, and deposition takes place *in situ*. Thus contamination by either particles or vapor species can be held to very low levels.

Vapor phase deposition techniques are highly effective for preparation of waveguides because iron and similar transition metal chlorides can be easily and effectively removed from $SiCl_4$, $GeCl_4$, and similar starting materials. These materials are commercially available in a highly purified state by virtue of their widespread use in semiconductor processing. A variety of ingenious purification techniques for these chlorides can be credited to H. Theuerer in the early 1940s. These methods are described in his patent (Theuerer, 1963). The materials served as sources for the growth

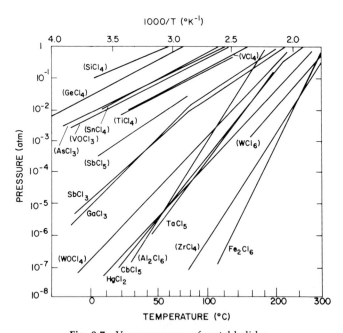

Fig. 8.7 Vapor pressure of metal halides.

of epitaxial silicon as well as silicon single crystals during the late 1940s and 1950s.

One property of the chlorides of silicon and germanium which greatly aids in their purification is the tremendous difference in vapor pressure between these species and those of the transition metal ion chlorides. Figure 8.7 shows vapor pressure temperature curves for selected species. One observes, for instance, that the vapor pressure of $SiCl_4$ is some twelve orders of magnitude greater than that of ferric chloride at room temperature. Liquid $SiCl_4$ is purified by vaporization; commercially obtained material typically contains less than 100 ppb iron. Also, the act of vaporizing the chloride via a carrier gas through the bubbler further reduces the iron concentration in the gas stream (as well as other transition metal contaminants) until presumably only a few parts per billion remain in the final deposited waveguide core.

Another advantage of vapor phase processes is that they allow the use of refractory compositions which cannot be easily prepared in bulk. For example, a silica composition containing 5% germanium would be very difficult to melt and homogenize by conventional glass melting techniques. Such a composition can be made easily, however, by the simultaneous oxidation of SiH_4 and GeH_4 or $SiCl_4$ and $GeCl_4$.

8.4.2 Flame Hydrolysis (Soot) Techniques

The first successful vapor deposition process to be applied to the manufacture of optical waveguides was developed at the Corning Glass Works. Its origins can be traced to work in the early 1930s by Hyde (1942). He patented a flame hydrolysis process intended to "produce articles containing vitreous silica at relatively low temperatures and, if desired, of a high degree of purity." There then followed work by Nordberg (1943) on silica containing TiO_2, which was originally intended as a low expansion coefficient glass. In both cases silica or doped silica particles were formed by introduction of silicon tetrachloride vapor and dopant chloride species into a gas burner flame (flame hydrolysis). These particles were collected and vitrified in a manner so as to form an optical quality glass.

Some 30 years later this same basic process was adapted by Keck and Schultz (1973) to make a waveguide having a core of fused silica doped so as to have a suitable refractive index relative to that of silica cladding. The flame hydrolysis technique was employed (Schultz, 1974) to implement this invention. The most versatile and convenient version uses a burner to generate particulates by hydrolysis of $SiCl_4$, and a dopant-ion-chloride chosen to increase the index of the glass. The burner flame is directed against a mandrel or "starting rod" (graphite, fused silica, or crystalline ceramic) which is rotated in a machine lathe and upon which an adherent but porous layer is formed. A second layer of pure silica or a doped silica, intended to exhibit a lower index than that of the first layer, is subsequently deposited. This serves as cladding of the eventual fiber waveguide.

Upon completion of the second layer, the porous layer is removed from the lathe and introduced slowly into a furnace. Here it is sintered (Flamenbaum *et al.*, 1974) in a dry inert atmosphere at temperatures above 1400°C. The formation of bubbles is avoided by the use of a helium atmosphere because entrapped gases can escape through the unsintered and porous portions of the preform. The mandrel is then removed and the hollow cylindrical preform is drawn into fibers under conditions which cause the central hole to collapse.

Initially, fibers with TiO_2–SiO_2 cores were produced, but titania doping proved to be disadvantageous; this was because at the high temperatures required for drawing, titania tends to reduce, yielding a small concentration of Ti^{3+} which strongly absorbs in the visible and near-infrared. To reduce this ion's concentration, fibers had to be annealed (DeLuca *et al.*, 1974) at temperatures of 800–1000°C. This treatment diminished the fibers' strength. Later fibers used GeO_2 as a dopant to increase the index of the core. These proved to be satisfactory and yielded losses as low as 2 dB/km.

An alternate approach was based upon the discovery by Van Uitert *et al.* (1973) that borosilicate compositions have lower indexes of refraction than that of pure SiO_2. Accordingly Dabby *et al.* (1974, 1975) deposited B_2O_3–SiO_2 particulates on a highly pure SiO_2 rod (Suprasil W, for instance). This was subsequently consolidated to a vitreous layer by sintering at 1000°C in a He atmosphere to yield cladding material. Fibers drawn from such preforms exhibited losses as low as 5 dB/km.

8.4.3 Chemical Vapor Deposition Techniques

The observation that vapor phase processes are inherently clean had been widely applied in the semiconductor industry. Indeed, the initial Bell Laboratories CVD waveguide fabrication experiments were patterned directly after the process used to produce SiO_2 protective films on semiconductor devices.

In a typical application, an easily oxidized reagent, such as SiH_4, is mixed with an oxidizing agent in a gas stream highly diluted with inert gas. This high dilution, together with the relatively low temperatures used, ensures that the oxidation reaction takes place only on the substrate surface (i.e., heterogeneous reaction). Should a gas phase or homogeneous reaction occur, a fine particulate material would form and become a source of defects in the film.

CVD methods were adapted to the formation of optical waveguides. The substrate was normally the inside surface of a fused quartz tube, upon which a higher refractive index glassy film was deposited. After sufficient film thickness was obtained, the tube was collapsed so that the deposited layer formed the high-index core and the tube provided the lower index cladding (see Fig. 8.8).

One adaptation of this method uses GeO_2, obtained from oxidation of germane GeH_4, to increase the index of the silica (MacChesney *et al.*, 1973b; Black, *et al.*, 1974). In another adaptation, fibers with silica cores and low-index borosilicate cladding were made by first depositing a borosilicate layer on the inner surface of a silica tube, then either depositing a silica layer (French *et al.*, 1973) or inserting a pure fused silica rod (MacChesney *et al.*, 1973a) before collapsing the tube.

These methods produced fibers with losses below 10 dB/km. However, they suffered from two major disadvantages: low deposition rates and the use of hydride reactants. The low deposition rate is a direct consequence of the low concentrations and temperatures needed to favor the heterogeneous (wall) reaction over the homogeneous reaction. Attempts aimed at increasing the rate by increasing reactant concentrations resulted in particulate incorporation of the films which caused high loss. An inevitable result of the oxidation of hydrides is the incorporation of OH groups in the deposited film, which therefore absorb light strongly at certain wavelengths, particularly, near 0.95 μm.

8.4.4 Modified Chemical Vapor Deposition

These problems have been overcome by use of a technique, developed at Bell Laboratories, called "modified chemical vapor deposition" (MCVD) to distinguish it from vapor deposition methods derived from semiconductor processing. The earlier CVD process results in deposition on a heated wall due to reaction in the immediate vicinity of the wall. Since

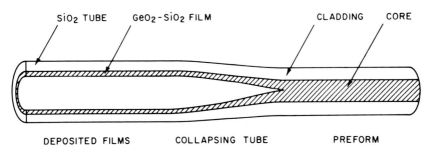

SiO₂ TUBE   GeO₂-SiO₂ FILM   CLADDING   CORE

DEPOSITED FILMS   COLLAPSING TUBE   PREFORM

Fig. 8.8 In the vapor deposition method, the high refractive index core is deposited inside a silica tube. This material becomes the core after the tube is collapsed.

reaction involves solid wall and gas phase precursor reactant, deposition is solely by the thermodynamically preferred heterogeneous route. In accordance with the Bell MCVD process, glass precursor reactants are heated within a tube under conditions such that a substantial part of the reaction is homogeneous—i.e., involves only one phase—in this instance, gas phase material. In accordance with MCVD, particulate material formed within the hot zone is seen to travel downstream and deposit on wall regions removed from the hot zone. In MCVD, as generally practiced, continuous movement of the hot zone results in layer-by-layer fusion of the deposited particles to result in the desired low-loss transparent glassy material. Because required temperatures (1400–1700°C) are usually above the softening point of the substrate tube, the latter is rotated between the synchronous chucks of a glass-working lathe as illustrated in Fig. 8.9. The rotating tube is heated by a flame which traverses its length. Reaction takes place in the hot zone of the torch, and a glassy film, from heterogeneous reaction, is deposited in this region. In addition, particulate material obtained from homogeneous reaction is deposited downstream from the hot zone. This material is fined to a vitreous deposit on a layer-by-layer basis as the torch moves along the tube. The torch is repeatedly traversed until sufficient deposit thickness is obtained. (This procedure differs from the flame hydrolysis process where vitrification occurs by a separate sintering step after deposition is completed.)

The increased reactant concentrations permitted by this method in-

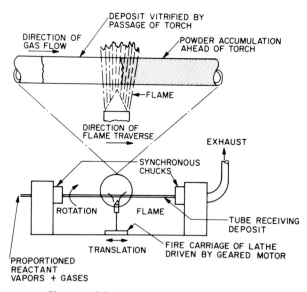

Fig. 8.9 Schematic of the MCVD process.

creased the deposition rates over that of CVD by more than a hundred-
fold, Furthermore, the high temperatures allow use of chloride reactants
such as $SiCl_4$:

$$SiCl_4 + O_2 \xrightarrow{>1200°C} SiO_2 + 2Cl_2$$

and

$$GeCl_4 + O_2 \rightarrow GeO_2 + 2Cl_2.$$

Use of halide reactants eliminates the large OH concentrations found in
films prepared by oxidation of hydrides or by hydrolysis reactions.

Several of the material composition systems which have been used in
the MCVD technique are listed in Table 8.2. The fabrication techniques for
each of them are similar. For example, in the fabrication of multimode
fibers having GeO_2–SiO_2–B_2O_3 cores and SiO_2 cladding (MacChesney,
1974; MacChesney et al., 1974; French et al., 1974), preparation begins by
mounting a commercial grade fused quartz tube in a glass working lathe.
Silicon and germanium tetrachloride vapors are transported to the tube
using a stream of oxygen bubbled through the chloride liquids. Boron
trichloride, which is gaseous at room temperature, is mixed with the other
constituents. The reaction is initiated by heating the tube with a moving
oxyhydrogen burner. As described above, the reaction produces both a
vitreous germania-doped borosilicate film and a stream of similarly doped
particles which accumulate on the downstream walls of the fused quartz
tube. These particles are fused to a clear glassy film as the torch moves
along the length of the tube. After sufficient film thickness is obtained, the
reactants are shut off, and the tube is heated strongly to cause it to collapse
into a solid rod. The deposited material then forms the core of the eventual
fiber while the fused quartz tube forms the lower index cladding.

Using this procedure, fibers having a graded index with numerical
apertures (NA) in the range 0.18–0.23 have been made for use in long-

TABLE 8.2
Material Composition Systems Used for Fibers

Core:Clad	Maximum NA	Minimum loss (dB/km)	Mode dispersion (nsec/km)
SiO_2–GeO_2–B_2O_3:SiO_2	0.4	1.3	0.5
SiO_2:SiO_2–B_2O_3	0.17	0.9	0.13
SiO_2–P_2O_5:SiO_2–B_2O_3	0.2	0.5	
Multicomponent glasses (e.g., SiO_2–B_2O_3–Na_2O)	0.2	5	1.0
SiO_2:Silicone	0.6	3	19

distance transmission. Preforms yielding multikilometer lengths of such a multimode fiber can be made in about 2 hr. The optical losses of these fibers are consistently low, as indicated in Fig. 8.10a. Typically, losses are below 5 dB/km at 850 nm and 2 dB/km at 1060 nm (DiMarcello and Williams, 1975).

In addition to the above fibers, a second class, intended for data bussing applications, can be made by the MCVD method. These fibers are aimed at switching applications where cost and reliability considerations require the use of LEDs as light sources. They have higher NAs (0.35–0.4), and larger cores (up to 125-μm diameter) in order to couple more efficiently to the incoherent light source. The fibers are prepared by the same procedure, except that the temperature of the preform is maintained above that of the glass transition temperature of the vitreous core deposit until final collapse to the rod preform. In this way, severe strain caused by thermal expansion mismatch between the highly doped core deposit and silica tube is avoided. The loss spectrum of such a fiber is shown in Fig. 8.10b (O'Connor *et al.*, 1976). Losses are somewhat higher than those of the other fibers shown. These are inherently due to higher scattering losses which occur because of concentration-induced fluctuation of the refractive index resulting from the higher doping level.

The OH absorption at 950 nm is small for most fibers and it is negligible in the case of certain of the high NA fibers. This is the result of heat treatment of the silica tube prior to core deposition so as to remove much of the incorporated OH. Additional improvement results from reduction in the

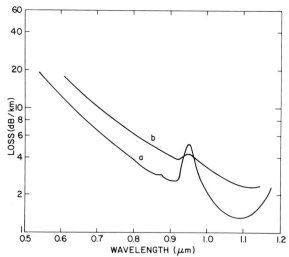

Fig. 8.10 Loss spectra of (a) a graded-index fiber with a $GeO_2 \cdot B_2O_3 \cdot SiO_2$ core and NA of about 0.2, and (b) a high NA (0.35) step-index fiber. Each fiber was made using the MCVD process.

quantity of hydrogen-bearing impurities in the deposition atmosphere, i.e., trichlorosilane impurities in $SiCl_4$ or leakage of water vapor from the atmosphere into the reactant stream.

Fibers comprising silica core and borosilicate cladding have been made by the same procedure (French *et al.*, 1976; Tasker and French, 1974). First, a borosilicate cladding is formed within a silica tube by reaction of boron trichloride and silicon tetrachloride and oxygen. Then, a silica core layer is formed by reaction of silicon tetrachloride and oxygen. Fibers drawn from such a preform have exhibited losses as low as 1.8 dB/km at 860 nm and less than 1 dB/km at 1060 nm. Unlike germania- or phosphorus-oxide-doped fibers, there is no problem with evaporation of the dopant during preform processing. This is advantageous in refractive index profile control, but the refractive index difference between the silica core and borosilicate cladding limits numerical apertures to less than 0.17.

The borosilicate composition is especially useful for the fabrication of single-mode fibers because, for these fibers, the cladding must have as low a loss as the core. Since the borosilicate cladding is made by the MCVD process, it has much lower loss than the fused quartz which is normally used as cladding for germania-doped fibers (French and Tasker, 1975; Tasker *et al.*, 1978). These single-mode fibers have been reproducibly fabricated with losses of less than 2 dB/km at 1060 nm and 3 dB/km at 850 nm.

Another materials system which has been shown to yield low-loss fibers is that of phosphosilicate, first reported by Payne and Gambling (1974a). The phosphosilicate core is formed inside a silica tube by oxidation of phosphorus oxychloride and silicon tetrachloride. Of particular interest is the broad low-loss window extending from 800 to 1300 nm (Payne and Gambling, 1976) exhibited by these fibers.

P_2O_5-doped silica core fibers with borosilicate claddings have achieved the lowest losses yet reported (Horiguchi and Osanai, 1976). The minimum loss at 1.2 μm (0.5 dB/km) is in a wavelength region where material dispersion is very low. When similar fibers can be routinely manufactured, and when optical transmitters and receivers in this wavelength region are developed, very high bandwidth systems with long repeater spacings will be possible. However, the use of phosphosilicate for commercial production is complicated by its increased sensitivity to iron contamination. Apparently iron is incorporated into such glass in the divalent state (Gambling *et al.*, 1976) which absorbs strongly in the infrared.

8.4.5 Graded-Index Fibers

Because the cores of fibers made by vapor deposition processes are built up in many layers, it is very simple to control the refractive index profile in a stepwise approximation to the power law profile which minimizes dispersion. Each of the methods described above has been used in this way.

An extensive examination of the dispersion dependence on index profile was carried out in the SiO_2–B_2O_3 system, and is illustrative of the method (French *et al.*, 1976; Cohen, 1976). In this study, the profile of the SiO_2–B_2O_3 core was controlled in the MCVD method by decreasing the BCl_3 concentration in the gas stream after each pass of the torch. The results of these experiments are shown in Fig. 8.11 as mode dispersion as a function of the refractive index profile parameter g, where

$$\Delta n = \Delta n_0[(1 - (r/a)^g]$$

describes the profile. As expected, there is a sharp minimum in the dispersion as an optimal g value is approached. The optimum value in this system, $g = 1.77$, is in excellent agreement with calculations based on Olshanski and Keck's (1976) theory when used with independent refractive index measurements (Fleming, 1976; Gloge *et al.*, 1975). The minimum

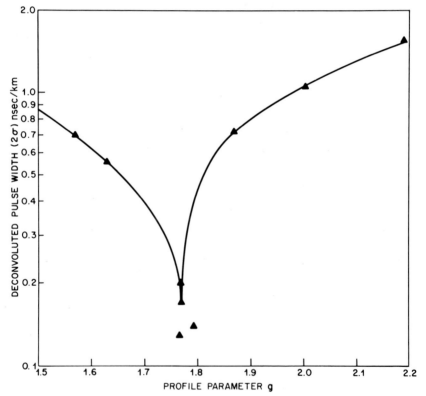

Fig. 8.11 Mode dispersion for a series of graded-index B_2O_3-SiO_2 fibers versus the profile parameter g.

dispersion in this system is about 130 psec/km, a factor of about 100 reduction from a step-index fiber with the same 0.15 NA (Cohen *et al.*, 1978).

8.4.6 Variants to the MCVD Technique

In an effort to provide alternate deposition methods, the MCVD technique has been modified to initiate the oxidation reactions by plasmas rather than external heating sources. The first efforts along these lines utilized a low-pressure microwave plasma (Küppers and Koenings, 1976; Geittner *et al.*, 1976). Because the silica tube does not have to be heated with the plasma, the reaction zone can be traversed rapidly. This can be applied to produce up to 2000 thin GeO_2–B_2O_3–SiO_2 layers. The resulting "smooth" profile exhibits improved mode dispersion. Losses as low as 1.4 dB/km at 1.06 μm are reported. Deposition rates approach those of MCVD but the deposition efficiency is higher, approaching 100%.

Another plasma process uses an RF heated argon plasma at 1 atm pressure (Jaeger *et al.*, 1978). To date losses are somewhat higher (6 dB/km at 1.06 μm) than those of either microwave plasma or ordinary MCVD. However, a three-fold increase in deposition rate over that of MCVD and near 100% efficiency was observed by this method.

Higher deposition rates using conventional MCVD have also been reported (Akamatsu *et al.*, 1976; O'Connor *et al.*, 1977). Nearly a fivefold increase over MCVD is permitted by adding He to the deposition atmosphere. The increased rate is alleged to result from improved fining of the particulate layer which accumulates downstream from the torch. Optical properties of fibers formed using He appear equivalent to those obtained by MCVD without the addition.

8.4.7 Summary

The current state of the art of fiber properties can be summarized. Low losses (<10 dB/km) can be attained by both multicomponent glass and high-silica fiber. Those multicomponent glass fibers exhibiting low loss generally have NAs of less than 0.2. However, they can be made in the step-index version with large cores for use with LED light sources. Graded-index profiles can be attained but it is still questionable whether fiber can be manufactured with optimal gradings to reduce pulse delay by a factor of 50 below that of the step-index fibers. Silica core–silicone clad fibers can be fabricated with NAs up to approximately 0.4, and losses below 10 dB/km for wavelengths from approximately 0.80 to 0.85 μm. They cannot be graded, and questions about their susceptibility to environmental attack are unresolved. To date fibers having germania-doped silica cores and silica or borosilicate claddings made by flame hydrolysis have generally exhibited NAs below 0.2. They exhibit low losses, have exhibited low dispersion, and when plastic coated are assumed to be relatively free from atmospheric attack. Graded-index silica borosilicate fibers produced by MCVD exhibit low loss and greatly reduced mode dispersion (150 psec/km), but are limited to NAs below 0.18. Higher NA fibers (up to 0.4) can be produced by MCVD; they consist of a GeO_2–SiO_2 core composition with small amounts of B_2O_3 or P_2O_5, and SiO_2 cladding. They exhibit losses generally below 5 dB/km in the 0.85 to 1.06-μm region. Grading of such fibers can be quite good (below 1 nsec/km impulse delay).

For the highest data rate systems, single-mode fibers could be used. The MCVD method has been successfully applied to the fabrication of single-mode fibers of borosilicate composition with losses of less than 2 dB/km.

REFERENCES

Akamatsu, T., Okamura, K., Ueda, Y., Inoue, K., and Unotoro, T. (1976). High deposition rate CVD method with helium gas. *Londen Elec.* **56,** 602.

Audsley, A., and Bayliss, R. K. (1969). Induced plasma torch as a high temperature chemical reactor. *J. Appl. Chem.* **19,** 33.

Beales, K. J., Duncan, W. J., and Newns, G. R. (1975). Sodium borosilicate glass for optical fibres. *Proc. Eur. Conf. Opt. Fiber Commun. 1st,* 1975 IEEE Conf. Publ. No. 132, p. 27.

Beales, K. J., Day, C. R., Duncan, W. J., Midwinter, J. E., and Newns, G. R. (1976). Preparation of sodium borosilicate glass fiber for optical communication. *Proc. IEE* **123,** 591.

Black, P. W., Irven, J., Byron, K., Few, I. S., and Worthington, R. I. (1974). Measurements on waveguide properties of GeO_2–SiO_2 cored optical fibres. *Electron. Lett.* **10,** 239.

Brückner, R. (1970). Properties and structure of vitreous silica. *J. Non-Cryst. Solids* **5,** 123 and 177.

Cohen, L. G. (1976). Pulse transmission measurements for determining near optimal profile gradings in multimode borosilicate optical fibers. *Appl. Opt.* **15,** 1808.

Cohen, L. G., DiMarcello, F. V., Fleming, J. W., French, W. G., Simpson, J. R., and Weissmann, E. (1978). Pulse dispersion properties of fibers with various material constituents. *Bell Syst. Tech. J* **57,** 1653.

Dabby, F. W., Pinnow, D. A., Ostermayer, F. W., Van Uitert, L. G., Saifi, M., and Camlibel, I. (1974). Borosilicate clad fused silica core fiber optical waveguide with low transmission loss prepared by a high efficiency process. *Appl. Phys. Lett.* **25,** 714.

Dabby, F. W. *et al.* (1975). A technique for making low loss fused silica core-borosilicate clad fiber optical waveguides. *Mater. Res. Bull.* **10,** 425.

DeLuca, R. D., Keck, D. B., and Maurer, R. O. (1974). Heat treating optical waveguides for OH ion removal. U.S. Patent 3,782,914.

DiMarcello, F. V., and Williams, J. C. (1975). Reproducibility of optical fibers prepared by a chemical vapor deposition process. *Proc. Eur. Conf. Opt. Fibre Commun. 1st,* 1975 IEEE Conf. Publ. No. 132.

Dumbaugh, W. H., and Schultz, P. C. (1968). Vitreous silica. *Kirk-Othmer Encycl. Chem. Technol.,* Vol. 18, p. 77.

Faulstich, M., Krause, D., Neuroth, N., and Reitmayer, F. (1975). Highly transparent glasses for producing optical fibers for telecommunication. *Tech. Dig., Top. Meet. Opt. Fiber Trans., 1975,* p. TuB5-1.

Flamenbaum, J. S., Schultz, P. C., and Voorhees, F. W. (1974). Method for producing high quality fused silica. U.S. Patent 3,806,570.

Fleming, J. W. (1976). Material and mode dispersion in $GeO_2 \cdot B_2O_3 \cdot SiO_2$ glasses. *J. Am. Ceram. Soc.* **59**, 503.

French, W. G., and Tasker, G. W. (1975). Fabrication of graded index and single mode fibers with silica cores. *Proc. Top. Meet Opt. Fiber Transm. 1st, 1975* p. TuA2-1.

French, W. G., Pearson, A. D., Tasker, G. W., and MacChesney, J. B. (1973). A low loss fused silica optical waveguide with borosilicate cladding *Appl. Phys. Lett.* **23**, 338.

French, W. G., MacChesney, J. B., O'Connor, P. B., and Tasker, G. W. (1974). Optical waveguides with very low losses. *Bell Syst. Tech. J.* **53**, 951.

French, W. G., Tasker, G. W., and Simpson, J. R. (1976). Graded index fiber waveguides with borosilicate composition: Fabrication techniques. *Appl. Opt.* **15**, 1803.

Gambling, W. D., Payne, D. N., Hammond, C. R., and Norman, S. R. (1976). Optical fibers based on phosphosilicate glass. *Proc. Inst. Electr. Eng.* **123**, 570.

Geittner, P., Küppers, D., and Lydtin, H. (1976). Low loss optical fibers prepared by plasma-activated chemical vapor deposition (CVD). *Appl. Phys. Lett.* **28**, 645.

Gloge, D., Kaminow, I. P., and Presby, H. M. (1975). Profile dispersion in multimode fibers: Measurement and analysis. *Electron. Lett.* **11**, 469.

Horiguchi, M., and Osanai, H. (1976). Spectral losses of low-OH content optical fibers. *Elect. Lett.* **12**, 310.

Hyde, J. F. (1942). Making a transparent article of silica. U.S. Patent 2,272,342.

Ikeda, Y., and Yoshiyagawa, M. (1976). "Development of low-loss glasses for SELFOC *fibers.* *Proc. Eur. Conf. Opt. Fiber Commun., 2nd, 1976* p. 27.

Ikeda, Y., Yoshiyagawa, M., and Furuse, Y. (1974). Low-loss glasses for light-focusing fiber waveguides. *Proc. Int. Congr. Glass, 10th, 1974* Vol. 6, p. 82.

Imagawa, H., and Ogino, N. (1977). Low-loss silicate glass optical fibers with small (0.15) and medium (0.3) numerical apertures. *Tech. Dig., Int. Conf. Integr. Opt. Opt. Fiber Commun., 1977* p. 613.

Jaeger, R. E., and Miller, T. J. (1974). Preparation of ceramic oxide powders by liquid drying. *Am. Ceram. Soc., Bull.* **53**, 855.

Jaeger, R. E., MacChesney, J. B., and Miller, T. J. (1978). The preparation of optical waveguide preforms by plasma deposition. *Bell Syst. Tech. J.* **57**, 205.

Kaiser, P., and Astle, H. W. (1974). Low-loss single-material fibers made from pure fused silica. *Bell Syst. Tech. J.* **53**, 1021.

Keck, D. B., and Schultz, P. C. (1973). Method of producing optical waveguide fibers. U.S. Patent 3,711,262.

Kikuchi, B., Okamura, K., and Arima, T. (1975). Preparation of pure silica bulk for optical fiber. *Fujitsu Sci. Tech. J.* **11**, 99.

Koizumi, K., Ikeda, Y., Kitano, I., Furukawa, M., and Sumimoto, T. (1974). New light-focusing fibers made by a continuous process. *Appl. Opt.*, **13**, 255.

Küppers, D., and Koenings, J. (1976). Preform fabrication by deposition of thousands of layers with the aid of plasma activated CVD. *Proc. Eur. Conf. Opt. Fibre Commun. 2nd, 1976* p. 49.

Leko, V. K., Meshcheryakova, E. V., Gusakova, N. K., and Lebedeva, R. B. (1974). "Investigation into the viscosity of domestically manufactured fused silica. *Sov. J. Opt. Technol. (Engl. Transl.)* **41**, 600.

MacChesney, J. B. (1974). Preparation of low loss optical fibers using simultaneous vapor phase deposition and fusion. Proc. Int. Congr. Glass, 10th, 1974 Vol. 6, p. 40.

MacChesney, J. B., Jaeger, R. E., Pinnow, D. A., Ostermayer, F. W., Rich, T. C., and Van Uitert, L. G. (1973a). Low loss silica core-borosilicate clad fiber optical waveguide. *Appl. Phys. Lett.* **23**, 340.

MacChesney, J. B., O'Connor, P. B., Simpson, J. R., and DiMarcello, F. V. (1973b). Multimode optical waveguides having a vapor deposited core of germania doped borosilicate glass. *Am. Ceram. Soc., Bull.* **52**, 704.

MacChesney, J. B., O'Connor, P. B., and Presby, H. M. (1974). A new technique for preparation of low-loss and graded-index optical fibers. *Proc. IEEE* **62**, 1280.

Macedo, P. B., Simmons, J. H., Olson, T., Mohr, R. K., Samanta, M., Gupta, P. K., and Litovitz, T. A. (1976). Molecular stuffing of phasil glasses for graded index optical fibers. *Proc. Eur. Conf. Opt. Fibre Commun., 2nd 1976* p. 37.

Miller, T. J., and Sigety, E. A. (1975). Chemical preparation of multicomponent glass powders for plasma fusion. *Am. Ceram. Soc., Bull.* **54**, 430.

Mitchell, J. W., and Riley, J. E. (1976). Practical application of neutron activation analyses in optical waveguide research and development. *Proc. Int. Conf. Mod. Trends Activ. Anal., 1976* **2**, 1348.

Miyashita, T., Edahiro, T., Horiguchi, M., Masuno, K., Tokimoto, T., and Isawa, J. (1974). Vitreous silica optical fibers. *Proc. Int. Cong. Glass. 10th, 1974* Vol. 6, p. 52.

Nagel, S. R., Pearson, A. D., and Tynes, A. R. (1976). Compound glass waveguides fabricated by a metal evaporation technique. *J. Am. Ceram. Soc.* **59**, 47.

Nassau, K., and Shiever, J. W. (1975). Plasma torch preparation of high purity, low OH content fused silica. *Am. Ceram. Soc., Bull.* **54**, 1004.

Nassau, K., and Shiever, J. W. (1976). Three anisotropies in vitreous silica. *J. Am. Ceram. Soc.* **59**, 253.

Nassau, K., Rich, T. C., and Shiever, J. W. (1974). Low loss fused silica made by the plasma torch. *Appl. Opt.* **13**, 744.

Nassau, K., Shiever, J. W., and Krause, J. T. (1975). Preparation and properties of fused silica containing alumina. *J. Am. Ceram. Soc.* **58**, 461.

Newns, G. R., Pantelis, P., Wilson, J. L., Uffen, R. W. J., and Worthington, R. (1973). Absorption losses in glasses and glass fibre waveguides. *Opto-electronics* **5**, 289.

Newns, G. R., Beales, K. J., and Duncan, W. J. (1974). Low-loss glass for optical transmission. *Electron. Lett.* **10**, 201.

Newns, G. R., Beales, K. J., and Day, C. R. (1977). Development of low-loss optical fibres. *Tech. Dig., Int. Conf. Integr. Opt. Opt. Fiber Commun., 1977* p. 609.

Nordberg, M. E. (1943). Glass having an expansion lower than that of silica. U.S. Patent 2,326,059.

O'Connor, P. B., MacChesney, J. B., DiMarcello, F. V., Kaiser, P., Burrus, C. A., Presby, H. M., and Cohen, L. G. (1976). Large numerical aperture, germania-doped fibers for LED applications. *Proc. Eur. Conf. Opt. Fibre Commun., 2nd. 1976* p. 55.

O'Connor, P. B., MacChesney, J. B., and Melliar-Smith, C. M. (1977). Large-core high N.A. fibres for data-link applications. *Electron. Lett.* **13**, 170.

Olshanski, T., and Keck, D. B. (1976). Pulse broadening in graded index optical fibers. *Appl. Opt.* **15**, 483.

Payne, D. N., and Gambling, W. A. (1974a). New silica-based low loss optical-fibre. *Electron. Lett.* **10**, 289.

Payne, D. N., and Gambling, W. A. (1974b). Preparation of water-free silica-based optical-fibre waveguide. *Electron. Lett.* **10**, 335.

Payne, D. N., and Gambling, W. A. (1976). A borosilicate-cladded phosphosilicate-core optical fibre preparation of very low loss optical waveguides. *Opt. Commun.* **13**, 422.

Pearson, A. D. (1974). Progress in compound glass optical waveguides. *Proc. Int. Congr. Glass, 10th, 1974* Vol. 6, p. 31.

Pearson, A. D., and French W. G. (1972). Low-loss glass fibers for optical transmission. *Bell Lab. Rec.* **50**, 103.

Pearson, A. D., and Northover, W. R. (1978). Preparation and properties of ultrapure low-loss Na borosilicate glass. *Am. Ceram. Soc. Bull.* **57**, 1032.

Reed, T. B. (1961a) Induction-coupled plasma torch. *J. Appl. Phys.* **32**, 821.

Reed, T. B. (1961b). Growth of refractory crystals using the induction plasma torch. *J. Appl. Phys.* **32**, 2534.

Schultz, P. C. (1973). Preparation of very low loss optical waveguides. *J. Am. Ceram. Soc.* **57**, 383 (abstr.).

Schultz, P. C. (1974). Optical absorption of the transition elements in vitreous silica. *J. Am. Ceram. Soc.* **57**, 309.

Scott, B., and Rawson, H. (1973a). Techniques for producing low loss glasses for optical fiber communication systems. *Glass Technol.* **14**, 115.

Scott, B., and Rawson, H. (1973b). Preparation of low-loss glasses for optical fibre communication systems. *Opto-electronics* **5**, 285.

Stewart, C. E. E., and Black, P. W. (1974). Optical losses in soda-lime/silica-cladded fibres produced from composite rods. Electron. Lett. **10**, 53.

Stewart, C. E. E., Tyldesley, D., Scott, B., Rawson, H., and Newns, G. R. (1973). High-purity glasses for optical-fibre communication. Electron. Lett. **9**, 482.

Takahashi, S., and Kawashima, T. (1977). Preparation of low loss multi-component glass fiber. *Tech. Dig., Int. Conf. Integr. Opt. Opt. Fiber Commun. 1977* p. 621.

Tasker, G. W., and French, W. G. (1974). Low-loss optical waveguides with pure fused SiO_2 cores. *Proc. IEEE* **62**, 1281.

Tasker, G. W., French, W. G., Simpson, J. R., Kaiser, P., and Presby, H. M. (1978). Low loss fiber with a different B_2O_3–SiO_2 compositions. *Appl. Opt.* **17**, 1836.

Theurer, H. C. (1963). Preparation of purified semiconductor material. U.S. Patent 3,071,444.

Van Uitert, L. G., Pinnow, D. A., Williams, J. C., Rich, T. C., Jaeger, R. E., and Grodkiewicz, W. H. (1973). Borosilicate glasses for optical fibre waveguides. *Mater. Res. Bull.* **8**, 469.

Verneuil, A. (1904). Reproduction artificielle du rubis par fusion. *Nature (Paris)* **32**, 77.

Yamazaki, T., and Yoshiyagawa, M. (1977). Fabrication of low loss, multi-component glass fibers with graded index and pseudo step index distribution. *Tech. Dig., Int. Conf. Integr. Opt. Opt. Fiber Commun., 1977* p. 617.

Chapter 9

Fiber Drawing and Control

RAYMOND E. JAEGER
A. DAVID PEARSON
JOHN C. WILLIAMS
HERMAN M. PRESBY

9.1 INTRODUCTION

9.1.1 Prior Art

This chapter addresses the task of drawing lengths, greater than a kilometer, of low-loss optical fibers, or lightguides, intended for long-distance telecommunication systems. Since this is a relatively new technology, large-scale lightguide production facilities are only beginning to evolve. This discussion is based mostly on experience acquired using a laboratory-type apparatus.

The ability to form glasses into fine filaments or fibers is attributed to their behavior as Newtonian liquids in which viscosity decreases gradually as temperature increases. Although the phenomenon is not understood fully, it has been recognized as a property of glass for ages so that the concept of drawing fibers from a source of softened glass is not new. Man has practiced the art of drawing glass into fibers for at least 2000 years. There are two major categories of glass fibers: nonoptical and optical. Regardless of the differences between the two types, drawing is the common processing step.

9.1.1.1 Nonoptical Glass Fibers. Most nonoptical glass fibers are drawn from a given single-glass composition. In the first century B.C. (Revi, 1957), Palestinian artisans created miniature mosaic portraits comprised of a bundle of glass fibers. Venetians produced intricately decorated glassware (Dickson, 1951) from finely drawn glass fibers during the thirteenth

263

century. Early technical applications include lamp wicks in 1822 and lamp shades as exhibited at Chicago during the Columbian Exposition of 1893. Glass fibers were substituted for asbestos during World War I for use as thermal insulation. The first plant for the production of short lengths, or staple fibrous glass, was founded in the United States in 1934. Shortly after the Owens–Corning Fiberglas Corporation was formed in 1939 the continuous filament process for the production of glass fibers became established (Diamond, 1953). Glass-fiber reinforced plastics emerged from special products developed during World War II. Many details about the manufacture of nonoptical glass fibers together with citations of the patent literature in that field have been published for the first time only recently (Lowenstein, 1973).

9.1.1.2 Optical Glass Fibers. Optical glass fibers generally consist of a central core surrounded by a cladding that necessitates drawing two or more glass compositions simultaneously. In addition to selecting compositions with the proper index of refraction difference, consideration must also be given to their differences in coefficients of thermal expansion, softening points, and viscosities for success in drawing.

Early applications of optical glass fibers include image transmission in 1926 and flexible fiber optics by surgeons for illuminating internal regions of the body in the 1930s (Maloney, 1968). The basic techniques, such as the rod-tube preform and double-crucible methods, and equipment for drawing optical fibers were developed in the 1950s (Kapany, 1967) Other authors have discussed similar techniques as well as procedures for fabricating fiber optic bundles (Bagley *et al.*, 1964; Lisita *et al.*, 1972; Allan, 1973). Commercial drawing of glass optical fibers has attained speeds of hundreds of miles per hour and a diameter less than 1 μm for a single fiber within a multifiber array (Straka, 1968). The production of glass fibers had advanced considerably in the 80 years since drawing a thin fiber of glass by shooting a straw arrow, attached to a glob of molten glass, from a crossbow was reported (Boys, 1887).

9.1.1.3 Fiber Lightguides. The previous remarks account for only part of the legacy of the art and technology of drawing glass fiber with which investigators, many of whom were without previous experience, embarked on developing lightguides once feasibility was predicted (Kao and Hockham, 1966).

The initial lightguide material candidates at Bell Labs were high softening point fused quartz in the form of liquid-filled hollow fibers (Stone, 1972) and the single material, "Unifiber" (Kaiser *et al.*, 1973), as well as low softening point compound glass compositions (Pearson and French, 1972). These were drawn mostly from preforms of the rod-tube type and other configurations. During that time emphasis on developing glasses of sufficiently low loss for long-distance lightguide applications foreshadowed the significance of the drawing process upon other qualities of the lightguide. The low loss threshold of <20 dB/km at Bell Labs was passed with the development of the modified chemical vapor deposition (MCVD) process (MacChesney et al., 1974). Subsequent efforts in cabling and splicing focused attention on the importance of uniform geometry, dimensions, and strength. Some of the approaches taken to accomplish these are discussed later. They emphasize work on drawing lightguides in the Bell System using high-silica glass performs prepared by the MCVD process and requiring heat sources capable of providing temperatures of at least 2000°C.

9.1.2 Essential Components

One can manage to draw a glass fiber using a relatively simple apparatus comprised of four basic components: (1) a glass feed system, (2) a heat source sufficient to soften the glass, (3) a pulling or drawing device, such as a capstan or winding drum, and (4) a structure for supporting and aligning the other three.

The early laboratory type facilities for lightguide drawing in the Bell System were relatively simple designs. They ranged from a portable, horizontal bench-top unit with a screw-driven perform feed, an array of hypodermic needles forming an O_2-H_2 torch heat source, and plastic spools converted from reels for wire serving as winding drums, to a vertical metal frame with improved preform feed mechanism, furnace or gas torch heat sources, and winding directly on precision metal drums (Pearson and French, 1972).

The requirements of lightguides posed new challenges to the drawing process. Consequently, more sophisticated and taller designs have been developed to accommodate improved feed mechanisms, graphite resistance furnaces (DiMarcello and Williams 1975, 1978), CO_2 laser (Jaeger, 1976; Paek, 1974; Saifi and Borutta, 1976), and RF inductively heated zirconia (Runk, 1977) heat sources, faster drawing speeds employing capstans for applying the drawing force, coating systems, and instrumentation for monitoring temperature and controlling on-line fiber and coating diameters (Presby, 1976). Investigators at universities and research organizations in Europe and Japan have applied improved double-crucible designs (Kawamura et al., 1973; Koizumi et al., 1974; Beales et al., 1976; Inoue et al., 1976; van Ass et al., 1976), high-temperature electric resistance (Payne and Gambling, 1976), and RF inductively heated graphite furnaces (Chiccacci et al., 1976). An on-line method for continuously monitoring optical loss during lightguide manufacturing has been reported (Cannell, 1977).

9.1.3 Process–Product Relationship

The drawing process affects three important properties of lightguides: dimensional, strength, and optical. Solutions for achieving the ultimate among these properties in large volume production have not been established, but certain trends and correlations relating their behavior to the drawing process have been observed. When designing a fiber-drawing apparatus, it is good practice to eliminate vibrations that may induce perturbations to the fiber diameter, to align all components on axis as well as to avoid exposing the preform and fiber to sources of contamination and surface flaws originating from the heat source, ambient atmosphere, and mishandling. Many drawing facilities are located in ordinary laboratory and plant environments but it is anticipated that some degree of "clean room" philosophy and conditions will generally be found advantageous.

9.1.3.1 Dimensions.   Typical multimode lightguides have core and cladding diameters in the range of 50 to 125 μm and 100 to 220 μm, respectively. Except for slight enhancement of the circularity of the cladding during drawing, the circularity and concentricity of the core and cladding as well as the core/cladding diameter ratio are established in the MCVD preforms by the deposition and collapsing phases. Uniformity of fiber diameter, on the other hand, is dependent upon the drawing process (DiMarcello and Williams, 1975, 1978). The outer fiber diameter usually serves as the reference surface for alignment during splicing. For short lengths, transverse displacement splicing tolerances are greater for graded-index fibers than for step-index fibers. In long lengths, however, the situation is reversed (Kitayama, and Ikeda, 1977). It is imperative, therefore, to minimize offset in fiber and cable splices by providing control and uniformity of fiber diameter.

Factors affecting diameter during drawing include uniformity of temperature distribution around the preform, proper balance between preform feed rate and drawing speed to comply with the conservation of mass, and vibrations that perturb the profile of the neck-down region of the preform (see Fig. 9.1). Thermal fluctuation resulting from drafts or the "chimney effect" within the heat source are also contributory. Diameter fluctuations due to the heat source are worst in the case of gas torches with the exception of the design used at Philips (van Ass *et al.*, 1976). Depending upon the heat source, various sealing techniques have been improvised in an attempt to reduce the "chimney effect," such as "O" ring seals around the preform, iris-type diaphragms, and counterflow gas curtains.

Control of diameter is also dependent upon the sensitivity of the monitoring device and response time of the feedback system. Diameter control

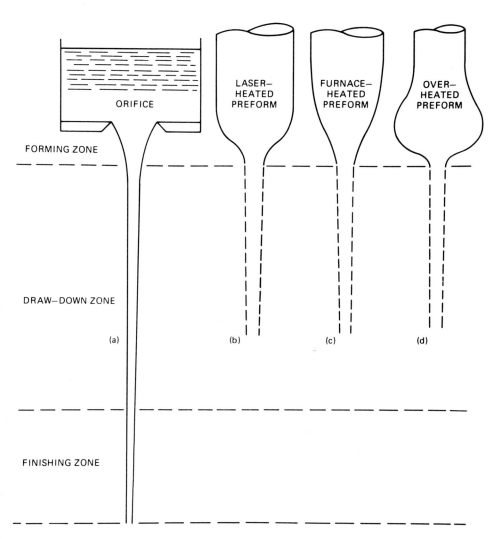

Fig. 9.1 Fiber-forming profiles for glass melt and preform feed methods (after Geyling, 1976). Reprinted with permission from the Bell System Technical Journal, copyright 1976, the American Telephone and Telegraph Company).

of $\pm 2\%$ has been obtained on experimental fibers using a Milmaster* monitor and manual feedback. A precision of $\pm 1\%$ has been reported using a laser monitoring device with electronic feedback to a capstan puller (Cohen and Klaiber, 1977).

Cyclic changes in the drawing force due to out-of-round direct-draw drums and capstan devices can also vary fiber diameter. Lightguide drawing speeds are still relatively slow (1 to 10 m/sec) compared to conventional glass fibers (100 m/sec) and metal wires (5 to 15 m/sec) (Dove, 1968). The ultimate drawing speed will depend in part on the on-line coating system. As drawing speeds increase, all process parameters require adjustment in order to maintain diameter uniformity.

9.1.3.2 Strength.   A freshly drawn glass fiber in the pristine state, free of surface defects, is capable of exhibiting strengths in excess of 7 GN/m^2 (10^6 psi). Although kilometer lengths of experimental coated fibers have recently approached tensile strengths of 7 GN/m², many difficulties have beset investigators in achieving reproducibility of high strength, kilometer lengths of lightguides (see Chap. 12).

A major difficulty has been the inability to rigorously limit the size or completely eliminate the presence of surface flaws on the fiber (Takayama *et al.*, 1977). Surface flaws may originate from a number of sources. They may exist prior to, and persist after, drawing, as in the case of residual contaminants and defects from the support tube used for preform preparation, and damage induced in the surface of the preform during processing and handling. Other sources include the presence of devitrification at the surface of the preform, and contact with any foreign substance during and after drawing, including particulate debris and condensates from furnace refractories and heating elements, dust particles from the forming atmosphere (Maurer, 1977), misalignment of coating applicators, and foreign or even coarse particles in the coating system. Condensation of the preform upon itself by volatilization from overheating (Li *et al.*, 1969) during drawing may also create surface defects. Measurements of the electrostatic surface charge on fused silica fibers from drawing with a graphite resistance furnace or CO_2 laser heat source is so low ($< 10^{-11}$ C/cm²) that it is unlikely to be the force responsible for attracting dust particles to the fiber prior to on-line coating (L. L. Blyler, unpublished data, 1976). Improvements in achieving higher strength lightguides continue, but more consideration of all aspects of the drawing process is required for enhancing reproducibility of high strengths.

9.1.3.3 Optical.   Defects induced into the silica network as the result of drawing have been observed to cause coloration of fused silica fibers at the

---

* Model SSE-5R Milmaster, Electron Machine Corp., Umatilla, Florida.

wavelength of about 630 μm (Kaiser, 1974). Fortunately this wavelength is below that of the light sources currently of interest for lightguides. The presence of contaminants of the type noted earlier also contribute to scattering losses along the core–cladding (glass/air) interface particularly in the case of unclad fibers (Kaiser, 1977) and fibers drawn from rod-tube type preforms (French et al., 1975). Similarly, fluctuations in core diameter that accompany changes in fiber diameter may also cause scattering losses. On the other hand, core diameter fluctuations of the proper spatial frequency have proved beneficial in promoting mode coupling, thereby reducing dispersion in multimode lightguides (Marcuse and Presby, 1975; Personick, 1971).

9.1.4 Dynamics

Lightguides are drawn by heating the source material to soften it while applying a pulling stress sufficient to attenuate the material into a fiber as illustrated in Fig. 9.1. Attenuation initiates in the forming of the neck-down zone and continues at a more gradual rate in the draw-down zone before solidifying into the finished fiber. The fiber undergoes a thermal quench in which its temperature drops hundreds of degrees within a few millimeters upon exiting the heat source. In the case of preforms the profile in the forming zone is markedly dependent upon the geometry of the heating source as shown in Fig. 9.1b,c, and d. The short profile for the CO_2 laser drawn preform is due to the narrow heat zone, short dwell time at temperature, and dominance of surface heating. Typical diameter draw-down ratios with the use of preforms range from 10:1 to 100:1. Figure 9.1a is the profile for material drawn from an orifice of a crucible or bushing.

The ultimate objective, forming a lightguide of uniform specified diameter, requires among other conditions that the principle of conservation of mass be satisfied. Under steady-state drawing conditions the volume of material fed into the heat source must be equivalent to the volume of fiber being removed. This relationship is represented in the following expression:

$$s = S(D^2/d^2), \tag{9.1}$$

where D and d are the preform and fiber diameters, and S and s are the preform feed and fiber draw speeds, respectively, in centimeters per second. To compensate for variables such as fluctuations in the preform diameter and temperature changes of the source material, the practice in most drawing operations is to maintain a constant preform feed rate and temperature setting while adjusting the draw speed to control fiber diameter. The drawing of glass fibers is simple to describe qualitatively, but it is a complex fluid dynamic process of viscous flow subject to a variety of instabilities resulting from changes in temperature and viscosity profiles, vibrations and dynamic fluctuations at higher drawing speeds, as well as tensile and surface tension stresses. An analysis of the process has been attempted (Geyling, 1976) by developing one-dimensional models of an isothermal base state for comparison to models that account for the interaction of the dynamic forces. The capillary "pinch-off" and tensile "neck-down" mechanisms, illustrated in Fig. 9.2, are proposed as the instabilities that dominate in the forming and draw-down zones, respectively. In the capillary case, surface tension overcomes viscous stresses to cause fluid from both directions to form "beads." In the tensile model, tensile stresses reduce the cross-sectional area which, in the worst case, lead to a runaway tensile separation when surface tension and viscous stress are exceeded by the draw force. Related studies (J. A. Lewis, unpublished data, 1977) have modeled steady furnace drawing as well as the effect of heat transfer mechanisms (P. H. Krawarik, unpublished data, 1973) in which radiation heating at 3 μm is claimed to be preferable for heating fused silica. The latter work confirms earlier findings (Kutukov and Khodakovskii, 1964) that dynamic instabilities at higher drawing

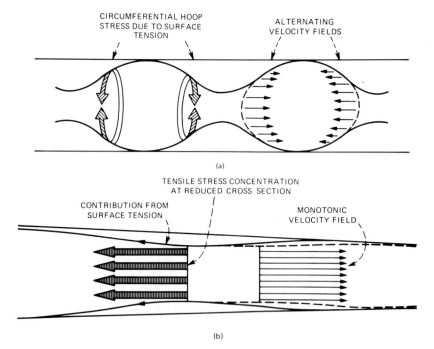

Fig. 9.2 Comparison of (a) capillary and (b) tensile instabilities (after Geyling, 1976. Reprinted with permission from the Bell System Technical Journal, copyright 1976, the American Telephone and Telegraph Company).

speeds (10–50 m/sec) cause the neck-down region to pulsate and result in diameter fluctuations. Thermal pulses, using either a laser or flame heat source, have been found at times to produce a "sausage effect" with periodic diameter variations of 2%.

Process control is complicated further when on-line coating systems are incorporated into the apparatus. The use of a sensitive dynamometer for measuring the drawing tension is becoming valuable both as a diagnostic tool and as a means of monitoring the complex interaction of forces. Typical drawing tensions range from <4 to 100 gm with preference for the lower end. At a given temperature, tension increases with drawing speed (Runk, 1977); while fiber strength decreases with increasing drawing tension. In other work (Peyches, 1945) the term "drawing index" was used as a measure of the drawability of glass fibers. The index is the ratio of viscosity (in poise) to surface tension. In the case of fused silica at a temperature of 2000°C (Li *et al.*, 1969) the drawing index value of 2000 is obtained from 600,000 P/300 dyn cm^{-2}.

9.2 GLASS FEED

9.2.1 Introduction

There are two general methods for feeding glass into a fiber-drawing machine for the manufacture of clad glass fibers. These depend on the softening temperatures of the glasses involved. In the case of the very high softening temperature doped silica glasses, the material is introduced as a solid rod or tube which softens at the furnace temperature so that it can be drawn into fiber. The multicomponent glasses, which soften at much lower temperatures, are usually drawn from double-crucible devices in which the glass is held in the molten state. The glass flows through an orifice in the bottom of the device and is then drawn into fiber.

9.2.2 Solid Preform Technique

Several types of "preforms" or "blanks" can be drawn by this method. These include the solid composite rod preforms made by the MCVD and Phasil processes, and the hollow tubular blanks prepared by the "soot" method. In this last mentioned technique, the fiber drawing is carried out at temperatures and rates such that the central hole or bore of the blank closes up during drawing so that the finished fiber has a solid core. Also, rod-and-tube preforms can be employed (Kapany, 1967; Pearson and French, 1972). Here, a polished rod of the core glass is inserted into a polished tube of the cladding glass, and the composite is fed into the fiber-drawing furnace. No low-loss fibers have been reported from this method, largely due to the trapping of gas bubbles at the core–cladding interface. These give rise to unacceptable interfacial scattering losses in the finished fiber.

The feed mechanism itself is fairly simple, although it should be precisely and sturdily constructed. The preform holder consists of either a clamp or a chuck which is used to hold the preform rigidly in a vertical position. Horizontal X and Y adjusters are usually provided to allow for centering the preform in the core of the drawing furnace. The holder is driven in the vertical direction by a lead screw connected to a variable-speed electric motor. It is important that the lead screw runs smoothly and that the motor speed, at any given setting, be constant. If these conditions are not met, the rate of introduction of the preform into the furnace will vary, with resultant effects on the fiber diameter.

The major advantage of this method is that the glass preform is never melted to a low viscosity, but only softens enough to be drawn into fiber. This is important because during preform preparation the core–cladding diameter ratio and the refractive index gradient of the core are carefully established. These are maintained during the solid preform drawing method. The main disadvantage is that the technique is a batch process. This means that after each preform is drawn the machine must be stopped and reloaded before drawing can continue. This entails extra costs for the process. In addition there is some material wastage in the start-up and shut-down of each drawing operation.

9.2.3 Double Crucible Technique

This method has been described by a number of workers (Kapany, 1967; Pearson and French, 1972). The basic crucible design for the preparation of step-index fibers is shown in Fig. 9.3. It consists of two containers, usually made from platinum. Each has an orifice in its bottom, and these orifices or nozzles are lined up concentrically. The whole arrangement fits inside a vertical muffle furnace capable of heating the double crucible and its contents to temperatures in the range 800–1200°C. The core and cladding glasses are placed in the inner and outer crucibles, respectively. At the fiber-drawing temperature they flow simultaneously under gravity from the double nozzle and are drawn directly into fiber. The rate of flow of each glass through its respective nozzle depends upon the liquid "head" above the nozzle and the glass viscosity, which is in turn dependent on temperature. These factors, in conjunction with the diameters of the core and cladding nozzles, determine the core to cladding diameter ratio in the finished fiber.

A modification of this technique for the production of fibers with graded refractive index cores has been described by several authors (Koizumi *et al.*, 1974; Inoue *et al.*, 1976; Beales *et al.*, 1976). The modification

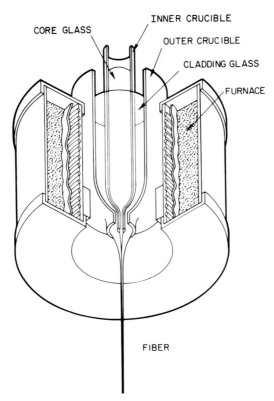

Fig. 9.3 Double-crucible preparation of step-index fibers. After Pearson and French (1972). Copyright 1972, Bell Laboratories, Inc. Reprinted by permission, editor, Bell Laboratories Record.

consists of extending the nozzle of the outer crucible as shown in Fig. 9.4. Thus the core glass comes into contact with the cladding glass some time before it reaches the end of the cladding nozzle from which the fiber is drawn. As the two molten glasses flow together toward the drawing neck, ionic diffusion takes place across the interface between them, producing a graded refractive index profile. The two crucibles can be fed with rods of core and cladding glass so as to maintain constant liquid heads and facilitate fiber diameter control. Also, under these conditions, fiber can be drawn continuously.

If a double crucible is operated with an oxygen-rich atmosphere, scattering sites can be formed at the core–cladding interface. It was postulated that these were minute gas bubbles produced by the electrolytic cell formed between the molten core and cladding glasses and the tip of the core nozzle (Cowan *et al.*, 1966). It was also suggested (Beales *et al.*, 1976) that the cell-driving force was the diffusion of oxygen from the atmos-

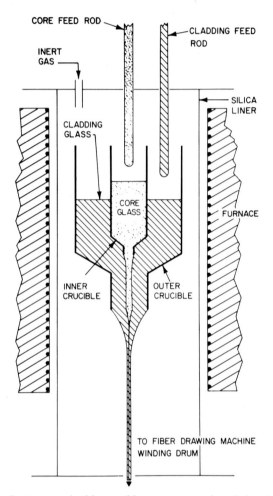

Fig. 9.4 Continuous double-crucible preparation of graded-index fiber.

phere through the molten cladding glass. The effect was eliminated by maintaining an inert gas atmosphere ($P_{O_2} \approx 10^{-5}$ atm) inside the crucible (Beales *et al.*, 1976).

Although most fibers drawn by these techniques are multimode, the preparation of single-mode fibers from a double crucible has been reported (Kawamura *et al.*, 1973).

The double-crucible method represents the only continuous process for drawing clad optical fibers, and herein lies its potential for cost savings compared to batch processes. On the other hand, the losses of double-crucible fibers are not as low as the doped high-silica compositions, and perhaps even more important, the control of refractive index profile is difficult, which will probably limit the bandwidths that can be obtained. Finally, the reliance on diffusion processes to form the profile severely limits the rate at which fiber can be drawn. The presently reported rates are around 20 m/min (Yamazaki and Yoshiyagawa, 1977; Takahashi and Kawashima, 1977).

9.3 HEAT SOURCES

Lightguides exhibiting the lowest optical losses are normally drawn using the preform technique. The most frequently used heat sources are gas burners, a variety of resistance and inductively heated furnaces, and CO_2 lasers. Very precise temperature control in the drawing zone is necessary, since the viscosity of glass in the softening range is essentially a monotonic function of temperature (Rawson, 1967) (Fig. 9.5). Fluctuations in the thermal gradient in the draw-down zone may stimulate instability affecting diameter control.

9.3.1 Gas Burners

Flame burners have historically been used for the drawing of high temperature glasses into fiber. These are normally oxy-hydrogen or oxy-gas torches which possess inherent turbulence problems resulting in relatively poor diameter control. One such burner design used in early fiber development by Pearson and Tynes (French et al., 1975) is shown in Fig. 9.6. The design consists of 16 oxygen jets converging on the preform in a conical fashion. Hydrogen is injected from four lower radial ports. A dif-

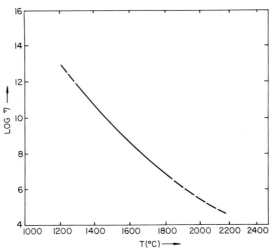

Fig. 9.5 Viscosity versus temperature characteristics for fused silica.

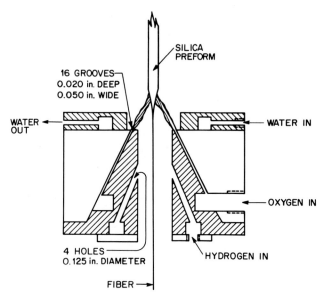

Fig. 9.6 Oxy-hydrogen burner. After French *et al.* (1975). Copyright 1975, Annual Reviews, Inc. Used with permission.

ferent configuration is described by Zvanut (1972) which uses an additional noncombustible gas to control the temperature gradient across the flame. Both of these designs are capable of producing nominal diameter uniformity of 5%, but are subject to intermittent large excursions due to flame turbulence.

9.3.2 Furnaces

A variety of laboratory and commercial high-temperature furnaces have been developed in recent years for use in drawing high-silica glasses. The high-temperature (graphite or metal) element is typically heated by direct electrical resistance or by high-frequency induction. The major drawback of such heat sources in the past has been contamination of the glass through evaporation of the heating element and/or the furnace insulation.

DiMarcello has successfully used a graphite resistance furnace in the preparation of kilometer lengths of low-loss waveguides (DiMarcello and Williams, 1975, 1978) and in the evaluation of protective coatings (Kurkjian *et al.*, 1976). P. Kaiser (unpublished data, 1975) has effectively reduced the contamination imparted to the fiber surface in a tungsten resistance furnace by providing a high-purity laminar gas flow in the hot zone through the use of special manifolds. The gas in this case must be inert in order to prevent oxidation of the tungsten. Through the use of a stabilized zirconium oxide RF susceptor, the required temperatures have been achieved in an oxidizing and still environment (Runk, 1977).

A schematic diagram of the zirconia induction furnace is shown in Fig. 9.7. The larger thermal mass of such heating sources produces a relatively long neck-down region (several preform diameters) compared with gas burners or lasers. This presents problems in drawing rod and tube type preforms as a result of entrapped defects at the core–clad interface. Excess losses greater than 100 db/km reportedly resulted from drawing this type of preform configuration with a furnace as the heat source. No such problems occur with preforms prepared by the chemical vapor deposition process described in Chapter 8.

Drawing speeds up to 10 m/sec have been achieved with the zirconia induction furnace. The best diameter control to date, however, has been obtained at much lower draw speeds in the order to 2 m/sec. Through the use of a diameter feedback control, excellent fiber uniformity (standard deviation less than 0.5%) has been obtained. Figure 9.8 is a power spectral density plot of fiber diameter variation (see p. 278), indicating the magnitude of diameter variation versus spatial wavelength for such a fiber.

9.3.3 Laser

The use of high-powered lasers for cutting, scribing, and drilling ceramic oxides is well known (Epperson *et al.*, 1966; Cohen, 1967; Longfellow, 1971). In terms of process evaluation, this technique for heating the glass can be expected to be more flexible than conventional heat sources with regard to ultimate temperature capability, ease of adapting atmosphere control, overall cleanliness, temperature response and control, and geometry control in the fluid zone. A typical setup is shown schematically in Fig. 9.9. The output laser beam passes through a lens which is rotated at a rate in excess of 50 Hz. The lens axis is laterally displaced from the

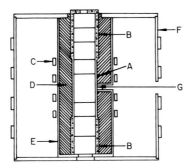

Fig. 9.7 Vertical section of zirconia induction furnace. A. Zirconia susceptor rings. B. Zirconia support rings. C. Induction coil. D. Zirconia insulation. E. Fused quartz. F. Copper shell. G. Temperature monitoring.

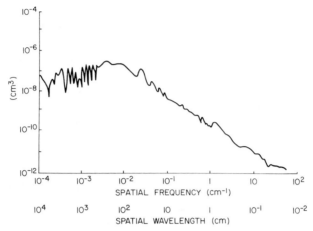

Fig. 9.8 Power spectrum of a closed-loop furnace-drawn fiber.

rotation axis by 6.4 mm and the faceted conical reflector is positioned at a distance of twice the focal length of the lens. These conditions produce an annular path of the beam at the reflector surface having a diameter of approximately 25 mm and a width of approximately 7 mm. The glass preform is fed into the rear of the conical reflector and the fiber is drawn through the hole in the 45° mirror.

A variety of fused silica, high-silica, and multicomponent glasses have been drawn into fibers using this technique. By comparison with the two

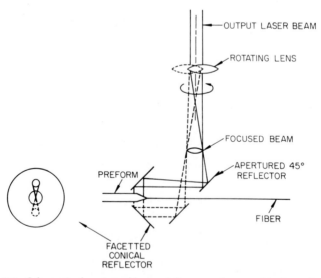

Fig. 9.9 Schematic drawing of beam delivery system used in laser drawing.

other types of heating sources, the fluid neck region assumes a markedly different geometry, and there is a decided reduction in the amount of vaporization from the preform surface. Both effects result from the very localized heating achieved through the use of the laser. This is desirable from the standpoint of suppressing phase separation, devitrification, diameter variation, and interface scatter loss in fibers drawn from rod and tube preforms.

The laser technique was the first to achieve less than 1% standard deviation in fiber diameter in long lengths without the aid of control loops. Figure 9.10 gives the power spectral density of diameter variation for a laser-drawn fiber. The peak which occurs at a spatial wavelength of 1.6 cm is the result of an intentional periodic variation in diameter induced by a 50 Hz lens rotation rate. Such controlled periodicities in diameter variation have been used experimentally by Jaeger and Lazay (unpublished data, 1976) to enhance mode mixing in plastic clad silica fibers.

The cleanliness of the drawing environment inherent in the laser process has been used to advantage in the evaluation of protective coatings applied in-line to the fiber surface to maintain the strength characteristics of the pristine fiber. Tensile strengths greater than 500,000 psi, in kilometer lengths, have been reported for epoxy-acrylate coated fused silica fibers drawn with a CO_2 laser heat source (Schonhorn et al., 1976).

The cleanliness of the process has also been demonstrated in the preparation of low loss polymer clad fibers as described in Chapter 10. In drawing low OH content fused-silica core polymer clad fibers, it has been determined that laser drawing produces consistently lower absorption at 0.95 μm than similar fibers drawn with a high-purity flame burner. The difference is of the order of several decibels per kilometer.

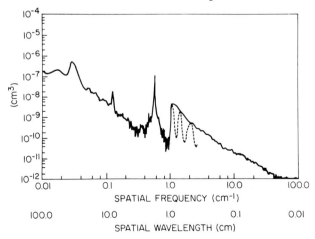

Fig. 9.10 Power spectrum of an open-loop laser drawn fiber.

9.3.4 Combined Heat Sources

In laboratory drawing equipment, gas burners, furnaces, and lasers have been used to draw high-silica optical waveguides at speeds of 1–3 m/sec. Table 9.1 gives the laser power required to draw a variety of glass composition and preform geometries at speeds of 1 m/sec. For further production needs, however, higher drawing speeds will more than likely be required. Considering only total available power, there is a limit to the speed of a laser system, whereas it appears that drawing with induction heating can be engineered to attain very high drawing speeds by virtue of the almost unlimited amount of available power. Overall efficiency of black-body radiative heating is comparable to the efficiency of present CO_2 laser systems, however, since black-body heating is more efficient at lower temperatures, the best configuration in terms of efficiency and control might well be a combination of the two methods.

9.4 DRAWING MECHANISMS

A third essential feature in a fiber-drawing facility along with a glass feed mechanism and heat source is some sort of fiber-drawing and spooling system. As indicated in Section 9.1.4, the dynamics of the process are such that in order to obtain a peak-to-peak variation in diameter of not more than 1% requires control of feed and draw speeds to within a total variation of 2% (Eq. 9.1). This degree of control can be reached by two approaches completely different in principle. In the first case, a high inertia mechanism is employed which essentially acts as a fly wheel in resisting forces which would tend to vary the drawing speed. This approach requires that all other process perturbations be reduced to a level which will provide the desired diameter stability. Small perturbations are diffi-

TABLE 9.1
Laser-Drawing Conditions for a Variety of Preform Materials

Material	Preform diam. (mm)	Fiber diam. (μm)	Donut i.d. $\times$ o.d. (mm)	Laser power (W)
Corning 7971	8	140	22 $\times$ 31	250
Soda-lime-silicate				
rod	10	160	22 $\times$ 31	95
rod-	6	60 $\times$ 110	22 $\times$ 31	95
in tube	8 $\times$ 12			
7971 rod-	3	79 $\times$ 180	21 $\times$ 30	175
SS-WII tube	4 $\times$ 7			
SS-I	8	175	21 $\times$ 30	300

cult to compensate for with this type of drawing mechanism because of the large mass purposely built into the system.

Another technique, gaining in acceptance, uses a low inertia mechanism driven by a powerful motor capable of high bandwidth speed response. Such systems generally employ rubber-coated pulling tractors or pinch wheels (capstan) as opposed to a massive drum which would normally be used in the method described above. The use of pulling tractors results in several other advantages, in addition to eliminating the need for a transversing drive incorporated into the actual drawing facility. These include the ability for sampling the fiber without interruption of the process, ease of in-line proof testing to establish minimum tensile strength, and capability of spooling the fiber for storage under a controlled winding tension that is less than the draw tension.

A photograph of a drawing system incorporating all these features and an in-line tensiometer is shown in Fig. 9.11. In-line sampling of the fiber is convenient for determining coating concentricity, fiber geometry, and optical characteristics. Proof testing is becoming increasingly important to ensure survivability of the fiber in subsequent cabling and installation. Monitoring of draw tension gives an independent indication of the temperature control in the forming zone. This is important to maintain diameter stability and high strength. The effect of draw tension on fiber strength appears less critical and Jaeger and Schonhorn (unpublished data, 1976) have determined that the tensile strength of the fiber begins to degrade as draw tensions greater than 30 gm are approached.

9.5 DIAMETER UNIFORMITY

9.5.1 Introduction

The diameter of an optical fiber is an important parameter in determining the transmission behavior and handling characteristics of the fiber. The diameter must be large enough to facilitate reliable splicing and fiber-to-source coupling as well as to ensure, in multimode fibers, the propagation of a sufficient number of modes and a reasonable insensitivity to microbending loss. In addition, single-mode fibers must possess a sufficient cladding thickness to ensure proper mode confinement. On the other hand, the fiber must not be so large as to waste fabrication time and material, be too stiff to bend, nor take up valuable space in cables and ducts. The outside diameters of typical optical fibers currently fabricated satisfying the above requirements are about 100 μm.

Variations of the fiber diameter from the desired value can have serious consequences. Large long-scale changes can make splicing and precise

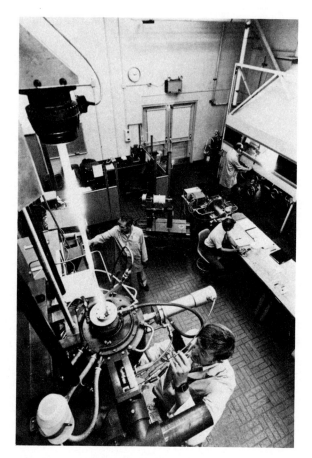

Fig. 9.11 Example of a pilot plant lightguide drawing apparatus (Bell Lab News, 7-12-76).

handling difficult and rapid short-range variations, if not carefully controlled, can lead to scattering loss and unpredictable mode-coupling effects (Marcuse, 1974).

It is necessary, therefore, to have some means of continuously monitoring the fiber diameter during the pulling operation to enable the operator, or an automatic control system, to compensate for unwanted variations. It is also desirable to have high-resolution, easy-to-use off-line techniques which can measure changes in geometry as well as diameter to evaluate various fibers and pulling arrangements.

In this section we present methods to measure the diameters and determine the geometries of optical fibers in both on-line and laboratory situations, and discuss implications for feedback control systems.

9.5.2 Diameter Measurement by Contact Methods

There are two very broad classifications of diameter measurement techniques, *viz.* (1) contacting or destructive methods and (2) noncontacting and nondestructive methods.

Among the former are measurements by micrometers and optical microscopes. Micrometer measurements, while simple and cheap, are severely limited in accuracy and resolution. Improvements in this method can be realized with the use of a transducer gauge which converts the displacement of its tip by the fiber, above a reference surface, into a voltage. This voltage can then be transformed into a diameter reading, accurate to better than 1 μm. If the fiber is rotated in increments and the measurements repeated, an estimate of the geometry can also be obtained. An on-line contact method of measurement has been proposed in which the fiber passes through two precise, spring-loaded cylinders, the separation of which is measured by optical means (Kapany, 1967). While a high degree of sensitivity is achievable in principle, the method is limited in accuracy by the mechanical motions, vibrations, and surface imperfections of the rollers. But both of these techniques involve physical, and probably damaging, contact with the fiber, and are thus not well suited to obtaining the long lengths of high-quality lightguides needed for telecommunications.

To determine a fiber's diameter accurately by microscopic means a flat cross section of the fiber is generally measured with an image-shearing eyepiece. Accuracies of about 0.2 μm are obtainable with this method. It is also possible to observe microscopically the fiber perpendicular to its axis and thus avoid breaking it. However, strong diffraction and focusing effects make an accurate measurement difficult.

9.5.3 Diameter Measurement by Noncontact Methods

Those diameter measuring methods possessing the greatest sensitivity and adaptability to on-line use are optical in nature. As such they are inherently noncontacting and nondestructive. The techniques are based either on shadowgraph or light-scattering principles. Some of the methods have been developed into commercially available instruments for on-line use and others have proven valuable in their ability to statically detect minute deviations in a fiber's geometry.

9.5.3.1 Shadowgraph-Based Methods. Shadowgraph-based methods process a signal arising from the blockage of a beam of light by the fiber. In one commonly used embodiment of this principle (Milmaster), the shadow of the fiber is compared against an illuminated preset slit opening. The size of the fiber relative to the slit opening is indicated on a nullmeter having a precision of about ± 0.15 μm. Calibration of the device, when necessary, is achieved by drawing a number of fibers of different diameters and using known slit widths, measuring the diameters independently, and plotting a calibration curve. A continuous strip chart recorder is generally coupled with the instrument to provide a permanent record of the diameter.

In on-line use diameter variations occurring in distances along the fiber of less than the approximately 5 mm measuring beam width will not be detected. Also, since the response time of the device is about 0.25 sec, diameter variations occurring in a shorter time period will be smoothed over and attenuated. For example, if the fiber is being drawn at 1 m/second, then only changes in diameter occurring in portions of the fiber 25 cm apart will be faithfully recorded, while variations occurring between these portions will be smoothed.

In another shadowgraph method (Cohen and Glyn, 1973) an oscillating mirror is used to deflect a laser beam with constant velocity across the fiber. The time interval during which the fiber intercepts the beam and casts a shadow on a photodetector is measured and is, to a first approximation, linearly related to the outside diameter of the fiber. Measurements are made every 0.06 sec and are repeatable within a spread of $\pm 1\%$ for stationary fibers with diameters greater than 20 μm.

9.5.3.2 Forward Light Scattering. Forward-scattered radiation patterns, generated when a collimated beam of laser light is incident at right angles to an object's axis, have been successfully used to measure the diameters of opaque cylinders such as wires (Gagliano *et al.*, 1969). The extensions of this technique to optical fibers (transparent cylinders) involves the theoretical solution of a complicated scattering problem in which both the diameter, refractive index, and combinations thereof appear (Kerker, 1969; Watkins, 1974; Marcuse and Presby, 1975). Experiments have shown close agreement between theoretical characteristics and actual measurements of the scattered light (Watkins, 1974). The scattered light forms an interference pattern in which the number of fringes in a predetermined angular range is proportional to the fiber's outer diameter. In addition, measurements of a fringe modulation gives relatively sensitive determinations of the core diameter. Total-diameter determinations are reported to have accuracies of ± 0.2 μm and core diameters of ± 0.5 μm for 0.02 refractive index difference between core and cladding (Watkins, 1974).

In instrumentation that has been developed to implement this method (Watkins, 1975; Smithgall *et al.*, 1977), a beam from a continuous wave laser is incident on a fiber perpendicular to its axis (Fig. 9.12) and po-

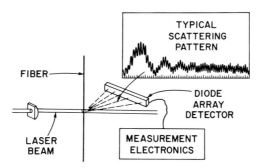

Fig. 9.12 Arrangement for measurements on forward scattering pattern. A typical pattern is shown in the insert.

larized parallel to the fiber's axis. The scattering pattern is focused onto a self-scanned diode array using a lens, and a variable density filter is used in front of the array to compensate for the intensity decrease with increasing scattering angle. The number of maxima and minima (half-fringes) are counted by electronic conditioning and counting circuits, the first of which enable counting to be achieved even in the modulation region. The self-scanned array completes a scan in 0.5 msec and then repeats giving 2000 determinations of core ratio and fiber diameter each second, a rate suitable for on-line use in high-speed drawing machines.

The above system has been greatly enhanced by the addition of a spatial–spectral analysis system (Krawarik, 1975). The system consists of a real-time spectrum analyzer and a computerized data acquisition system to obtain spatial power spectra from the fiber diameter data. Using this technique, the amplitude of the time-varying analog signal representing the fiber diameter variation is separated into its frequency components. The units of power spectral density are then rms amplitude^2 per Hz. In this case, the amplitude is equal to a voltage proportional to an rms diameter variation in centimeters. The bandwidth of the analysis filter is fixed by the instrument but affected by the fiber draw speed according to the equation

$$\frac{f \text{ (sec}^{-1})}{v \text{ (cm sec}^{-1})} = \text{spatial frequency (cm}^{-1}) \text{ or cycles}$$

per unit length of fiber.

The units of the power spectral density, therefore, are in square centimeters per centimeter of spatial frequency or cubic centimeters. If one subtracts the mean value of fiber diameter from the input data, then the power spectral density gives the distribution of the variance with spatial frequency.

Figure 9.13 compares the power spectra of fibers drawn without the aid

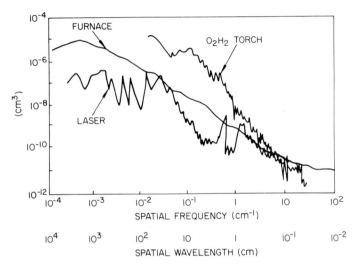

Fig. 9.13 Power spectra for open-loop furnace, laser, and torch drawn fibers.

of external control using a furnace, laser, and gas burner as heat sources. The data indicate several features common to the process of drawing high-silica fiber regardless of the drawing system used. For example, all spectra when extended into the low-frequency range show a broadband peak attributed to the self-regulation of the process. The spectra may decrease slightly toward lower frequencies, but in general, maintain approximately the same magnitude such that below the process peak the spectrum can be regarded as essentially random. Thus, even in the absence of a diameter control system, the process is regulated by the law of conservation of mass. Such regulation, however, applies to spatial wavelengths longer than the section of fiber which can be drawn from that volume of glass in the neck-down region of the preform at any given time. It is for this reason that the leveling off of the spectrum for the furnace-drawn fiber appears at longer wavelengths than in the laser-drawn fiber. In the latter case, the thermal mass in the neck-down region is much smaller.

For lengths of fiber that are short in comparison with the mass of molten material available, the fiber diameter variation steadily decreases. In the region around one cycle per centimeter, the magnitude of the diameter variation for unity bandwidth is between 0.1 and 1.0 μm which correlates well to measured values. One can infer from these data that the fiber-drawing process is essentially a low-pass filtered random process with the filter cutoff frequency proportional to the size of the molten zone.

Figure 9.14 compares fiber diameter spectra for two fibers drawn under identical conditions using a laser system. One was drawn with a drum mechanism, and the other with a tractor wheel type of capstan. A periodic

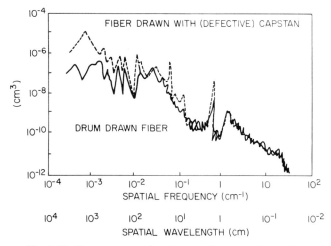

Fig. 9.14 Power spectra for drum and capstan drawn fiber.

diameter variation at a spatial wavelength of 15 cm is evident in the fiber drawn with the capstan device. In the same fiber, an additional perturbation associated with the feed mechanism appears at a wavelength of approximately 15 m. The general similarity of the curves over the entire frequency range, however, is strong evidence that the process is indeed a stationary random one.

9.5.3.3 Backscattered Light Analysis. An elegant method to measure both the diameter and geometry of optical fibers is by back-scattered light analysis (Presby, 1974). The basis of this technique is the observation that the complete back-scattered fringe pattern is localized in a range of angular deviation, ϕ, on the order of $\pm 20°$ from the incident direction. The spacing of certain interference fringes is a sensitive measure of the diameter, and the position of the sharp cutoff for a perfectly circular fiber depends only on the index of refraction. For an elliptical fiber the location of the cutoff also depends on the degree of ellipticity (Presby, 1976a). Diameter variations on the order of 0.1 μm and ellipticities smaller than 0.997 have been measured in static situations (Presby, 1977).

An experimental arrangement to observe the backscattered light pattern in static situations is shown in Fig. 9.15. Light from a CW He–Ne laser strikes plane mirror M1 which reflects it to oscillating mirror M2. This serves to transform the ~1-mm circular beam into a line 1-mm wide, with length determined by the amplitude of oscillation. Observations can thus be made on an extended length of fiber (typically 7 cm) upon which the light impinges after passing through a slit in the observation screen. The fiber is held in a rotatable mount and the back-scattered light falling on the screen is photographed. Several back-scattered light patterns are shown in

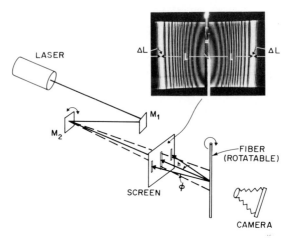

Fig. 9.15 Arrangement for observing back-scattered light pattern. A typical pattern from an unclad fiber is shown in the insert.

Fig. 9.16. Figure 9.16a shows the pattern from a 7-cm length of very uniform fiber pulled in an electric furnace. The section of the fiber shown in Fig. 9.16b contains a slight perturbation as indicated by the arrow. The extent of the disturbance is about 1.4 cm and the change in diameter over this region is approximately 2 μm. The sinusoid-like perturbation in the fiber of Fig. 9.16c, introduced by vibrations of the pulling machine, has a period of approximately 1.2 cm and corresponds to a maximum diameter change of 3 μm. A pattern for a torch-drawn fiber, indicating a high degree of geometrical nonuniformity, is shown in Fig. 9.16d. The fiber of Fig. 9.16e was pulled with a CO_2 laser beam focused upon a rotating preform. A combination of this rotation and the uneven preform heating gives rise to the very regular perturbation seen.

The core diameters of step-index and graded-index fibers are also subject to analysis by this technique (Presby and Marcuse, 1974; Marcuse and Presby, 1975).

9.5.3.4 Other Diameter Measuring Methods.   Several other noncontact techniques have been proposed to measure fiber diameters. In one of these (Rzepecka and Stuchly, 1974), the external cavity of a Gunn diode oscillator is utilized as a basic sensor. The fiber passes through the oscillator cavity affecting its frequency of operation which is being compared with the frequency of an identical but unloaded oscillator. The frequency difference between the two oscillators, being a measure of the test fiber diameter, is monitored by an RF frequency counter. Accuracies better than 1% are reported.

Another noncontact technique (Holly, 1976) makes use of a laterally

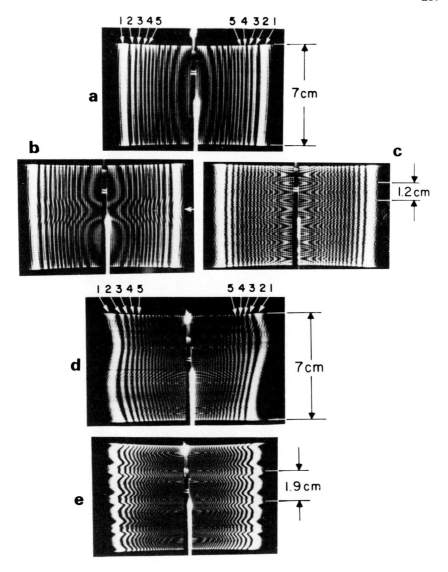

Fig. 9.16 Back-scattered light patterns from sections of fibers pulled with various sources: (a) electric furnace—uniform fiber. (b) electric furnace—single perturbation. (c) electric furnace—regular perturbation. (d) oxy-hydrogen torch. (e) CO_2-laser.

moving optical fringe pattern that is generated by two laser beams converging at a small angle. The fringe spacing is used as a "yardstick" to compare its space to the diameter of the fiber. A resolution better than 0.1% was obtained for fibers with a diameter ranging between 45 and 260 μm.

A coherent-optical method based on Moiré interference for measuring fiber diameters has also been described (Fink and Schneider, 1974). In this technique, the optical diffraction pattern of the fiber is sampled by a special spatial filter so that the fiber diameter corresponds to the position of a straight Moiré line of least-modulation across the filter.

Two optical-based techniques have been proposed to check the dimensions of hollow glass fibers as used in the fabrication of liquid-core waveguides. In one of the methods (Burkhtiarova, 1975), converging cylindrical waves are focused onto the capillary such that the focal line is perpendicular to its axis. A structure with a small spatial period, carrying information about the diameter, superimposed on a structure with a large spatial period, caused by the wall thickness, appears in the interference pattern formed by the superposition of the four waves reflected from the surfaces of the fiber. In the case of specimens with varying thickness a Moiré pattern is observed, characterizing the variation in wall thickness.

The second nondestructive (Horton and Williamson, 1973) method for studying capillaries makes use of the interference patterns obtained by illuminating the fiber normal to its axis with a laser beam polarized perpendicular to the plane of incidence. Characteristics of these patterns, explained by using a ray-vector analysis, permit the evaluation of the external radius and the ratio of radii.

9.5.3.5 Measurements on Coated Fibers.   Plastic jackets are used directly as the cladding for fused silica cores, creating large-numerical-aperture fibers and also as an overcoat to glass-clad fibers to help control microbending loss and to provide mechanical protection. It is necessary that the coating be applied uniformly and concentrically around the fiber, and that its diameter be maintained constant for routine handling and splicing, as well as to ensure optimum strength and transmission characteristics.

The jackets are usually applied to the fibers during the pulling operation, and the constancy of their diameters can generally be monitored by many of the methods previously discussed. An additional factor to be monitored however, is now present, viz., that of coating concentricity. This is presumably controlled by the position of the fiber relative to the exit orifice of the coating applicator, and since there is the possibility that this alignment may vary, due to such factors as preform variations or vibrations, some means to monitor concentricity must be incorporated into the drawing operation.

Recently a sensitive, nondestructive and noncontacting method to do this has been described (Presby, 1976b; Marcuse and Presby, 1977). The technique is based on observations of the back-scattered light pattern arising when a beam of laser light is incident at right angles to the axis of the fiber. Eccentricity between the fiber and the jacket is manifested by a very sensitive dependence of the back-scattered light pattern on fiber orientation. If a perfectly symmetric fiber is rotated around its axis, the back-scattered light pattern does not change. If an eccentrically coated fiber is rotated, bright bands shift around, broadening or narrowing in width and becoming asymmetrically positioned around the center line; in addition, bright bands appear and disappear depending on fiber orientation. This fact is utilized in the monitoring arrangement by producing two back-scattered patterns simultaneously, using light incident from two different directions. Observations of differences between the two light distributions give a strong indication of fiber eccentricity in the coating. A necessary requirement for the implementation of this technique is that the coating material be fairly transparent.

9.5.3.6 Control Systems.   Once the diameter of an optical fiber is determined as the fiber is being pulled, the fiber fabricator can in principle make adjustments to the apparatus to maintain the diameter at a constant value if variations are found to occur. A preferred situation would be to have the variations compensated for automatically through a feedback control system. The essential elements of such a system consist of the diameter measurement sensor located some distance downstream from the forming zone of the preform, appropriate electronics comprising filtering and amplification, and the drawing mechanism, the speed of which is controlled by the processed diameter signal. The inclusion of other elements in the control system, such as the preform feed mechanism or the oven temperature, is generally not warranted due to their very slow response.

In implementing a practical control system, many factors affecting its dynamic response characteristics must be taken into account. The measurement sensor, as previously discussed, will have a finite response due to both its spatial and time resolution, limiting the range of small scale variations that can be detected. In addition, if the mean fiber velocity is V and the neck-to-sensor distance is L, there will be a delay $T = L/V$ between the creation of a diameter error and the time that it is sensed. Of course, this measurement delay time can be shortened by moving the measurement system closer to the heat zone but some finite delay will still exist. The electronics and control circuitry will also possess a finite response time, though this, in general, will be small.

Of critical importance is the dynamic response of the drawing mechanism. This consists of the motor controlling the take-up drum or the drawing capstan and the mechanism itself. The inertia of various components of the drawing mechanism will limit its speed of response.

In principle, all of these factors can be taken into account in linear perturbation models used in control theory (Kuo, 1967; D'Azzo and Houpis, 1966) and the results utilized to design a feedback control system. In practice, an accurate analysis involves a detailed characterization of the response of each of the components of the system, which have to be obtained either theoretically or experimentally. In addition, trade-off's have to be made between the desired degree of diameter control and the system dynamics. The latter can actually enhance undesirable diameter variations at specific frequencies of instability where oscillations may build up. It may generally be expected that control systems will be effective in counteracting large, long-scale variations but not rapid variations as may be caused by random vibrations or air current and temperature variations in the neck-down region of the preform. The latter must be reduced by a careful design of the drawing apparatus and its environment. With such systems, fibers have been produced consistently that have a standard deviation of diameter of less than ±1% (Cohen and Klaiber, 1977; Runk, 1977; Irven *et al.*, 1977; Kobayashi *et al.*, 1977).

A novel method of diameter control has been reported in which an additional flow rate of gas is superimposed on a steady gas flow in the muffle tube of a resistance furnace (Imoto *et al.*, 1977). The additional flow is changed by a gas-flow controller according to signals processed for the diameter detected. The additional increase or decrease in gas flow rate induces a corresponding transient increase or decrease in the fiber diameter. The diameter-change response time is below 0.1 sec/μm, and resultant diameter fluctuations have been reduced to ±0.5% for fiber lengths of over 1 km.

9.6 COATING AND JACKETING

In-line coating of fibers is important to provide a protective buffer on the fiber, which will help in maintaining the pristine strength of the fiber and protect it from atmospheric attack, mechanical abuse, and microbending. The relationship of these properties to the coating materials and methods of application is discussed in detail in Chapter 10 and will be only briefly noted here.

The method of application depends largely on the coating material. The best results to date have been obtained with polymer coatings applied from the melt, or in monomer form followed by in-line thermal or ultraviolet activated curing (Schonhorn *et al.*, 1976). These coatings have been successfully applied using a flexible silicone applicator (Hart and Albarino, 1977) which, due to its compliant nature, applies a concentric coating even with slight lateral movement of the fiber during drawing. In addition, the applicator material is chosen so as to minimize damage to the fiber surface during operation. To date, these polymer coatings have been cured at drawing rates of up to 1 m/sec. It may be, however, that the curing operation will ultimately limit maximum production draw speeds.

9.7 COMBINED APPARATUS

9.7.1 General Principles

The ways in which the basic components are combined in an apparatus for drawing glass lightguides may differ in detail among organizations and even within the same organization, but the overall appearance of the apparatus varies little. Most systems are designed for drawing the fiber downward in a vertical path. While horizontal drawing is possible it has the disadvantages of the neck-down profile becoming asymmetrical, and it poses greater difficulty in maintaining alignment and geometric uniformity. Drawing vertically upward results in similar disadvantages plus added difficulties with coating applicators.

The apparatus should be free-standing with ample floor space to accommodate platforms for ease of servicing and operating the unit. Most consist of sturdy metal frames ranging from precision optical slabs to all varieties of metal angles, posts, and brackets. The apparatus should be constructed, and located in an area free from vibrations and with sufficient ceiling height to accommodate all of the components, especially on-line coating systems. In addition to providing the required services, such as electrical power and cooling water, it is important to have good ventilation for exhausting vapors that may emanate from the coating units and preform heat source.

A typical laboratory apparatus for use with glass preforms is shown in Fig. 9.17. The preform support and feed mechanism is located at the top, followed by a high temperature-graphite resistance furnace, a fiber diameter monitoring unit, an experimental coating system, and finally a take-up drum and winding mechanism. The assembly is mounted on a heavy aluminum frame measuring ~1.2 × 1.8 × 2.4 m. The feed mechanism has a variable speed motor and control unit capable of less than 1% variation and full torque at low feed speeds of 0.5–1.5 cm/min. The winding mechanism is driven by a variable speed motor connected by a gear belt to a lead screw that traverses the drum to give a 0.3 mm (0.012″) pitch to the fiber winding. With this apparatus a fiber of 110-μm diameter is drawn from a 8-mm diameter preform at a speed of 1 m/sec using a preform feed rate of 1.2 cm/min.

The same apparatus has been used with alternate heat sources: a Pt re-

Fig. 9.17 Typical laboratory glass fiber lightguide drawing apparatus (IEE Conf. Publication No. 132, p. 38).

sistance furnace, an O_2–H_2 torch, and an RF inductively-heated zirconia furnace. One advantage of laser heating compared to furnace heating is the ability to view the preform during the drawing operation.

Component alignment becomes critical when applying coatings online. The path of the fiber between the tip of the preform and the fiber puller should be plumb. The use of x–y stages with the preform support and coating applicators simplifies the alignment problem.

A variety of plastics and metals have served as materials for winding drums in sizes from 10 to 60 cm in diameter. The winding drum shown in Fig. 9.17 is a modified deep-drawn aluminum can with a black anodized finish measuring 28 cm diameter by 45 cm. Coated and uncoated fibers up to 0.25 mm (0.010″) diameter have been wound in a single layer and in lengths of 1100 m at speeds up to 1.5 m/sec.

Winding directly onto a take-up drum used for applying the drawing force places the fiber under stress. One method for avoiding this is to use a capstan device for the drawing force, and to take up subsequent slack in the fiber by a separate winding unit. Another advantage to this method is that the low inertia of a properly designed capstan enables it to respond faster to diameter control signals than a drum winding mechanism.

Other apparatuses have incorporated improved instrumentation for better diameter control and monitoring coating thickness, capstan pullers, and winding mechanisms designed with variable pitch and reversible traverse for preparing multilayer windings of coated fibers.

An example of a drawing apparatus used in a lightguide pilot production line is shown in Fig. 9.11. Compared to the laboratory unit, the major difference is the height of ~ 6 m which adds flexibility for accommodating various on-line coating systems and process monitoring instruments.

9.7.2 Storage and Handling

Coated fibers exhibit good durability, including resistance to mechanical stresses encountered in automatic cabling operations, and are easily handled. Uncoated fibers exhibit increasing fragility and degradation of strength with handling and aging. Storage of fibers for long periods of time on drums of small diameters (<25 cm) subjects them to bending stresses that could lead to premature static fatigue failures. When laying down multilayers of fiber on a take-up drum, the insertion of a plastic or tissue sheet between layers assists in unreeling the fiber.

Kilometer lengths of coated and uncoated lightguides have been shipped without damage over considerable distances on drums of the type shown in Fig. 9.17. One packaging scheme consists of wrapping a plastic film over the fiber and enclosing the drum in a cylindrical fiberboard carton containing wooden inserts as drum supports at either end. The first carton is enclosed in a similar outer carton containing foam cushion pads inside each end. Good housekeeping in all work areas is recommended for reducing the risk of fiber damage.

REFERENCES

Allan, W. B. (1973). "Fibre Optics—Theory and Practice." Plenum, New York.

Bagley, W. G., *et al.* (1964). "A Special Report on Fiber Optics." Nimrad Press, Inc., Boston, Massachusetts.

Beales, K. J., Day, C. R., Duncan, W. J., Midwinter, J. E., and Newns, G. R. (1976). Preparation of sodium borosilicate glass fibre for optical communication. *Proc. Inst. Electr. Eng.* **126,** 591–596.

Boys, C. V. (1887). On the production, properties, and some suggested uses of the finest threads. *Philos. Mag.* **23,** 489–499.

Burkhtiarova, T. V., Dyachenko, A. A., Zhabotinsky, M. Ye., and Shushpanov, O. Ye.

(1975). An interference method of checking the geometrical dimensions of capillary fibers and tubes. *Radio Eng. Electron. Phys. (Engl. Transl.)* **20,** No. 8, 1.

Cannell, G. J. (1977). Continuous measurement of optical-fibre attenuation during manufacture. *Electron. Lett.* **13,** 125–126.

Chiccacci, P. F., Sheggi, A. M., and Brenci M. (1976). R. F. induction furnace for silica fibre drawing. *Electron. Lett.* **12,** 265–266.

Cohen, L. G., and Glyn, P. (1973). Dynamic measurement of optical fiber diameter. *Rev. Sci. Instrum.* **44,** 1749.

Cohen, M. I. (1967). Laser beams and integrated circuits. *Bell Lab. Rec.* **45,** 246–252.

Cohen, M. I., and Klaiber, R. J. Drawing of smooth optical fibers *Tech. Dig. Top. Meet. Opt. Fiber Transm., 2nd. 1977* Paper TuB4.

Cowan, J. H., Buehl, W. M., and Hutchins, J. R., III (1966). An electrochemical theory for oxygen reboil. *J. Am. Ceram. Soc.* **49,** 559.

D'Azzo, J. J., and Houpis, C. H. (1966). "Feedback Control System Analysis and Synthesis." McGraw-Hill, New York.

Diamond, F. (1953). "The Story of Glass." Harcourt, New York.

Dickson, J. H. (1951). "Glass—A Handbook for Students and Technicians," Chapter XIII. Chem. Publ. Co., New York.

DiMarcello, F. V., and Williams, J. C. (1975). Reproducibility of optical fibers prepared by a chemical vapor deposition process. Inst. Electr. Eng. **132,** 36–38.

DiMarcello, F. V., and Williams, J. C. (1978). *Bell Syst. Tech. J.* **57,** Part I, 1723–1734.

Dove, A. B., ed. (1968). "Steel Wire Handbook," Vol. 2, Chapter 2. Wire Assoc. Inc., Branford, Connecticut.

Epperson, J. P., Dyer, R. W., and Grzywa, J. C. (1966). The laser now a production tool. *West. Electr. Eng.* **10,** 9–17.

Fink, W., and Schneider, W. (1974). A coherent-optical method for measuring fibre diameters. *Opt. Acta* **21,** 151.

French, W. G., MacChesney, J. B., and Pearson, A. D. (1975). Glass fibers for optical communications. *Annu. Rev. Mater. Sci.* **5,** 373–393.

Gagliano, R. P., Lumley, R. M., and Watkins, G. S. (1969). Lasers in industry *Proc. IEEE* **57,** 114.

Geyling, F. T. (1976). Basic fluid-dynamic considerations in the drawing of optical fibers. *Bell Syst. Tech. J.* **55,** 1011–1056.

Hart, A. C., and Albarino, R. V. (1977). An improved fabrication technique for applying coating to optical fiber waveguides. *Proc. Opt. Fiber Transm., 2nd, 1977* p. Tu B2-1.

Holly, S. (1976). Lateral interferometry monitor fiber diameter. *Laser Focus* **12,** 58.

Horton, R., and Williamson, W. J., (1973). Interference patterns of a plane-polarized wave From a hollow glass fiber. *J. Opt. Soc. Am.* **63,** 1204.

Imoto, K., Aoki, S., and Sumi, M. (1977). Novel method of diameter control in optical-fibre drawing process *Electron. Lett.* **13,** 726.

Inoue, T., Koizumi, K., and Ikeda, Y. (1976). Low-loss light focusing fibers manufactured by a continuous process. *Proc. Inst. Electr. Eng.* **123,** 577–580.

Irven, J., Black, P. W., Harrison, A. P., Hearn, V., Lamb, J. G., and Scott, B. J. (1977). Control techniques for producing silica fibers to high specifications. *Tech. Dig., Int. Opt. Opt. Fiber Comm.,* Paper B9-2, p. 323.

Jaeger, R. E. (1976). Laser drawing of glass fiber optical Waveguides. *Am. Ceram. Soc., Bull.* **55,** 270–273.

Kaiser, P. (1974). Drawing-induced coloration in vitreous silica fibers. *J. Opt. Soc. Am.* **64,** 475–481.

Kaiser, P. (1977). Contamination of furnace drawn silica fibers. *Appl. Opt.* **16,** 701–704.

Kaiser, P., Marcatili, E. A. J., and Miller, E. (1973). A new optical fiber. *Bell Syst. Tech. J.* **52,** 265–269.

Kao, K. C., and Hockham, G. A. (1966). Dielectric-fibre surface waveguides for optical frequencies. *Proc. Inst. Electr. Eng.* **113,** 1151–1158.

Kapany, N. S. (1967). "Fiber Optics, Principles and Applications." Academic Press, New York.

Kawamura, Y., Arima, T., and Syogaki, T. (1973). Single mode fiber production by double crucibles. *Fujitusu Sci. Tech. J.* **9,** 121–140.

Kerker, M., (1969). "The Scattering of Light and Other Electromagnetic Radiation." Academic Press, New York.

Kitayama, K., and Ikeda, M. (1977). Leaky modes in spliced graded-index fibers. *Appl. Phys. Lett.* **30,** 227–228.

Kobayashi, T., Osanai, H., Sato, M., Takata, H., and Nakahara, M. (1977). Tensile strength of optical fiber by furnace drawing method *Tech. Dig., Int. Opt. Opt. Fiber Comm., 1977* Paper B9–4, p. 331.

Koizumi, K., Ikeda, Y., Furukawa, M., and Surmimoto, T. (1974). New light-focusing fibers made by a continuous process. *Appl. Opt.* **13,** 225–229.

Krawarik, P. H. (1975). Power spectral measurements for optical fiber outer diameter variation. *Top. Meet. Opt. Fiber Transm. 1st. 1975* Paper PD1.

Kuo, B. C. (1967). "Automatic Control Systems." Prentice-Hall, Englewood Cliffs, New Jersey.

Kurkjian, C. R., Albarino, R. V., Krause, J. T., Vazirani, H. N., DiMarcello, F. V., Torza, S., and Schonhorn, H. (1976). Strength of 0.04–50 m lengths of coated fused silica fibers. *Appl. Phys. Lett.* **28,** 588.

Kutukov, S. S., and Khodakovskii, M. D. (1964). Investigating the movement of glass in the forming of continuous glass fibers by high speed filming. *Keramika* **21,** 3–10.

Li, P. C., Pontarelli, D. A., Olson, O. H., and Schwartz, M. A. (1969). Fused quartz fiber optics for ultraviolet transmission. *Am. Ceram. Soc., Bull.* **48,** 214–220.

Lisita, M. P., Berezhinski, L. I., and Volakh, M. Ya. (1972). "Fiber Optics." Isr. Program Sci. Transl., New York.

Loewenstein, K. L. (1973). "The Manufacturing Technology of Continuous Glass Fibers." Am. Elsevier, New York.

Longfellow, J. (1971). High speed drilling in alumina substrates with a CO_2 laser. *Am. Ceram. Soc., Bull.* **50,** 251–253.

MacChesney, J. B., O'Connor, P. B., DiMarcello, F. W., Simpson, J. R., and Lazay, P. D. (1974). Preparation of low-loss optical fibers using simultaneous vapor phase deposition and fusion *Proc. Int. Congr. Glass, 10th, 1974* Paper 6-40.

Maloney, E. J. (1968). "Glass in the Modern World." Doubleday, Garden City, New York.

Marcuse, D. (1974). "Theory of Dielectric Optical Waveguides." Academic Press, New York.

Marcuse, D., and Presby, H. M. (1975). Mode coupling in an optical fiber with core distortions. *Bell Syst. Tech. J.* **1,** 3.

Marcuse, D., and Presby, H. M. (1977). Optical fiber coating concentricity: Measurement and analysis. *Appl. Opt.* **16,** 2383.

Maurer, R. D. (1977). Effect of dust on glass fiber strength. *Appl. Phys. Lett.* **30,** 82–84.

Paek, U. C. (1974). Laser drawing of optical fibers. *Appl. Opt.* **13,** 1383–1386.

Payne, D. N., and Gambling, W. A. (1976). A resistance heated high temperature furnace for drawing silica based fibers for optical communications. *Am. Ceram. Soc., Bull.* **55,** 195–197.

Pearson, A. D., and French, W. G. (1972). Low loss glass fibers for optical transmission. *Bell Lab. Rec.* **50,** 102–109.

Personick, S. D. (1971). Time dispersion in dielectric waveguides. *Bell Syst. Tech. J.* **50,** 843–859.

Peyches, I. (1945). Physical problems in the production of glass fibers. *Mem. Soc. Ing. Civ. Fr.* **98,** 411–422.

Presby, H. M. (1974). Refractive index and diameter measurements of unclad optical fibers. *J. Opt. Soc. Am.* **64**, 280.

Presby, H. M. (1976a). Ellipticity measurement of optical fibers. *Appl. Opt.* **15**, 492.

Presby, H. M. (1976b). Geometrical uniformity of plastic coatings on optical fibers. *Bell Syst. Tech. J.* **55**, 1525.

Presby, H. M. (1977). Detection of geometric perturbations in optical fibers. *Appl. Opt.* **16**, 695.

Presby, H. M., and Marcuse, D. (1974). Refractive index and diameter determinations of step index optical fibers and preforms. *Appl. Opt.* **13**, 2882.

Rawson, H. (1967). "Inorganic Glass Forming Systems." Academic Press, New York.

Revi, A. C. (1957). Miniature portraits in glass rods. *Glass Ind.* **38**, 323–330.

Runk, R. B. (1977). A zirconia induction furnace for drawing precision silica waveguides. *Tech. Dig., Top. Meet. Opt. Fiber Transm., 2nd, 1977* Paper TuB5.

Rzepeka, M. A., and Stuchly, S. S. (1974). A microwave system for measurement of the diameter of thin dielectric fibers. *IEEE Trans. Instrum. Meas.* March, 100.

Saifi, M. A., and Borutta, R. (1976). Split beam system for laser drawing of optical fibers *Dig. Tech. Pap. Conf. Laser Electro-Opt. Syst., 1976* TuB3, pp. 6–8.

Schonhorn, H., Kurkjian, C. R., Jaeger, R. E., Vazirani, H. N., Albarino, R. V., and DiMarcello, F. V. (1976). Epoxy acrylate coated fused silica fibers with tensile strengths greater than 500 ksi (3.5 GN/m²) in 1 kilometer gauge lengths. *Appl. Phys. Lett.* **29**, 712.

Smithgall, D. H., Watkins, L. S., and Frazee, R. E., Jr. (1977). High-speed noncontact fiber-diameter measurement using forward light scattering. *Appl. Opt.* **16**, 2395.

Stone, J. (1972). Optical transmission in liquid-core quartz fibers. *Appl. Phys. Lett.* **20**, 239.

Straka, E. R. (1968). Survey of the present technology of fiber optics. *SPIE Semin. Proc.* pp. 31–41.

Takahashi, S., and Kawashima, T. (1977). Preparation of low loss multi-component glass fiber. *Tech. Dig. Int. Conf. Integr. Opt. Opt. Fiber Commun. 1977* p. 621.

Takayama, K., Susa, N., Hirai, M., and Uchida, N. (1977). Observations of surface flows in fused silica optical fibers. *Appl. Phys. Lett.* **30**, 155–157.

van Ass, H. M. J. M., Geittner, P., Gossink, R. G., Küppers, D., and Severin, P. J. W. (1976). The manufacture of glass fibers for optical communications. *Philips Tech. Rev.* **36**, 182–189.

Watkins, L. S. (1974). Scattering from side-illuminated clad glass fibers for determination of fiber parameters. *J. Opt. Soc. Am.* **64**, 769.

Watkins, L. S. (1975). Instrument for continuously monitoring fiber core and outer diameters. *Top. Meet. Opt. Fiber Transm. 1st. 1975* Paper TuA4-1.

Yamazaki, T., and Yoshiyagawa, M. (1977). Fabrication of low loss multi-component glass fibers with graded index and pseudo step index distribution. *Tech. Dig. Int. Conf. Integr. Opt. Opt. Fiber Commun. 1977* p. 617.

Zvanut, C. M. (1972). Manufacturing development: High strength continuous silica filaments. *Am. Ceram. Soc., Bull.* **51**, No. 9.

Chapter 10

Coatings and Jackets

LEE L. BLYLER, JR.
BERNARD R. EICHENBAUM
HAROLD SCHONHORN

10.1 THE ROLES OF COATINGS

The roles coatings can play in the performance of an optical fiber system fall broadly into five categories.

(a) Abrasion protection
(b) Microbending loss protection
(c) Static fatigue protection
(d) Identification
(e) Cladding

We will focus our attention on each category in turn. In the later parts of the chapter, the need for coatings, requirements on coatings, surface treatments, and techniques of application will be discussed.

10.1.1 Abrasion Protection

Glass strength characteristics are dominated by the brittle nature of glass materials. Chapter 12 deals with the relevant theories and data in detail and here we will merely point out the key factors.

The lack of ductility in the glass fiber optic medium does not permit relaxations of any stress concentrations. We see in Figure 10.1 a fiber viewed from the side with a flaw in the surface. The fiber is under tensile stress with the longitudinal lines spaced to represent the local level of stress; the closer the lines, the higher the stress. Stress concentration exists at the tip of the flaw. If plastic flow were an available response of the

Fig. 10.1   Stress concentration at a crack tip.

medium, the flaw could then distort into a more rounded shape with a resultant decrease in the degree of stress concentration. But glass cannot respond this way and the stress concentration remains. The larger the flaw, the greater the degree of stress concentration. When the local stress is sufficiently high, catastrophic crack growth begins and results in glass fracture.

A fiber freshly drawn from a furnace has the best surface flaw distribution it will ever realize. Abrasive contact of the fiber with other surfaces during transport, take-up, or any subsequent stage of handling can only aggravate existing flaws or generate new ones. Therefore, not only must the coating protect the fiber from abrasion during subsequent handling but also during the transport of the fiber from the coating applicator to the take-up. Furthermore, the fiber must be transported from the furnace to the applicator without damage and the applicator itself must not be a source of abrasion.

To fully appreciate the deleterious effects of surface flaws in a quantitative manner we refer the reader to Chapter 12.

10.1.2 Microbending Loss Protection

Microbending loss is the name applied to radiation losses arising from mode coupling generated by random bends of the optical fiber. Its theoretical foundation is covered in Section 3.12. The term microbending is appropriate because the short period (~ 1 mm) random bends which most efficiently couple the modes for typical fibers need merely micron-like amplitudes to severely degrade transmission loss properties of low-loss fibers.

The cable environment itself can impose these random bends on the fiber. In this context, the term packaging loss is often used.

Gloge (1975) pointed out that the coating can offer protection from microbending loss in two ways.

(1) It can act as a buffer, a soft compliant enclosure which masks the peaks and valleys of adjacent surfaces in a mechanical way. We can intuitively see that in the limit of a sufficiently thick and very soft coating, the fiber path is mechanically oblivious to its environment.

(2) It can act as a stiffener, a hard enclosure which, by stiffening the fiber-coating structure, makes the fiber resistant to conforming to environ-

mental surfaces. Again, a limiting case can illustrate the effect. We can consider a fiber encased in a thick steel sheath so stiff that it cannot respond with the spatial bending frequencies which cause microbending losses.

Ideally, both of these protective mechanisms can be incorporated into one coating by using a hard outer shell and a soft inner layer. Finally, the coating itself should be as uniform and smooth as possible to avoid imparting microbends to the fiber from coating irregularities.

10.1.3 Identification

In multifiber cables, there is a need to distinguish the fibers from one another. Geometrical orientation can serve this purpose to some degree, but a color code plus geometry identification system is useful for quality control during manufacture, and ambiguities are less likely in the field. In the case of a ribbon structure cable, a color code can differentiate one fiber from another and also one ribbon from another.

Dyes are used for coloring the coatings. Pigments add particulate matter to the coating which hardens it (reduces its microbending—buffering action) and which may also be a source of abrasion during application. Dyes which have a high molecular weight or, more advantageously, bond chemically to the coating polymer, lessen the chance of dye migration.

10.1.4 Cladding

Polymer coatings used as claddings carry some of the optical signal through them because they are an active part of the waveguide. In this particular application, polymer optical properties are especially important since (1) signal power loss can occur as a result of coating absorption or scattering, and (2) the refractive index of the polymer must be lower than the core so that a waveguide is indeed formed.

Power loss can be limited by choosing highly transparent polymers and also by constructing the waveguide so that a minimal amount of the signal power propagates in the cladding. If the core material is pure fused silica, then two broad categories of polymers satisfy the refractive index requirement—fluorocarbons and silicones. Polymer-clad fibers are discussed in detail in Section 10.5.

10.2 REQUIREMENTS ON COATINGS

A variety of parameters relating to dimensions and material characteristics are at the disposal of the coating designer to optimize the coating for its specific use. While these parameters are interrelated, a separation into three broad categories may be helpful for setting perspectives in the discussions. The design areas are: (a) dimensional considerations, (b) mechanical characteristics of the coating material, and (c) chemical characteristics of the coating material.

10.2.1 Coating Dimensions

In the context of the various roles coatings play listed in Section 10.1, coating dimensions relate mainly to abrasion protection, microbending loss protection, and the special case of the cladding role. We will concentrate here on the first two and leave cladding considerations to Section 10.4.5 where polymer-clad fibers are discussed in detail.

Certainly, the thicker a coating of a particular material, the more abrasion protection it affords the fiber. Also, there is more microbending loss protection since the buffering action of a soft coating improves with greater thickness and the stiffening action of a hard coating likewise increases with thickness. Weighed against these arguments for applying a thicker coating are some countervailing considerations.

A basic consideration is space efficiency. Ultimately, if the fibers will be emplaced in a cable, the capacity of the cable will be partly dependent on the coated outside diameter (o.d.) of each fiber. Chapter 13 explores various cable designs. Depending on the design, the cross section taken up directly by the coated fibers influences the total cross section of the cable, sometimes with a multiplier effect. For some applications a minimization of cable diameter is desired, or alternatively, the cable o.d. is limited. In such cases, a thinner coating means a greater number of fibers and fixed costs are amortized over a greater total bandwidth.

For splicing ease, fibers must be well-centered in their coatings. The contiguous ribbon design (see Chapter 13) relies on the coating uniformity to establish a first-order level of fiber alignment for insertion into the connector. Relations between coating thickness and coating eccentricity are investigated in Section 10.4.3 for the pressureless die applicator. In general, a thinner coating gives better centering. Therefore, in addition to considerations of space efficiency and spacing regularity, we have a third argument against extraordinarily large coating thickness, at least when using a pressureless die.

Splicing requirements affect coating thickness in yet another way. The hardware for stacked array splices limits the closeness between adjacent fibers. Current alignment fixture construction becomes too fragile if it is scaled down to too small a fiber spacing. Thus stacked-array splicing hardware demands a fiber well centered in its coating which suggests thin coatings; however, the same limits how thin the coatings can be.

10.2.2 Mechanical Characteristics of the Coating Material

Again, in the context of the various roles coatings play listed in Section 10.1, there are overall two broad ways in which the mechanical characteristics of the coating material influence fiber properties: (1) abrasion protection and (2) microbending loss protection.

The main relationship of mechanical characteristics to abrasion protection is direct and obvious—the tougher the coating material the more abrasion protection available for a given thickness. On the other hand, there are more subtle considerations related to glass substrate adhesion, filler particles within the coating, and draw-down conditions in the case of tubing extrusion. Fillers may abrade because of flow shear during application from a melt, prepolymer, or solution or conceivably, particles can abrade the glass under conditions of weak adhesion and high shear stress. A speed differential between fiber and tubing in tube extrusion can also cause abrasion depending on mechanical properties of the molten polymer.

The relations between mechanical properties and microbending loss protection are even more involved. We have already discussed in Section 10.1.2 how microbending can be controlled through proper choice of coating properties. Gloge (1975) has calculated the effects of various coating moduli thicknesses, and multiple-layer constructions and Gardner (1975) has compared these predictions to actual cases of fibers coated by a variety of materials. Attention was focused on the buffering action of soft coatings.

The fibers were wound under varying degrees of tension onto an acrylic drum both before and after coating. The resulting microbending caused by the fiber conforming to the slightly irregular drum surface was measured for various tensions both before and after application of the coating. With increased winding tension, the fiber conformed to surface irregularities more and more and the transmission loss increased. The slope of the loss-vs-tension curve changed from bare fiber to coated fiber, the decrease in slope showing that the coating provides microbending loss protection. The data show (Figure 10.2) that different coatings produce differing degrees of microbending loss protection. In general, the softer the coating, the more the slope decreases, indicating better the buffering action as predicted by the Gloge theory for the range of coatings studied.

Whereas the Young's modulus describes an elastic property of material, the viscous response of a coating can be characterized by cold flow or creep. This property, a slow inelastic deformation, also has important implications with respect to microbending. Indeed, coated fibers wound on a drum under tension show a decrease in transmission loss with time as material flows to relax microbends. Time scales depend on temperature as

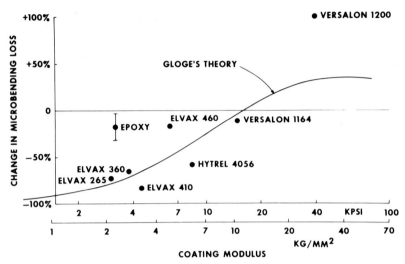

Fig. 10.2 Experimental and theoretical effects on microbending loss by varying the Young's modulus of the coating.

well as the particular material. Hundreds of hours are typical relaxation times for ethylene–vinyl acetate copolymers (EVA) at room temperature.

In addition to reeled-fiber tests just described, relaxation has been observed in fiber ribbon arrays (Standley, 1974). Fibers coated with EVA were assembled into ribbons by two techniques:

(1) sandwiching between two tapes with pressure-sensitive adhesive (adhesive sandwich ribbon or ASR) (Saunders, 1977), and

(2) welding the coatings together with solvents and/or heat (solvent welded ribbon, heat welded ribbon or SWR, HWR).

For ASRs, a progressive decrease in packaging loss was noted as time passed after manufacture. Figure 10.3 illustrates the typical behavior for

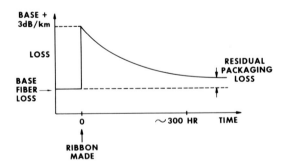

Fig. 10.3 General form of packaging loss time dependence.

early ASRs which showed high added loss upon ribboning and a decay of this added loss to some residual value, usually $\frac{1}{2} \rightarrow 1$ dB/km (Buckler *et al.*, 1978). Higher-molecular-weight EVAs showed consistently longer decay time constants than lower-weight grades. Time constants were much longer for SWRs but their relaxation rates could be vastly accelerated with heat. Initial added loss when making SWRs was sometimes extraordinarily high for some fibers, several tens of decibels per kilometer. While monitoring transmission, hot air was directed onto the SWR to provide a dramatic illustration of microbend relaxation. In the span of 2 min, an added loss of 40 dB/km was totally annealed away. Not only can relaxation of the coating relieve microbends, but relaxation in other cable materials, such as the adhesive in ASRs, can also provide relief.

10.2.3 Chemical Characteristics—Strippability

When fibers are made ready for splicing, their coatings are stripped off. Splicing requires such precise alignment (see Chapter 14) that fiber dimensions should be kept within ± 2 μm to minimize loss. Coating technology is nowhere near this accuracy so that coatings must be removed before alignment is obtained.

To remove many polymers from the glass fiber substrate, a bath of hot propylene glycol is particularly effective. It removes a variety of coatings including solution-applied EVA, hot-melt-applied EVA, EVA-based hot melts, Kraton®-based* hot melts, uv-cured epoxy acrylate, and uv-cured polyester thiol. The procedure is to dip the coated fibers in a reservoir of propylene glycol at 160°C for about 30 sec. At the top of the reservoir the fiber is clamped between pads of Scott Felt®,[†] a soft polyurethane foam (see Section 10.4.1). Upon withdrawal from the bath the fiber is pulled up through the clamp and the coating is grabbed by the pad and slides off.

The technique can even be used with ASRs without first removing the tape. A hot wire is used to burn the polyester tape at approximately the point where stripping is to start. Applying heat from only one side is sufficient. The ASR is then dipped into the hot reservoir up to the burn and clamped as before. After 3 min, the ASR is pulled through the clamp and the tape and coatings slide off.

10.3 SURFACE TREATMENT OF SILICA

Although the fracture strengths for long lengths ($\sim$ km) of coated optical fibers have been reported to approach values of 7 GN/m² (10^6 psi) (Schonhorn *et al.*, 1976; Kurkjian *et al.*, 1976), there is a serious degradation in

---

* Kraton®—trademark of Shell Chemical Co.
† Scott Felt®—trademark of Scott Paper Co.

strength when these same coated fibers are exposed simultaneously to high humidity and tensile stress (Wang *et al.*, 1977). The diffusion of water vapor or bulk water through the coating may possibly displace the coating from the glass, exposing the pristine surface to environmental effects. While a variety of procedures have been adopted to prevent this premature failure, among them the use of silane coupling agents, it is not clear that this effect has been explained adequately.

In this section we shall focus attention on the chemical, physical, and geometric characterization of a silica surface and propose some generalizations regarding specific mechanisms whereby this type of surface can interact with polymeric coatings. Further, we shall consider specific types of interfacial reactions which may prevent or retard bond disruption after exposure to water, humid air, or other hostile environments in the presence of an applied stress.

10.3.1 Nature of Silica Surface

Although a considerable effort has been expended in examining the hydrated nature of oxide surfaces (Anderson and Wickerscheim, 1964) little, if any, information exists concerning the nature of the surface of a freshly drawn silica fiber. However, when considering the permanance of a composite structure and the aging characteristics of the interface, i.e., possible rehydration of the silicon oxide in the presence of the ambient, the following considerations may become of importance. Debye and Van Beek (1959), Bowden and Throssell (1951), Bowden (1962), and Bowden and Tabor (1963) have found that there is a reasonably thick layer of water adsorbed on uncoated silica at ambient temperatures and humidities. Even though the humidity is reduced appreciably, it is virtually impossible to remove the last few adsorbed monolayers of water. Only after exposing silica to high temperatures can the surface be dehydrated (Shafrin and Zisman, 1967). In the case of amorphous silica fibers, caution must be exercised in not annealing at too high a temperature for too long a period of time since devitrification processes would ensue and the strength would be markedly reduced. The experiments of Irwin (1967), concerning the fracture of glass and the cleavage experiments of Orowan (1932) and Bowden and Tabor (1963) with mica, clearly show that the crack propagation velocities in glass and the energy to cause cleavage in mica depend strongly on the activity of water in the surrounding phase. Orowan (1932), measured a factor of 10 decrease in strength from high vacuum to cleavage in air or water vapor, illustrating the extreme rapidity with which silica surfaces hydrate and the profound ability of these adsorbed surface layers to reduce the surface energy of a high-energy solid.

The attractive forces attributable to the surface hydroxyl groups on silica depend strongly on the electron density of the oxygen atom in the group

$\equiv$Si—OH. One manner of gauging the reactivity of solid surfaces has been to consider the concept of the isoelectric point of the solid oxide (IEPS). The IEPS is defined as the pH at which the net surface charge is zero. Analysis of the factors that influence the IEPS of silicon oxide in water, should therefore, also serve as a qualitative guide to the factors that govern the ability of these surfaces to lose, donate, attract, or bind protons in contact with nonaqueous liquids.

Parks (1965) has compiled, and has critically reviewed, the literature for the isoelectric points of a large number of solid oxides, including SiO_2 (quartz). Fully hydrated surfaces give a higher IEPS than freshly prepared (or drawn in our case), reflecting the time dependence of the hydroxyl formation reaction.

10.3.2 Hydrolysis of a Silica Surface

A description of a scheme for the hydrolysis of Si–O–Si linkages by various species can be considered within the framework of electrophilic or nucleophilic mechanisms of attack. Obviously, analogies are related to the relative acidity or basicity of the oxide and the attacking moiety. In the scheme of Budd (1961), attacking species are classified into electrophilic reagents which always seek to attack positions where there is an excess of electrons and nucleophilic reagents which always seek to attack positions where there is a deficiency of electrons. Water is capable of being both nucleophilic and electrophilic. Since the surface of SiO_2 is acidic (IEPS $\cong$ 2), water acts more strongly in a nucleophilic manner, viz.

$$
\begin{array}{cc}
\text{H} \qquad \text{H} & \\
\diagdown \diagup & \\
\text{O} \cdots & \quad (10.1)\\
\diagup \qquad \diagdown & \\
\text{O—H} \qquad \text{H} & \\
| \qquad\quad | & \\
\text{Si} \qquad\quad \text{O} & \\
\diagup | \diagdown \qquad | & \\
\qquad\qquad \text{Si} & \\
\qquad\quad \diagup | \diagdown &
\end{array}
$$

Nucleophilic attack by water will cause disruption of Si–O–Si linkages leading to formation of microcracks, which may result in premature fracture. This is schematically illustrated in Fig. 10.4. While oxide surfaces can be dehydrated at elevated temperatures, at normal ambient conditions the outermost surface oxygens hydrate to form a high density of hydroxyl groups. Brooks (1958, 1960), Zettlemoyer (1965), and Ter-Minassian-Saraga (1964), for example, have estimated that about one silanol group is present, per 60 $Å^2$ on the surface of a variety of glasses and silicas. Of course, the freshly drawn fiber will not contain such a large number of silanol groups but if inadequate protection is afforded the interface, re-

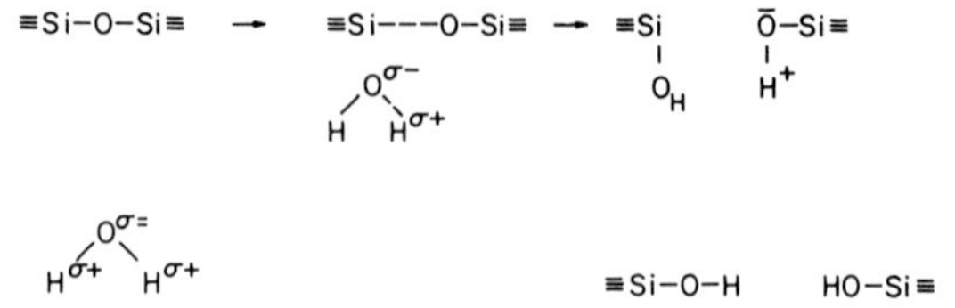

Fig. 10.4 Schematic of hydrolysis of an $-S_i-O-S_i-$ linkage.

hydration will proceed spontaneously, ultimately leading to a reduction in strength. Once hydroxyl groups are formed on an unprotected silica surface, or if inadequate coating procedures fail to provide protection, the hydroxyl rich surface that forms under ambient conditions or elevated humidity adsorbs and strongly retains several monomolecular layers of water.

Since the IEPS of silica in water is 2.2, it is expected that a somewhat lower value would exist for the freshly prepared surface. If there exists the possibility of an interfacial reaction between a solid and liquid the interfacial energy change in establishing contact will be large and negative (or the work of adhesion, large and positive). Typically, as in the case of silica, when a surface containing acidic groups is brought into contact with a liquid containing basic groups we would expect considerable chemisorption. After solidification of the liquid phase, delamination of these two solids would require either the separation of opposing electrostatic charges, or (in the presence of water) the regeneration of free acid and free base—both of which are high-energy demand processes. Thus, the strength of this interfacial bond would be much greater than that expected if only nonspecific dispersion interactions were operative across the interface. From the above considerations, it would appear that the silica surface can interact with polar organic groups by a variety of mechanisms.

Infrared measurements indicate that the number of hydroxyl groups per unit surface area depends on oxide composition, level and type of impurities, thermal history, time and temperature of exposure to water vapor, oxide crystallinity, and other factors. Attractive forces which may be attributable to surface hydroxyl groups depend strongly on the electron density of the oxygen atoms in the $\equiv$Si—OH group. If the electron density of the oxygen atom is low, the strength of the hydrogen bonds which can be formed with polarized hydrogen atoms in a coating phase is reduced and the probability of ionization may be increased. Based on the analysis of Parks (1965), one finds that as the valence of the surface metal atoms increases, the M—O bond strength increases, the —OH bond strength decreases, the electronegativity (ability of the oxygen atom to attract electrons or to donate a proton) increases and the overall basicity of the $-$MOH group decreases. Apparently, $\equiv$SiOH ionizes via the dissociative reactions.

$$\equiv SiOH + H_2O \rightleftarrows -SiO^- + H_3O^+ \tag{10.2}$$

and

$$\equiv SiOH + H_2O \rightleftarrows SiOH_2^+ + OH^- \tag{10.3}$$

and the IEPS of the solid is equal to the IEP of the dissolved hydroxo complexes.

10.3.3 Interactions of Coatings with Silica Surfaces

Typically, a silica surface containing silanol groups may interact with a basic amine group in a possible coating system to give either dipole or ionic bonds:

$$-Si-OH + R_3N \rightarrow SiOH \cdots NR_3 \quad \text{or} \quad -SiO^-HNR_3. \tag{10.4}$$

The silica surface may react with strongly cationic groups to yield an ionic bond.

$$-SiOH + R_4N^+Cl^- \rightarrow -SiONR_4 + HCl. \tag{10.5}$$

To predict the interaction for a silicon oxide surface in contact with polar groups we may write the equilibrium constants for both acid and base type reactions as

$$K_A = \frac{[SiOH_2^+][X^-]}{[SiOH][X]}, \tag{10.6}$$

$$K_B = \frac{[SiO^-][HX^-]}{[SiOH][X]}, \tag{10.7}$$

$$\Delta_A \equiv \log K_A, \qquad \Delta_B \equiv \log K_B.$$

By considering the possible surface reactions of the oxide acting as an acid and base, Bolger and Michaels (1968) showed that

$$\Delta_A = IEPS - pK_{A(A)}, \tag{10.8}$$

$$\Delta_B = pK_{A(B)} - IEPS, \tag{10.9}$$

where $pK_{A(A)} \equiv -\log K_{A(A)}$ and $pK_{A(B)} \equiv -\log K_{A(B)}$. The equilibrium constants $K_{A(A)}$ and $K_{A(B)}$ are derived from the ionization of typical acids or bases, $viz.$,

$$K_{A(A)} \equiv \frac{(H^+)(X^-)}{HX} \tag{10.10}$$

and

$$K_{A(B)} \equiv \frac{(H^+)(X)}{(HX^+)}. \tag{10.11}$$

TABLE 10.1
Polar Surface Interactions with an Organic Acid—$\Delta_A \equiv$ IEPS $-$ $pK_{A(A)}$

Organic acid	$pK_{A(A)}$	SiO_2 (IEPS = 2)
Dodecyl sulfonic acid	−1	3
Trichloroacetic acid	0.7	1.3
Chloroacetic acid	2.4	−0.4
Phthalic acid	3.0	−1
Benzoic acid	4.2	−2.2
Adipic acid	4.4	−2.4
Acetic acid	4.7	−2.7
Hydrogen cyanide	6.7	−4.7
Phenol	9.9	−7.9
Ethyl mercaptan	10.6	−8.6
Water	15.7	−13.7
Ethanol	16	−14
Acetone	20	−18
Ethyl acetate	26	−24
Toluene	37	−35

TABLE 10.2
Polar Surface Interactions with an Organic Base—$\Delta_B \equiv pK_{A(B)} -$ IEPS

Organic base	$pK_{A(B)}$	SiO_2 (IEPS = 2)
Trimethyl dodecyl amonium hydroxide	12.5	10.5
Piperdine	11.2	9.2
Ethylamine	10.6	8.6
Triethylamine	10.6	8.6
Ethylenediamine	10	8
Ethanolamine	9.5	7.5
Benzylamine	9.4	7.4
Pyridine	5.3	3.3
Aniline	4.6	2.6
Urea	1.0	−1.0
Acetamide	−1	−3
Water	−1.7	−3
Tetrahydrofuran	−2.2	−4.2
Ethyl ether	−3.6	−5.6
t-Butanol	−3.6	−5.6
n-Butanol	−4.1	−6.1
Acetic acid	−6.1	−8.1
Phenol	−6.7	−8.7
Acetone	−7.2	−9.2
Benzoic acid	−7.2	−9.2

If Δ_A or Δ_B is positive, ionic reactions should predominate, whereas if Δ_A or Δ_B is negative, dipole reactions should predominate. Furthermore, the total interaction energy should increase if Δ_A or Δ_B increases, that is, as the ionic contributions to the interfacial bond increases. At large negative values of Δ, ionic-type interactions are minimal and the corresponding dipole force weak. Nonpolar polymers (such as polyethylene) are essentially weak acids ($pK_{A(A)}$ is large and positive) and very weak bases ($pK_{A(B)}$ is large and negative), giving large negative values of Δ_A and Δ_B with any oxide surface, leading to the prediction that only London dispersion forces are operative across the interface. This does not imply that polyethylene will not adhere well to other nonpolar surfaces, or polar surfaces which have been modified to behave as nonpolar surfaces. Noller (1965) has compiled pK_A values for a large number of organic acids and bases, including water. Bolger and Michaels have tabulated the Δ_A and Δ_B values for these acids and bases in contact with SiO_2 (IEPS = 2), Al_2O_3 (IEPS = 8), and MgO (IEPS = 12). Portions of these tables are reproduced in Tables 10.1 and 10.2. Typically, water reactions are nonionic, although water may act as a weak acid or a weak base with surfaces of both low and high IEPS. Whether a polar group acts as an acid or a base can be predicted from the IEPS of the oxide. Figure 10.5 illustrates the interaction type and orientations for a variety of liquid types based on the Δ_A and Δ_B values.

10.3.4 Stability of Silica–Organic Interface

Interaction of the environment with coatings may be visualized by considering the magnitude of the thermodynamic work of adhesion, that is, the free energy change to create an organic–water and a substrate–water interface and to eliminate the original organic–substrate interface.

Consider a system drawn schematically in Fig. 10.6a,b for the bonded composite A–B. It will be assumed that formation of the interface has been maximized with respect to interfacial contact. The situation with coated glass fibers may be quite different since we are concerned with dynamic wetting. For simple liquids, water or alcohol, the equilibrium contact angle on solids is reached in a relatively short time ($\sim$ seconds). However,

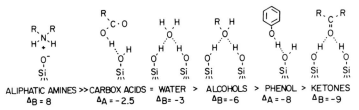

Fig. 10.5 Dipole orientations, dominant interaction modes, and relative bond strengths predicted for low IEPS surfaces such as S_iO_2.

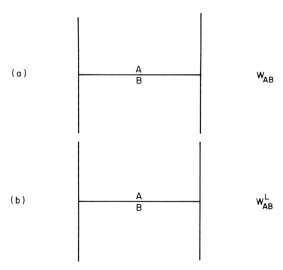

Fig. 10.6 Schematic of AB composite in the presence and absence of water.

for highly viscous liquids similar to polymer melts or the uv curable epoxy–acrylate, extended periods of time may be required to reach equilibrium contact angles. The dynamics of wetting is associated with viscosity, relative surface energetics of the solid and liquid, and velocity of the liquid flow. In the fiber-coating process, the contact angle appears to be associated with the dimensionless parameter $(\eta v/\gamma)$, when η is the viscosity of the liquid, v is the velocity of fiber drawing, and γ is the surface tension of the liquid (Hoffman, 1975).

In the absence of a liquid phase, but in the presence of air (V), the thermodynamic work of adhesion is

$$W_{AB} = \gamma_{AV} + \gamma_{BV} - \gamma_{AB}, \tag{10.12}$$

where the γ's represent the solid surface tensions of the members of the composites. In the presence of a liquid phase, the thermodynamic work of adhesion becomes

$$W^L_{AB} = \gamma_{AL} + \gamma_{BL} - \gamma_{AB}. \tag{10.13}$$

Since the work of adhesion for substrate A in contact with liquid L, is,

$$\gamma_{AL} = \gamma_{AV} + \gamma_{LV} - W_{AL} \tag{10.14}$$

and for substrate B, the work of adhesion in contact with liquid L is

$$\gamma_{BL} = \gamma_{BV} + \gamma_{LV} - W_{BL} \tag{10.15}$$

we obtain by combining Eqs. (10.12)–(10.15),

$$W^L_{AB} = 2\gamma_{LV} + W_{AB} - W_{AL} - W_{BL}. \tag{10.16}$$

TABLE 10.3

The Thermodynamic Work of Adhesion of the Silica Polymer Composite in the Presence of Water[a]

Polymer	W_{AB} (ergs/cm²)	W_{BL} (ergs/cm²)	W_{AB}^{L} (ergs/cm²)
Nylon 6,6	268.8	108.8	−162.4
Polyethylene	177.6	67.7	−212.5
Polystyrene	211.8	71.6	−182.2
Polydimethylsiloxane	163.8	57.7	−216.3
Polyethyleneterphthalate	219.6	84.1	−186.9

[a] $W_{AB}^{L} = 2\gamma_{LV} + W_{AB} - W_{AL} - W_{BL}$, where $W_{AL} = 468$ ergs/cm² and $\gamma_{LV} = 72.88$ ergs/cm².

Employing the techniques of Fowkes (1965) and Kaelble (1971) values of W_{AB}^{L} are computed and presented in Table 10.3. What is striking is that for a variety of thermoplastic polymer coatings ranging from the rather polar nylon 6,6 to (nonpolar) polyethlene, the W_{AB}^{L} values are large and negative, indicating a tendency to delaminate spontaneously in the presence of water. Clearly, if that situation existed, the composite could be viewed as in Fig. 10.7. Under these circumstances the silica surface would be under direct attack by the water.

To reduce the effect of the large value for W_{AL}, two approaches may be taken.

(a) to form covalent bonds across the interface,

(b) to modify the silica by covalently bonding species onto the surface to reduce its interaction with water (coupling agents). These species could be of the nonreactive type, just to lower the γ_S value, or the reactive type which could covalently bond to the coating (acid–base reaction).

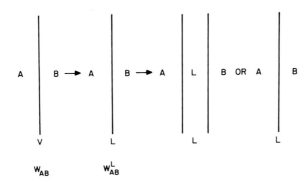

IF $W_{AB} \simeq W_{AB}^{L}$, NO SPONTANEOUS DELAMINATION HOWEVER, IF $W_{AB} \sim 0$ OR NEGATIVE, SPONTANEOUS DELAMINATION IS FAVORED.

Fig. 10.7 Schematic of possible delamination in the presence and absence of hostile environment (water).

10.3.5 Silane Coupling Agents in Ultraviolet Cured
Epoxy Acrylate

Providing for covalent bonding cross the interface should establish a stronger adhesive joint and (because of the electronic charge symmetry about a covalent bond) generate a bond insensitive to rupture or displacement by polar species such as water. Coupling agents afford the possibility of modifying the interfacial region so as to preclude the reaction of water with the interface. Maintaining an intact interfacial region, one in which the organic phase is not displaced from the glass, would preclude the hydrolytic attack on the Si–O–Si linkages when the interface is stressed. Coupling agents generally have associated with them two functionalities, one capable of interacting with the resin phase and one favoring reaction with the substrate.

$$R_x SiR'_{4-x}$$

If the silicon in the silane is bonded to a strongly electronegative atom, such as chlorine, the bond is highly polarized and hydrolysis proceeds rapidly and exothermically to yield (initially) a silanol group. As the electronegitivity of the group attached to the silicon atom decreases, the hydrolytic stability increases. Thus, alkoxy compounds (Si–OR, where R is an alkyl group) hydrolyze rather slowly. Silicon atoms bonded to still weaker electronegative groups (Si–C, or Si–O–Ar, where Ar is an aromatic group) are relatively stable to hydrolysis.

Although the rate of reaction is slow at neutral pH and ambient temperatures, silanol groups can condense to yield a network structure. Clearly, the additives (silane) must diffuse to, and selectively adsorb at, the interface prior to solidification of the coating. In addition, the additive must condense with itself and react with the resin phase.

Several possible choices are suggested for the application of a suitable coupling agent to provide environmental protection to the silica surface. A two-stage application is feasible whereby the silane is applied and reacted to the silica surface prior to the resin application, or the suitable silane could be an integral part of the coating resin and migrate to the interface. For initial studies (Schonhorn *et al.*, 1976; Kurkjian *et al.*, 1976), the latter procedure was chosen, although it is probably a less desirable choice for the high-speed coating envisioned in optical fiber technology.

Although a large number of coating systems and methods of application are available, (see Section 10.2) initial studies were made using a uv curable epoxy acrylate that not only preserves the initial strength (Schonhorn *et al.*, 1976), but provides for environmental stability. Table 10.4 gives the components used in the preparation of the prepolymer. The constituents are reacted until an acid value less than one is reached. The prepolymer

TABLE 10.4
Prepolymer Composition

Components	Percent
LY8161: $(H_2C\overset{\diagdown}{\underset{O}{\diagup}}CH-CH_2)R'$—Ciba–Geigy Co.	45.4
RD-2: $(H_2C\overset{\diagdown}{\underset{O}{\diagup}}CH-CH_2)_2R''$—Stauffer Chemical Corp.	27.6
Acrylic Acid: $CH_2{=}CH-COOH$	
where $R' = -O-\langle\bigcirc\rangle-\overset{CH_3}{\underset{CH_3}{C}}-\overset{Br}{\langle\bigcirc\rangle}-O-$	27.0
and $R'' = -O-(CH_2)_4-O-$	100.0

contains approximately 90% of the following species:

$$R'(-CH_2-\overset{OH}{\overset{|}{CH}}-CH_2-O-\overset{O}{\overset{\|}{C}}-CH{=}CH_2)_2$$

and

$$R''(-CH_2-\overset{OH}{\overset{|}{CH}}-CH_2-O-\overset{O}{\overset{\|}{C}}-CH{=}CH_2)_2.$$

The remaining 10% consists of

$$H_2C\overset{\diagdown}{\underset{O}{\diagup}}CH-CH_2-R'-CH_2-\overset{OH}{\overset{|}{CH}}-CH_2-O-\overset{O}{\overset{\|}{C}}-CH{=}CH_2$$

and

$$H_2C\overset{\diagdown}{\underset{O}{\diagup}}CH-CH_2-R''-CH_2-\overset{OH}{\overset{|}{CH}}-CH_2-O-\overset{O}{\overset{\|}{C}}-CH{=}CH_2.$$

The above prepolymer blend may be polymerized to a flexible film by electron or uv radiation. Extremely rapid curing can be obtained by uv radiation when a sensitizer such as benzoin or its ether derivative is added to the formulation. In an initial study (Schonhorn et al., 1976; Kurkjian et al., 1976), the isobutyl ether derivative of benzoin, Vicure 10 (Stauffer Chem. Corp.) was employed.

Currently, a styrene-functional amine hydrochloride silane (Plueddemann, 1972) designated Z-6032 (Dow Corning Corp., Midland, Mich.) has been selected; it has the structural formula:

$$(CH_3O)_3 - Si(CH_2)_3NH(CH_2)_2 \overset{\overset{\displaystyle HCl}{|}}{NH} - CH_2 \langle O \rangle CH=CH_2.$$

The wet strength data of Plueddemann (1972) provided the impetus for the selection of this particular coupling agent.

In assessing the stability of a bonded coating in the presence of water to which additional acid or base is added, other considerations become important. The foregoing arguments indicate that the region of maximum stability with respect to delamination by water would be in the range

$$pK_{A(B)} > pH > IEPS.$$

Under strongly acidic or alkaline conditions, bonds should become relatively weak owing to the development of mutual repulsion forces or to suppression of ionization in the surface or organic groups.

$$RSi - (OH)_3 \rightarrow \begin{bmatrix} R & R \\ -Si-O-Si- \\ O & O \end{bmatrix}_{n/2} + 1.5 \text{ m } H_2O. \tag{10.17}$$

Recent work by Gent and Hsu (1974) and Koenig (1975) has shown that the silanol groups do condense on the surface of silica to yield Si–O–Si covalent bonds. The overall (although admittedly idealized) reaction for the formation of a cross-linked surface zone can therefore be visualized as

$$
\begin{array}{ccc}
-\overset{|}{Si}-OH & & -\overset{|}{Si}-O-\overset{|}{Si}-R \\
| & & | \quad\quad | \\
O & & O \quad\quad O \\
| & & | \quad\quad | \\
-\overset{|}{Si}-OH + 3R-Si(X_3) + 3H_2O \longrightarrow & -\overset{|}{Si}-O-\overset{|}{Si}-R + 9H_2O \\
| & & | \quad\quad | \\
O & & O \quad\quad O \\
| & & | \quad\quad | \\
-\overset{|}{Si}-OH & & -\overset{|}{Si}-O-\overset{|}{Si}-R
\end{array}
\tag{10.18}
$$

If the nonhydrolyzable group R contains a functional organic moeity such as a vinyl group, these may then form covalent bonds with corresponding reactive groups in the resin phase, leading to a bond which should, in practice, provide a high degree of water resistance.

10.4 TECHNIQUES OF COATING APPLICATION

Common to all techniques of coating application are two requirements:

(1) the process itself must not abrade the fiber surface and thereby weaken the fiber,

(2) the coating must be smooth so as not to introduce microbending losses when in contact with its environment.

Coatings can be applied in-line at the fiber drawing machine by means of a wick. Other techniques of application include those that require that the fiber pass through a hole or die somewhere in the coating fixture. Because contact with the die wall can result in abrasion, much attention has been directed to limit that abrasion, through alignment of the fiber path and die position, through exploiting the lubricant properties of the coating fluid itself, and through choosing die wall materials which would cause as little abrasion as possible should contact be made. To these ends, alignment tools and techniques have been developed, flexible dies which follow small drifts in fiber position designed, die contours chosen, and soft materials selected for molding applicator parts.

Coating smoothness is related to the coating material, especially via the dynamics by which the coating fluid sets up to its solid state. For example, a drying mechanism may cool the coating surface to the point where water vapor condenses to yield an "orange peel" texture. This is especially evident in acetone-based systems where evaporation is rapid and the water is miscible with the acetone.

Coating with a low viscosity liquid can lead to beading of the coating before set-up, independent of the particular set-up dynamics. The essentially cylindrical column of liquid placed on the fiber by the die is inherently unstable. The preferred state of lower free energy has the geometry of a series of beads. A model of the phenomenon was developed by Roe (1975). Bead growth rate depends on the viscosity (higher viscosity → slower growth) of the liquid and the ratio of coating thickness to glass fiber radius (greater ratio → faster growth). An analytical representation is especially difficult for real situations since the liquid viscosity is time-dependent. Indeed, the best way to avoid a beaded coating is to increase viscosity quickly after application, that is, have a rapid set.

Bead-type irregularities can have drastic effects on transmission loss. The beading period is typically highly regular. If the period is close to the mode coupling period for graded-index fibers, then the fiber becomes especially prone to microbends. This critical period is approximately 1 mm.

10.4.1 Wick Coating Application

A wick wetted with coating liquid has been used at Bell Labs to apply Kynar®* 7201 coating from solution and hexamethyldisilazane as a surface treatment. The Kynar was an early coating which provided some degree of abrasion protection. A wick, consisting of a felt pad sandwich, applied up to 12 μm of Kynar from an acetone solution. A greater thickness could not dry quickly enough to prevent beading. There were a variety of problems associated with the wick–Kynar coating system. The cellulosic felt abraded the fiber. Coating thickness, primarily controlled by wick satura-

---

* Kynar®—trademark Pennwalt Corp.

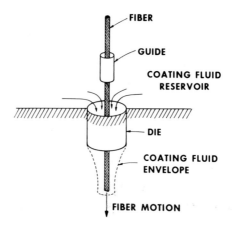

Fig. 10.8 Schematic of pressureless die with guide.

tion and solution concentration, had poor repeatability, and also varied during a run. The thin coating, typically about 5 μm, afforded limited abrasion protection. Polymer wick coating was abandoned for die coating.

At a much later stage in coating development, the wick was reintroduced to apply a very thin coating of low viscosity organo-silanes to bare fiber as a hydrophobic surface treatment. Scott Felt®,[†] a controlled porosity urethane foam from Scott Paper Company, was used instead of felt as the wick material. The intrinsic softness of the urethane polymer resulted in much less abrasion of the fiber. Indeed, no degradation of fiber strength has ever been conclusively attributed to lubricated Scott Felt contact. However, foreign particles which may become entrapped in the foam are a potential source of abrasion that is hard to evaluate but reasonable to assume exists.

10.4.2 Pressureless Die with Guide Applicator

The simplest embodiment of the pressureless die with guide applicator is shown in Fig. 10.8. At the bottom of a reservoir of coating fluid is a die. After passing through the reservoir and die, the fiber exits with an envelope of coating fluid adhered. It later dries or cures to a solid coating. The centering of the fiber within the die determines the geometry of the coating envelope and the resultant centering of the fiber in the coating. To ensure the centering of the fiber in the die, a guide is placed in the reservoir, close to the die. The geometry is analogous to a pressure extrusion head with the guide in the role of the core tube. Here, however, the guide is filled with coating fluid and therein lies one advantage: the coating fluid itself provides lubrication to limit fiber abrasion when passing through

[†] Scott Felt®—trademark of Scott Paper Co.

the guide. Also, the pressureless nature of the process avoids problems of transverse pressure differentials seen in pressure extrusion. The only pressures are those resulting from the flow generated by the moving fiber. For most geometries, these pressures tend to center the fiber in the guide and are predicted by lubrication theory. A closer look at the theory is given later in Section 10.4.3.1.

10.4.2.1 Materials Used with the Pressureless Die/Guides. Much experience was obtained with the pressureless die/guide applicator while applying coatings of ethylene-vinyl acetate copolymer (EVA). The EVA was applied from a hot solution. For example, an 8-melt index, 18%-vinyl acetate grade was dissolved in 1,1,1-trichloroethane to a concentration of 28.3% W/V and applied at 51°C. Upon existing the die, the solution cooled and gelled in about 1 sec to a gel tough enough to traverse a capstan-tractor without damage. From there, the coated fibers were taken up on a reel under low tension and the solvent evaporated while on the reel.

Because of solvent evaporation, the final solid coating thickness was about half the initial gel thickness. For a given final coating thickness, then, solvent systems in general and the EVA-trichloroethane system in particular need a larger die diameter than 100% solids systems. As we shall see later, the centering forces generated by the flow decrease as die diameter increases. Two conclusions are (1) centering forces in dies for solvent systems will be less than for 100% solids systems, and (2) a narrow guide can have a high centering force. Therefore, the die/guide applicator is particularly appropriate for solvent systems.

Fibers coated with EVA from the die/guide applicator by Western Electric in Atlanta passed a 0.24-giga Pascal proof test stress over kilometer lengths in the great majority of the cases. The die was of stainless steel and 0.74 mm in diameter. Final coating thickness was 56 μm on a 110-μm fiber.

10.4.3 Self-Centering Coating Applicators

As a fiber passes through a die, fluid dynamic forces tend to center it within the die to a degree dependent on die design and fluid properties. However, the forces are characteristicly so small that a minor misalignment of the die away from the free path of the fiber can upset the centering and result in grossly eccentric coatings. There are, then, two approaches to minimizing eccentricity:

(1) design the die to maximize the fluid dynamic centering forces, and
(2) maintain accurate alignment of the coating die.

A full mathematical modeling of the fluid flow to predict the centering forces within the die requires a three-dimensional solution of the Navier–Stokes equation under appropriate boundary conditions. That is a difficult numerical analysis problem for a Newtonian fluid and even more so for non-Newtonian fluids.

Several models have been proposed (Lenahan, 1978; Lewis, 1978; France *et al.;* 1977) but experimental results are scarce which makes it difficult to evaluate the models. All, however, have certain conclusions in common. The usual Newtonian fluid assumption leads to a centering force linear in both coating fluid viscosity and line speed. Also, the centering force is found sensitive to the ratio of die exit diameter to fiber diameter. To show how this arises, a simple, modified one-dimensional model is sketched in Section 10.4.3.1 and then the implications for drawing line performance are covered in Section 10.4.3.2. There, the centering force predictions are used to place requirements on the lateral compliance of a flexible die mount. It is shown that with this type of mount, alignment of the die with the fiber's free path is made less sensitive to mechanical drift.

10.4.3.1 Forces in the Coating Die, An Illustrative Model. If we look at the coating die geometry from an idealized point of view, we see an endless cylinder (the fiber) moving axially through a tube (the die, see Fig. 10.9). The inner walls of the die tube take the form of a body of rotation with a monotonic trend of decreasing radius in the direction of fiber motion. The monotonic nature is not a design restriction but rather a statement of current practice. The fiber is centered in the die when its axis and the die axis are coincident. Coating fluid completely fills the gap between the fiber and die.

For any specific geometry within this general class, centering forces are generated. That is, if the fiber axis moves off to one side, the resultant hydrodynamic force on the fiber will be toward the center. This behavior is inferred from the two-dimensional case where the pressure profile has been calculated for a variety of sliding pad bearing geometries (see, for example, Purday, 1949, pp. 60–62). The result for one geometry is illustrated in Figure 10.10. The procedure involves an integration of the Reynolds equation in a simplified one-dimensional form.

For a fluid, bounded by two surfaces separated by $h(x, y)$, the elimination of inertia terms in the Navier–Stokes equation leads to the Reynolds equation.

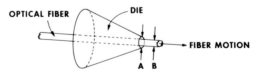

Fig. 10.9 General die contour.

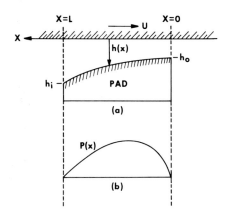

Fig. 10.10 (a) Geometry of a pad near a surface moving with speed U. The distance between the pad and the surface varies from h_i to h_0. A fluid, filling the space between the wall and pad, obtains a pressure profile nominally given by $P(x)$ in (b). In general, $P(x)$ would increase if the pad is positioned closer to the surface. This response provides an intuitive understanding of why a centering force arises in a tapered die.

$$\frac{\partial}{\partial x}\left(\frac{\rho h^3}{\mu}\frac{\partial P}{\partial x}\right) + \frac{\partial}{\partial y}\left(\frac{\rho h^3}{\mu}\frac{\partial P}{\partial y}\right) = -6\,\frac{\partial(\rho h U)}{\partial x} + 12\,\frac{\partial(\rho h)}{\partial t}. \quad (10.19)$$

where U is the relative surface speed (in the $-x$ direction), and $P(x, y)$, $\rho(x, y)$, and $\mu(x, y)$ are the fluid pressure, density, and viscosity, respectively.

After a variety of assumptions more or less appropriate to practical coating conditions, (P = constant, μ = constant, $P(x, y) = P(x)$, $\partial h/\partial t = 0$, $\partial\mu/\partial x = 0$), Eq. (10.19) can be reduced to one dimension and integrated. We insert an edge-loss factor (Purday, 1949) into the solution to account for the narrow width of the fiber. When the fluid pressures are calculated for the sides of an offset fiber, the side nearer the die wall is found to have a greater pressure. This pressure differential, when integrated over the surface, gives an expression for the total fluid dynamic force on the fiber, F, as a function of its offset, x, from the central position:

$$\Delta F(x) = \frac{q\mu U B^3}{6}$$

$$\times \left\{\frac{1}{(C - x)^2\,\ln(h_i/(C - x)e^2)} - \frac{1}{(C + x)^2\,\ln(h_i/(C + x)e^2)}\right\}, \quad (10.20)$$

where $e = 2.718$, $C = (B - A)/2$, and h_i is the fiber-to-wall distance at the die extrance and assumed much greater than the die exit clearance C, B is the fiber diameter, and q is a geometrical factor normalized to a die with linear longitudinal taper.

Now, if we expand $\Delta F(x)$ in a Taylor series around the center, $x = 0$, then $\Delta F(0) = 0$ and the first term is linear in x. To first order, then,

$$\Delta F(x) = kx, \tag{10.21}$$

where

$$k = \frac{\mu U B^3 q}{3 C^3 \ln(h_i/Ce^2)} \left\{ 2 + \frac{1}{\ln(h_i/Ce^2)} \right\}. \tag{10.22}$$

We have assumed $\ln h_i$ to be constant since it has a very weak dependence on x.

From (10.21) and (10.22), we see that the linear restoring force term, which is the dominant term for small eccentricities, has a sensitive C^{-3} factor (indeed, the nonlinear terms vary inversely with higher powers of C). This finding makes a strong argument for using thinner coatings to obtain better centering. The considerations are identical to the case of a tapered-land thrust-bearing design (see, for example, Fuller, 1956). Another, subtler consideration, is that stability of the fiber position within the die may be improved with a smaller die. There are no organized data available on this point, though.

For a given final coating thickness, a solvent system coating fluid requires a larger die than a 100% solids system. Therefore, in terms of fiber centering, the 100% solids systems have an advantage. This is in spite of any scaling-down of eccentricity as the coating dies.

The centering force described here must counter the offset force caused by coating line misalignment. The degree of misalignment is dependent on initial positioning accuracy and on drift over a period of time. We will see in Section 10.4.3.2 that if the die is on a passive flexible support, it can compensate for drift to an extent dependent on die flexibility, and line tension and length.

10.4.3.2 External Forces on the Applicator and the Role of Flexibility.   If the drawing tension at the capstan is T and the geometry of the fiber part is as shown in Fig. 10.11 then the horizontal force, in the plane of the figure, exerted by the die on the fiber is given by

$$F = T \sin \theta_p + T \sin \theta_c \tag{10.23}$$

or, in terms of a linearized fluid force, (Equation (10.21),

$$F = kx, \tag{10.24}$$

where F is physically the force needed to deflect the fiber by an amount a from its straight-line path, AB. If we assume the angles are sufficiently small, then

$$x = TLa/kl_p l_c, \tag{10.25}$$

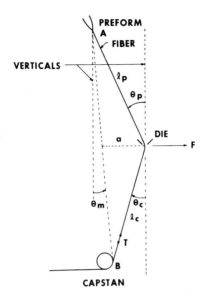

Fig. 10.11   General geometry of a vertically oriented fiber drawing and coating line.

where

$$L \equiv l_\mathrm{p} + l_\mathrm{c}. \qquad (10.26)$$

From Eq. (10.22) we can evaluate k.

For typical fiber drawing line values, we find

$$x \sim 0.1a. \qquad (10.27)$$

Experiments relating die exit clearance to mean coating thickness for some 100% solids systems have indicated a ratio of about 1.3:1. Upon scaling Eq. (10.27) by this factor, we have

$$\epsilon \sim 0.08a, \qquad (10.28)$$

where ϵ is the eccentricity of the set-up coating.

If the die is on a compliant support, it can respond to a drift in line position and thereby tend to compensate for the change in alignment. In practice, changes in fiber path result most often from a shift in position of the draw-down zone; preforms are not necessarily straight. It can be shown that a shift of δ at Point A in the plane of Fig. 10.11 (a worst-case type of shift) effects a change in x given by

$$\Delta x = \frac{1/k}{1/k + 1/j + (l_\mathrm{p} l_\mathrm{c}/TL)} \frac{l_\mathrm{c}}{L} \delta, \qquad (10.29)$$

where j is the stiffness or spring constant of the flexible die support. In the limit of a highly compliant support,

$$\Delta x \cong \frac{j}{k} \frac{l_c}{L} \delta. \tag{10.30}$$

In a practical situation,

$$\Delta \epsilon \sim 0.02\delta \tag{10.31}$$

is readily achieved through proper design. For compliant die supports, then, centering is relatively insensitive to line position drift. A guide can be dispensed with. One design achieves high compliance by making the die and its support out of elastomeric materials (Hart and Albarino, 1977). Simultaneously, the elastomeric construction makes the problem of fiber-die wall abrasion much less severe, an important result. This construction was a valuable contribution to the technology that made possible high fiber strengths over long lengths.

Does better centering improve fiber strength? There are no organized studies addressing this question. About all that can be said empirically is that fiber weakness has often been associated with highly eccentric coatings. A major study obstacle has been that the preform change from one run to the next introduces hidden variables; well-centered fibers have been weak too.

From a theoretical tribology viewpoint, two abrasion mechanisms relate to coating eccentricity. First, there is less chance of a transient excursion pushing the fiber against the die wall when the fiber is well-centered. More energy would then be needed to rupture the lubricating film of coating fluid. This lower chance of contact abrasion would have special significance in heavy machinery environments, such as factories, where vibration levels are high. Second, if we assume that the fiber carries along particles attached to its surface, then wall contact would not be necessary for abrasive action. As the fiber approaches the wall, the fluid shear rate increases at the fiber surface and the chance of dislodging a particle also increases. Once free, the particle can then abrade the surface. Increased fluid shear rates would likewise make particles already suspended in the fluid more likely to abrade the fiber. Thus, at least two mechanisms are theoretically less damaging when the fiber is well centered in the die.

10.4.4 Melt Extrusion Coating

The technology of melt extrusion coating is highly developed in the wire and cable industry. It is natural, therefore, to examine the strengths and limitations of melt extrusion as applied to the coating of optical fibers.

The process of melt extrusion involves melting a polymer and forcing the molten liquid through a die which has an opening shaped to produce a desired cross section. Two types of dies may be considered for applying the polymer to a fiber. These types are pressure dies and tubing dies. The pressure die shown in Fig. 10.12 is commonly used for insulating wire. In it the polymer contacts the fiber under pressure with the coated fiber existing the die at nearly the dimensions of the desired product. This method ensures a tight coating but requires that the core tube bore through which the fiber passes be only slightly larger than the fiber diameter to prevent the molten polymer from leaking back past the fiber. Such a process is extremely difficult to carry out without damaging the fiber surface by scraping against the core tube wall. Additionally since pressure dies have nearly the same size as the coated fiber diameter, very high shear rates are generated when applying thin coatings to small diameter fibers, even at modest line speeds. Distorted coatings due to viscoelastic flow instabilities usually occur under these circumstances. For these

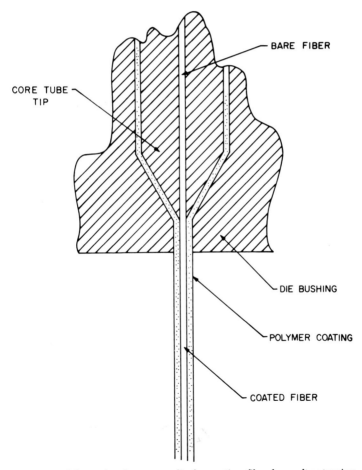

Fig. 10.12 Schematic of pressure die for coating fiber by melt extrusion.

reasons the pressure method is more useful for jacketing previously coated fibers than for coating bare fibers as they are drawn.

The tubing method of extrusion has distinct advantages for coating fibers as they are drawn. The fiber passes through a core tube which extends to the die exit as shown in Fig. 10.13. Sufficient clearance is allowed to enable the fiber to pass through the core tube without contact. The polymer is extruded completely independently as a large diameter, thick-walled tube about the fiber and is drawn down to form a tight thin coating on the fiber.

It is most important that defects in the extruded coating such as tears, bumps, and other irregularities be avoided. The polymer itself must be capable of withstanding the large extensional deformation or draw-down without failure by tearing. Many polymers such as those melt spun into synthetic fibers, e.g., nylon, polypropylene, and certain thermoplastic elastomers, are suitable candidates. Beyond this the homogeneity of the melt delivered to the die exit is critical. Inhomogeneities such as gel particles, particulate inclusions, and the like produce defects in the coating by acting as stress concentrators during draw-down. It is therefore crucial that the material be clean, particularly if thin coatings (25 to 75 μm) are desired.

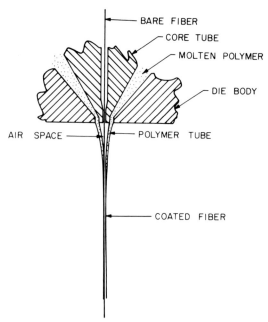

Fig. 10.13 Schematic of tubing die for coating fiber by melt extrusion (Blyler and Hart, 1977).

Uniform output and thermal homogeneity of the melt at the die exit is also important for uniform draw-down. Good melt mixing in the extruder is necessary for delivering a thermally homogeneous melt to the die. Screw extruders are therefore to be preferred over ram extruders for tubing coatings on fibers. However, many fiber-coating operations require very low output rates as, for example, the application of thin coatings on small diameter fibers at low drawing speeds. In these cases ram extruders can be used to advantage because the heated volume of polymer in the extruder can be kept small, thereby reducing the likelihood of polymer degradation.

Polymers selected for extrusion coating on optical fibers may be chosen to provide a number of critical properties such as abrasion resistance, microbending loss isolation, strippability, etc. A qualitative evaluation of selected extrudable polymers as regards these properties is given in Table 10.5. None of these properties, however, is more important than maintenance of fiber strength. It is particularly important than the extrusion process itself not significantly lower the strength of the fiber. Maintaining fiber alignment through the core tube to avoid contact is crucial. In addition, however, the polymer itself cools and begins to solidify as it is drawn onto the fiber. This contact between the fiber and drawing polymer may cause significant damage and lowered strength. There is also little opportunity for the polymer to wet the fiber surface and consequently there is little or no interfacial bonding. The fiber surface may be more prone to attack by water and other corrosive agents in this situation than when the polymer is bonded to the fiber.

TABLE 10.5
Properties of Polymers for Extrusion Coating of Optical Fibers

Polymer	Young's modulus (psi)	Influence on microbending loss	Abrasion resistance	Solvents for stripping
Nylon	200,000	increases	very high	Propylene glycol 160°C
Plasticized nylon	50,000	increases slightly	high	Propylene glycol 160°C
Perfluorinated ethylene–propylene copolymer (FEP) and perfluorinated alkoxy copolymer (PFA)	90,000	increases	high	none
Polyester thermoplastic elastomer	8,000	decreases	moderate	Methylene chloride
Polyurethane thermoplastic	5,000	decreases	moderate	Ketones, tetrahydrofuran

The above concerns are illustrated in Fig. 10.14 which compares histograms of the strength distributions of two fibers drawn with a CO_2-laser from the same TO-8 preform rod. One fiber was simply extrusion coated with nylon 6-12. The other fiber was first coated with a thin ($\sim 5\ \mu$m) layer of an epoxy resin from solution and thermally cured just ahead of the nylon extrusion die. The flexible die applicator described in Section 10.4.3 is especially useful for producing thin coatings without damaging the fiber.

As seen in Fig. 10.14 a histogram of tensile strengths for samples tested in 24-in. gauge lengths shows that the strength distribution for the fiber coated with nylon alone is multimodal and complex, with a rather low median value. Conversely the distribution for the epoxy/nylon combination is unimodal with a high median strength. The obvious conclusion is that the extruded nylon damages the fiber as it is applied. This damage is caused by shear at the interface as the solidifying polymer is drawn onto the silica surface. The epoxy resin offers mechanical protection to preclude damage in this step.

Melt extrusion coating of fibers is therefore most efficacious when coupled with the initial application of a thin primary coating. Many pri-

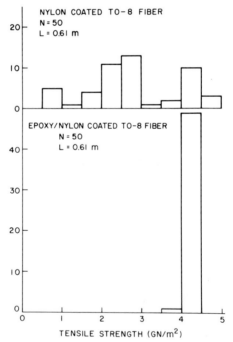

Fig. 10.14 Histograms of tensile strengths of nylon 6-12 coated, CO_2-laser-drawn TO-8 fibers with and without primary epoxy coating.

mary coatings are suitable for providing high initial strength, including epoxies, organo-silanes, and silicone resins. The principal requisites are that the coating liquid be free of particulates, capable of application from a noncontacting die, and have the ability to be solidified ahead of the inline extruder.

In this scheme long-term passivation of the silica surface against the effects of water becomes the function of the primary coating. Its ability to provide such protection depends upon its interaction with the silica surface as discussed earlier in this chapter. Indeed given the wide range of materials which meet the requisites outlined in the preceding paragraph the choice of primary coating for system implementation may rest most heavily upon static fatigue requirements.

10.5 POLYMER-CLAD FIBERS

10.5.1 Introduction

For many years grades of fused silica with very low bulk optical losses have been available commercially at reasonable prices. This situation has spurred interest in utilizing fused silica as the core of a fiber waveguide and has stimulated a search for an effective cladding material. The low refractive index of fused silica complicates this problem but two approaches which have met with success are the utilization of certain borosilicate glass compositions as the cladding material (MacChesney et al., 1973; Dabby et al., 1974) and the development of the single-material fiber structure (Kaiser et al., 1973b) discussed earlier. A third possible approach is to use a low refractive index organic polymer as the cladding (Blyler and Hart, 1977).

10.5.2 Polymer Optical Properties

Most commonly used polymers are based upon the hydrocarbon chain and have refractive indices of 1.480 or greater, too high to be of use for cladding fused silica. Fortunately, however, two special classes of organic polymers exist which provide suitably low indices—fluorocarbon polymers and silicones. In fluoropolymers the substituent fluorine atoms lower the molar refraction of the repeat unit of the polymer chain, whereas with silicone polymers the low index derives from the siloxane chain. A list of available low index polymers is provided in Table 10.6. When used as a cladding on a fused silica core fiber these polymers yield theoretical NAs ranging from 0.33 to 0.58. It is, in fact, the potential of high NA, coupled with the low losses of fused silica, that make the polymer-clad fused silica core fiber of special interest.

The utility of the polymers listed in Table 10.6 as cladding materials de-

TABLE 10.6
Low Refractive Index Polymers for Cladding Fused Silica Core Fibers

Polymer	Refractive index	Approximate minimum loss (dB/km)
Perfluorinated ethylene–propylene copolymer (FEP)	1.338	5×10^5
Perfluorinated alkoxy copolymer (PFA)	1.350	5×10^5
Vinylidene fluoride-hexafluoropropylene copolymer (VDF-HFP)	1.39	10^4
Vinylidene fluoride-tetrafluoroethylene copolymer (VDF-TFE)	1.40	5×10^5
Polydimethyl siloxane (PDMS)	1.40	10^3
Vinylidene fluoride-chlorotrifluoroethylene copolymer (VDF-CTFE)	1.417	10^4
Polyvinylidene fluoride (PVDF)	1.42	5×10^5
Polychlorotrifluoroethylene (PCTFE)	1.425	10^5

pends to a large extent on their bulk optical losses which contribute to fiber loss through evanescent field penetration. Polymers exhibit absorption and scattering losses which may be very large. The principal contributors to absorption losses in polymers in the wavelength region from 0.70 to 1.10 μm are overtones of various CH absorption bands. These absorption peaks may be several thousand decibels per kilometer in height, but with specific compositions windows may exist at wavelengths of interest. Perfluorinated polymers contain no CH groups and are therefore devoid of large absorption peaks in this wavelength region.

Contributors to bulk scattering losses in polymers are crystallinity and solid particle inclusions such as dirt, catalyst residues, etc. Semicrystalline polymers are generally translucent to opaque solids with scattering losses near 10^6 dB/km. Amorphous polymers avoid these enormous losses, but careful preparation and handling are still required to minimize particulate concentrations.

10.5.3 Fiber Loss Considerations

In addition to bulk optical loss effects, waveguide losses arise in polymer-clad fibers from scattering at the core–cladding interface. The process for applying the polymer to the fiber core is therefore crucial for attaining reasonable loss. Since the polymer is applied while in the liquid phase (melt, solution, pre-polymer, etc.) wetting of the core surface by the polymer liquid is a principle concern. Other problems of interfacial void formation (particularly in solvent systems), beading and physical damage to the core surface by the coating operation must be dealt with. It is also most important to deliver a high quality contamination-free fiber core to be coated with the polymer.

If the core–cladding interface is assumed to be free of imperfections, it is possible to estimate the total fiber loss from the bulk losses of the core and cladding materials. According to Gloge (1973),

$$\alpha_{\text{total}} = \alpha_{\text{core}} + \frac{\alpha_{\text{clad}} - \alpha_{\text{core}}}{4\,\pi a\,[\text{NA}]} \left[\frac{\text{NA}_{\text{eff}}}{\text{NA}}\right]^2 \lambda, \tag{10.32}$$

where α_{total} is the total fiber loss, α_{core} and α_{clad} are the bulk losses of the core and cladding materials, a is the core radius, λ is the wavelength, $[\text{NA}]$ is the theoretical numerical aperture of the fiber, and $[\text{NA}_{\text{eff}}]$ is the effective numerical aperture at steady state, in the presence of mode coupling. By measuring or estimating the loss spectra of candidate core and cladding materials we may use this relationship to calculate expected fiber losses at specific wavelengths of interest.

An example of such a calculation is shown in Fig. 10.15. Equation (10.32) has been used to construct curves of constant total fiber loss on a grid of cladding loss vs cladding refractive index for a fused silica core fiber with $\alpha_{\text{core}} = 5\,\text{dB/km}$. Superimposed on this grid are data points representing commercially available polymers for which α_{clad} has been measured or estimated (see Table 10.6). It may be noted that fiber losses less than 10 dB/km (representing less than 5 dB/km excess loss contributed by the cladding) occur only when the cladding polymer loss is 10^3 dB/km or lower. Only one polymer displayed in Fig. 10.15, Silicone A, a cross-linkable polydimethyl siloxane resin, has the requisite transparency to

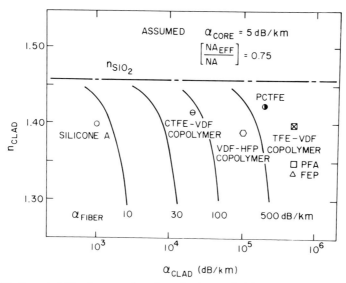

Fig. 10.15 Expected fiber loss as a function of polymer cladding loss and refractive index for a fused silica core fiber with $\alpha_{\text{core}} = 5$ dB/km (Blyler and Hart, 1977).

produce this low-loss fiber. The crystalline fluoropolymers (PFA, FEP, PCTFE, TFE-VDF copolymer) have scattering losses greater than 10^5 dB/km and therefore yield estimated fiber losses of more than 500 dB/km. Amorphous elastomeric fluoropolymers (CTFE-VDF copolymer, VDF-HFP copolymer) have relatively high losses presumably caused by particulate inclusions and impurities. It is likely that diligent effort in monomer and polymer preparation and handling could significantly lower these values. Nonetheless with the exception of silicone resins which have been developed in clean, transparent grades for use in electronics packaging, presently available commercial polymers are not highly satisfactory to the fiber engineer whose goal is a low-loss waveguide.

10.5.4 Early Work

Perhaps the earliest major development of polymer clad fiber waveguides began with the duPont Crofon® plastic fiber (Hager *et al.*, 1967) which features a polymethylmethacrylate core ($n = 1.488$) and a proprietary amorphous fluorocarbon cladding ($n = 1.390$). This development is an example of innovative materials and process engineering, but because of the high loss of the core material the fibers produced are too lossy (several hundred decibels per kilometer at 0.8 μm) for use in all but the shortest telecommunication links. More recent work (Schleinitz, 1977) using poly(perdeuteromethyl-methacrylate) as the core material shows promise of providing reduced loss.

Other efforts utilizing fused silica as the core material produced moderately high-loss fibers, principally due to deficiencies of the cladding materials and coating methods. Suzuki and Kashiwagi (1974) coated silica fibers from solution with a copolymer of hexafluoropropylene and vinyl fluoride ($n = 1.415$) and obtained losses in the range of 70 dB/km. Yoshimura *et al.* (1974) attained losses of 17 dB/km at 1.07 μm for a low-OH content silica fiber coated from a dispersion of a fluorocarbon resin. Buyken and Kriege (1972) and Dislich and Jacobsen (1973) produced FEP-clad silica fibers for use as lightguides in the uv spectral region. They achieved losses of 360 dB/km at 0.546 μm using uv-grade fused silica coated with an aqueous dispersion of FEP particles which were subsequently fused together through application of heat. It is likely that the high losses realized in these fibers resulted from a combination of the high bulk scattering of the cladding materials owing to their semicrystalline structure and void formation near the core–cladding interface occurring during solvent or water removal after coating.

10.5.5 Loose Polymer Claddings

Because of the high losses of the vast majority of low index polymers structures have been explored to minimize the problem of field penetration of transmitted light from the core into the lossy cladding. One approach involves the fabrication of a structure in which the fiber core is loosely surrounded by an extruded cladding tube of Teflon FEP or PFA (Kaiser *et al*, 1975; Blyler *et al.*, 1975). A photomicrograph of such a structure is shown in Fig. 10.16.

The fabrication of these loosely clad fibers is carried out by drawing the silica core from a preform rod and passing the pristine fiber through the crosshead die of a polymer melt extruder. A tubing die as depicted in Fig. 10.13 is used to form the loose tube. As in the case of extrusion coating of fibers sufficient clearance in the core tube must exist to allow the fiber to pass through it without contact or else surface damage to the fiber core and high loss will result. As the molten tube extruded around the fiber is drawn down around the fiber, it begins to solidify in the ambient air as the polymer temperature falls below the crystallization temperature. By adjusting such design and processing variables as die dimensions, melt temperature, extrusion rate, and line speed, the final dimensions of the polymer tube may be controlled.

Fig. 10.16 Cross-sectional photomicrograph of PFA-clad, fused silica fiber (Blyler and Hart, 1977).

In addition to the polymer tubing step, the fabrication of a low-loss fiber is critically dependent on the production of a uniform diameter, contamination-free silica fiber core. Oxy-hydrogen torch drawing yields a clean fiber but random diameter variations are usually severe, causing coupling of lower order guided modes to lossy radiation modes. Drawing with a high-temperature resistance furnace gives uniform fiber diameter but unless extreme measures are taken, contamination of the silica surface with impurities causing both scattering and absorption losses are a problem. An ideal heating source is the CO_2 laser which produces very clean, uniform diameter fibers (Jaeger, 1976).

The lowest loss polymer-clad fibers have been fabricated using low-OH content silicas for the core. The spectral loss curves for a 98 μm-diameter Spectrosil WF core fiber, loosely clad with Teflon PFA are shown in Fig. 10.17. The angular power distributions propagating in the fiber as a result of changing the NA of the injected beam were measured by scanning the far-field radiation patterns, at the near and far ends of the fiber (see Chapter 11). The approximate steady state corresponds to a launching condition for which the near and far end patterns most closely coincide, giving an $NA_{eff} = 0.32$ based on the 20-dB points of the radiation patterns, with losses of 5.8, 5.7, and 4.6 dB/km at wavelengths of 0.82, 0.91, and 1.10 μm, respectively. This NA_{eff} is considerably lower than the theoretical

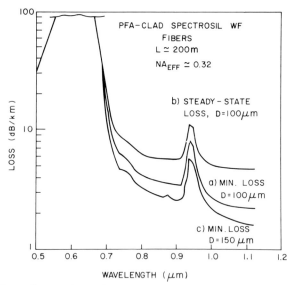

Fig. 10.17 Spectral transmission losses of two approximately 200-m-long PFA-clad Spectrosil WF-core fibers for (a) small-angle, and (b) steady-state ($NA_{eff} \approx 0.32$) excitation of a 100-μm core fiber, and (c) small-angle excitation of a 200-μm diameter fiber (Blyler et al., 1975).

value of 0.55, and reflects the selective attenuation of higher order modes caused by intermittent core contact with the lossy cladding tube.

Minimum losses of 2.2 dB/km near 1.1 μm are measured for low-order mode excitation (Fig. 10.17, curve a) and even lower losses of 1.6 dB/km are obtained for a fiber made with the same material, but having a larger core diameter of 150 μm (curve c). These losses are believed to be characteristic of the bulk material losses of Spectrosil WF.

Despite the achievement of low losses in these loosely clad fiber structures, problems exist for their successful implementation. The interior of the tube provides an air space in which atmospheric moisture can exist as both vapor and condensate. Diffusion of water vapor through the FEP or PFA tube is moderately rapid with diffusion coefficients of order 10^{-7} cm^2/sec, so that the interior of the tube comes to equilibrium with the atmosphere surrounding the exterior of the fiber in a relatively short time. Rapid temperature decreases can cause condensation within the tube, thereby increasing fiber loss due to scattering. The long-term mechanical reliability of the fiber core is also a concern when it is exposed to such atmospheric conditions.

Another problem with the loosely clad fiber structure arises because the cladding tube and silica core are decoupled from one another over short lengths, but remain coupled over long lengths owing to frictional forces. Differential thermal expansion and contraction between the core and cladding, occurring either during the fabrication process or as a result of temperature cycling, can cause the core to buckle and undulate in a helical path within the tube. This effect results in loss increases up to several 10 dB/km and represents a type of microbending in which the buckled silica core is pressed against the tube wall and forced to conform to its surface perturbations.

Because of the aforementioned problems, interest has shifted from the loose tube structure to tight polymer claddings.

10.5.6 Silicone-Clad Fibers

As indicated earlier the high optical losses of the vast majority of commercial low refractive index polymers makes them unsuitable for use as an intimate cladding for producing low-loss fibers. However, in 1975 Tanaka et al. reported on a fused silica core fiber with low transmission losses which utilized a silicone resin cladding. This commercial resin is specifically prepared for high transparency. It is a two-part addition cure system consisting of a base resin and a curing agent.

The optical transmission properties of the silicone material have been determined by Tanaka et al. (1975), and its spectral loss curve is shown in Fig. 10.18. A number of prominent absorption peaks are present at wave-

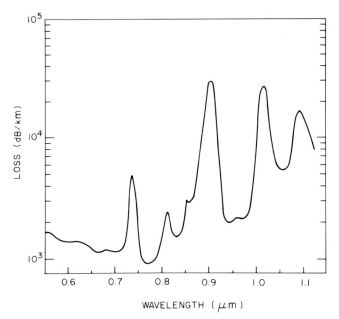

Fig. 10.18 Spectral transmission loss curve for silicone resin after Tanaka *et al.* (1975).

lengths of 0.74, 0,91, 1.02, and 1.08 μm. These are overtones of the CH absorption bands discussed above. Minimum losses are observed at 0.77 μm and amount to 900 dB/km.

Silicone-clad, fused silica core fibers are produced by applying the liquid silicone mixture (base resin and curing agent) to the silica core fiber as it is drawn. It is essential to carry out this operation without contacting the core against any solid surface to avoid damage and attendant high loss. The flexible die applicator discussed earlier is a convenient device for accomplishing this process. It is noncontacting and produces concentric coatings of the silicone resin with only moderate care in fiber alignment during drawing.

After the liquid resin is applied by the coating die the fiber is immediately passed through a furnace maintained at a temperature between 200 and 500°C to cross link the silicone. The optimum temperature depends on drawing speed, coating thickness, fiber diameter, and other drawing parameters. It is important to cure the resin before it starts to bead on the fiber core. Fortunately the curing reaction is very fast at elevated temperatures and this step is not difficult. The fiber emerges from the furnace with a completely cured, nontacky silicone cladding. Because this cladding is not abrasion resistant it is usually necessary to coat the fiber for handling protection. An extruded nylon jacket tubed over the silicone in the manner described in Section 10.4.4 provides a tough, abrasion-resistant coating.

Spectral loss curves for a CO_2 laser-drawn, 900-m-long silicone-clad fiber with a 100-μm-diameter core of Suprasil 2 are shown in Fig. 10.19. [The intense absorption bands visible in these curves result from the high-OH content of this synthetic silica (Kaiser *et al.*, 1973a).] Minimum losses occur at 0.79 μm and amount to 6 dB/km for small angle excitation and 7.6 dB/km for the steady state. The steady state NA of 0.40 is nearly identical to the theoretical value of 0.41 based on values of $n_{core} = 1.458$ and $n_{cladding} = 1.40$. This result indicates that the silicone–silica interface of laser-drawn fibers is essentially free of contamination and imperfections which would selectively attenuate the higher order modes through interfacial scattering and reduce the NA_{eff} below the theoretical value.

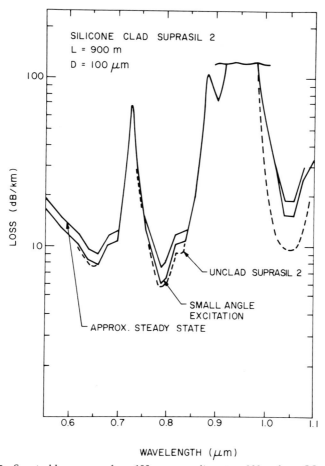

SILICONE CLAD SUPRASIL 2
L = 900 m
D = 100 μm

UNCLAD SUPRASIL 2

SMALL ANGLE EXCITATION

APPROX. STEADY STATE

LOSS (dB/km)

WAVELENGTH (μm)

Fig. 10.19 Spectral loss curves for a 100-μm core diameter, 900-m-long CO_2-laser-drawn silicone-clad Suprasil 2 core fiber for (a) small-angle and (b) steady-state ($NA_{eff} = 0.40$) excitation. The dashed curve is typical of unclad Suprasil 2 fibers (Blyler and Hart, 1977).

Furnace-drawn silicone-clad fibers typically exhibit a lower NA_{eff} ($\cong 0.3$) due to surface contamination. Also plotted for comparison in Fig. 10.19 is the spectral loss curve for a typical Suprasil 2 unclad fiber. It may be noted that the excess steady-state loss contributed by the cladding at 0.79-μm amounts to only about 1.5 dB/km, while at 1.06 μm it is considerably higher at about 10 dB/km. The CH-overtones contributed by the resin and centered at 1.02 and 1.09 μm (Fig. 10.18) are responsible for this result, as they elevate the cladding loss to approximately 600 dB/km at 1.06 μm.

In Fig. 10.20 a spectral loss curve is shown for a silicone-clad fiber utilizing a low-OH content synthetic silica, Suprasil W-2 as the core. Using a 0.25-μm NA launch condition the minimum losses occur at 0.78 μm and

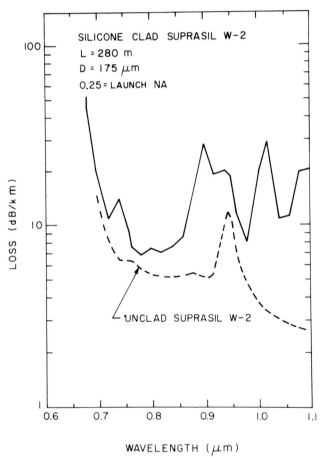

Fig. 10.20 Spectral loss curve for a 175-μm diameter, 280-meter-long CO_2-laser-drawn silicone-clad Suprasil W-2 core fiber using a 0.25 NA launch condition. The dashed curve represents a typical unclad Suprasil W-2 fiber.

amount to 8.3 dB/km. At longer wavelengths (0.86 to 1.10 μm) the silicone-clad Suprasil W-2 fiber exhibits considerably higher losses than does an unclad fiber of the same core material. The influence of the absorption overtones contributed by the resin is strongly felt in this region and the absorption peaks are prominently displayed in the spectral loss curve of the fiber. Thus, the low transmission losses of Suprasil W-2 at longer wavelengths, e.g., 2.2 dB/km at 1.06 μm are effectively swamped by the high losses of the silicone cladding.

10.5.7 Future Directions

Because of their large numerical aperture, low losses in certain wavelength regions and ease of fabrication in any core size desired, polymer-clad fused silica fibers may prove to be very useful for data bussing links which employ LED sources. Future work will be directed toward achieving low losses in these fibers at wavelengths specific to particular sources. It is likely that this activity will require the development of cladding polymers tailored in their chemical structure to provide absorption windows in these regions. A general reduction in scattering losses in these polymers through a lowering of particulate inclusions would aid in this endeavor.

REFERENCES

Anderson, J. H., and Wickerscheim, K. A. (1964). Solid surfaces. *Proc. Int. Congr. Phys. Chem. Solid Surfaces, 1964* p. 252.

Blyler, L. L., Jr., and Hart, A. C., Jr. (1977). Polymer claddings for optical fiber waveguides. *35th ANTEC Prepr., Soc. Plast. Eng., 1977* pp. 383–387.

Blyler, L. L., Jr., Hart, A. C., Jr., Jaeger, R. E., Kaiser, P., and Miller, T. J. (1975). Low-loss, polymer clad silica fibers produced by laser drawing. *Top. Meet. Opt. Fiber Transm., 1st, 1975* p. TuA5-1.

Bolger, J. C., and Michaels, A. S. (1968). Molecular structure and electrostatic interactions at polymer-solid interfaces. *In* "Interface Conversion for Polymer Coatings" (P. Weiss and G. D. Cheever, eds.), p. 3. Am. Elsevier, New York.

Bowden, F. P. (1962). The adhesion of metals and the influence of surface contamination and topography. *In* "Adhesion and Cohesion" (P. Weiss, ed.), p. 121. Elsevier, Amsterdam.

Bowden, F. P., and Tabor, D. (1963). The adhesion of solids. *In* "Structure and Properties of Solid Surfaces" (R. Gomes and C. H. Smith, eds.), Chapter VI. Univ. of Chicago Press, Chicago, Illinois.

Bowden, F. P., and Throssel, W. R. (1951). Adsorption of water vapor on solid surfaces. *Nature (London)* **167**, 601.

Brooks, C. S. (1958). Competitive adsorption of aliphatic compounds and water on silica surfaces. *J. Colloid Sci.* **13**, 522.

Brooks, C. S. (1960). Free energies of immersion for clay minerals in water, ethanol and *n*-heptane. *J. Phys. Chem.* **64**, 532.

Buckler, M. J., Santana, M. R., and Saunders, M. J. (1978). Lightguide cable manufacture and performance. *Bell Syst. Tech. J.* **57**, 1745.

Budd, S. M. (1961). "The mechanisms of chemical reaction between silicate glass and attacking agents. I. Electrophilic and nucleophilic mechanisms of attack. *Phys. Chem. Glasses* **2**, No. 4, 111.

Buyken, H., and Kriege, W. (1972). Properties of UV-fibers. *Proc. Soc. Photo-opt. Instrum. Eng.* **31**, 37–43.

Dabby, F. W., Pinnow, D. A., Ostermayer, F. W., Van Uitert, L. G., Saifi, M. A., and Camlibel, I. (1974). Borosilicate-clad fused silica core fiber optical waveguide with low transmission loss prepared by a high efficiency process. *Appl. Phys. Lett.* **25**, 714.

Debye, P. J. W., and VanBeek, L. K. H. (1959). Effect of adsorbed water on the optical transmission properties of isotropic powders. *J. Chem. Phys.* **31**, 1595.

Dislich, H., and Jacobsen, A. (1973). Lightguide systems for the ultraviolet region of the spectrum. *Angew. Chem.* **12**, No. 6, 439.

Fowkes, F. M. (1965). Attractive forces at interfaces. *In* "Chemistry and Physics of Interfaces" (D. E. Gushee, ed.), Chapter I. Am. Chem. Soc., Washington, D.C.

France, P. W., Dunn, P. L., and Reeve, M. H. (1977). On line fibre coating using tapered nozzles. *Proc. Eur. Conf. Opt. Commun., 3rd, 1977* p. 90.

Fuller, D. D. (1956). "Theory and Practice of Lubrication for Engineers," pp. 159–162. Wiley, New York.

Gardner W. B. (1975). Microbending loss in coated and uncoated optical fibers. *Top. Meet. Opt. Fiber Transm. 1st, 1975* Paper WA3-1.

Gent, A. N., and Hsu, E. C. (1974). Coupling reactions of vinyl silanes with silica and poly(ethylene-co-propylene). *Macromolecules* **7**, No. 6, 933.

Gloge, D. (1973). See Miller, S. E., Marcatili, E. A. J., and Li, T. (1973). Research toward optical-fiber transmission systems. *Proc. IEEE* **61**, 1703.

Gloge, D. (1975). Optical fiber packaging and its influence on fiber straightness and loss. *Bell Syst. Tech. J.* **54**, No. 2, 245.

Hager, T. C., Brown, R. G., and Derick, B. N. (1967). Plastic fiber optics. *SAE Trans.* **76**, 75.

Hart, A. C., Jr., and Albarino, R. V. (1977). An improved fabrication technique for applying coatings to optical fiber waveguides. *Top. Meet. Opt. Fiber Trans., 2nd, 1977* Paper TuB2-1.

Hoffman, R. L. (1975). A study of the advancing interface. I. Interface shape in liquid–gas systems. *J. Colloid Interface Sci.* **50**, No. 2, 228.

Irwin, G. R. (1967). Fracture mechanics applied to adhesive systems. *Treatise Adhes. Adhes.* **1**, 233.

Jaeger, R. E. (1976). Laser drawing of glass fiber optical waveguides. *Am. Ceram. Soc., Bull.* **55**, No. 3, 270.

Kaelble, D. H. (1971). "Physical Chemistry of Adhesion. "Wiley (Interscience), New York.

Kaiser, P., Tynes, A. R., Astle, H. W., Pearson, A. D., French, W. G., Jaeger, R. E., and Cherin, A. H. (1973a). Spectral losses of unclad vitreous silica and soda-lime silicate fibers. *J. Opt. Soc. Am.* **63**, No. 9, 1141.

Kaiser, P., Marcatili, E. A. J., and Miller, S. E. (1973b). A new optical fiber. *Bell Syst. Tech. J.* **52**, No. 2, 265–269.

Kaiser, P., Hart, A. C., Jr., and Blyler, L. L., (1975). Jr. Low loss FEP-clad silica fibers. *Appl. Opt.* **14**, 156.

Koenig, J. L. (1975). Application of fourier transform infrared spectroscopy to chemical systems. *Appl. Spectrosc.* **29**, 293.

Kurkjian, C. R., Albarino, R. V., Krause, J. T., Vazirani, H. N., DiMarcello, F. V., Torza, S., and Schonhorn, H. (1976). Strength of 0.04-50m lengths of coated fused silica fibers. *Appl. Phys. Lett.* **28**, 588.

Lenahan, T. A. (1978). Viscous centering force on a fiber in a coating applicator. Abstr. Summer Mtg. Am. Math. Soc., p. A546.

Lewis, J. A. (1979). Centering force in the fiber coating applicator. *Bell Syst. Tech. J.,* in press.

MacChesney, J. B., Jaeger, R. E., Pinnow, D. A., Ostermayer, F. W., Rich, T. C., and Van Uitert, L. G. (1973). Low-loss silica core-borosilicate clad fiber optical waveguides. *Appl. Phys. Lett.* **23,** 340.

Noller, C. R. (1965). *"Chemistry of Organic Compounds."* Saunders, Philadelphia, Pennsylvania.

Orowan, E. (1932). Bemorkung zu den Arbeiten von F. Zwicky über die strucktrurder
Realkristalle. *Z. Phys.* **79,** 573.

Parks, G. A. (1965). The isoelectric points of solid oxides, solid hydroxides and aqueous hydroxo complex systems. *Chem. Rev.* **65,** 177.

Pleuddemann, E. P. (1972). Reactive silanes as adhesion promoters to hydrophilic surfaces.
27th Annu. Tech. Conf. Reinf. Plast., SPI Sect. 21-B.

Purday, H. F. P. (1949). "An Introduction to the Mechanics of Viscous Flow," p. 37 pp.
16–18, 28, 48–49, and 57–59. Dover, New York.

Roe, R. J. (1975). Wetting of fine wires and fibers by a liquid film. *J. Colloid Interface Sci.* **50,**
70.

Saunders, M. J. (1977). Adhesive sandwich optical fiber ribbons. *Bell Syst. Tech. J.* **56,** 1013.

Schleinitz, H. M. (1977). "Ductile Plastic Optical Fibers With Improved Visible and Near
Infrared Transmission," Proc. 26th Int. Wire & Cable Symp. CORADCOM, Cherry Hill,
New Jersey, p. 352.

Schonhorn, H., Kurkjian, C. R., Jaeger, R. E., Vazirani, H. N., Albarino, R. V., and Di-
Marcello, F. V. (1976). Epoxy acrylate coated fused silica fibers with tensile strengths
>500 ksi (3.5 GN/m²) in 1-km gauge lengths. *Appl. Phys. Lett.* **29,** 712.

Shafrin, E. G., and Zisman, W. A. (1967). Effect of adsorbed water on the spreading of
organic liquids on soda-lime glass. NRL Report No. 6496.

Standley, R. D. (1974). Fiber ribbon optical transmission lines. *Bell Syst. Tech. J.* **53,** 1183.

Suzuki, H., and Kashiwagi, H. (1974). Polymer-clad fused-silica optical fiber. *Appl. Opt.* **13,**
No. 1, 1.

Tanaka, S., Inada, K., Akamoto, T., and Kozima, M., 1975. Silicone-clad fused silica core
fiber. *Electron Lett.* **11,** 153.

Ter-Minassian-Sarage, L. (1964). Chemisorption and dewetting of glass and silica. *Adv.
Chem. Ser.* **43,** 232.

Wang, T. T., Zupko, H. M., Vazirani, H. N., and Schonhorn, H. (1977). UV cured epoxy-
acrylate coatings on optical fibers. III. Effect of environment on long term strength. *Top.
Meet. Opt. Fiber Transm., 2nd, 1977.* Paper TuA5-1.

Yoshimura, K., Nakahara, T., Tsukamoto, A., and Isomura, A. (1974). Low-loss plastics-
cladding fibre. *Electron. Lett.* **10,** No. 25/26, 534.

Zettlemoyer, A. C. (1965). Immersional wetting of solid surfaces. *In* "Chemistry and Physics
of Interfaces" (D. E. Gushee, ed.), Chapter XIII.

Chapter 11

Fiber Characterization

LEONARD G. COHEN
PETER KAISER
PAUL D. LAZAY
HERMAN M. PRESBY

11.1 INTRODUCTION

Central to gaining insight into the operational characteristics of optical fibers is a measurement capability with which the transmission properties of glass-fiber waveguides can be determined. In this chapter we will concentrate on the evaluation of the transmission losses, pulse dispersion, refractive-index difference, and index profile. Among the various techniques available we will emphasize those which in our opinion have proven most useful. While the techniques discussed are generally applicable to fibers of different kind and composition, we present results obtained with low-loss, high-silica fibers for illustrative purposes, and to indicate the quality of the data.

An optical characterization typically serves two different purposes: on one hand, the acquired data form the basis upon which the material scientist attempts to improve the properties of the glasses used for fiber fabrication. On the other hand, the systems engineer requires data which ensure that realistic fiber parameters are used in design calculations. The characterization techniques should be sufficiently flexible to accommodate both sets of requirements as will become clearer in subsequent discussions.

343

11.2 TRANSMISSION LOSS

11.2.1 Loss Mechanisms

Perhaps the single most important characteristic of glass-fiber optical waveguides is their transmission loss, and it is the spectacular reduction of this loss in recent years which has made telecommunication via light waves a reality. In order to facilitate the interpretation of data acquired in fiber loss measurements, we first summarize the various mechanisms responsible for such loss.

Attenuation in optical fibers is due to absorption and radiation processes, with the latter being either caused by material scattering or waveguide effects. Material absorption is caused by (a) atomic transitions in the uv and molecular vibrations in the ir wavelength regions; (b) by atomic and network effects, being either intrinsic, or caused by irradiation or during the drawing process; (c) by impurities. Amongst the latter, transition metal ions like Fe^{2+} and Cu^{2+} are responsible for losses of about 1 dB/km/ppb contamination in certain portions of the spectrum (Smith and Cohen, 1963) (Fig. 11.1), while water via the harmonics of the 2.73 μm O–H stretching vibration and combinational overtones with the 12.5 μm SiO_2 vibration, is responsible for numerous absorption bands throughout the wavelength region of interest for optical communication (Jones and Kao, 1969; Keck et al., 1973; Kaiser et al., 1973). Even in low-OH-content silica fibers, the residual water content typically still creates absorption bands near 0.95, 1.24, and 1.39 μm, with corresponding intensities of about 1, 2, and 40 db/km/ppm OH, respectively (Fig. 11.2).

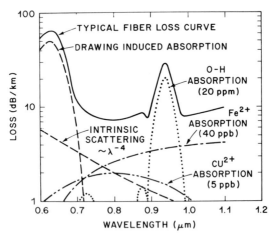

Fig. 11.1 Loss spectrum of a low-OH-content silica fiber synthesized from typical loss components. (Cohen, 1975.)

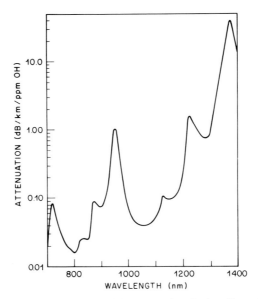

Fig. 11.2 OH-related spectral loss contribution in fused silica fibers (D. B. Keck, in Barnoski and Personick, 1978).

Intrinsic radiation losses are associated with Rayleigh scattering due to frozen-in thermal and compositional fluctuations of the glass structure. Rayleigh scattering is characterized by a $(1 + \cos^2 \theta)$ angular dependence (with θ representing the angle of radiation relative to the fiber axis) and a λ^{-4} wavelength dependence. It determines the lowest losses achievable with a particular glass composition. Scattering losses are also caused by material defects such as elongated bubbles and inclusions, as well as inhomogeneities in the refractive-index distribution, with the radiation being primarily forward directed (Tynes *et al.*, 1971; Rawson, 1972). A resonance-scattering band found near 0.63 μm in low-OH content, pure-fused-silica fibers is associated with a drawing-induced absorption band (Kaiser, 1974).

Both the core and cladding contribute to the material-related loss α according to their relative power concentrations p_{core} and p_{clad},

$$\alpha = p_{\text{core}} \alpha_{\text{core}} + p_{\text{clad}} \alpha_{\text{clad}} \tag{11.1}$$

with α_{core} and α_{clad} being the attenuation coefficients of the respective materials.

Radiation losses due to waveguide effects result from the loss of high-order guided modes and particularly those close to cutoff, from leaky modes, from variations of the core cross section, and from a curvature of the guide axis (Snyder and Mitchell, 1974; Marcuse and Presby, 1975b;

Marcatili, 1969; Gloge, 1972b). Small-scale, random perturbations of the guide axis, as for example caused by an intimate contact of the fiber with a rough drum surface, or in the cabling process, introduce what is commonly referred to as microbending losses which are discussed more extensively in Chapter 6 of this book (Gardner, 1975; Gloge, 1975). Using the wavelength-dependent contributions of some of the more important loss mechanisms which may be encountered in high-silica fibers, a loss spectrum is semiquantitatively synthesized in Fig. 11.1 which approximates actually measured spectra of some low-OH-content synthetic silica fibers (Kaiser, 1973).

11.2.2 Preliminary Fiber Evaluation

The evaluation of the propagation characteristics of optical fibers is greatly facilitated because of their high transparency in the visible portion of the spectrum. This permits a qualitative visual examination via the scattered light intensity when the fiber is, for example, excited with the intense 0.6328 μm light of a HeNe laser: isolated waveguide defects are easily recognizable through their excess scattering. Interface and cladding defects are distinguishable from core imperfections because the former scatter preferentially for higher order mode excitation. Mode loss due to cutoff phenomena, and microbending losses manifest themselves as excess forward scattering. In contrast, only Rayleigh scattering is present if the light scattered in the forward and backward direction is identical. A high-quality fiber is thus characterized by a uniform scattering level with gradually decreasing intensity along its length.

While an analysis of the light scattered from the fiber provides insight into its loss characteristic, information concerning the numerical aperture, the degree of mode coupling, and the quality of the fiber end break, can be deduced from an inspection of the radiation pattern at the end of the fiber. A good break, as for example obtained by scoring the fiber with a diamond scribe while under tension, or with other, more sophisticated techniques (Gloge et al., 1973), produces a symmetric radiation pattern. The fiber end section itself appears dark under this condition since no power is reflected into fast escaping higher modes as is the case with an arbitrarily fractured end surface. Provided all fiber modes are excited, the sine of the half-angle θ_c of the radiation pattern (baseline width) yields a numerical aperture (NA) which is in good agreement with the theoretical NA_{th} computed from

$$NA_{th} = (n_1^2 - n_2^2)^{1/2} = n_1(2\Delta)^{1/2} \tag{11.2}$$

with n_1 and n_2 being the refractive indices of the core and cladding, respectively, and $\Delta \cong n_1/n_2 - 1$ (Cherin et al., 1974). The index difference can thus be obtained from the measured half-angle θ_c according to

$$\Delta = \tfrac{1}{2}(\sin \theta_c/n_1)^2. \tag{11.3}$$

For a multimode fiber, the intensity distribution of the radiation pattern is a measure of the modal power distribution, $N(\theta)$, given approximately by

$$N(\theta) \simeq N_{tot}(\theta/\theta_c)^2 \tag{11.4}$$

with N_{tot} being the total number of modes propagating in the fiber. Hence, one can get an estimate of the degree of mode coupling by selectively exciting only a few fiber modes, or equivalently, by exciting the fiber with a narrow beam of light (relative to θ_c): mode coupling is small if the output pattern closely resembles the input distribution (Marcuse, 1973; Keck, 1974). On the other hand, extensive mode mixing causes a broadening of the input pattern. A steady-state distribution is reached if an equilibrium exists between the selective loss of higher modes and their replenishing by mode coupling from lower order modes (Gloge, 1972a; Kaiser et al., 1975). For a lossy cladding or core–cladding interface with its accompanying selective loss of the higher modes, the steady-state NA is smaller than NA_{th}. A steady-state mode distribution and associated loss cannot establish itself if mode coupling is absent. Assuming that the loss increases with mode number, an initially large mode distribution will gradually narrow along the fiber, and the incremental losses will become the smaller, the longer the fiber. While the knowledge of the steady-state losses is necessary for determining the loss of arbitrary fiber lengths, in many cases it is merely important to determine the insertion loss of a given fiber length associated with a specific excitation condition used.

11.2.3 Transmission Loss Measurements

11.2.3.1 Destructive Technique. The transmission losses of optical fibers are commonly determined by measuring the total power at two points of separation L (Tynes, 1970; Keck and Tynes, 1972; Kaiser and Astle, 1974). With P_{FE} representing the power at the output or far end, and P_{NE} the power at the broken-off near-end point, the normalized loss α is given by

$$\alpha = \frac{10}{L(km)} \log \left(\frac{P_{NE}}{P_{FE}}\right) \text{ (dB/km)}. \tag{11.5}$$

The schematic of a loss measuring apparatus is shown in Fig. 11.3. The collimated light of either a laser, an LED, or a filtered white-light source, passes through a variable attenuator, a chopper, a beam splitter, a variable aperture wheel, and a focusing lens. Chopped light enables low-noise ac or phase-sensitive detection of weak signals without the interference of ambient light. A reference detector is desirable for monitoring source

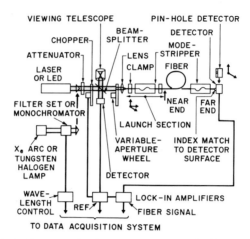

Fig. 11.3 Transmission loss measuring set. (Cohen, 1976.)

instabilities. It also permits their elimination through ratioing, provided they are caused by amplitude, and not spatial variations of the light beam such as those occurring in arc lamp systems. The size and the alignment of the focused beam on the fiber end face is observable via its reflection through a beamsplitter and a viewing telescope. The NA of the launch beam can be changed with a variable-aperture wheel in front of the focusing lens.

The fiber input end is held in place by a clamping device or vacuum chuck mounted on a micropositioner, whereby care has to be taken that the mode spectrum launched is not altered by excessive clamp pressure and accompanying bends of the fiber axis. The short launch- or reference-fiber section immediately following the launch point is typically a few meters long to allow for mechanical isolation of the input end, as well as for cladding mode stripping with an index-matching liquid. A curvature in the reference fiber section reduces the intensity of highest order and leaky modes. Because of its effect on the modal power distribution, the layout of the reference section typically affects the loss of the fiber, and its influence must be carefully analyzed for a particular fiber type investigated. A cladding mode stripper may also be required near the output end under those circumstances where the power scattered from the core into the cladding can accumulate. This will happen, for example, if the fiber is jacketed with a low-index silicone resin. On the other hand, no cladding mode stripper is required at either end, if the fiber is surrounded with a lossy or higher index jacket.

In order to obtain reproducible data it is good practice to index-match the fiber output end to the detector surface. In addition, the same spot on the detector surface should be illuminated for the far- and near-end power measurements in order to minimize the influence of absolute and spectral sensitivity changes across the detector surface. The far-field patterns, whose knowledge is required for steady-state loss measurements as we shall see later, is measured with a pinhole detector or a scanning diode array, which for this purpose replaces the total power detector.

While silicon diodes are generally employed as detectors for loss measurements below 1.1 μm, germanium, lead sulfide, and some more recently developed detectors (Wagner *et al.*, 1974) extend the range of the measuring set to about 1.8 μm. Suitable white-light sources for spectral loss measurements are, for example, the high-brightness Xe-arc lamp, which is particularly useful for small-NA and single-mode fiber measurements, and the more stable, yet less intense, tungsten–halogen lamp. Both sources can be used either with interference filters, which have a high power transfer but are limited to the discrete wavelengths chosen, or monochromators, which allow a continuous scan, but have an inferior coupling efficiency to the circular fiber geometry.

For constant launch conditions, the loss α as calculated above can be determined with a precision of about ± 0.1 dB/km for 1-km-long fiber sections. However, different launch conditions may result in significantly different loss values because of the mode dependence of the fiber losses. Specifically, higher modes suffer increased losses because of mode cutoff phenomena, lossy claddings, tunneling through a finite cladding layer, core–cladding imperfections, geometrical variations of the guide cross section, as well as micro- and macrobends of the guide axis. Lower modes, too, may experience selective attenuation if the material losses are higher near the guide axis, as may be caused by the addition of dopants in graded-index fibers (Ostermayer and Pinnow, 1974; O'Connor *et al.*, 1976), or if waveguide imperfections exist near the guide axis such as a pronounced index dip which may couple lowest modes to radiation modes. As a consequence, the mode distribution changes along the fiber until a dynamic equilibrium or steady-state distribution establishes itself after a coupling length L_c (Cohen and Personick, 1975) as a result of the selective loss of the various modes and the replenishing of their power via mode coupling from other modes. Mode coupling may be caused by waveguide imperfections, core diameter variations, or microbends. In general, mode coupling is associated with excess loss due to coupling to radiation modes unless special precautions are taken to prevent such coupling (Marcuse, 1974b). Typical mode-coupling lengths range from tens of meters to more than 10 km (Cohen, 1976).

Provided the reference point lies beyond L_c, the measured α represents the length-independent steady-state loss which can be extrapolated to arbitrary fiber lengths. In general, though, L_c is too long to render the above approach practical, and approximations to the steady-state mode distribution have been generated in separate, long, reference fibers having similar transmission characteristics (Murata, 1976), or through the use of mode scramblers (Ikeda *et al.*, 1976; Eve *et al.*, 1976). In both cases, though, it is not known to what degree the steady-state mode distribution has actually been established.

This ambiguity is avoided by scanning the radiation patterns and using them as a measure of the mode distribution. While it is difficult, even when known, to launch the steady-state distribution exactly, an approximation can be achieved by launching beams with different NA's and measuring the corresponding radiation patterns both at the far- and near-end points. The steady-state losses are associated with that set of radiation patterns whose distribution remains essentially unchanged (Kaiser *et al.*, 1975).

While changing the launch NA is sufficient to determine the steady-state losses of step-index fibers, the more complex excitation behavior of graded-index fibers requires in addition a restriction of the spot size (relative to the core diameter) in order to avoid the excitation of higher order and leaky modes even for small-angle launch beams. This can be accomplished by incorporating a 25-μm pinhole imaging system in the collimated beam of the loss measuring set of Fig. 11.3, if it is to be used, for example, to determine the steady-state losses of a 55-μm core size, 0.23 NA, Ge-doped, graded-index fiber. The NA-dependent near- and far-end radiation patterns of such a fiber, together with the corresponding losses, are shown in Fig. 11.4. As is typically observed, the radiation patterns widen along the fiber for under-excitation, and narrow for over-excitation. Only when the launch NA in this particular example corresponded to the NA of the fiber did the near- and far-end patterns essentially coincide. The associated steady-state losses were 0.57 dB/km higher than the minimum losses observed for small-angle excitation.

In general the steady-state losses are higher than the small-angle losses. However, depending on the type and quality of the graded-index fiber, they may be identical or even lower than the small-angle losses (Kaiser, 1977, 1978). Since a central index dip precludes the excitation of lowest modes, fibers having such a dip tend not to exhibit the small-NA excess loss behavior.

The differential mode attenuation of graded-index fibers has been studied by Olshansky and Oaks (1978), and by Yamada *et al.* (1977). The principal-mode-number-dependent total and scattering losses of a 0.28 NA, Ge-P-doped graded-index fiber (Fig. 11.5) also indicate lower attenuation for intermediate-mode excitation, and a decrease of this differential loss with wavelength due to diminished compositional-fluctuation scattering.

The spectral losses of a 0.22 NA, Ge-doped, graded-index fiber for

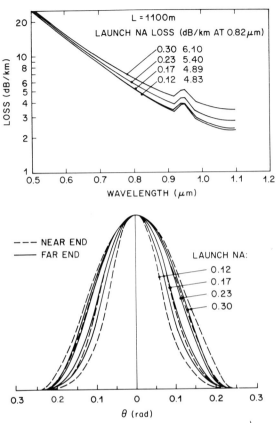

Fig. 11.4 Launch-NA-dependent loss spectra of a 0.23 NA, Ge–B-doped graded-index fiber, including associated far- and near-end radiation patterns. The steady-state losses are obtained for a 0.23 launch NA.

which identical losses have been measured for steady-state incoherent, and laser excitation are shown in Fig. 11.6. The losses of this fiber follow closely the λ^{-4} dependence characteristic for Rayleigh scattering, which for high-silica fibers comprises the dominant loss mechanism at shorter wavelengths. In confirmation of this, scattering losses measured independently at 0.6328 μm using a 4-cm-long integrating cube which will be described later, approached the total loss value within 1 dB/km.

The λ^{-4} dependence of intrinsic Rayleigh scattering at shorter wavelengths, and the tails of the dopant-dependent infrared absorption bands of the glass constituents at longer wavelengths, create regions of highest transparency in the 1 to 1.5 μm wavelength region (Fig. 11.7) (Osanai *et al.*, 1976). Future telecommunication systems are therefore expected to operate in that part of the spectrum, while presently preference is being

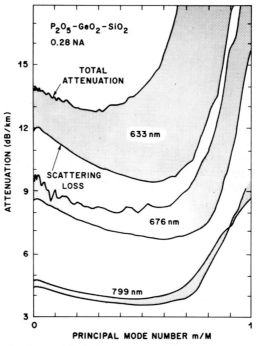

Fig. 11.5 Absorptive losses (shaded areas) of a 0.28 NA, graded-index fiber at three wavelengths are shown as the difference between total loss (upper curve) and scattering loss (lower curve). (Olshansky, 1977; Olshansky and Oaks, 1978).

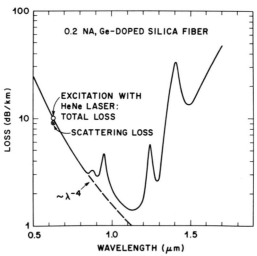

Fig. 11.6 Steady-state spectral losses of a 0.22 NA, GeO_2/B_2O_3-doped silica fiber, including small-angle excitation and scattering losses measured at 0.6328 μm. (Cohen *et al.*, 1977.)

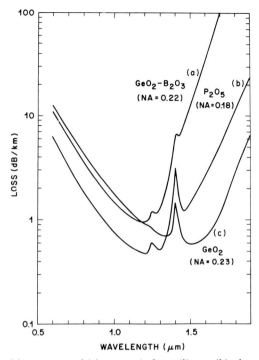

Fig. 11.7 Spectral loss curves of (a) germania borosilicate, (b) phosphosilicate, and (c) germania-doped fibers (Osanai *et al.*, 1976).

given to the 0.8- to 0.9-μm-wavelength region, where reliable sources and fast detectors are already available.

The fact that the losses obtained for small-angle excitation are a close approximation to the core material losses, is utilized in the determination of glass bulk losses via unclad-fiber measurements (Kaiser, 1973; Kaiser *et al.*, 1973). For this purpose, 10- to 60-m-long sections of unclad fibers drawn under high-purity conditions were freely suspended between low-index Teflon®-FEP clamps. Spectral loss curves of commercial-fused-quartz as well as high- and low-OH-content synthetic silica fibers are shown in Fig. 11.8. The fiber losses are considered to be close approximations of the bulk material losses with the exception of the 0.63 μm, drawing-induced loss band of the Spectrosil WF fiber (Kaiser, 1974). Instead of using unclad fibers, similar results have been obtained with fibers which were loosely jacketed with Teflon® FEP or PFA (Blyler *et al.*, 1975). It is interesting to note that, although available for some time, the high transparency of synthetic fused silica was unknown, until sensitive fiber-loss-measuring-techniques had been developed.

While the losses of single-mode fibers are primarily due to material

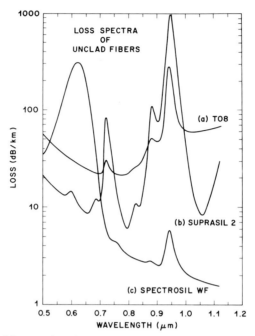

Fig. 11.8 Spectral losses of unclad silica fibers drawn from (a) commercial-grade fused quartz (TO8), (b) a high-OH-content synthetic silica, Suprasil 2, and (c) a low-OH-content synthetic silica, Spectrosil WF. (Personick, 1977.)

losses when they are operated in their single-mode domain, mode cutoff phenomena introduce loss bands of variable intensity if they are operated in the multimode domain (Kaiser et al., 1977; Tasker et al., 1978; Reeve et al., 1976). As a consequence of a limited cladding thickness, the cutoff wavelengths of the higher modes are shifted to shorter wavelengths, and a rapid increase of the fundamental mode loss occurs beyond a critical wavelength, in aggreement with theoretical predictions (Fig. 11.9).

Recently, an elegant graphic method has been proposed in which the spectral losses are plotted versus λ^{-4} (Inada et al., 1976; Yoshida et al., 1978). If α is represented by

$$\alpha = (A/\lambda^4) + B + C(\lambda), \tag{11.6}$$

A is the so-called Rayleigh scattering coefficient; B is a constant loss contribution which is affected by launch conditions, waveguide imperfections, and microbends; and $C(\lambda)$ represents wavelength-dependent loss contributions originating from impurities (including OH), drawing-induced effects, and uv and ir absorption tails. Above representation is valuable in as much as it facilitates the identification and separation of these different loss mechanisms.

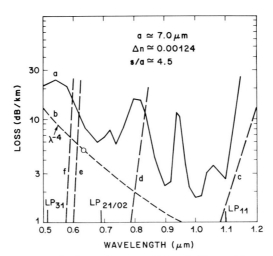

Fig. 11.9 Loss spectrum of a borosilicate single-mode fiber with pure silica core; core radius $a = 7$ μm, fiber diameter $= 170$ μm, $\Delta n_{0.9\,\mu m} = 0.00124$, cladding/core ratio $= 4.5$. $\circ$ = scattering loss measured at 0.6328 μm. (a) loss curve of a 254-m-long fiber, (b) Rayleigh scattering of pure silica. Theoretical leakage losses of the (c) LP_{01}, (d) LP_{11}, (e) LP_{21}, and (f) LP_{02} modes.

11.2.3.2 Nondestructive Techniques.

The breaking method exclusively discussed thus far has the disadvantage of being a destructive technique. Several nondestructive techniques exist which may be preferable under certain circumstances. In the simplest case of an insertion-loss-type measurement, after ascertaining the magnitude of the near-end power level, and assuming reasonable source stability, the fiber losses can be deduced from the far-end reading alone. Similarly straightforward is the loss measurement of a fiber containing end connectors, if their contribution, together with the reflection losses, can be included.

In a novel approach, pulse-amplitude measurements in single-reflection or shuttle-pulse arrangements yield the transmission losses, provided mirror reflection losses are subtracted as discussed in more detail in Section 11.4.2. In an extension of this technique, fiber losses have also been determined via backward Rayleigh scattering, using a highly sensitive, gated detection scheme as described in Section 11.4.4.

11.2.4 Scattering and Absorption Loss Measurements

Total transmission loss data may not provide sufficient information to indicate whether scattering or absorption is the main loss mechanism. Techniques have therefore been developed that allow an independent measurement of these losses.

In an indirect approach to scattering loss measurements, the spectrum of the scattered light is analyzed with the Brillouin scattering technique which employs a Fabry–Perot spectrometer to separate the elastic (Rayleigh) scattering from the inelastic (Brillouin) scattering. From a foreknowledge of the inelastic scattering cross section it is possible to determine the elastic scattering loss (Rich and Pinnow, 1974). The technique is not simple and has not been widely adopted.

The usual approach is to collect the scattered light that exits a short length of fiber and compare it to the power traveling in the fiber. The fiber is surrounded with an index-matching fluid so that light scattered into all angles (other than the near forward and back-scattered light guided in the core) is detected. Practical scattering cells use either an enclosure lined with solar cells, or an integrating sphere and photodetector (Tynes, 1970; Ostermayer and Benson, 1974). A measuring set employing a modified Tynes cell is shown in Fig. 11.10. After measuring the scattered power P_{sc}, the fiber end is placed into the cell and the total power P_{tot} is determined. The scattering loss can then be calculated from

$$\alpha_{sc} = \frac{4.34 \cdot 10^5}{l(cm)} \frac{P_{sc}}{P_{tot}} \quad (dB/km), \tag{11.7}$$

with l being the fiber length sampled. Measured scattering losses of different Ge-B-doped fibers whose NA varied between 0.13 and 0.38 are presented in Fig. 11.11 (Kaiser, 1977, 1978). The loss data are normalized with the Rayleigh scattering losses of pure fused silica which were measured to

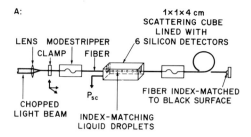

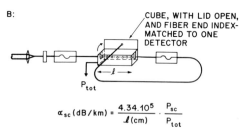

$$\alpha_{sc}(dB/km) = \frac{4.34 \cdot 10^5}{l(cm)} \cdot \frac{P_{sc}}{P_{tot}}$$

SCATTER-LOSS MEASUREMENT SETUP

Fig. 11.10　Scattering loss measuring setup.

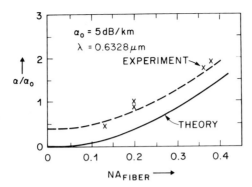

Fig. 11.11 Excess scattering losses of Ge–B-doped silica fibers at 0.6328 μm and comparison with theoretical prediction for Ge-doped fibers.

be 5.08 dB/km (at 0.6328 μm) using the same integrating cube and plastic-clad silica fibers. While the quadratic increase of the scattering loss was well predicted by theory, there was a residual constant excess loss which was attributed to the compositional-fluctuation scattering caused by the boron dopant, and to a lesser degree on residual waveguide imperfections.

While relative accuracies of ± 0.2 dB/km can be achieved in scattering loss measurements using laser sources, the absolute accuracy depends on a careful calibration of the scatter cube and the mode spectrum propagating in the fiber. Although a laser is the preferred light source because of the low scattering levels involved, meaningful data have also been obtained for incoherent excitation (Kaiser, 1974; Midwinter and Reeve, 1974).

In general, the absorption losses are determined as the difference between the total and scattering losses. However, they can also be measured directly as, for example, via the temperature rise caused by the absorbed power. While direct-contact schemes have been considered, the usual approach is to place the fiber in a thin-walled glass capillary (Witte, 1972; R. L. Cohen, 1975). The capillary can be air- or liquid-filled (White, 1976). The temperature rise at the exterior capillary surface is measured either with thermocouples or with a fine Pt wire wound on the capillary. In the first of two approaches taken, after the equilibrium temperature rise, ΔT, is measured, the incident beam is blocked and the thermal decay time, τ, determined (Pinnow and Rich, 1975; White, 1976). The absorption loss α_{abs} is given by

$$\alpha_{\text{abs}} = CA\Delta T/\tau P_o, \tag{11.8}$$

where A is the fiber cross-sectional area, C is the fiber heat capacity per unit volume, and P_o is the guided optical power. In a second approach, a

Pt sensing wire is placed in an ac bridge circuit (R. L. Cohen *et al.*, 1975). Calibration is accomplished either by a known self-heating of the Pt winding, or by using the measurable power dissipation in a fine Constantan wire that is substituted for the fiber. The power in the fiber is measured with a calibrated thermopile or radiometer. The techniques are limited to wavelengths available from laser sources as 10 to 100 mW of incident power are usually required. Losses of less than 1 dB/km have been measured.

Most absorption cells are constructed so that the scattered light does not affect the measurement. However, scattering losses can be included by using opaque absorbing coatings. By threading the fiber simultaneously through opaque and transparent capillary tubes, and by making a difference temperature measurement, it is also possible to measure the scattering loss alone (R. L. Cohen, 1975; Stone, 1978). Other proposals include the independent determination of the scattering and absorption losses via their different thermal response times (Zaganiaris, 1974); and through the use of an acousto-optic cell (R. L. Cohen *et al.*, 1975; Pinnow and Rich, 1975; Lewis, 1976).

11.2.5 Measuring Accuracy

While comments regarding the accuracy of fiber loss measurements have already been made where appropriate, it is necessary to add some general remarks. Principally we distinguish between relative accuracy, or sensitivity, which is primarily affected by instrumental factors, and the absolute accuracy, which is mainly determined by mode spectrum considerations, with the fiber length influencing both types of accuracies. Primarily limited by source instabilities and detector responsivity changes, relative accuracies of about ± 0.1 to 0.2 dB/km can be achieved with 1-km-long fiber sections. While the measurement error increases for shorter fiber lengths, reliable results can still be obtained with relatively short fiber sections using statistical averaging. With the dynamic range of typical measuring setups being on the order of 30 to 40 dB, suitably short fiber sections have to be measured if the losses of a given length exceed this value.

In contrast to a high relative accuracy, the absolute losses depend on the launch conditions and other factors of influence, and differences of many decibels per kilometer for different measurements of the same fiber may not be uncommon unless precautions are taken to standardize the launch conditions as discussed earlier.

11.2.6 Automated Data Acquisition

Because of the large volume of data involved, loss measurements lend themselves readily to automated data acquisition and analysis. Without going into a detailed discussion of a particular automated system, we will consider the merits of the two general approaches available, namely, off-line and real-time data acquisition.

In the off-line system the data are collected and recorded on paper or magnetic tape, and the tape is read into a central computing center at a later time for analysis and reduction of the data to a desired form such as graphical output (Kaiser and Astle, 1974). This method has the disadvantage that little or no automated control of the experiment is possible and results are not immediately available. Real-time or on-line systems employ a dedicated computer or calculator (including terminal, printer, and file storage) that collects the data, controls the equipment, and generates results during and after data acquisition. Until the recent adoption of the 488 standard interface bus the interconnection of instruments was one of the main obstacles to automation (IEEE, 1975). This standard bus permits the interfacing of minicomputers and calculators with a large number of instruments which are commercially available (Hewlett-Packard, 1975; see also Traifari, 1976).

11.3 REFRACTIVE-INDEX DISTRIBUTION

11.3.1 Significance of Index Distribution

Glass optical fiber waveguides currently under investigation possess a wide variety of refractive index profiles ranging from the simple dielectric rod waveguide or unclad fiber to the more sophisticated graded-index fiber. In the former case a discontinuous change in index of refraction occurs at the fiber surface, from the dielectric's value n_1 to the ambient index n_0 and in the latter case the composition of the glass is varied in such a way as to develop an index of refraction with a maximum on the axis and decreasing gradually in the radial direction until it merges into the constant cladding index (Gloge and Marcatili, 1973).

The refractive-index distribution plays an important role in characterizing the fiber and in determining its transmission behavior (Miller et al., 1973). It provides information on the numerical aperture of the fiber, on the number of modes a given fiber can propagate, and, if measured as a function of wavelength, on the profile dispersion (Burrus et al., 1973; Gloge 1971; Gloge et al., 1975). The distribution of refractive-index also determines the multimode delay distortion of the waveguide and possible mode-coupling and radiation loss factors if it has short-range variations along the axis of the fiber (Marcuse, 1976).

The problem of tailoring the refractive-index profile of the fiber to achieve a minimum of pulse dispersion has received considerable attention. The reason is that theoretical considerations show that an optimum profile exists for which the transit time of all modes is very nearly equa-

lized, and a considerable increase in bandwidth results (Gloge and Marcatili, 1973; Olshansky and Keck, 1976). To achieve this, however, the
index variation must be controlled with great precision, and techniques to
measure it with high accuracy are required.

There are several methods to measure the refractive index distribution,
most of which have only recently been proposed. They can basically be
considered to be of an interferometric or noninterferometric type. Among
the former are measurements by single- or double-pass transmission
interference microscopy, and included in the latter are the techniques of
near- or far-field power distribution measurements, end-face power reflection measurements, immersion methods, and microprobe analysis.
We will concentrate on those methods which have been more fully developed and have provided reliable and consistent profiling information. This information, as noted, not only allows a prediction of the expected transmission behavior of the waveguide, but also serves as a valuable direct guide to the fiber fabricator in making changes or corrections in
his process to achieve the most desirable index profile.

11.3.2 Interferometric Techniques for Index Measurement

If a very thin transverse slice is taken from an optical fiber or preform
rod, and the end faces are ground and polished to be flat and parallel, the
sample acts as a phase object, and techniques of phase-sensitive detection
can be utilized to study its properties. One of the most sensitive of these
techniques, also amenable to quantitative analysis, is interference microscopy (Krug et al., 1964). In this method, the refractive-index distribution
of the fiber is determined either from the interference of light transmitted
once through the fiber with light transversing an independent reference
path in a Mach–Zehnder type arrangement, or by the interference of light
reflected from the bottom surface of the sample and a reference beam in a
Michelson-type arrangement.

The thickness required of the sample to ensure its phaselike behavior
depends on the maximum index difference Δn between the cladding and
the core and on the radius a of the core (Stone and Burrus, 1975). If the
sample is too thick, rays traversing it are bent and focused, thus producing curved wavefronts which can lead to erroneous results.

Care must also be taken that the phase shift observed in interferometric
analysis is due only to the refractive-index differences in the sample and
not to variations in thickness of fiber caused by the polishing procedure
(Stone and Derosier, 1976). Composition-dependent thickness variations
can occur in soft-lap polishing, causing errors as large as 50% in index
measurements. This error may be avoided by additional polishing of the
sample for several minutes on a hard lap such as tin. In practice, the

sample itself is obtained by potting one or several ~5-cm lengths of fiber in a glass capillary tube several millimeters in diameter with an epoxy resin, curing the epoxy, and cutting slabs about 1-mm thick from the composite rod followed by the polishing procedure.

11.3.2.1 Mach–Zehnder Interferometry. The instrument most generally used for transmitted-light interferometry on fiber samples is the LEITZ interference microscope. Its application to measure refractive-index distributions has been described (Kaminow and Carruthers, 1973; Burrus and Standley, 1974), and extended to the profiling of graded-index optical fibers (Stone and Burrus, 1975; Presby *et al.*, 1976). The method has also been used by Cherin *et al.* (1974) and Burrus *et al.* (1973) for similar measurements. It is essentially a combination of a microscope and an interferometer in such a way that the magnified image of the object appears to be superimposed by interference fringes.

Interference is achieved by the Mach–Zehnder arrangement shown in Figure 11.12. The light emitted by a white or monochromatic light source is split into two coherent beams (object and reference beam) by a beam splitter, T_1, and reunited after a certain path length by means of a second

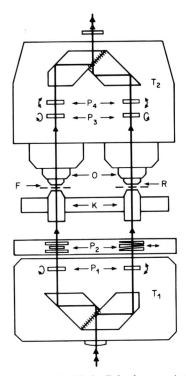

Fig. 11.12 Optical arrangement in Mach–Zehnder type interference microscope.

beam splitter, T_2. Between the two beam splitters, the object and reference beams pass through two identical microscopes consisting of condensor lenses K and objective lenses O and various plate compensators P_1-P_4. The plate compensators adjust for path differences of the object and reference beams which are initially made similar by the insertion of a flat sample of pure fused silica (R) with thickness nearly equal to the polished fiber sample (F), into the reference beam. They also serve to orient the interference fringes and to vary their width.

In an interference microscope, maximum lateral resolution and exact phase measurements cannot be combined in a single instrument (Ingelstam and Johansson, 1958), and a condensor aperture of 0.5 is chosen as a compromise. Larger apertures reduce the contrast of the interferences and smaller apertures reduce the lateral resolving power. Spatial index resolution of about 1 μm, and index values accurate to about two parts in 10^4 can be realized (Presby et al., 1976). By electronically processing the image of the microscope, index measurements with an accuracy of one part in 10^5 have recently been reported (Presby and Kaminow, 1976; Presby and Astle, 1978).

In white light, the bright empty field of the interference microscope contains a number of parallel, colored, evenly spaced interference fringes; in monochromatic light a system of parallel, equidistant dark fringes appears on the bright background whose color corresponds to that of the light.

An object in the measuring beam of the interference microscope displaces the phase of the wave passing through it according to its thickness and refractive index. This phase displacement is visible in the interference fringe field as a displacement of the interference fringes and in the homogeneous field (infinite fringe width) as a change in the color or in the brightness at the site of the image.

Upon observation in the interference fringe field, a graded-index fiber will appear as in Fig. 11.13a. The fiber is shown at a magnification of $500\times$, although the interference microscope can be used at magnifications of from $50\times$, for preform samples, to $1000\times$, for smaller fibers or for greater spatial resolution.

The fiber is shown in Fig. 11.13b for the homogeneous field case in which points of equal optical path differences are connected by fringes. This is a particularly sensitive way to detect asymmetries in the geometry as demonstrated in Fig. 11.13c for a slightly asymmetric sample.

The index profile is determined across a diameter of the fiber with care taken to locate the center of the core (Presby et al., 1976; Wonsiewicz et al., 1976). This is readily achieved on most chemical-vapor-deposited fibers due to a slight index depression in the center which appears as a dark or bright spot in the interference microscope and uniquely locates this point (Presby et al., 1975). If the fiber is not circularly symmetric, the re-

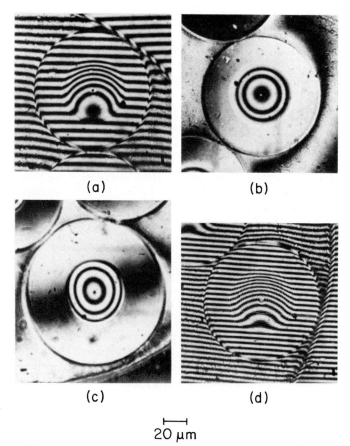

(a)   (b)

(c)   (d)

20 μm

Fig. 11.13   Microinterferograms of graded index optical fibers, (a) observed with microscope set for interference fringe field, (b) observed with microscope set for homogeneous
field case, (c) slight geometrical asymmetry, (d) photograph prepared for data extraction.
Magnification 500 × .

fractive-index profile can be different depending on the choice of diameter, and an analysis along several diameters is required for more complete
information.

To extract profile information from the microinterferogram, circles on
the scale of magnification of the photograph with radii increasing in 1- or
2-μm increments (depending upon required resolution) are drawn with a
divider by lightly scratching the surface of the photograph. A straight line
through the midpoints of the fringe in the cladding which is displaced to
the center of the core is also scribed. One photograph prepared in this way
is shown in Fig. 11.13d.

The microinterferogram is then mounted on the movable stage of a low-power binocular microscope, and the displacement from the cladding level of the fringe which passes through the center of the core is measured at each of the scribed radii with the aid of a micrometer scale in the eyepiece. The ratio of the results of these measurements with the distance between the uniformly spaced parallel fringes observed in the cladding gives the band or fringe fraction q.

The difference in refractive index Δn between the cladding and a point measured in the core is then given by

$$\Delta n = q\lambda/t , \qquad (11.9)$$

where λ is the wavelength at which the observation is made (usually $\lambda = 0.546$ μm) and t is the thickness of the sample. The thickness is most accurately measured with an electronic transducer gauge, in which the displacement of the tip of the gauge by the sample above a reference level is translated into an electrical signal which is amplified, converted, and digitally displayed.

The refractive-index profile can now be plotted as Δn-versus-radius, or further computer-analyzed to provide information on how closely the profile of the fiber approaches the ideal distribution (Presby et al., 1976).

The previous data reduction procedure can be automated resulting in the fast and accurate interpretation of the refractive-index profile (Wonsiewicz et al., 1976). To achieve this, a transparency of the microinterferogram is made and encoded with a transmission-type scanning microdensitometer. A computer program then locates the precise center of the fringes to be measured, with care taken to ignore extraneous information such as spots or blemishes on the image. The index profile is calculated from the fringe data and by using a least-square routine, a best-fit g curve is predicted. A typical refractive-index profile obtained in this manner is shown in Fig. 11.14. The solid line is the best fit to the data points indicated by the circles. The resultant g value is 1.97.

Techniques have also been developed to automatically process the output field of the interference microscope directly with a video-based, computer-controlled system (Presby et al., 1978a). The experimental reproducibility of the measured profile is limited mainly by noise on the video signal and is within ± 5 parts in 10^5. With this system it is possible to obtain detailed profiles at different wavelengths and different sample orientations in less than 3 minutes.

An alternate means of obtaining profile information directly from the microscope involves inserting a known variable path change in one arm of the interferometer, and following a single circular fringe as it expands radially outward through the sample (Stone and Burrus, 1975). The Leitz instrument has a built-in provision for this mode of operation, which has the disadvantage, however, that the fringe position and path change data

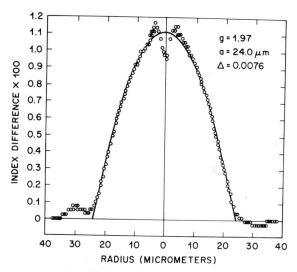

Fig. 11.14 Refractive-index profile obtained by automatic analysis of interferogram. The solid line is the least-squares fit to the data shown by the circles.

must be read step-by-step from the instrument, which is less convenient than measurement of a permanent photographic record.

A variation of the above method which reduces measurement time and increases accuracy is offered by a technique called phase-locked interference microscopy (Sommargren and Thompson, 1973). In this approach one of the mirrors in a dual-beam interferometer is vibrated by approximately $\lambda/4$ at the illumination wavelength. A point detector is placed in the image plane, and at positions other than the fringe extremes there will be a component of the detected signal at the driving frequency proportional to the distance of the detector from the fringe peak. Processing this signal yields a voltage proportional to changes in the optical path which can thus be measured directly.   •

Mach–Zehnder-type interference microscopy has also been used to obtain valuable profile information on the preforms from which modified-chemical-vapor-deposited optical fibers are pulled (Presby *et al.*, 1975). The objective of this experiment was the correlation of the deposited material structure in the preform to that in the resulting optical fiber. A view of one-half the preform sample as observed by interference microscopy is shown in Fig. 11.15a. The fused-silica cadding, a borosilicate layer, having a lower index of refraction than pure fused silica, and the first three germania deposited steps are labeled. Comparison of this and the composite scanning-electron-beam microscope photograph of a segment of one-half the corresponding fiber (Fig. 11.15b) indicates the preservation of features present in the preform through the drawing process.

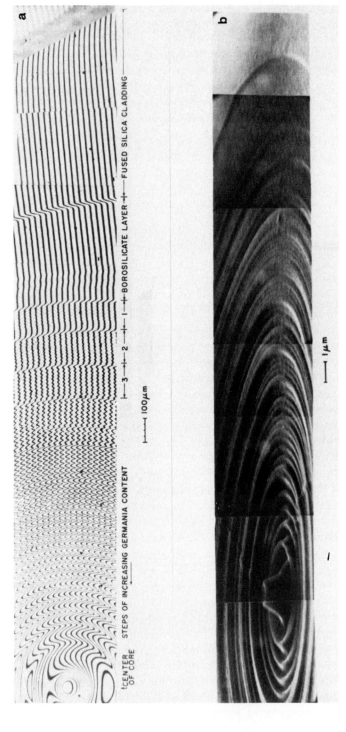

Fig. 11.15 (a) Composite microinterferogram of graded-index perform sample, (b) composite scanning electron-beam-microscope photograph of corresponding fiber.

This lends confidence to the fact that the same distribution of refractive index that is introduced into the preform by changing the material composition of the deposited layers exists in the fiber pulled from this preform.

The use of interference microscopy as described in this section requires rather elaborate sample preparation because a thin slice has to be cut out of the fiber and polished to a high degree of flatness and parallelism. The net time involved in this procedure can be reduced to a great extent by simultaneously processing many samples and utilizing an automatic analysis scheme (Presby *et al.*, 1978a) in the profile evaluation. However, it would still be highly desirable to avoid any sample preparation. Techniques based on interferometry requiring no sample preparation have been proposed in which a sample of the whole fiber is immersed in index-matching oil and illuminated perpendicular to its axis. These techniques are treated in Section 11.3.3.6 of this chapter.

11.3.2.2 Michelson Interferometry.   A micro-Michelson optical configuration normally utilized in a Zeiss interference microscope for surface profile measurements by reflection, can be adapted to obtain refractive-index profiles of optical fibers by aluminizing one side of the sample to act as the sample mirror in the interferometer (Martin, 1974). The same objective is focused through the sample on the aluminized back face. Light from the sample objective then passes through the slice, is reflected by the aluminum surface, and returns to the objective. The reference leg of the microscope contains a matched objective and a plane mirror located at the objective focal plane.

Photographic negatives of the sample and the reference fringes are processed with a microphotometer to give the location of a given interference fringe versus position, from which the refractive-index profile can be deduced.

A sensitive interferometric technique which yields information on the departure of the ideal radial distribution of the dielectric constant of graded-index rods has been reported (Rawson and Murray, 1973). A collimated laser beam is focused on one surface of the sample which is cut to such a length that an object on one surface is reimaged at the second surface (one-half period). The partially reflected light from the first surface constitutes the reference beam and that light partially reflected from the second surface forms the sample beam. Both beams return through a beam splitter and interfere on photographic film. Any fringes which appear on the film are related to departures of the index profile from the assumed ideal value, and provide information on the magnitude of this departure.

Other applications of Michelson-type interferometers for refractive-index profiling can be found in the study of glass rods by ion exchange techniques (Pearson *et al.*, 1969) and in the study of waveguides for integrated optical devices (Izawa and Nakagowe, 1972).

11.3.3 Noninterferometric Techniques

11.3.3.1 Near- and Far-Field Radiation Pattern Measurements. The near-field scanning technique provides a simple and rapid method for obtaining detailed refractive index profiles (Burrus *et al.*, 1973). It is based on the fact that in a fiber with all modes equally excited, a close resemblance exists between the near-field intensity distribution and the refractive-index profile (Gloge and Marcatili, 1973). In refinements of this theory, account has been taken of the presence of tunneling leaky modes. These modes contribute additional power to the observed near-field intensity distribution, resulting in a possible error in the inferred refractive-index profile if not accounted for (Sladen *et al.*, 1976).

An experimental arrangement is shown in Fig. 11.16. A Lambertian source such as a tungsten-filament lamp or an LED is used to equally excite all fiber modes. The source is focused into the end of the fiber and a magnified image of the flat fiber output face is displayed in the plane of an apertured photodiode which is arranged to scan the field transversely. Amplification is by a phase-sensitive detection system, and the intensity profile is plotted directly by an $x-y$ recorder. The fiber is typically less than 1 m long. The use of the near-field technique to determine profiles has also been reported by (Olshansky and Keck, 1976), and a modified version, in which a small area of the fiber core is illuminated and the total transmitted power measured by (Arnaud and Derosier, 1976).

A further modification of the near-field technique has been devised in which no leaky mode correction is necessary (Stewart, 1977). This is achieved by using power not trapped by the fiber core. This embodiment also has improved resolution and enables the absolute indices to be determined since both core and cladding are profiled.

The far-field radiation pattern of an optical fiber is useful in determining the fiber's numerical aperture (Cherin *et al.*, 1974). The experimental arrangement is similar to that of Fig. 11.16 except for the omission of the

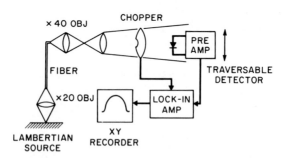

Fig. 11.16 Experimental arrangement for making near-field radiation pattern measurements.

image-forming objectives, and the use of long fiber lengths. The numerical aperture is simply obtained from a knowledge of the distance of the flat fiber end to the detector and the extent of the radiation pattern. Results are found to be in good agreement with values calculated from interferometrically measured core–cladding refractive-index differences (L. G. Cohen *et al.*, 1975).

11.3.3.2 Reflection Measurements. Refractive-index profiles of optical fibers have been obtained based on the fact that the reflected power from the surface of the fiber is directly related to its refractive index (Ikeda *et al.*, 1975; Eickhoff and Weidel, 1975). This technique measures the reflectivity as a function of position across an end-face of the fiber. A block diagram of a reflection apparatus is shown in Fig. 11.17. The circularly polarized spatially filtered laser beam is focused to a spot on the end-face of the fiber. The fiber end is scanned past the focal spot and the reflected power is recorded. The index profile $\Delta n(r)$ is (for small Δn) given by

$$\frac{\Delta n(r)}{n(0)} = \frac{\Delta R(r)}{R(0)} \frac{(R(0))^{1/2}}{1 - R(0)}, \tag{11.10}$$

where $R(0)$ is the reflectivity at $r = 0$, $\Delta R(r)$ is the change in reflectivity at radial position r, and $n(0)$ is the index at the fiber core axis $r = 0$. In practice, $n(0)$ is not known and $R(0)$ is difficult to measure with great absolute precision. To avoid these problems, a sample of known index is substituted for the fiber and a calibration factor, C, is determined so that

$$\Delta n(r) = C\Delta R(r). \tag{11.11}$$

Thus, $\Delta n(r)$ can be plotted directly. The spatial resolution tends to be slightly higher than that of the interference microscope because of the

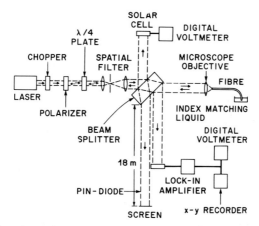

Fig. 11.17 Experimental arrangement for making reflected power measurements.

large numerical aperture (~ 0.8) objectives used to focus the beam. The index resolution is determined by a number of factors. The large range of incident angles means that Eq. (11.10), which is based on the expression for the reflectivity at normal incidence, is not quite correct. The finite focal spot size causes the profile to be slightly averaged. The end surface of the fiber may not be exactly normal to the beam axis and may not be in the lens focal plane. The incident beam has amplitude fluctuations that results in noise in the detector signal. Finally, the electronics contributes to the final noise observed at the recorder. These effects result in an error of about 2% in the measured reflectivity which is reduced by about a factor of five in Δn. These figures would indicate an index accuracy of $\sim 15 \times 10^{-4}$, or about an order of magnitude worse than the interference microscope technique. The errors associated with the calibration factor are of course in addition to those already discussed.

There are other factors that may lead to a further degradation in the accuracy of the technique. The reflection technique is very sensitive to the condition of the end-face of the fiber (Stone and Earl, 1976). If the end surface is polished, then the surface layer density (and thus the surface index) is altered by compaction. Also, for glass compositions containing reactive species, the surface index may be substantially altered by reaction with either the polishing compounds or ambient water vapor. Good results were obtained with fractured samples of germania or phosphorous-doped silica, and very poor results, due to rapid atmospheric attack, were evidenced with borosilicate fibers in a normal room environment.

11.3.3.3 Immersion Measurement. Oil immersion techniques, one of the most widely employed methods to determine refractive indices of solids in minerology and crystallography, have been utilized to measure the refractive indices of unclad (Bateson, 1958) and cladded fibers (Liu, 1974).

In one method, the fiber is mounted in a special hot stage refractometer controlled with a variac, and the temperature of the index liquid adjacent to the fiber is read by means of a thermocouple. Knowing the temperature coefficient of the liquid, the refractive index is determined by reading the endpoint temperature at which the fiber disappears from the field.

In the case of cladded fibers, the index of the matching liquid is gradually increased until the boundary between the core and the cladding, as observed in a microscope, is not visible. Then, knowing the refractive index of the cladding glass and the radii of the core and the cladding, the refractive index of the core can be calculated. An accuracy greater than 10^{-4} is reported.

A combination of index-matching and multiple-beam interference methods employing fringes of equal chromatic order has also been utilized in making refractive-index measurements on optical fibers (Kapany, 1957). In this method the fiber is immersed in a liquid of known index and its index is matched with that of the specimen by the Becke line method. The immersion liquid is chosen so that its dispersion curve intersects that of the glass in the middle region of the visible spectrum. The specimen and immersion liquid are then placed between the two parallel plates of a Fabry–Perot interferometer illuminated with white light and projected on the slit of a spectrograph with a microscope. By adjusting the temperature of the liquid the observed fringes can be made to suffer no displacement at a given wavelength indicating that at this wavelength the index of the fiber and the liquid are matched. A dispersion curve of the sample can be determined in this way from the known dispersion curve for the liquid at different temperatures. The index is reported to be measured with an accuracy of ±0.0001.

11.3.3.4 Scanning Electron Microscopy and X-Ray Microprobe Technique.   Because of its high spatial resolution and great depth of field, the scanning electron microscope (SEM) is a valuable diagnostic tool for studying optical fibers. When equipped with an X-ray analyzer, the SEM can provide quantitative data on composition variations (Wells, 1974).

For qualitative studies of graded-index fibers or preforms, the index profile is converted into a contour profile and the SEM is used to produce a microphotograph of the contour (Burrus and Standley, 1974). The contour is produced by taking advantage of the fact that a fiber end surface will etch at a rate that is dependent on the local composition of the glass. The etched surface is overcoated with a conductor to prevent charge accumulation. Figure 11.15b shows a microphotograph of the etched and overcoated end surface of a $GeO_2:SiO_2$ core fiber made by modified chemical vapor deposition (Presby et al., 1975). In this fiber, the GeO_2 concentration has been increased in several steps and these steps are apparent in the photograph. The composition variations produced by the individual depositions at fixed GeO_2 concentration are also visible. The rise at the core center is a region depleted of Ge during the preform collapse. This photograph illustrates the power of the SEM as a diagnostic tool in the development of fiber fabrication techniques and process control.

This technique is limited by the fact that the etch rate may depend on factors other than composition (i.e., strain) or that more than one dopant determines the etch rate. Also, since the etch rate may not be a linear function of concentration, the extraction of accurate quantitative data requires further study.

Quantitative studies of the composition profile of fibers can be made by using an X-ray microprobe analyzer (McKinley et al., 1966). The incident beam in the SEM is used to generate characteristic X rays in the sample.

These X rays are energy analyzed and the unique relationship between the characteristic X-ray energy (or wavelength) and the atomic number Z is used to identify the element or elements present in the beam target area of the sample.

The composition at a fixed position can be determined by scanning the X-ray energy analyzer or plots of the variation of a single element can be made by fixing the X-ray energy accepted and scanning the incident electron beam. The limitations of the technique are (1) the reduced spatial resolution (~ 1 μm) due to the increased electron beam diameter at the higher required beam currents, and the X-ray source size due to electron scattering in the sample, (2) the difficulty of accurately differentiating between characteristic and fluorescent X rays, and (3) the difficulty in converting an X-ray count rate at a particular energy into concentration of an element having atomic number Z. This last problem can be overcome if suitable standard samples of known concentration are available. Also, it is not possible to detect elements lighter than Be, due to the Z dependence of the X-ray cross section. This means that boron cannot be measured except by differentiating from data on all other elements present.

The technique has been used to obtain the refractive-index profile of glass fibers and rods (Kita *et al.*, 1971) and comparative results of profiles obtained by transmission interference microscopy and electron microprobe analysis are in reasonably good agreement (Burrus *et al.*, 1973).

11.3.3.5 Scattering Methods. If a beam of light is incident upon an optical fiber at right angles to the fiber axis, an analysis of the resulting scattered radiation can be utilized in limited cases to determine the refractive-index profile of the fiber.

The back-scattered fringe pattern is uniquely useful for this application (Presby, 1974a) in that refractive-index and diameter information can be separated for certain cases. It is based on the fact that the back-scattered pattern is localized in a range of angular deviation on the order of $\pm 25°$ from the incident direction, and the sharp cutoff of the pattern, for unclad fibers, depends only on the index of refraction and not the diameter of the fiber. The relationship of the cutoff angle ϕ and the index n is given by

$$\phi = 4 \sin^{-1} \left[\frac{2}{n(3)^{1/2}} \left(1 - \frac{n^2}{4} \right) \right]^{1/2} - 2 \sin^{-1} \left[\frac{2}{(3)^{1/2}} \left(1 - \frac{n^2}{4} \right) \right]^{1/2}. \quad (11.12)$$

The method has been used to determine the variation of index with wavelength over a limited range of laser wavelengths (Presby, 1974b) and over a broader range using incoherent light in a preform comparative study (Presby, 1974c).

The technique is also applicable to both step-index fibers (Presby and Marcuse, 1974; Chu, 1976) and fibers with arbitrary refractive-index distributions (Marcuse and Presby, (1975a).

The arrangement to observe the back-scattered light is shown in Fig. 11.18 along with a photograph and an x–y chart plot of a typical pattern from a step-index fiber. Information on the refractive index and the diameter of the core and the cladding is obtained from the location of the major peaks. In addition, the location of the maxima of intensity is a sensitive function of any ellipticity which may exist in the fiber and must be taken into account (Marcuse, 1974a; Presby, 1976).

A method based on the analysis of the forward-scattered radiation (Watkins, 1974) has also been implemented to compute the refractive-index distribution of an optical fiber (Okoshi and Hotate, 1976a,b).

11.3.3.6 Abel Transform Techniques.   In the scattering techniques, information on the phase of the scattered wavefront is not directly available for measurement. It is possible to measure the phase of the transmitted wavefront and determine the core-index profile from the phase data under certain circumstances. The restriction that must be satisfied is that the fiber be a weak phase object (phase shift $<2\pi$). This can be accomplished by removing the large phase shift associated with the air–cladding interface through the use of index-matching fluids. The fiber-plus-index matching fluid now appears as a weak phase object if the core index is not much greater than the cladding index (the usual case for optical fibers), and the core size is such that the total phase change is less than 2π. The geometry is quite similar to that used in the scattering technique. The wavefront phase distortion can be measured by interfering the transmitted wavefront with a plane reference wavefront in a Mach–Zehnder interferometer. By tilting the reference wavefront, a system of straight parallel fringes is produced except in regions where the sample wave-

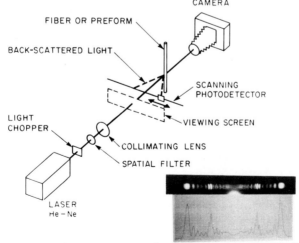

Fig. 11.18 Experimental arrangement to observe back-scattered light distribution with photograph of pattern from step-index fiber, and x–y chart plot of same pattern.

front has been distorted by the core-index profile. In these regions, the
fringes are displaced and the wavefront phase can be determined from the
fringe displacement. The phase measurement proceeds in a nearly identical manner to that used in the interference microscope technique. From
the phase data the index profile is determined via Abel's integral equation. The wavefront phase distortion and the index profile form an Abel
transform pair (Vest, 1975). Additional experiments to prove the feasibility
of this technique have been performed on liquid-filled capillaries and a
Vycor clad silica core fiber (Marhic et al., 1975). The accuracy with which
an index profile can be determined will be limited by the ability to measure the fringe displacements and thus the phase. In the above experiment
index variations of the order of 10^{-3} were determined.

These measurements have been extended to graded-index profile fibers.
In one application (Saunders and Gardner, 1977), index distributions have
been deduced for fibers with power-law profiles which can be characterized by two constants: the refractive-index difference (between the maximum value at the core center and the cladding value) and the power law
coefficient. In another (Kokubun and Iga, 1977), the accuracy of this
method has been increased by correcting for refraction due to the index
gradient of the fiber. In this last reference, a transverse shearing interferometer is utilized, and a simulation shows that the measurement error is in
principle within 0.3% of the index difference between the core center and
the cladding. In a further embodiment, this method has been successfully
applied to fibers with an arbitrary index profile, and the measurement
procedure has been fully automated so that profiles, accurate to a few parts
in 10^4, can be produced within 3 min of fiber fabrication (Presby et al.,
1978b).

This "whole-fiber" method has thus shown itself to be a viable technique to measure index profiles, although with somewhat less inherent
attainable precision than the slab method. Its main advantage of not requiring sample preparation and thus providing information within
minutes should make it a very valuable diagnostic tool. Complicating
factors are the index-matching fluid and asymmetric fiber geometries.

11.3.4 Profile Dispersion Measurements

The techniques previously presented generally determine the
refractive-index profile of the optical fiber at a single wavelength. The
variation of the profile with wavelength, the profile dispersion, however,
is an important factor in determining the radial-index distribution that
maximizes information-carrying capacity (Olshansky and Keck, 1976).
Calculations, based on refractive-index data for a titania-doped silica, indicate that the index gradient parameter g, which characterizes the op-

timal profile shape, should be altered by 10–20% to compensate for dispersion in the core and cladding glasses.

The index difference Δn between the core and cladding as a function of wavelength must be measured with extreme accuracy to calculate profile dispersion effects which depend on the first derivatives of Δn.

Measurements of the variation of index of refraction with wavelength have been made on bulk glass samples in the form of prisms to determine the profile dispersion (Olshansky and Keck, 1976; Fleming, 1975). However, in addition to the dopant material, the index profile may also depend on the process by which the glassy matrix is prepared, and on the temperature history of the fiber which affects the stress distribution in its cross section. It is therefore desirable to measure dispersion as a function of composition in the fiber itself, rather than in the bulk samples which simulate the composition at certain points in the fiber.

An extremely sensitive technique for performing the measurement in fibers has been developed (Gloge et al., 1975), and applied to measure profile dispersion in the binary silica optical fibers: GeO_2–SiO_2, B_2O_3–SiO_2, TiO_2–SiO_2, P_2O_5–SiO_2, Al_2O_3–SiO_2, and Cs_2O–SiO_2 (Presby and Kaminow, 1976). The experimental apparatus is illustrated in Fig. 11.19a. It consists of a Leitz transmission interference microscope, a light source that covers the range 0.5 to 1.1 μm by means of a set of 200 Å-wide filters, a silicon camera tube, and an oscilloscope. The microscope field is presented on a monitor and the amplitude of two individual scan lines are displayed on the oscilloscope. These displays, with labeled fringes, are shown in Figs. 11.19b and 11.19c for wavelengths of 0.5 and 0.9 μm, respectively. The displacements of respective fringes between core and cladding are measured from photographs of the oscilloscope display with a precision of about $\frac{1}{100}$ of a fringe. This measurement procedure has been further refined by the addition of a video line selector and digital timer with signal averaging features to make the fringe displacement measurements directly from the oscilloscope (Presby and Astle, 1978). An accuracy of $\frac{1}{1000}$ of a fringe, not including systematic errors, is reported.

Predictions of the optimum profile parameter g and of material dispersion effects are in good agreement with measurements made by pulse propagation techniques (Cohen, 1976; Luther-Davies et al., 1975). The results of these measurements have also been utilized to design multicomponent glass optical fibers having several optimized transmission characteristics (Kaminow and Presby, 1977). These fibers are fabricated by independently specifying the concentration profiles of each of the components.

Profile-dispersion measurements have also been made by measuring the wavelength dependence of a fiber numerical aperture directly (Sladen et al., 1977). This yields the difference in dispersion between the core and cladding and hence the profile dispersion. The total output power from a

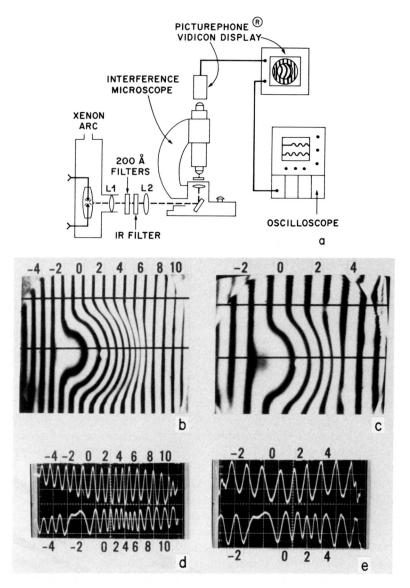

Fig. 11.19 (a) Experimental apparatus to perform profile dispersion measurements; (b), (c) graded-index optical fiber and electronic fringe display at $\lambda = 0.5$ μm; (d), (e) same for $\lambda = 0.9$ μm.

short length of straight fiber is measured at each wavelength for both apertured and nonapertured excitation, the use of the ratio of the two transmitted powers, which is proportional to the square of the numerical
aperture, provides compensation of the Lambertian light source, and
eliminates the effect of fiber attenuation. The measured $\Delta n(\lambda)$ is then
fitted by the least-squares technique to an expanded Sellmeier representation of $\Delta n(\lambda)$ with six physically significant coefficients.

11.4 DISPERSION AND BANDWIDTH

11.4.1 Causes of Pulse Dispersion and Bandwidth Limitation

As a consequence of dispersion, pulses of light traveling in a fiber become wider as the transmission distance increases. The limit of
information-carrying capacity, or bit rate, of a fiber waveguide depends
on how close together light pulses may be spaced before successive overlapping pulses smear out the information. Pulse-broadening in single-
mode fibers is caused by intramodal material dispersion due to the functional variation of refractive index with wavelength, and by waveguide
dispersion because the group velocity of a single mode is slightly dependent on frequency. For silica fibers, this dispersion approaches zero in the
vicinity of $\lambda = 1.3$ μm wavelength. It is approximately 0.07 nsec/km for
every nanometer of source spectral width near $\lambda = 0.9$ μm.

Additional pulse distortion occurs in multimode fibers because of intermodal dispersion due to group delay variations between propagating
modes at a single frequency. The magnitude of the time delay, τ, between
the ray with the longest and shortest path is proportional to the difference
in refractive index, Δn, between the core and cladding. For example, for
a fiber with $\Delta n = 0.015$, $\tau/L \approx \Delta n/c = 50$ nsec/km, which implies that
information rates of 100 Mbit/sec could only be transmitted about 200 m.
Fortunately, intermodal dispersion effects can be reduced by fabricating
fibers with graded-power-law refractive-index profiles approximating the
form $n = n_1[1-2\Delta (r/a)^g]^{1/2}$ (Gloge and Marcatili, 1973). The optimal profile, at a particular wavelength, is one for which the speed variation
between rays traveling different zigzag paths most nearly compensates the
corresponding path length variation. Dispersive refractive-index differences between material constituents, such as B_2O_3, GeO_2, and SiO_2
cause modal group velocities to depend not only on the index profile but
also on the wavelength (profile dispersion) (Olshansky and Keck, 1976).
Consequently, the optimal profile is wavelength-dependent and its g
exponent differs from two. Theory predicts reductions in intermodal
dispersion by approximately three orders of magnitude from what it
would be in a step- or discontinuous-index multimode fiber and reduction

factors of two orders of magnitude have already been achieved (Cohen *et al.*, 1978b; Keck and Bouillie, 1978).

Other mechanisms besides refractive-index profiling may influence intermodal delay distortion. Loss differences between modes modify the shape as well as the width of transmitted pulses and the effect changes with fiber length. Axially varying imperfections in the refractive index profile or the geometry of an optical waveguide and microbends in an optical fiber cable would induce power mixing among the guided modes which can reduce signal distortion by changing the length dependence (Personick, 1971; Cohen and Personick, 1975). Unfortunately, decreased signal distortion is usually accompanied by increased radiation loss.

Dispersion effects on optical fiber transmission are characterized by making measurements of impulse response in the time domain or power transfer function in the frequency domain. The methods assume that the fiber response behaves quasilinearly in power (Personick, 1975). The implication is

$$P_{out}(f) = H(f)P_{in}(f) \tag{11.13}$$

or

$$p_{out}(t) = h(t) * p_{in}(t), \tag{11.14}$$

where $H(f)$ is the power transfer function at the baseband frequency f, $h(t)$ is the power impulse response, and the asterisk (*) denotes convolution.

11.4.2 Time Domain Measurements

(a) *Impulse response measuring techniques.* Time domain measurements of pulse dispersion are normally made by injecting a sequence of narrow impulses of light into a fiber and measuring the temporal pulse spread at the output with a square-law (power-sensitive) detector (Gloge *et al.*, 1972a). A power impulse response function, $h(t)$, is used to determine the fiber output response, p_{out}, to the envelope of a general power modulated carrier.

$$p_{out} = \int_{-T/2}^{T/2} p_{in}(t - \tau)h(\tau) \, d\tau, \tag{11.15}$$

where T is the period between input pulses and is wider than the broadened output pulsewidths.

Impulse response measurements through a fixed fiber length cannot generally be linearly extrapolated to arbitrary lengths because mode-mixing and differential attenuation tend to influence the rate of pulse spreading. However, pulse transmission properties can be measured for lengths longer than the physical length with the shuttle pulse arrangement illustrated in Figure 11.20a (L. G. Cohen, 1975). Light pulses from a GaAs injection laser ($\lambda = 0.908$ μm) are coupled into the fiber and allowed to shuttle back and forth between partially transparent mirrors pressed in contact with the input and output ends of the fiber. Dispersion is measured by comparing the widths of pulses, transmitted through the output end mirror, which return from successive round trips through the fiber. In order to record those pulses on a sampling oscilloscope, the scope trigger is delayed to coincide with the arrival of each shuttle pulse at the output.

The fiber and mirrors are mounted in cylindrical holders (Fig. 11.20b) which have grooves precisely milled perpendicular to machined faces. The dielectric coated mirrors have reflectivity spectra with flat peaks centered approximately at the source wavelength. Measurements in visible light ($\lambda = 0.63$ μm) confirmed that scattering caused by reflecting mirrors is small compared to the scattering within the fiber (Stone *et al.*, 1974).

Insertion and reflection coefficients of the mirrors reduce the dynamic range of a measurement. The first pulse is attenuated by (T^2), the transmission coefficient of the two mirrors. Each successive pulse is attenuated by the square of the mirror reflectivity (R^2) since there are two reflections per round trip. A 56-dB dynamic range is attained by using a high-peak

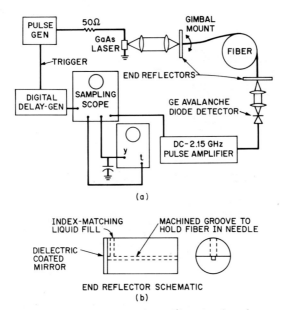

(a)

END REFLECTOR SCHEMATIC
(b)

Fig. 11.20 (a) Experimental arrangement for making shuttle pulse measurements with a pulsed GaAs injection laser ($\lambda = 0.908$ μm). (b) Schematic of the holder that is used to press a fiber against a reflecting mirror. The holder mates with a gimbal mount that can be tilted at an angle relative to the injected laser beam in order to emphasize the launching of higher order modes.

power laser (~ 1 W) and a sensitive detector consisting of a germanium avalanche diode followed by a wideband (0–2 GHz) pulse amplifier. The detectable number of shuttle pulses can be predicted from the following equation which governs the system's dynamic range, D, in decibels:

$$D + 20 \log T - \alpha L_0 = (-20 \log R + 2\alpha L_0)(N - 1) \qquad (11.16)$$

The extrapolated fiber length L corresponding to N pulses is

$$L = (2N - 1)(L_0). \qquad (11.17)$$

For example, consider an $L_0 = 1$ km fiber with $\alpha = 1$ dB/km loss measured on a shuttle pulse test set with $R = 0.9$ reflecting mirrors at each end. Equations (11.16) and (11.17) predict that 10 pulses can be detected and pulse spreading can be extrapolated to 10 km. Figure 11.21 illustrates results of shuttle pulse measurements in a 106-m length of CGW-Bell-10 fiber, which was the first graded-index fiber (NA $\cong 0.09$) which had a reasonably low 13 dB/km loss ($\lambda = 0.9$ μm). The GaAs laser is driven by a mercury reed relay, switching a variable length of delay line, which produces pulses of controllable duration. Pulsed lasers using fast solid-state avalanche switches to discharge transmission lines, or ceramic chip capacitors, have also been described in the literature (Andrews, 1974; Dannwolf et al., 1976). The photograph in Fig. 11.21a shows the first five shuttle pulse returns when mirrors with $R \approx 0.9$ were pressed against each fiber end. Pulse

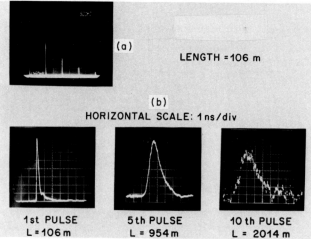

Fig. 11.21 (a) Shuttle pulse train in a 106-m length of CGW-Bell-10 fiber with nominally 90% reflecting mirrors against each end. Pulse spreading effects were made negligible by injecting 30-nsec-wide laser pulses that were broad compared to the fiber dispersion. (b) Pulse spreading is observed when narrow impulses are injected into the fiber. The photographs are sampling scope displays of the first pulse received after $L = 106$ m, the fifth pulse after $L = 954$ m and the tenth pulse after $L = 2014$ m.

area is proportional to the pulse energy. Therefore, peak pulse amplitudes traveling along a fiber decrease due to transmission loss and broadening by dispersion effects. Transmission loss can be measured by using laser pulses that are wide compared to the spreading caused by the fiber. The trip-to-trip loss between successive shuttle pulses in Fig. 11.21a was 3 dB. Approximately 0.9 dB was due to the mirror reflectivity per round trip (-20 log R), and the remainder was due to the fiber transmission loss. Pulse dispersion is measured by injecting narrow subnanosecond impulses of laser light into the fiber and comparing the output pulse widths that return from successive line traversals. The three photographs in Fig. 11.21b show the evolution of pulse broadening from a subnanosecond width into a 3-nsec width for pulse propagation lengths between one (0.106 km) and 19 times (2.014 km) the fiber length.

(b) *Evaluation of intermodal dispersion.* When mode-mixing effects are small, the structures of fiber impulse responses can be related to refractive-index profile exponents so that transmission measurements can be used to guide the fabrication of new fibers with closer-to-optimal profile gradients. Relative time delays between the on-axis ray, which has the shortest path down the fiber axis, and off-axis rays, which travel longer zigzag paths are measured by blocking the fiber output far field with circular irises or annular rings (Cohen, 1976). Figure 11.22 shows spatially filtered impulse response shapes in two fibers with NA ≈ 0.14 output numerical aperture after 1 km of propagation. Pulses in Figure 11.22a resulted from an overcompensated profile ($g < g_{opt}$) because small angle rays, corresponding to low-order modes, progagating through a small iris with NA < 0.038 arrived after larger angle rays, corresponding to high-order modes, passed through an annular ring with $0.112 <$ NA < 0.156. Figure 11.22b resulted from an undercompensated profile ($g > g_{opt}$) because small angle rays peaked, in time, before larger angle rays. The relative delay time between large and small angle ray peaks is directly related to the deviation of the profile exponent from its optimal value, g_{opt}.

Impulse response shapes are characterized by computing their full rms pulse widths (2σ) defined by

$$\sigma^2 = \frac{1}{A} \int_{-\infty}^{\infty} p_{out}(t) \, (t - \bar{t})^2 \, dt, \tag{11.18}$$

where

$$A = \int_{-\infty}^{\infty} p_{out}(t) \, dt \quad \text{and} \quad \bar{t} = \frac{1}{A} \int_{-\infty}^{\infty} p_{out}(t) t \, dt.$$

Shuttle pulse measurements have been used to determine the optimal profile grading which minimizes intermodal dispersion in graded-index fibers with different material constituents (Cohen *et al.*, 1978b). Mode-

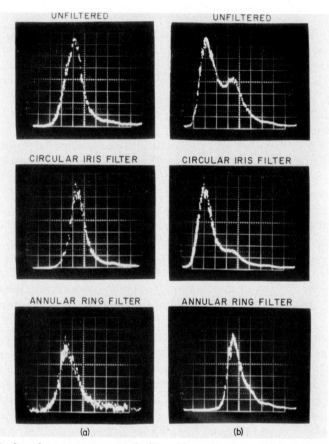

Fig. 11.22 Impulse responses spatially filtered by a circular iris (NA < 0.038) and an annular ring (0.112 < NA < 0.156). (a) Fiber with an overcompensated refractive-index profile. (b) Fiber with an undercompensated profile. Horizontal scale: 0.5 ns/div.

mixing effects were relatively small in these fibers since pulsewidths broadened with an almost-linear length dependence for multikilometer pathlengths. Therefore, output pulse broadening was primarily caused by intermodal dispersion, profile dispersion, and material dispersion effects due to relative time delays between the semiconductor laser source spectral components within its 2.5 to 3 nm bandwidth. Material dispersion effects are reduced by a narrow-band interference filter which has a 1.4-nm bandwidth approximately centered about the laser line peak at $\lambda = 0.908$ μm. The filtered source spectral bandwidth should cause $2\sigma_s \sim 0.09$ nsec/km full rms pulse spreading in silica fibers doped primarily with boron and 0.11 nsec/km pulse spreading in fibers doped primarily with germanium (Fleming, 1976). These material dispersion values are deconvolved [$\sigma_{dec} = (\sigma^2 - \sigma_s^2)^{1/2}$] from the total measured rms pulse spreading in order to evaluate intermodal dispersion data.

Figure 11.23 contains a plot of the bandwidth improvement factor or equivalently, the rms pulse width reduction factor, $\sigma(step)/\sigma(graded)$, relative to step-index fibers with the same numerical aperture, as a function of g for fibers with different profile gradients. The strongly peaked behavior of the curves shows that the amount of pulse broadening is a very sensitive function of the profile gradient. The two sets of curves apply to fibers with different material constituents. Nearly optimal gradients correspond to the peak data points which are characterized by $g \sim 2.03$ in germanium borosilicate (GBS) fibers doped primarily with germanium and by $g \sim 1.78$ for borosilicate (BS) fibers doped primarily with boron. The optimal g values for each type of fiber are consistent with values from refractive-index measurements on bulk materials and thinly polished fiber samples (Fleming, 1976; Presby and Kaminow, 1976). The theoretical minimum pulse broadening in an optimally graded-index fiber is characterized by (Arnaud and Fleming, 1976)

$$2\sigma(min) \approx 141(\Delta n)^2 \tag{11.19}$$

and the theoretical maximum pulse width reduction factor is given by

$$\left(\frac{\sigma(step)}{\sigma(graded)}\right)_{max} = \frac{1}{12^{1/2}} \frac{\Delta nL/c}{\sigma(min)L} = \frac{14}{\Delta n} \tag{11.20}$$

which is inversely proportional to Δn, the maximum core-to-cladding index difference. Therefore, the ratio of two between pulse width reduc-

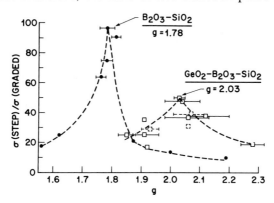

Fig. 11.23   Root-mean-square pulse width reduction factor

$$\frac{\sigma(step)}{\sigma(graded)} = \left[\frac{\Delta nL}{(12)^{1/2}c}\right] \frac{1}{\sigma(graded)}$$

is plotted versus g for germanium borosilicate and borosilicate fibers. Data points □ for GBS fibers; ● for BS fibers.

tion factors for the borosilicate and germanium borosilicate fibers are consistent with the fact that $\Delta n(\text{GBS}) \sim 2\Delta n(\text{BS})$. However, the measured pulse widths are approximately 20 times greater than the minimum values predicted by Eq. (11.19) [$2\sigma(\text{min}) \sim 0.007$ nsec/km for BS fibers with $\Delta n \sim 0.007$; $2\sigma(\text{min}) \sim 0.024$ nsec/km for GBS fibers with $\Delta n \sim 0.013$] and it indicates the difficulty in fabricating perfect profiles. Results in Fig. 11.23 apply at the wavelength ($\lambda = 0.908$ μm) of a GaAs semiconductor laser. However, profile dispersion makes the optimal profile gradient depend on wavelength. Consequently, even if a fiber is optimized for operation at one wavelength the fiber may not have a large bandwidth at another wavelength (Section 11.4.3a).

(c) *Evaluation of intramodal dispersion.* Ultimately, the information carrying capacity of a fiber may be limited by intramodal material and waveguide dispersion which exist even in a single-mode fiber. Wavelength-dependent measurements of refractive index in bulk materials (Fleming, 1976; Mallitson, 1965) and thinly polished fiber samples (Presby et al., 1976) have been used to calculate material dispersion. Intramodal pulse broadening; $\Delta\lambda \, d\tau/d\lambda$, due to an optical source with spectral width, $\Delta\lambda$, centered about λ has been evaluated by measuring time delay, τ, as a function of λ for a light pulse propagating in a fiber of length L. Emission wavelengths over the range $\lambda = 0.7$–0.93 μm have been generated by pulsed krypton lasers and various gallium aluminum arsenide injection lasers (Gloge and Chinnock, 1972; Gloge et al., 1974) and by Raman generation in liquids (Luther-Davies et al., 1975). A recently developed fiber Raman laser has been used to make the first transmission measurements of intramodal dispersion over the wavelength range $\lambda = 1.06$–1.55 μm which covers both sides of the zero intramodal dispersion wavelength in a variety of long germanium and phosphorus borosilicate single-mode and multimode fibers (Cohen and Lin, 1977; Lin et al., 1978). Similar measurements have been done with a dye laser pumped optical parametric oscillator (Payne and Hartog, 1977).

The measuring apparatus described by Cohen and Lin is illustrated in Figure 11.24a. Subnanosecond pulses are generated by efficient multiple-order stimulated Raman scattering in a single-mode "generating" fiber pumped by a Q-switched and mode-locked Nd:YAG laser at $\lambda = 1.06$ μm. With approximately 1-kW peak pump power in the "generating fiber," six orders of stimulated Stokes emission bands are generated near $\lambda = 1.12, 1.18, 1.24, 1.31, 1.41$, and 1.51 μm. The multiwavelength output beam can be spatially dispersed by a grating and displayed as an array of spots on an ir phosphor card. Output light wavelengths from the "generating" fiber are also selected by a monochromator, injected into a long "test" fiber, detected by a germanium photodiode and displayed on an oscilloscope. When the input wavelength is changed, intramodal dispersion

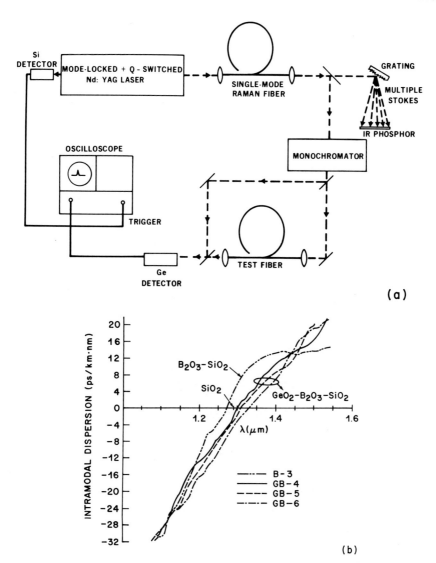

Fig. 11.24   (a) Experimental arrangement for measuring transmission time delays versus wavelength. (b) Intramodal dispersion versus wavelength in doped silica fibers. Fiber B-3 has a graded borosilicate core. The other germanium borosilicate fiber cores have a small amount of B_2O_3 and a larger graded GeO_2 concentration. The dopant concentrations were increased from fiber GB-4 to GB-6.

causes time delay shifts of the mode locked pulse train at the "test" fiber output. Increasing the input wavelength causes propagation delay shifts, τ, relative to a marker pulse at $\lambda = 1.06$ μm, to decrease to a minimum value at λ_0 where intramodal dispersion $d\tau/d\lambda$ (psec/km-nm), becomes zero. Propagation delay shifts increase for increasing wavelength longer than λ_0. Intramodal dispersion, plotted versus wavelength in Figure 11.24b, is evaluated by taking derivatives of transmission delay versus wavelength curves for a borosilicate (B_2O_3–SiO_2) and several germanium borosilicate (GeO_2–B_2O_3–SiO_2) fibers with different material concentrations (Cohen and Lin, 1977). The borosilicate fiber has zero intramodal dispersion at $\lambda_0 = 1.27$ μm while the germanium borosilicate fibers have zero intramodal dispersion in the range $\lambda_0 = 1.3$–1.33 μm. Pulse spreading is proportional to the square of the source spectral bandwidth, $(\Delta\lambda)^2$, at the wavelength where intramodal dispersion vanishes (Kapron, 1977). This imposes a fundamental limit of about 1.25×10^{-2} psec/km-nm^2 on the pulse broadening rate through doped silica fibers.

11.4.3 Frequency Domain Measurements

These measurement techniques assume that the fiber modes may be considered as independent baseband channels so that the Fourier transform of the impulse response yields

$$H(f) = \int_{-\infty}^{\infty} h(t)e^{-i2\pi ft}\, dt = \sum_{\nu=1}^{N} |C_\nu|^2 e^{-i2\pi f\tau_\nu} dt. \tag{11.21}$$

where f is the baseband frequency of the envelope of the modulated optical carrier,

$$H(f) = |H(f)|e^{-i2\pi f\tau_0}e^{i\theta(f)} \tag{11.22}$$

is the complex power transfer function, $|C_\nu|^2$ is the output power in a mode, and τ_0 is the (large) average delay common to all modes. Present measurements yield only $|H(f)|$ but phase-sensitive detection may be used to measure the phase angle $\theta(f)$ (Boisrobert et al., 1976, 1977) which is required to obtain the impulse response $h(t)$ from

$$h(t) = \frac{1}{2\pi} \int_{-\infty}^{\infty} H(f)e^{-i2\pi ft}\, dt. \tag{11.23}$$

One advantage of measurements in the frequency domain is that their precision is better than in the time domain because the deconvolution process, for removing signal distortion caused by the limited detector bandwidth, is simply an arithmetic division of output-by-input frequency response rather than a cumbersome deconvolution of pulse shapes as required in the time domain.

(a) *Wavelength-dependent measurements.* Measurements can be made by externally modulating a cw light carrier or directly modulating a (LED) light-emitting diode with a frequency tunable sinusoidal signal (Auffret *et al.*, 1975; Personick *et al.*, 1974). Then the baseband frequency response is the ratio of the sine wave amplitudes at the beginning and end of the fiber. Figure 11.25a illustrates an experimental arrangement for directly measuring fiber baseband frequency response in spectrally filtered incoherent light suitable for wavelength-dependent studies of profile dispersion over a wide range of wavelengths without resorting to a multitude of

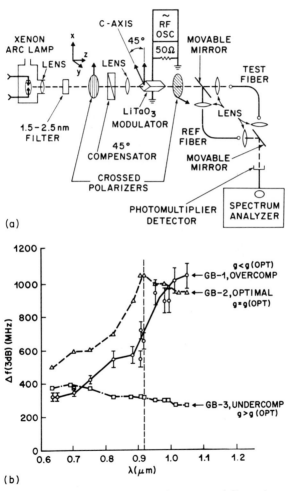

Fig. 11.25   (a) Experimental arrangement for making spectral dispersion measurements in the frequency domain with an incoherent light source and an electrooptic modulator. (b) Transmission bandwidth Δf (3 dB) is plotted versus wavelength λ for three germanium borosilicate fibers GB-1, $L = 1.12$ km, GB-2, $L = 0.92$ km; GB-3, $L = 1.03$ km which are numbered in order of increasing g values.

mode-locked laser sources (Cohen *et al.*, 1976). The xenon arc lamp light passes through one of a set of narrow-band interference filters and is focused into a LiTaO$_3$ electrooptic crystal which intensity modulates the incoherent cw light carrier on a frequency tunable sinusoidal envelope (Chen, 1970). The modulated beam is focused into either the fiber under test or a short (2 m) length of reference fiber. Fiber output power is then detected by a sensitive broadband crossed-field photomultiplier tube (Varian/LSE Division, Palo Alto, Calif.) and the baseband modulation components are displayed by a spectrum analyzer. The component at modulating frequency f from the reference fiber is taken as $P_{in}(f)$ and the same component from the test fiber as $P_{out}(f)$.

The electro-optically modulated input is

$$P_{in}(t) = (1 + M \cos 2\pi ft)e^{2\pi f_0 t}, \tag{11.24}$$

where f_0 is the optical carrier frequency and M is the modulation index at the modulating frequency f.

The fiber output power is divided among N modes and only the terms at the fundamental frequency, f, are measured with the spectrum analyzer. The reduction of sine-wave envelope intensity, $P_{out}(f)/P_{in}(f)$, due to transmission in a fiber gives:

$$\frac{P_{out}(f)}{P_{in}(f)} = |H(f)| = Re \left\{ \sum_{\nu=1}^{N} |C_\nu^2| e^{-2\pi f\tau_\nu} \right\} \tag{11.25}$$

which is the magnitude of the baseband frequency response or power transfer function defined in Eq. (11.22). This function is approximately equivalent to the magnitude of the Fourier transform of the fiber impulse response $h(t)$.

Input-to-output power transfer functions have been measured from dc to 1 GHz in a variety of germanium- and boron-doped kilometer length fibers. Wavelength-dependent measurements are made through a series of interference filters whose center wavelengths range from $\lambda = 0.65 \mu$m to 1.1 μm. In order to reduce modulator errors due to smearing of the optical bias and material dispersion effects, the filter bandwidths are kept less than 1.5 nm between 0.65 μm $< \lambda < 0.908 \mu$m, 2.5 nm for 0.92 μm $< \lambda < 0.98 \mu$m and 10 nm for 1.04 μm $< \lambda < 1.1 \mu$m.

Frequency domain data are characterized by the transmission bandwidth Δf (3 dB) for which the power transfer function is attenuated by 3 dB relative to its dc value. The rate at which transmission bandwidths change with wavelength depends on $dg(opt)/d\lambda$. Borosilicate fibers exhibit relatively small profile dispersion effects since $|\Delta g(opt)| < 0.05$ in the range $0.65 < \lambda < 1.05 \mu$m. This was confirmed by frequency domain measurements which showed that the transmission bandwidth in a nearly optimally graded fiber, with Δf (1 dB) $\approx 0.63/\pi 2\sigma \approx 1100$ MHz-km, changed by less than 25% in the range $0.65 < \lambda < 1.1$ μm. However, the addition of germanium dopant into a silica fiber significantly increases its profile dispersion because $|\Delta g(\text{opt})| > 0.2$ for GeO_2–SiO_2 material constituents in the range $0.65 < \lambda < 1.1$ μm.

Figure 11.25b clearly illustrates dispersion effects on transmission bandwidths as a function of wavelength for three different kilometer length fibers. The qualitative behavior of each curve depends on whether the fabricated profile g is larger (undercompensated) or smaller (overcompensated) than the optimum value for particular wavelengths and individual bandwidth curves peak at the wavelength for which the fabricated profiles are optimal (Cohen et al., 1978a). In quantitative terms, the location of the bandwidth peak shifts from $\lambda_p > 1.05$ μm for fiber GB-1 with $g \sim 1.94$ to $\lambda_p < 0.65$ μm for fiber GB-3 with $g \sim 2.3$ and the bandwidth of an individual fiber can change by more than a factor of three within the range 0.65 μm $< \lambda < 1.05$ μm. For example, Δf (3 dB) in fiber GB-1 increases monotonically with wavelength from 300 MHz at $\lambda = 0.65$ μm to more than 1000 MHz at $\lambda = 1.05$ μm. The bandwidth at 1.05 μm is approximately 40% wider than at $\lambda = 0.908$ μm where pulse dispersion measurements were made.

(b) *Baseband frequency response to laser excitation.* Free-running lasers can be used to measure fiber frequency response by comparing their beat spectra before and after transmission through a fiber (Gloge et al., 1972b). Many signal frequencies are fed into the fiber simultaneously by beating the longitudinal laser mode spectrum to produce a spectrum consisting of N components spaced by the roundtrip frequency $f = c/2L$ which is 100 MHz for a 1.5-m laser cavity length. Although the technique does not require a pulsed source, it does require that the input and output signals be constant for at least the delay time through the fiber. Most free-running lasers are subject to external disturbances which change the (random) phase relations between the longitudinal modes within fractions of a microsecond. Therefore measurements have to be averaged over times longer than the correlation time of the fluctuations mentioned above. The experimental arrangement is very similar to Figure 11.25 with a laser source replacing the incoherent arc lamp and electrooptic modulator. A variable beam splitter divided the output from the laser in such a way that the dc current from a high speed photodiode detector was the same for the reference laser beam and the fiber output beam. The rf current fed into the spectrum analyzer exhibited fluctuations caused by external disturbances affecting the cavity but their effect was eliminated by photographing the display with a 4 second time constant.

Figure 11.26a shows a photograph of a logarithmic plot of the beat spectrum (power spectrum of the optical intensity fluctuations) between 0.5

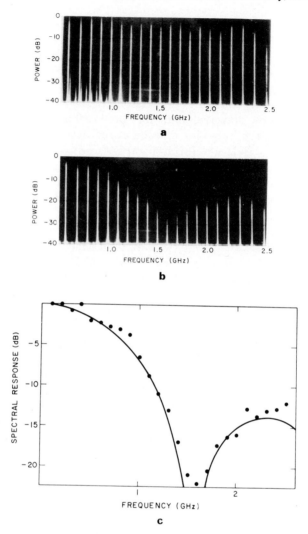

Fig. 11.26 Power spectra of the optical intensity fluctuations in decibels of longitudinal modes from a free-running laser (a) before transmission through a fiber, (b) after transmission through a fiber, (c) logarithmic plot of the baseband frequency response of the fiber tested.

and 2.5 GHz produced by the laser. Beat components appear at multiples of 100 MHz and the additional components at the left of the figure represent spurious responses from the spectrum analyzer which should be ignored. Figure 11.26b shows the beat spectrum of the same signal after it has propagated through a step-index fiber. The high-frequency components are significantly reduced and the spectrum has a minimum at 1.6

GHz. The magnitude of the transfer function is shown in Figure 11.26c after computing the logarithmic difference of Figs. 11.26a and (b) in decibels. The test fiber was short (10 m) enough to ensure that there were no differences in the attentuation of the modes and no coupling between them. The fiber impulse response $h(t)$ should be a rectangular pulse with width τ. The real part of its Fourier transform has the form

$$|H(f)| = \sin(\pi\tau)/\pi\tau f. \qquad (11.26)$$

Equation (11.26) is plotted in Fig. 11.26c so that the parameter τ can be fit to the measurement. This fit suggests a value of $\tau = 0.6$ nsec which is only 10% larger than the expected value, $\tau = 1/2nc\ (NA)^2 = 0.67$ nsec, for the group delay difference between the fastest and slowest mode in the step-index fiber.

11.4.4 Optical Ranging and Fault Location

Time domain reflectometer techniques used for testing conventional cables have been successfully used to locate faults in optical fibers (Boisrobert, 1975; Guttman and Krumpholz, 1975; Ueno and Shimizu, 1976). The types of faults of interest are breaks, misaligned connectors and splices, and regions of high scattering due to bubbles or other discrete defects in the fiber core.

The experimental arrangement is very similar to that of Figure 11.20a without external shuttle pulse mirrors. The roundtrip travel time t of a narrow optical pulse from the fiber input face to the fault and back to the input by reflection from a beam splitter onto a detector at the fiber input is measured. The location of the fault L (measured from the input end of the fiber) is then determined from

$$L = \tfrac{1}{2}(c/n_1)t, \qquad (11.27)$$

where n_1 is the core refractive index. To locate a fault to within one meter of its true position requires an optical pulse width, $\Delta t < 5$ nsec.

The power reaching the detector, P_d, is given by

$$P_d = TSP_0\ 10^{-(2\alpha L/10)}, \qquad (11.28)$$

where P_0 is the output power of the source, T is the transmittance of the optical system including fiber coupling efficiency and beam splitter losses, α is the total transmission loss of the fiber in decibels per kilometer, and S is the effective reflectivity of the fault which is 0.04 for a flat perpendicular break, but would be considerably smaller for a jagged nonperpendicular break (Marcuse, 1975). A typical fault location apparatus (Ueno and Shimizu, 1976) has detected breaks at a range of 1.4 km in a fiber of 17 dB/km loss with a GaAlAs laser diode operating at 0.83 μm and an avalanche photodiode detector. A typical effective reflectivity was found to be 0.005. The factor T in Eq. (11.28) was -16 dB (-6 dB due to the beam splitter and -10 dB coupling loss).

Recently a crossed-field photomultiplier detector was used to increase the dynamic range (Varian) (Personick, 1976). Gating the detector off during each laser pulse is desirable to protect the tube from saturating due to the intense reflection off the fiber input face. The recovery time saturation or gating determines the minimum range that can be probed. The reflection from the input end can be avoided by injecting light into a tapered fiber section (Barnoski and Jensen, 1976; Barnoski et al., 1977), or it can be substantially reduced by using a beam splitter with index-matched fiber (Costa and Sordo, 1977). The high signal-to-noise ratio obtained with a photomultiplier tube in conjunction with a boxcar averager, allows the Rayleigh and Brillouin back-scattering from the core glass to be observed. In this mode of operation, the coefficient S in Eq. (11.28) represents the amount of light being scattered and trapped in the reverse direction. Its magnitude is approximately $3 \cdot 10^{-5}$, or 30 dB below the 3.5% reflection of an ideal break (Neumann, 1978).

Figure 11.27 depicts the decay of the back-scattered light observed in a 1500-m-long, low-loss fiber. The length-dependent transmission losses, including splice and connector losses, and losses due to any other type of fiber imperfections, are directly obtainable from such displays. Furthermore, by measuring the width of a narrow pulse reflected from the far-end of the fiber, dispersion data can be obtained with the same setup. Since

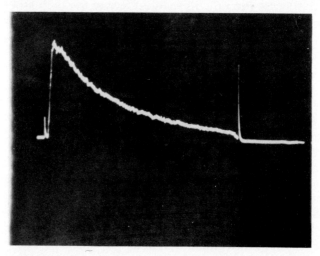

Fig. 11.27 Detector signal versus time delay from a 1500-m length of low-low-loss fiber showing the ability to detect the Rayleigh–Brillouin back-scattering. Signal-to-noise enhancement by use of boxcar averaging was employed (Personick, 1976).

only one end of the fiber has to be available, optical time domain reflectometry (OTDR) is well-suited for field use. As a consequence, it may evolve as a powerful fiber characterization technique.

REFERENCES

Andrews, J. R. (1974). Inexpensive laser diode pulse generator for optical waveguide studies. *Rev. Sci. Instrum.* **45,** 22.

Arnaud, J. A., and Derosier, R. M. (1976). Novel technique for measuring the index profile of optical fibers. *Bell Syst. Tech. J.* **55,** 1489.

Arnaud, J. A., and Fleming, J. W. (1976). Pulse broadening in multimode optical fibers with large $\Delta n/n$ numerical results. *Electron. Lett.* **12,** 167.

Auffret, R., Boisrobert, C., and Cozannet, A. (1975). Wobulation technique applied to optical fibre transfer function measurement. *Tech. Dig. Eur. Conf. Opt. Fibre Commun. 1st, 1975* p. 60.

Barnoski, M. K., and Jensen, S. M. (1976). Fiber waveguides: A novel technique for investigating attenuation characteristics. *Appl. Opt.* **15,** 2112.

Barnoski, M. K., and Personick, S. D. (1978). Measurements in fiber optics. *Proc. IEEE* **66,** 429.

Barnoski, M. K., Rourke, M. D., Jensen, S. M., and Melville, R. T. (1977). Optical time domain reflectometer. *Appl. Opt.* **16,** 2375.

Bateson, S. (1958). Critical study of the optical and mechanical properties of glass fibers. *J. Appl. Phys.* **29,** 13.

Blyler, L. L., Jr., Hart, A. C., Jr., Jaeger, R. E., Kaiser, P., and Miller, T. J. (1975). Low-loss, polymer-clad silica fibers produced by laser drawing. *Top. Meet. Opt. Fiber Transm., 1st, 1975* Paper TuA5-1.

Boisrobert, C. Y. (1975). New engineering consideration for fiber optic digital transmission systems. *Top. Meet. Opt. Fiber Transm. 1st, 1975* Paper ThB3.

Boisrobert, C., Cozannet, A., and Vassallo, C. (1976). Sweep frequency transfer function measurement applied to optical fibre. *IEEE Trans. Instrum. Meas.* **IM-25,** 294.

Boisrobert, C., Cozannet, A., Lamouler, P., Martin, L., and Diraison, H. (1977). Mesure des distortions de temps de propagation de groupe sur les fibres optiques à saut d'indice. *URSI Proc. Int. Symp. Meas. Telecommun.* p. 445.

Burrus, C. A., and Standley, R. D. (1974). Viewing refractive-index profiles and small-scale inhomogeneities in glass optical fibers: Some techniques. *Appl. Opt.* **13,** 2365.

Burrus, C. A., Chinnock, E. L., Gloge, D., Holden, W. S., Tingye Li, Standley, R. D., and Keck, D. B. (1973). Pulse dispersion and refractive index profiles of some low-loss multimode optical fibers. *Proc. IEEE* **61,** 1498.

Chen, F. S. (1970). Modulators for optical communications. *Proc. IEEE* **58,** 1440.

Cherin, A. H., Cohen, L. G., Holden, W. S., Burrus, C. A., and Kaiser, P. (1974). Transmission characteristics of three Corning multimode optical fibers. *Appl. Opt.* **13,** 2359.

Chu, P. L. (1976). Determination of diameters and refractive indices of step-index optical fibers. *Electron. Lett.* **12,** 155.

Cohen, L. G. (1975). Shuttle pulse measurements of pulse spreading in an optical fiber. *Appl. Opt.* **14,** 1351.

Cohen, L. G. (1976). Pulse transmission measurements for determining near optimal profile gradings in multimode borosilicate optical fibers. *Appl. Opt.* **15,** 1808.

Cohen, L. G., and Lin, C. (1977). Pulse delay measurements in the zero material dispersion wavelength region for optical fibers. *Appl. Opt.* **16,** 3136.

Cohen, L. G., and Personick, S. D. (1975). Length dependence of pulse dispersion in a long multimode optical fiber. *Appl. Opt.* **14,** 1357.

Cohen, L. G., and Presby, H. M. (1975). Shuttle pulse measurements of pulse spreading in a low-loss graded-index fiber. *Appl. Opt.* **14,** 1361.

Cohen, L. G., Kaiser, P., MacChesney, J. B., O'Connor, P. B., and Presby, H. M. (1975). Transmission properties of a low-loss near-parabolic-index fiber. *Appl. Phys. Lett.* **26,** 472.

Cohen, L. G., Astle, H. W., and Kaminow, I. P. (1976). Frequency domain measurements of dispersion in multimode optical fibers. *Bell Syst. Tech. J.* **55,** 1509.

Cohen, L. G., Astle, H. W., and Kaminow, I. P. (1977). Frequency domain measurements of dispersion in multimode optical fibers. *Appl. Phys. Lett.* **30,** 17–19.

Cohen, L. G., Kaminow, I. P., Astle, H. W., and Stulz, L. W. (1978a). Profile dispersion effects on transmission bandwidths in graded index optical fibers. *IEEE J. Quantum Electron.* **QE-14,** 37.

Cohen, L. G., DiMarcello, F. V., Fleming, J. W., French, W. G., Simpson, J. R., and Weiszmann, E. (1978b). Pulse dispersion properties of fibers with various material constituents. *Bell Syst. Tech. J.* **57,** 1653.

Cohen, R. L. (1975). Loss measurements in optical fibers. 1. Sensitivity limit of bolometric techniques. *Appl. Opt.* **13,** 2518.

Cohen, R. L., West, K. W. Lazay, P. D., and Simpson, J. R., Jr. (1975). Loss measurements in optical fibers. 2. Bolometric measuring instrumentation. *Appl. Opt.* **13,** 2522.

Costa, B., and Sordo, B. (1977). Experimental study of optical fiber attenuation by modified backscattering technique. *Conf., Dig., Eur. Conf. Opt. Commun. 3rd, 1977* p. 69.

Dannwolf, J. W. Gottfried, S., Sargent, G. A., and Strun, R. L. (1976). Optical fiber impulse response measurement system. *IEEE Trans. Instrum. Meas.* **IM-25,** 401.

Eickhoff, W., and Weidel, E. (1975). Measuring method for the refractive index profile of optical glass fibers. *Opt. Quantum. Electron.* **7,** 109.

Eve, M., Hill, A. M., Malyon, D. J., Midwinter, J. E., Nelson, B. P., Stern, J. R., and Wright, J. V. (1976). Launching-independent measurements of multimode fibers. *Conf. Dig., Eur. Conf. Opt. Commun. 2nd, 1976* p. 143.

Fleming, J. W. (1975). Measurements of dispersion in GeO_2–B_2O_3 SiO_2 glasses. *Am. Ceram. Soc. Bull.* **54,** 814.

Fleming, J. W. (1976). Material and mode dispersion in $GeO_2 \cdot B_2O_3 \cdot SiO_2$ glasses. *J. Am. Ceram. Soc.* **59,** 503.

French, W. G., and Tasker, G. W. (1975). Fabrication of graded-index and single-mode fibers with silica cores. *Top. Meet. Opt. Fiber Trans. 1st, 1975* TuA2-1.

Gardner, W. B. (1975). Microbending loss in optical fibers. *Bell Syst. Tech. J.* **54,** 457.

Gloge, D. (1971). Weakly guiding fibers. *Appl. Opt.* **10,** 2252.

Gloge, D. (1972a). Optical power flow in multimode fibers. *Bell. Syst. Tech. J.* **51,** 1767.

Gloge, D. (1972b). Bending loss in multimode fibers with graded and ungraded core. *Appl. Opt.* **11,** 2506.

Gloge, D. (1975). Optical fiber packaging and its influence on fiber straightness and loss. *Bell Syst. Tech. J.* **54,** 245.

Gloge, D., and Chinnock, E. L. (1972). Fiber dispersion measurements using a mode-locked krypton laser. *IEEE J. Quantum Electron.* **QE-8,** 852.

Gloge, D., and Marcatili, E. A. J. (1973). Multimode theory of graded-core fibers. *Bell Syst. Tech. J.* **52,** 1563

Gloge, D., Chinnock, E. L., and Lee, T. P. (1972a). Self-pulsing GaAs laser for fiber dispersion measurements. *IEEE J. Quantum Electron.* **QE-8,** 844.

Gloge, D., Chinnock, E. L., and Ring, D. H. (1972b). Direct measurement of the (baseband) frequency response of multimode fibers. *Appl. Opt.* **11,** 1534.

Gloge, D., Smith, P. W., Bisbee, D. L., and Chinnock, E. L. (1973). Optical fiber end preparation for low-loss splices. *Bell Syst. Tech. J.* **52,** 1579.

Gloge, D., Chinnock, E. L., and Lee, T. P. (1974). GaAs twin-laser setup to measure mode and material dispersion in optical fibers. *Appl. Opt.* **13,** 261.

Gloge, D., Kaminow, I. P., and Presby, H. M. (1975). Profile dispersion in multimode fibres: Measurement and analysis. *Electron. Lett.* **11,** No. 19.

Guttman, J., and Krumpholz, O. (1975). Location of imperfections in optical glass-fibre waveguides. *Electron. Lett.* **11,** 216.

Hewlett Packard (1975). The Hewlett–Packard interface bus: current perspectives. *Hewlett Packard J.* **26,** 2.

IEEE (1975). "Digital Interface for Programmable Instrumentation," IEEE Std. 488-1975. IEEE Inc., New York.

Ikeda, M., Tateda, M., and Yoshikiyo, H. (1975). Refractive index profile of a graded index fiber: Measurement by a reflection method. *Appl. Opt.* **14,** 814.

Ikeda, M., Sugimura, A., and Ikegami, T. (1976). Multimode optical fibers: Steady-state mode exciter. *Appl. Opt.* **15,** 2116.

Inada, K., Akimoto, T., and Tanaka, S. (1976). Pulse spread mechanism of long length silicone clad optical fiber. *Conf. Dig., Eur. Conf. Opt. Fiber Commun. 2nd, 1976* p. 157.

Igelstam, E., and Johansson, I. P. (1958). Correction due to aperture in transmission interference microscopes. *J. Sci. Instrum.* **35,** 15.

Izawa, T., and Nakagowe, H. (1972). Optical waveguide formed by electrically induced migration of ions in glass plates. *Appl. Phys. Lett.* **21,** 584.

Jones, M. W., and Kao, K. C. (1969). Spectrophotometric studies of ultra-low-loss optical glasses. *J. Phys. E.* **2,** 331.

Kaiser, P. (1973). Spectral losses of unclad fibers made from high-grade vitreous silica. *Appl. Phys. Lett.* **23,** 45.

Kaiser, P. (1974). Drawing-induced coloration in vitreous silica fibers. *J. Opt. Soc. Am.* **64,** 475.

Kaiser, P. (1977). NA-dependent spectral loss measurements of optical fibers. *Tech. Dig., Int. Conf. Integr. Opt. Opt. Fiber Transm. 1977* p. 267.

Kaiser, P. (1978). NA-dependent spectral loss measurements of optical fibers. *Trans. IECE, Jpn.* **E61,** 225.

Kaiser, P., and Astle, H. W. (1974). Measurement of spectral total and scattering losses in unclad optical fibers. *J. Opt. Soc. Am.* **64,** 469.

Kaiser, P., Tynes, A. R., Astle, H. W. Pearson, A. D., French, W. G., Jaeger, R. E., and Cherin, A. H. (1973). Spectral losses of unclad vitreous silica and soda-lime silicate fibers. *J. Opt. Soc. Am.* **63,** 1141.

Kaiser, P., Hart, A. C., Jr., and Blyler, L. L., Jr. (1975). Low-loss, FEP-clad silica fibers. *Appl. Opt.* **14,** 156.

Kaiser, P., Tasker, G. W., French, W. G., Simpson, J. R., and Presby, H. W. (1977). Single-mode fibers with different B_2O_3–SiO_2 compositions. *Top. Meet. Opt. Fiber Transm. 1st 1977* Paper TuD3-1.

Kaminow, I. P., and Carruthers, J. R. (1973). Optical waveguiding layers in $LiNbO_3$ and $LiTaO_3$. *Appl. Phys. Lett.* **22,** 326.

Kaminow, I. P., and Presby, H. M. (1977). Profile synthesis in multicomponent glass optical fibers. *Appl. Opt.* **16,** 108.

Kapany, N. S. (1957). Fiber optics. Part I. Optical properties of certain dielectric cylinders. *J. Opt. Soc. Am.* **47,** 413.

Kapron, F. P. (1977). Maximum information capacity of fiber-optic waveguides. *Electron. Lett.* **13,** 96.

Keck, D. B. (1974). Spatial and temporal power transfer measurements on a low-loss optical waveguide. *Appl. Opt.* **13,** 1882.

Keck, D. B., and Bouillie, R. (1978). Measurements on high-bandwidth optical waveguides. *Opt. Commun.* **25**, 43.

Keck, D. B., and Tynes, A. R. (1972). Spectral response of low-loss optical waveguides. *Appl. Opt.* **11**, 1502.

Keck, D. B., Maurer, R. D., and Schultz, P. C. (1973). On the ultimate lower limit of attenuation in glass fiber optical waveguides. *Appl. Phys. Lett.* **22**, 307.

Kita, H., Kitano, I., Uchida, T., and Furukawa, M. (1971). Light-focusing glass fibers and rods. *J. Am. Ceram. Soc.* **54**, 321.

Kokubun, Y., and Iga, K. (1977). Precise measurement of the refractive index profile of optical fibers by a nondestructive interference method. *Trans. IECE, Jpn.* **60**, 702.

Krug, W., Rienitz, J., and Schultz, G. (1964). "Contribution to Interference Microscopy." Hilger & Watts. London.

Lewis, J. A. (1976). Laser-heating of an optical fiber. *Appl. Opt.* **15**, 1304.

Lin, C., Cohen, L. G., French, W. G., and Foertmeyer, V. A. (1978). Pulse delay measurements in the zero-material-dispersion region for germanium and phosphorus-doped silica fibers. *Electron. Lett.* **14**, 170.

Liu, Y. S. (1974). Direct measurement of the refractive indices for a small numerical aperture cladded fiber: A simple method. *Appl. Opt.* **13**, 1255.

Luther-Davies, B., Payne, D. N., and Gambling, W. A. (1975). Evaluation of material dispersion in low loss phosphosilicate core optical fibers. *Opt. Commun.* **13**, 84.

McKinley, T. D., Heimich, K. F. J., and Wittry, D. B., eds. (1966). "The Electron Microprobe." Wiley, New York.

Mallitson, I. H. (1965). Interspecimen comparison of the refractive index of fused silica. *J. Opt. Soc. Am.* **55**, 1205.

Marcatili, E. A. J. (1969). Bends in optical dielectric guides. *Bell Syst. Tech. J.* **48**, 2103.

Marcuse, D. (1973) Coupled mode theory of round optical fibers. *Bell Syst. Tech. J.* **52**, 817.

Marcuse, D. (1974a). Light scattering from elliptical fibers. *Appl. Opt.* **13**, 1903.

Marcuse, D. (1974b). Reduction of multimode dispersion by intentional mode coupling. *Bell Syst. Tech. J.* **53**, 1795.

Marcuse, D. (1975). Reflection losses from imperfectly broken fiber ends. *Appl. Opt.* **14**, 3016.

Marcuse, D. (1976). Mode mixing with reduced losses in parabolic-index fibers. *Bell Syst. Tech. J.* **55**, 777.

Marcuse, D., and Presby, H. M. (1975a). Light scattering from optical fibers with arbitrary refractive-index distributions. *J. Opt. Soc. Am.* **65**, 367.

Marcuse, D., and Presby, H. M. (1975b). Mode coupling in an optical fiber with core distortions. *Bell Syst. Tech. J.* **54**, 3.

Marhic, M. E., Ho, P. S., and Epstein, M. (1975). Nondestructive refractive-index profile measurements of clad optical fibers. *Appl. Phys. Lett.* **26**, No. 10.

Martin, W. E. (1974). Refractive index profile measurements of diffused optical waveguides. *Appl. Opt.* **13**, 2112.

Midwinter, J. E., and Reeve, M. H. (1974). A technique for the study of mode cut-offs in multimode optical fibers. *Opto-Electronics* **6**, 411.

Miller, S. E., Marcatili, E. A. J., and Tingye Li (1973). Research toward optical-fiber transmission systems. *Proc. IEEE* **61**, 1703.

Murata, H. (1976). Broadband optical fiber cable and connecting. *Conf. Dig., Eur. Conf. Opt. Commun. 2nd, 1976* p. 167.

Neumann, E. G. (1978). Optical time-domain reflectometer: Comment. *Appl. Opt.* **17**, 1675.

O'Connor, P. B., Kaiser, P., MacChesney, P. B., Burrus, C. A., Presby, H. M., Cohen, L. G., and DiMarcello, F. V. (1976). Large numerical aperture, germanium-doped fibers for LED applications. *Proc. Eur. Conf. Opt. Fiber Commun., 2nd, 1976* pp. 55.

Okoshi, T., and Hotate, K. (1976a). Computation of the refractive index distribution in an optical fiber from its scattered pattern for a normally incident laser beam. *Opt. Quantum Electron.* **8**, 78.

Okoshi, T., and Hotate, K. (1976b). Refractive-index profile of an optical fiber: Its measurement by the scattering-pattern method. *Appl. Opt.* **15**, 2756.

Olshansky, R. (1977). Differential mode attenuation in graded-index optical waveguides. *Tech. Dig., Int. Conf. Integr. Opt. Opt. Fiber Transm., 1977* p. 423.

Olshansky, R., and Keck, D. B. (1976). Pulse broadening in graded-index-optical fibers. *Appl. Opt.* **15**, 483.

Olshansky, R., and Oaks, S. (1978). Differential mode attenuation measurements in graded-index fibers. *Appl. Opt.* **17**, 1830.

Osanai, H., Shioda, T., Moriyama, T., Araki, S., Horiguchi, M., Izawa, T., and Takata, H. (1976). Effect of dopants on transmission loss of low-OH-content optical fibers. *Electron. Lett.* **12**, 549.

Ostermayer, F. W., Jr., and Benson, W. W. (1974). Integrating sphere for measuring scattering loss in optical fiber waveguides. *Appl. Opt.* **13**, 1900.

Ostermayer, F. W., Jr., and Pinnow, D. A. (1974). Optimum refractive index difference for graded-index fibers resulting from concentration fluctuation scattering. *Bell Syst. Tech. J.* **53**, 1395.

Payne, D. N., and Hartog, A. H. (1977). Determination of the wavelength of zero material dispersion in optical fibers by pulse-delay measurements. *Electron. Lett.* **13**, 627.

Pearson, A. D., French, W. G., and Rawson, E. G. (1969). Preparation of light focussing glass rod by ion-exchange techniques. *Appl. Phys. Lett.* **15**, 76.

Personick, S. D. (1971). Time dispersion in dielectric waveguides. *Bell Syst. Tech. J.* **50**, 843.

Personick, S. D. (1975). Baseband linearity and equalization in fiber optic digital communication systems. *Bell Syst. Tech. J.* **52**, 1175.

Personick, S. D. (1976). Photon-probe—an optical fiber time domain reflectometer. *Bell Syst. Tech. J.* **56**, 355.

Personick, S. D., Hubbard, W. M., and Holden, W. S. (1974). Measurements of the baseband frequency response of a 1-km fiber. *Appl. Opt.* **13**, 266.

Pinnow, D. A., and Rich, T. C. (1975). Measurements of the absorption coefficient in fiber optical waveguides using a calorimetric technique. *Appl. Opt.* **14**, 1258.

Presby, H. M. (1974a). Refractive index and diameter measurements of unclad optical fibers. *J. Opt. Soc. Am* **64**, 280.

Presby, H. M. (1974b). Optical dispersion of unclad fibers over limited wavelengths. *Appl. Opt.* **13**, 465.

Presby, H. M. (1974c). Variation of refractive index with wavelength in fused silica optical fibers and preforms. *Appl. Phys. Lett.* **24**, 422.

Presby, H. M. (1976). Ellipticity measurement of optical fibers. *Appl. Opt.* **15**, 492.

Presby, H. M., and Astle, H. W. (1978). Optical fiber index profiling by video analysis of interference fringes. *Rev. Sci. Instrum.* **49**, 339.

Presby, H. M., and Kaminow, I. P. (1976). Binary silica optical fibers: Refractive index and profile dispersion measurements. *Appl. Opt.* **15**, 3029.

Presby, H. M., and Marcuse, D. (1974). Refractive index diameter determinations of step index optical fibers and preforms. *Appl. Opt.* **13**, 2882.

Presby, H. M., Standley, R. D., MacChesney, J. B., and O'Connor, P. B. (1975). Material structure of germanium-doped optical fibers and preforms. *Bell Syst. Tech. J.* **54**, 1681.

Presby, H. M., Mammel, W., and Derosier, R. M. (1976). Refractive index profiling of graded index optical fibers. *Rev. Sci. Instrum.* **47**, 348.

Presby, H. M., Marcuse, D., and Astle, H. (1978a). Automatic refractive index profiling of optical fibers. *Dig. Tech. Pap., Conf. Laser Electroopt. Syst., 1978* Paper WFF5, p. 54.

Presby, H. M., Marcuse, D., and Boggs, L. M. (1978b). Rapid and accurate automatic index profiling of optical fibers. *Proc. Eur. Conf. Opt. Fibre Commun., 4th, 1978* p. 162.

Rawson, E. G. (1972). Measurement of the angular distribution of light scattered from a glass fiber optical waveguide. *Appl. Opt.* **11,** 2477.

Rawson, E. G., and Murray, R. G. (1973). Interferometric measurements of SELFOG dielectric constant coefficients to sixth order. *IEEE J. Quantum Electron.* **QE-9,** 1114.

Reeve, M. H., Brierley, M. C., Midwinter, J. E., and White, K. I. (1976). Studies of radiative losses from multimode optical fibers. *Opt. Quantum Electron.* **8,** 39.

Rich, T. C., and Pinnow, D. A. (1974). Evaluation of fiber optical waveguides using Brillouin spectroscopy. *Appl. Opt.* **13,** 1376.

Saunders, M. J., and Gardner, W. B. (1977). Nondestructive interferometric measurement of the delta and alpha of clad optical fiber. *Appl. Opt.* **16,** 2368.

Sladen, F. M. E., Payne, D. N., and Adams, M. J. (1976). Determination of optical fiber refractive index profiles by a near-field scanning technique. *Appl. Phys. Lett.* **28,** 255.

Sladen, F. M. E., Payne, D. N., and Adams, M. J. (1977). Measurement of profile dispersion in optical fibers: a direct technique. *Electron. Lett.* **13,** 212.

Smith, H. L., and Cohen, A. J. (1963). Absorption spectra of cations in alkali-silicate-glasses of high ultra-violet transmission. *Phys. Chem. Glasses* **4,** 173.

Snyder, A. W., and Mitchell, D. J. (1974). Leaky rays on circular optical fibers. *J. Opt. Soc. Am.* **64,** 599.

Sommargren, G. E., and Thompson, B. J. (1973). Linear phase microscopy. *Appl. Opt.* **12,** 2130.

Stewart, W. J. (1977). A new technique for measuring the refractive index profiles of graded optical fibers. *Tech. Dig. Pap. IOOC, 1977* C2-2.

Stone, F. T. (1978). Launch-dependent loss in short lengths of graded-index multimode fibers. *Appl. Opt.* **17,** 2825.

Stone, J., and Burrus, C. A. (1975). Focusing effects in interferometric analysis of graded-index optical fibers. *Appl. Opt.* **14,** 151.

Stone, J., and Derosier, R. M. (1976). Elimination of errors due to sample polishing in refractive index profile measurements by interferometry. *Rev. Sci. Instrum.* **47,** 885.

Stone, J., and Earl, H. E. (1976). Surface effects and reflection refractometry of optical fibers. *Opt. Quantum Electron.* **8,** 459.

Stone, J., Ramaswamy, V., and Cohen, L. G. (1974). An efficient end reflector for optical fibres. *Opto-Electronics* **6,** 181.

Tasker, G. W., French, W. G., Simpson, J. R., Kaiser, P., and Presby, H. W. (1978). Low-loss single-mode fibers with different B_2O_3–SiO_2 compositions. *Appl. Opt.* **17,** 73.

Traifari, J. (1976). Bus standard brings new power to bench-top instrumentation. *Electron. Prod.* **19,** 31.

Tynes, A. R. (1970). Integrating cube scattering detector. *Appl. Opt.* **9,** 2706.

Tynes, A. R., Pearson, A., and Bisbee, D. L. (1971). Loss mechanisms and measurements in clad glass fibers and bulk glass. *J. Opt. Soc. Am.* **61,** 143.

Ueno, Y., and Shimizu, M. (1976). Optical fiber fault location method. *Appl. Opt.* **15,** 1385.

Vest, C. M. (1975). Interferometry of strongly refracting axisymmetric phase objects. *Appl. Opt.* **14,** 1601.

Wagner, S., Shay, J. L., and Migliorato, P. (1974). $CuInSe_2$/CdS heterojunction photovoltaic detectors. *Appl. Phys. Lett.* **25,** 434.

Watkins, L. S. (1974). Scattering from side-illuminated clad glass fibers for determination of fiber parameters. *J. Opt. Soc. Am.* **64,** 767.

Wells, O. C. (1974). "Scanning Electron Microscopy." McGraw-Hill, New York.

White, K. I. (1976). A calorimetric method for the measurement of low optical absorption losses in optical communication fibers. *Opt. Quantum Electron.* **8,** 73.

Witte, H. W. (1972). Determination of low bulk absorption coefficients. *Appl. Opt.* **11,** 777.

Wonsiewicz, B. C., French, W. G., Lazay, P. D., and Simpson, J. R. (1976). Automatic analysis of interferograms: Optical waveguide refractive index profiles. *Appl. Opt.* **15,** 1048.

Yamada, T., Kashimoto, H., Inada, K., and Tanaka, S. (1977). Launching dependence of transmission losses of graded-index optical fiber. *Tech. Dig., Int. Conf. Integr. Opt. Opt. Fiber Transm., 1977* p. 263.

Yoshida, K., Sentsui, S., Shii, H., and Kuroha, T. (1978). Optical fiber drawing and its influence on fiber loss. *Trans. IECE, Jpn.* **E61,** 181.

Zaganiaris, A. (1974). Simultaneous measurement of absorption and scattering losses in bulk glass and optical fibers by a microcalorimetric method. *Appl. Phys. Lett.* **25,** 345.

Chapter 12

Fiber Characterization— Mechanical

DAVID KALISH
P. LELAND KEY
CHARLES R. KURKJIAN
BASANT K. TARIYAL
TSUEY TANG WANG

12.1 INTRODUCTION

One of the principal advantages of optical fibers for transmission of information is their large bandwidth and hence high information carrying capacity compared with that of a copper wire pair. For example, each fiber installed in the Chicago lightwave trial (Boyle, 1977) is being operated at a carrier rate of 44.7 Mbits/sec; equivalent to 672 simultaneous voice channels or the equivalent mix of voice, data, and video channels. By the same token, this obviously places very stringent requirements on the mechanical reliability of an optical fiber as compared to a copper wire pair carrying a few voice channels.

Optical fiber cables must be designed to withstand the loads associated with manufacture, installation, and service. Manufacture and installation loads may be either impulsive or gradually applied whereas service loads are usually slowly varying. The magnitude of these loads can vary with the application but, for the Chicago installation, stresses in the fibers as high as 175 MPa (25 ksi) were observed as the cable was pulled into ducts. In addition to loads, the environmental conditions of the application can play a significant role in mechanical reliability since glass suffers a deterioration due to stress induced reaction with water. In telecommunications applications, optical fiber cables could be immersed in water in manholes and underground ducts.

401

Mechanical reliability is not the only concern in designing a cable. Constraints imposed by other considerations such as weight, size, and flexibility as well as optical transmission properties lead to design trade-off situations. For example, an increased coating thickness on the fibers can lead to decreased microbending loss and improved moisture resistance but at the expense of bulkier, less flexible cable. These considerations are discussed further in the next chapter.

The response of glass to an applied stress consists of two components: an elastic strain and an inelastic strain. The elastic strain is produced by stretching and rotating the bonds between the atoms comprising the glass. The inelastic strain is produced by local rearrangement of stressed bonds and involves breaking and making of bonds. These local rearrangements are greatly facilitated by thermal vibration of the atoms leading to increased inelastic flow at elevated temperature. However, for the temperatures and glasses used in optical glass fibers, this mechanism of inelastic flow is ineffective. Thus, optical glass fibers should behave as an ideal elastic solid supporting loads up to the fracture strength of the atomic bonds binding the atoms in the glass together.

Any glass object such as an optical fiber always contains a number of flaws, particularly on its surface, that arise from manufacturing and handling. These flaws may be cracks or inclusions which act to concentrate the stress in their vicinity causing local failure of the atomic bonds while the average stress in the glass is still low. If the size and the geometry of the most severe flaw (the one producing the largest stress concentration) is known, then the fracture strength of the glass solid can be calculated. Real glass solids, such as optical fibers, can contain many flaws with a distribution of flaw sizes. Since there is, as yet, no way to determine the flaw distribution in a given piece of glass, the fracture strength can only be described statistically.

The observation that the severity of surface flaws can change when stressed in the presence of water further complicates the question of glass strength. Thus, a glass body, which initially does not fail at a given stress, may eventually fail if the stress is maintained for a long enough time for a flaw within it to grow to a critical size.

In this chapter, the mechanical behavior of glass fibers: elastic properties, fracture strength, and time-dependent fracture, are reviewed. Techniques for mechanical design of optical glass fiber structures are presented.

12.2 ELASTIC PROPERTIES

Glasses used in optical fiber applications behave as elastic bodies up to their breaking strength. The elastic strains are produced by elongations

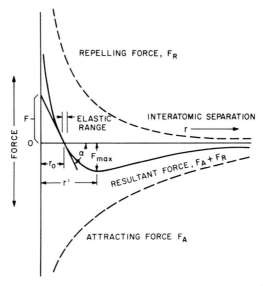

Fig. 12.1   Schematic force-displacement curve for atoms in an elastic solid.

and rotations of the bonds between the atoms comprising the glass. The relation between the elastic behavior and the interatomic deformations can be shown qualitatively by considering the resultant force between the atoms in the solid which can be assumed to result from the superposition of attractive and repulsive terms:

$$F = - (A/r^n) + (B/r^m) \tag{12.1}$$

where A, B, n, and m are empirical constants. This relation is sketched in Fig. 12.1. The equilibrium atomic spacing, r_0, corresponds to the distance at which the attractive and repulsive forces just balance. The macroscopic elastic strain is given by the average change in fractional spacing between atoms, $(r - r_0)/r_0$. The elastic stress is the force per unit area or F/r_0^2. Thus, with a change of coordinates, Fig. 12.1 can also represent the stress–strain relation for an elastic solid.

The tensile elastic modulus is the change in stress per unit change in strain or $d\sigma/d\epsilon$:

$$E = \frac{d\sigma}{d\epsilon} = \frac{1}{r_0} \frac{dF}{dr}, \tag{12.2}$$

which shows that the elastic modulus is proportional to the slope of the interatomic force curve. For small strains near r_0, this slope and thus the modulus is almost constant (Young's modulus).

In addition to Young's modulus; several other moduli are necessary to describe elastic deformation, the number of which depend on the symmetry of the structure of the elastic solid. For small strains, most glasses can be considered as homogeneous isotropic, linear elastic bodies characterized by Young's modulus, E, shear modulus, G, Poisson's ratio, ν, and bulk modulus, K. Only two of these four constants are independent with the relationship between them being:

$$G = \frac{E}{2(1 + \nu)} \quad \text{and} \quad K = \frac{E}{3(1 - 2\nu)}. \quad (12.3)$$

Typical values for several glasses are given in Table 12.1.

Since the breaking strength of glass fibers can be large (5 GPa), large elastic strains (>5%) can be obtained. Strains of this magnitude require consideration of nonlinear elastic effects (Novozhilov, 1953). Also, the large strains may lead to molecular orientation along the strain axis resulting in a transversely isotropic solid (and five elastic constants) rather than an isotropic solid.

Gradual increases in elastic modulus with increasing strains have been observed for fused silica (Brenner, 1951; Hillig, 1961; Mallinder and Proctor, 1964; Powell and Skove, 1968; Thomas and Brown, 1972) whereas decreases were observed for soda lime glass (Mallinder and Proctor, 1964). A typical nonlinear extension curve for fused silica is shown in Fig. 12.2. Mallinder and Proctor (1964) have reported their results in the form:

$$E/E_0 = 1 + \alpha\epsilon, \quad (12.4)$$

where E and E_0 are the tensile moduli $(d\sigma/d\epsilon)$ at a strain ϵ and at zero strain, respectively. The parameter α is given by 5.75 for fused silica and -5.11 for soda glass. Values of the other elastic properties for glasses such as Poisson's ratio ν, the shear modulus G, and the bulk modulus K also show variations with strain (Mallinder and Proctor, 1964).

The behavior of fused silica is especially interesting. The value of Poisson's ratio for silica is lower than for most solids and the modulus in-

TABLE 12.1
Typical Values of Elastic Constants for Glasses

	E (GPa)	G (GPa)	ν	K (GPa)
Fused silica fiber[a]	71.9	31.5	0.14	33.4
Aluminosilicate[b]	91	36.1	0.26	63.2
Borosilicate (Pyrex)[b]	61	25.0	0.22	36.3
Soda-lime-silica[b]	74	30.6	0.21	42.5
Lead-silicate[b]	61	25.2	0.21	35.1

[a] Mallinder (1964).
[b] Holloway (1973).

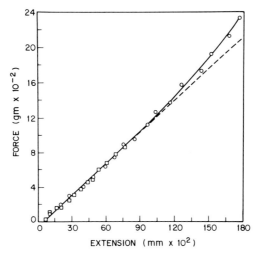

Fig. 12.2 Force-displacement curve of fused silica fibers at room temperature (Thomas and Brown, 1972). □ First run, ○ second run.

creases with strain which is also anomalous. Mallinder and Proctor (1964) suggest that this latter behavior results from deformation which occurs by pivoting of the silicon–oxygen tetrahedra about a corner in addition to bond stretching and bond angle changes.

Optical fibers will likely be coated to protect the glass surface. A coated glass fiber is actually a composite structure and the properties of both the glass and the coating must be considered. The fraction of the tensile load carried by a uniformly thick, linear elastic coating is given by

$$f_c = E_c A_c / (E_c A_c + E_g A_g), \tag{12.5}$$

where E is the elastic modulus, A is the cross-sectional area, and subscripts g and c refer to glass and coating, respectively. As an example, consider a 0.1-mm-diameter silica fiber with a 0.05-mm-thick coating of a polymer having a modulus of 350 MPa. The coating in this composite will carry 1.5% of the load. In considering this coating contribution, one should recognize that the properties of polymer coatings can be strongly strain rate-, time-, and temperature-dependent.

12.3 FRACTURE STRENGTH

12.3.1 Ultimate and Practical Strength

The theoretical strength of glass fibers is determined by the cohesive bond strength of the constituent atoms. A simple estimate of this may be made (Lawn and Wilshaw, 1975; Macmillan, 1972) by consideration of Fig. 12.1. The resultant force curve is approximated by a sine curve and the theoretical strength, σ_{th}, is taken as F_{max}. The area under the curve is then the work done per unit cross sectional area as the solid is separated by tensile forces and this must be at least equal to the energy of the new surfaces formed. This leads to an estimate of $\sigma_{th} \approx E/\pi \approx 20GPa$. Such strengths have been observed by several investigators. For example, Zhurkov (1935) obtained values up to 16 GPa, Hillig (1962), Morley *et al.* (1964), and Proctor *et al.* (1967) found values of approximately 6 and 14 GPa at room temperature and liquid nitrogen temperature, respectively.

Most glass objects, however, do not exhibit strengths anywhere near the theoretical value; strengths of 30 to 100 MPa (5000–15,000 psi) are more common. It had long been recognized that inhomogeneities and flaws could reduce the observed strength of materials (for a review, see Irwin and Wells, 1965). Griffith (1920) in his classic paper showed how this reduced strength could be calculated and studied experimentally. Using a thermodynamic argument, he suggested that a crack became unstable and led to fracture when the strain-energy release rate due to crack extension exceeded the rate of increase of surface energy. This leads to the familiar Griffith equation

$$\sigma = (2E\gamma/\pi a)^{1/2}, \tag{12.6}$$

where σ = fracture strength, E = Young's modulus, γ = surface energy, and a = half-crack length.

A similar relation is obtained if it is assumed that fracture occurs when the stress at the crack tip equals the theoretical value while the average stress in the body is still very low. Thus, the crack allows sequential fracture of the atomic bonds at the crack tip rather than the simultaneous fracture of the bonds across the entire fracture surface. Using the Inglis model (1913) for a straight surface crack which opens under an applied stress, σ, to an eliptical cross section with a semimajor axis of a and a radius at the crack tip of ρ, and setting the crack tip stress equal to the theoretical stress, σ_t, we have:

$$\sigma(a)^{1/2} = (\sigma_t/2)(\rho)^{1/2}. \tag{12.7}$$

The thermodynamic and the local stress approaches both lead to the expression, $\sigma(a)^{1/2}$ = constant for a given glass as a fracture criterion. The validity of this expression for glass has been established by several investigators as shown in Fig. 12.3.

The fracture criterion for a brittle solid was put in a somewhat more useful form by Irwin (1958) who showed that the stresses at a point in the vicinity of any crack in a linear elastic solid must have the form

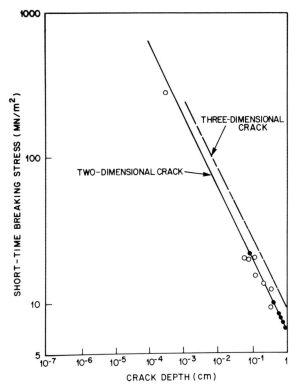

Fig. 12.3 Dependence of strength on crack size. Data from Shand (1954) and Anderson (1959).

$$\sigma = \frac{K_I}{(2\pi r)^{1/2}} f(r, \theta), \tag{12.8}$$

where K_I is a function of the loading and of the geometry of the solid and the embedded crack and r and θ are the polar coordinates of the point in question with the origin located at the crack tip. Applying the fracture criterion that the stresses in the vicinity of the crack must reach a critical, constant value dependent only upon the material results in the expression

$$K_I = K_{Ic}, \tag{12.9}$$

where K_{Ic} is the critical value of the parameter K_I. This formulation is part of the field known as fracture mechanics. K_I is called the stress intensity factor and K_{Ic}, the critical stress intensity factor or fracture toughness. For the straight crack discussed above:

$$K_I = \sigma(\pi a)^{1/2}. \tag{12.10}$$

Note that K_I is a function of both the applied stress and the crack size. Values for K_{Ic} for glass depend upon composition but have been found to lie in the range of 0.6 to 0.9 MN/m$^{3/2}$ (Wiederhorn, 1969). The advantage of the fracture mechanics form is that it facilitates consideration of various crack geometries and loading situations.

As an example of the use of fracture mechanics, the critical size of a crack in a glass fiber can be estimated. The stress intensity factor for a semicircular crack at the surface of a fiber with the crack plane normal to the fiber axis is given by (Irwin, 1958):

$$K_I = Y \sigma(a)^{1/2}, \tag{12.11}$$

where Y is a dimensionless geometric constant (1.24). For a fracture stress of 3.5×10^3 MPa (0.5×10^6 psi) and using a value of 0.8 MN/m$^{3/2}$ for K_{Ic} (Wiederhorn, 1969), the above relation predicts $a = 0.03$ μm when $K_I = K_{Ic}$.

12.3.2 Size Effects

As described above, fracture mechanics can be used to predict the failure of an elastic solid containing a crack of known size. However, to understand the strength of glass, one must also recognize that glass contains many flaws with a distribution of flaw sizes. Because of the monotonic dependency of K_I on crack size, the largest (i.e., most severe) flaw will reach the critical size first but there is generally no *a priori* means of locating and determining the size of this largest crack. In addition, the largest crack will usually be different in each glass body. Thus, glass strength is inherently a statistical quantity and one must apply statistics to describe glass strength. Assuming that the flaws are independent and randomly distributed in the solid, and that fracture occurs at the most severe flaw, it can be shown that the cumulative probability that a solid of volume V will fracture at an applied stress of σ or less can be written in the form:

$$F(\sigma, V) = 1 - \exp[-VH(\sigma)], \tag{12.12}$$

where $H(\sigma)$ represents the distribution of strengths (or most severe flaws) in otherwise identical bodies. This is simply a description of a "weakest link" model (Pierce, 1926). Weibull (1939) empirically postulated an "extreme value" relation (Fisher and Tippett, 1928):

$$H(\sigma) = (1/V_0)(\sigma/\sigma_0)^m, \tag{12.13}$$

where m, V_0, and σ_0 are constants. The resulting "Weibull" expression

$$F(\sigma, V) = 1 - \exp\left[-\frac{V}{V_0}\left(\frac{\sigma}{\sigma_0}\right)^m\right] \tag{12.14}$$

or

$$\ln \ln \frac{1}{1 - F} = \ln \frac{V}{V_0} + m \ln \frac{\sigma}{\sigma_0} \qquad (12.15)$$

has often been applied to the analysis of the fracture of brittle materials. In the case of glass, surface flaws are normally found to control the fracture strength and thus V in Eqs. (12.13)–(12.15) can be replaced by A, L or d, whichever variable principally characterizes the surface area. Thus, if the median strengths (σ at $F = 50\%$) are σ_1 and σ_2 for two sets of samples:

$$\frac{\sigma_1}{\sigma_2} = \left(\frac{A_2}{A_1}\right)^{1/m} = \left(\frac{L_2}{L_1}\right)^{1/m} = \left(\frac{d_2}{d_1}\right)^{1/m}, \qquad (12.16)$$

depending on the primary variable. It can be shown (Irwin, 1958) that m is inversely proportional to the coefficient of variation. For a fiber having a narrow strength distribution along its length characterized by a single mode Weibull distribution, a steep slope (large m) will result. In addition, the variation in the median strength with area (or L or d), e.g., the slope $\ln \sigma_m / \ln A = 1/m$, will be small.

   Griffith (1920) showed that, in practice, glass fibers displayed a sharply decreasing strength with increasing diameter when the gauge length was fixed, e.g., the coefficient of variation was very large. He postulated that the strengths were being controlled by surface flaws and that the probability of finding flaws increased with the amount of fiber under test as we have just shown. For many years following this work it was generally conceded that these flaws were intrinsic and that the fiber diameter effectively put an upper limit on the maximum flaw size and thus controlled strength. The careful experimental investigations of Otto in 1955 and Bartenev and Bovkunenko in 1956, followed by other work of Thomas (1960), Bartenev and Izmailova (1962), Metcalfe and Schmitz (1964), and Cameron (1965), showed conclusively that the observed diameter dependencies could be reduced or removed and therefore were the result of variations in the extrinsic flaw populations. The results of these investigations are shown in Fig. 12.4 (Kurkjian, 1977). In this figure, the original diameter or length dependence has been converted into a surface area dependence. This allows studies of length and diameter variations to be shown in the same figure. Otto (1955) and Thomas (1960) suggested that since fibers of different diameters were normally drawn at different temperatures, the apparent dependence of strength on diameter (or area) might not be intrinsic. By drawing at constant temperature, they were able to eliminate most of the variation in strength with diameter in the range of diameters where Griffith (1920) found the greatest change.

   Fibers with differing surface areas are more easily produced under constant drawing conditions if changes in length are studied rather than changes in diameter. This is illustrated by the early work of Anderegg

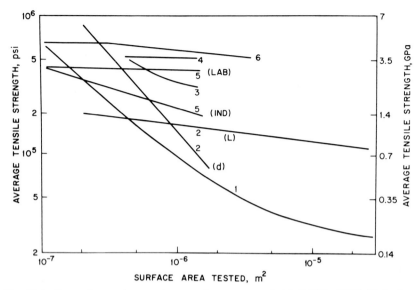

Fig. 12.4 Dependence of strength on surface area tested. (1) Griffith (1920) (2) Anderegg (1939) (3) Otto (1955) (4) Thomas (1960) (5) Bartenev and Izmailova (1962) (6) Metcalfe and Schmitz (1964). (From Kurkjian, 1977.)

(1939) (Fig. 12.4). His results show a substantially smaller area dependence when this change is brought about by varying the length rather than by varying the diameter. It is suggested then that additional practical variables may enter when diameter changes are made. Bartenev and Bovkunenko (1956) suggest that strength variations due to diameter and length changes result from different mechanism and thus are described by *two different expressions* (if $d < 100$ μm):

$$\sigma = A + B/d, \tag{12.17}$$

$$\sigma = C/L^{1/n}. \tag{12.18}$$

Although they have suggested that the effect of diameter changes on strength is the result of a quenched, oriented surface layer, this effect is difficult to show conclusively. In addition since this effect is not found when $d > 100$ μm, it will not be considered further.

The above results indicate that very major improvements in the *uniformity* of fiber strength can be accomplished by very strict control of the processing conditions. Unfortunately, the longest gauge length for which this very uniform behavior was observed was 20 cm and, since the number of samples tested was about 100, the total tested length was only about 20 m (Bartenev and Ismailova, 1964) rather than the 1-km lengths of interest for optical fiber applications.

This essentially was the status of the understanding in the early 1970s when work on the strength of optical waveguide fibers was initiated. There are obviously difficulties in predicting the strength of long waveguide fibers from this information.

(1) While the actual data taken covered two orders of magnitude in gauge length (0.1 to 10 cm), extrapolations of four orders of magnitude to kilometer lengths are required.

(2) Most of this early work was done on fibers drawn from crucible melts (soda-lime or E-glass) rather than high silica glass drawn from rod preforms.

In addition, in order to use fibers in optical cables without introducing flaws on the surface, some sort of protective coating obviously has to be applied in a manner which itself does not cause flaws. Considerable success with coatings and coating techniques has been achieved as described in Chapter 10. All of the results to be shown below will be on fibers which have been coated in-line with one of several coatings by various techniques of applications.

12.3.3 High Strength, Long Length Optical Fibers

The first works on the strength of fibers for optical communications were those of Maurer in 1974 and 1975 (Maurer et al., 1974; Maurer, 1975). He studied 125-μm-diameter furnace-drawn optical waveguide fibers with a 3-μm Kynar coating (Maurer, 1975). He made direct tensile strength measurements of 135 0.6-m-long sections selected in a "pseudorandom fashion from 1 km of fiber." In addition to these direct tensile strength measurements, Maurer recorded some data for proof testing (see Section 12.5.1) of 1-km-long fibers of the same type at three different stress levels. He found reasonable correlation between the proof test results (shifted to 0.6 m) and the low strength tail of the actual 0.6-m-length distribution. These data are shown in Fig. 12.5. While these results were encouraging and useful, the actual strengths for kilometer length fibers were quite low ($\sim$210 MPa at 50% failure probability). In fact, Maurer (1975) suggested that the strengths which he obtained were dominated by surface damage after fabrication.

Following Maurer's work, an attempt was made to (1) control the quality of the surface in these very long fibers more carefully, and (2) verify the weakest link and Weibull relationships more thoroughly. This is important in combining the strength data on different gauge lengths.

Recently (Kurkjian et al., 1976), fairly detailed studies of this sort have been performed on 100-μm-diameter silica fibers carefully drawn and coated. Silica rather than clad optical fibers were used in these studies, but

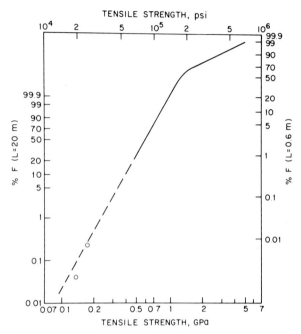

Fig. 12.5 Cumulative failure probability for Kynar-coated optical waveguide fibers. (Maurer, 1975).

since the critical outer surface in both cases is silica, similar behavior would be expected. In Figs. 12.6 and 12.7 data are plotted for three different lengths tested ($L = 0.04$, 0.75, and 20 m). In Fig. 12.7 the data for $L = 20$ m are plotted directly while the data for $L = 0.04$ and $L = 0.75$ have been shifted by ln L_1/L_2 according to Eq. (12.16). Except for the data for $L = 0.04$ m in which there were difficulties in gripping the fibers (subsequently overcome), the general continuity of the resulting curve verifies the weakest link hypothesis and the curve in the figure then represents $H(\sigma)$ in Eq. (12.12). The curvature of this plot, however, indicates that a single Weibull distribution is not sufficient.

The general procedure followed in this (Kurkjian *et al.*, 1976) and a subsequent investigation (Schonhorn *et al.*, 1976) was to draw and coat essentially a 1-km-long fiber under a given set of conditions. The entire fiber was broken in tension in 20-m gauge lengths. From the verification of the weakest link model above, the strength of the weakest of the 20-m lengths represents the strength of the kilometer length. Changes in the overall drawing and coating parameters were made and the process was repeated. Data for fibers drawn and tested in this way are shown in Figs. 12.8, 12.9, and 12.10. The actual strength level for these fibers is somewhat higher than that found above for E-glass, but is in line with that found by Hillig

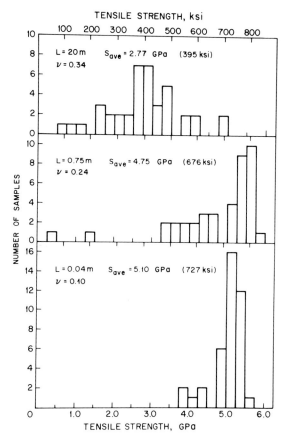

Fig. 12.6 Distributions of strength for different gauge lengths. Epoxy-acrylate-coated fused silica fibers (Kurkjian *et al.*, 1976).

(1961) and Morley *et al.* (1964) for fused silica fibers. Of much greater importance, however, is the fact that the laser-drawn, fire-polished Suprasil fibers shown in Fig. 12.8(d) exhibit a narrow, unimodal strength distribution with very high, long length strength.

Although fibers of this general sort (e.g., high strength, unimodal distribution in a 1-km length) have been produced many times, it is not simple to do so. All inputs to the drawing process must be controlled, e.g.,

(1) bulk glass quality—preferably synthetic silica.

(2) surface glass quality—the surface of the preform rod should be fire polished or etched.

(3) "clean" heat source, for example, a CO_2 laser.

(4) coating and coating applicator—nondamaging (chemical or mechanical) coating and application technique. The coating must at least af-

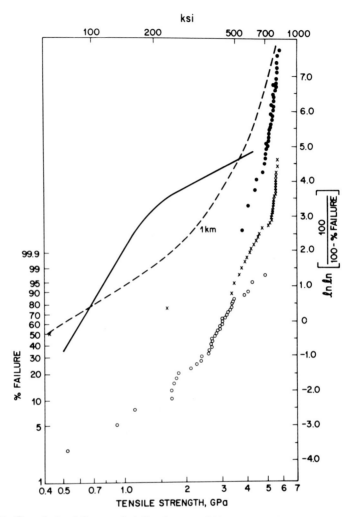

Fig. 12.7 Cumulative failure probability for coated fibers. Solid line from Maurer (1975). ○ = 20 m, × = 0.75 m, ● = 0.04-m gauge lengths. Data as shown are in terms of 20 m gauge length. Dashed curve represents data shifted to 1 km gauge length. (Kurkjian *et al.*, 1976).

ford mechanical protection. Although several satisfactory coatings have been used, a flexible tip applicator is the only satisfactory coating technique so far described.

The actual steps taken in consideration of the above factors are shown in Figs. 12.8, 12.9, and 12.10. It can be seen that each step is important. These kilometer long fibers, 100 μm in diameter were drawn from approximately 30 cm of a 0.7-cm rod. At 0.4 m/sec drawing rate the ~100 to 1 diameter

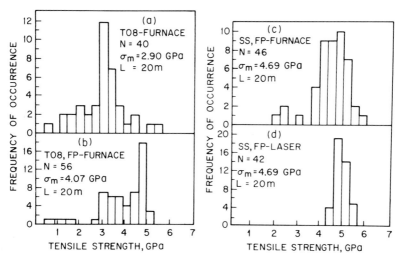

Fig. 12.8   Fiber strength distribution as a function of drawing and preform condition (Schonhorn *et al.*, 1976).

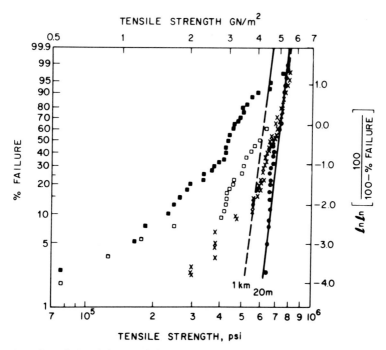

Fig. 12.9   Cumulative failure probability plot of data in Fig. 12.8 ■ 12.8(a), □ 12.8(b), × 12.8 (c), ● 12.8(d). (Schonhorn *et al.*, 1976).

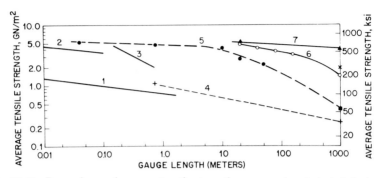

Fig. 12.10 Dependence of average tensile strength on gauge length tested. 1: Anderegg (1939). 2: Metcalfe and Schmitz (1964). 3 and 4: Maurer (1975). 5, 6, and 7 are a, c, and d, respectively, from Fig. 12.8.

attenuation requires ~35 min. Although some surface damage is repaired during melting (or removed by vaporization) some flaws may remain. It is thus important that the surface of the starting preform rod be of good quality. In addition, since a large amount of new surface is formed (~50 times that of the original rod surface), bulk glass quality is also important. The pronounced difference in strength between laser and furnace drawn fibers suggests that, during the ~35-min drawing, the fiber surface deteriorates in some way in the furnace atmosphere. Preliminary experimental and theoretical assessments of this process have been carried out (Holloway, 1959; Maurer, 1977).

After the initial drafts of this chapter had been written, a significant development in the ability to draw high strength fibers in an electric furnace was made by DiMarcello and Hart (1978). A new zirconia susceptor induction furnace (Runk, 1977) was used in place of the graphite resistance furnace previously used. Numerous kilometers of silica fiber exhibiting a unimodal strength distribution with a minimum strength of 3.5 GPa (essentially identical to the laser drawn fiber) have been prepared demonstrating the consistency with which high strengths can be achieved with this new furnace. In agreement with requirement (3) above, it is suggested that at least part of the reason for these high strengths is the cleaner, "static" atmosphere in the new furnace.

Several other investigators have studied fairly long lengths of optical fibers of the sort described above (Kao *et al.*, 1976; Tanaka *et al.*, 1976; Kalish *et al.*, 1977; Justice, 1977). In general, measurements have included both short (0.6 m) and long (≈1 km) tensile tests and long length proof tests as originally done by Mauer (1975). Kao *et al.* (1976) and Kalish *et al.* (1977) both found rather poor agreement between long and short length test results. One of the apparent problems was the multimode strength distributions obtained with the short length tests. Kalish *et al.* (1977) and Tariyal and Kalish (1977a) have suggested an analytical procedure for improving the ability to predict long length strength from short length data when bimodal distributions are present. These extrapolations are risky since it is difficult to assess the amount of short length testing which will provide sufficient statistical information to enable a reliable extrapolation to be made to long lengths. In general there is clearly no easy alternative to obtaining a complete probability distribution—the total amount of fiber about which information is required must be tested. The more complete the testing (e.g., number and length of samples) the more complete will be the information obtained. If the requirement is simply that a minimum value of strength be available in a given length, proof-testing can be used. Thus, a given minimum value of strength can be guaranteed even though nothing is known about $H(\sigma)$ (Tariyal *et al.*, 1977). On the other hand, if the overall strength distribution and its reproducibility have been previously well-characterized, a small number of short samples may be tested to assure that a given fiber conforms to this *general* characteristic.

12.4 TIME-DEPENDENT FRACTURE

The local environment at the surface of a glass can have a substantial detrimental effect on the strength. The basic phenomenon involved is the slow growth of flaws in an active environment (often referred to as "subcritical crack growth"). This slow flaw growth results in fracture at a lower stress than would occur in an inert environment where flaws are stable until the Griffith criterion is satisfied and unstable rapid fracture takes place. It is this slow flaw growth that leads to time-dependent fracture in glass.

There are two common manifestations of time-dependent fracture which relate to the manner in which the stress is applied. If the applied stress is maintained at a constant value below the inert strength then the glass can fracture after a period of time in an active environment and this behavior is known as "static fatigue." On the other hand, if the applied stress is continually increasing (such as in a standard tensile test), then the strength in an active environment will be lower than in an inert environment and this behavior is known as "dynamic fatigue."

Water in the environment is primarily responsible for slow crack growth in glass. Relatively low concentrations of water can cause long term failures by static fatigue at stresses well below one-half of the inert strength. Fused silica shows a time-dependent strength so that this phenomenon is of concern for the survival of optical fibers under the stresses and environments encountered during manufacturing, installation, and service.

12.4.1 General Characteristics and Mechanism

There are several mechanisms that have been proposed to explain static fatigue in glass which take into account the concentration of hydroxyl ions (the active chemical species), the rate of transport of the bulk environment to the flaw tip, and the details of bond-breaking reactions. These mechanisms are critically reviewed by Pavelcheck and Doremus (1976), Adams and McMillan (1977), and Wiederhorn (1978) and will not be discussed here. Also, the reviews of Hillig (1962), Otto (1965), Phillips (1965), Mould (1967), Wachtman (1974), and Wiederhorn (1975) present many details of time-dependent fracture phenomena in glass and contain extensive lists of references to original research papers.

The most pertinent observations about static fatigue in glass relative to the development of fused silica optical fibers are as follows:

(1) There is no slow flaw growth (and static fatigue) at sufficiently low temperatures (e.g., liquid nitrogen) and/or dry environments (e.g., vacuum); this defines the inert strength of glass.

(2) A number of investigators report that static fatigue does not occur in glass below a certain stress level, called the static fatigue limit. This limit appears to be in the range of about 0.15 to 0.25 of the inert strength, but this is still a matter of some controversy.

(3) Static fatigue occurs in various environments (alcohols, aqueous solutions of various pH and different relative humidities). An increasing rate of static fatigue occurs with an increase in hydroxyl ion concentration.

(4) Glass composition has a strong effect on static fatigue; fused silica is the most resistant of the glasses in water.

The preponderance of the experimental evidence to date supports the stress-corrosion mechanism, developed by Charles (1958a,b) and later by Charles and Hillig (1961), and Hillig and Charles (1965). The stress corrosion mechanism postulates a stress and temperature-dependent chemical reaction which controls the rate of growth of preexisting flaws. In the previous section it was noted that the maximum local stress in a glass component under stress, and hence the site for fracture, occurs at the tip of the maximum size flaw. In most glass components, and particularly in optical fibers because of the high quality of the bulk glass from which they are drawn, the flaws which control fracture are at the glass surface. The analytical descriptions of stress corrosion (Charles, 1958b; Charles and Hillig, 1961) also utilize the concept of an elastic stress concentration at the tip of an elliptical surface flaw which is accessible to the active environment and leads to relationships between the nominal applied stress (away from the flaw) and the time-to-fracture. The maximum size surface flaw gives the highest stress concentration and hence the fastest rate of stress assisted chemical breaking of primary atomic bonds.

12.4.2 Analytical Descriptions

The flaw growth model initially proposed by Charles (1958a,b) *assumes that the flaw growth rate is a power function of stress* and an exponential function of temperature:

$$V = C^1(\sigma_m)^n \exp(-Q_1/RT), \qquad (12.19)$$

where V = flaw growth velocity, C^1 and n are material/environment constants, σ_m = normal tensile stress at the crack tip, Q_1 = activation energy, R = gas constant, T = absolute temperature. In the case of a constant applied stress (σ_{fs}) at constant temperature, the time-to-failure (t_{fs}) in static fatigue derived by Charles (1958b) from Eq. (12.19) is

$$\log t_{fs} = -n \log \sigma_{fs} + \log C_1, \qquad (12.20)$$

where C_1 is material/environment constant. Similarly, for a constant applied stress rate (dynamic fatigue) Charles (1958c) showed that fracture strength (σ_{fd}) and the time-to-fracture (t_{fd}) are related by

$$\log t_{fd} = -n \log \sigma_{fd} + \log C_2, \qquad (12.21)$$

where C_2 is another constant. It can be shown (Davidge *et al.*, 1973; Charles, 1975) that the constants C_1 and C_2 are related by

$$\log C_2 = \log C_1 + \log(n + 1). \qquad (12.22)$$

The analytical relationship between static fatigue and dynamic fatigue is important because measurements of the latter are far less time-consuming than for the former, so that, in principal the characteristic static fatigue parameters of a glass fiber can be determined by dynamic fatigue measurements. Although in some cases (Krause and Kurkjian, 1977; Kalish and Tariyal, 1977) reasonable agreement is obtained, there is still controversy over whether the values of n and C_1 determined by dynamic fatigue measurements are always equivalent to those measured by static fatigue. Some evidence indicates that certain polymer coatings lead to a discrepancy in the correlation between dynamic and static fatigue (Tariyal and Kalish, 1978).

In two later papers (Charles and Hillig, 1961; Hillig and Charles, 1965), a thermodynamic description is given for a thermally activated rate of corrosion which depends upon the local stress state and curvature at the surface under attack. In this case the flaw growth rate is an *exponential function of stress and temperature*:

$$V = V_0 \exp(-Q_2 + V_2\sigma_m - V_M\gamma/\rho)/RT, \qquad (12.23)$$

where V_0 = stress-free rate of corrosion of a planar surface; Q_2 = stress-free activation energy, V_2 = activation volume, V_M = molar volume of glass, γ = interfacial surface energy, and ρ = radius of the flaw tip. In the case of a constant applied stress at constant temperature, the time-to-failure in static fatigue derived by Charles and Hillig (1961) from Eq. (12.23) is of the form:

$$\log t_{fs} = -n'\sigma_{fs} + \log C_3, \tag{12.24}$$

where n' and C_3 are material/environment constants. Consequently, there are two distinctly different models that relate the time-to-failure to the strength: one model, Eq. (12.20) gives a log–log relationship while the other, Eq. (12.24), gives a semilogarithmic relationship.

12.4.3 Some Experimental Results

Static fatigue test results typically show a large scatter in failure times for replicate fiber specimens tested under a given set of conditions, (Fig. 12.19 (Wang and Zupko, 1978)) due to the inherent statistical distribution of initial maximum flaw sizes in the otherwise identical specimens. Consequently, in order to test the validity of Eqs. (12.20) and (12.24), failure times of equal probability must be analyzed as a function of strength; this should give a comparison of equivalent initial flaw sizes tested at different stress levels. The median time-to-failure for the data in Fig. 12.11 is plotted in Fig. 12.12 and fit to Eqs. (12.20) and (12.24) with linear regression analyses. Both the power law and exponential flaw growth models lead to excellent correlations with the data, which extends to median t_{fs} of 40 days, but these two models give significantly different predictions of long-term static fatigue performance. For example, the log t_{fs} vs log σ_{fs} equation predicts a median 100-yr life on 0.6-m length specimens at a stress of 1.20 GPa (174 ksi) while the log t_{fs} vs σ_{fs} equations predicts that such a stress level would give a median life of only 10 yr. Consequently,

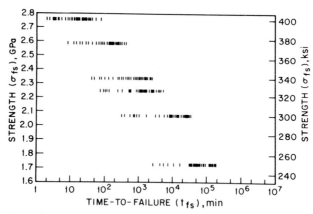

Fig. 12.11 Static fatigue of furnace-drawn TO8 clad optical fiber at 90% RH and 32°C. Fiber coating: uv cured epoxy acrylate. Gauge length: 0.6 m (Wang and Zupko, 1978).

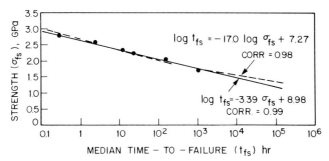

Fig. 12.12   Median time-to-failure by static fatigue of TO8 clad optical fiber at 90% RH and 32°C. Fiber coating: uv cured epoxy acrylate. Gauge length: 0.6 m.

while either model appears to be adequate for comparisons among such variables as fiber coatings, test environment or glass composition (Fig. 12.13), the more conservative semilogarithmic relationship, Eq. (12.24), must be used for lifetime predictions until some other relationship is firmly established.

12.4.4 Fracture Mechanics Analyses

Fracture mechanics provides an alternative approach to describing the time-dependent strength behavior of glass. Wiederhorn and Bolz (1970) presented a fracture mechanics analog to the Charles and Hillig (1961) stress corrosion model where the stress intensity factor (K_I) (Eq. (12.11)) at a crack tip is the factor governing the rate of crack growth. Specifically, substituting the expression for crack size given by Eq. (12.11) into the definition of crack velocity, V,

$$V = da/dt \tag{12.25}$$

results in

$$V = \frac{2K_I}{\sigma^2 Y^2} \frac{dK_I}{dt}. \tag{12.26}$$

The time-to-failure under constant applied stress (σ_{fs}) is obtained by integrating Eq. (12.26):

$$t_{fs} = \left(\frac{2}{\sigma_{fs}^2 Y^2}\right) \int_{K_{Ii}}^{K_{Ic}} \left(\frac{K_I}{V}\right) dK_I, \tag{12.27}$$

where Y = geometric constant, K_{Ii} = initial stress intensity factor, and K_{Ic} = critical stress intensity factor as defined in Eq. (12.9), and

$$K_I = \alpha + \beta \ln V \tag{12.28}$$

where α and β can be related to the parameters in Eq. (12.23). Integration

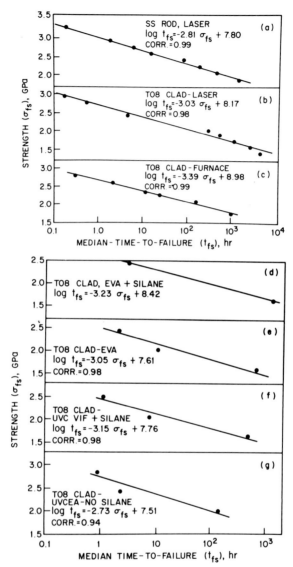

Fig. 12.13 Median time-to-failure by static fatigue for various silicas, coatings, drawing furnaces and environments. [(a–c): Wang and Zupko (1978), (d–f): Tariyal and Kalish (1978), (g): D. Kalish and B. K. Tariyal (personal communication, 1978).]

of Eq. (12.27), using Eq. (12.28), and making use of the fact that for a given initial flaw:

$$K_{Ii}/\sigma_{fs} = K_{Ic}/\sigma_{Ic} \qquad (12.29)$$

leads to

$$t_{fs} = [C_4/\sigma_{fs}^2](C_5\sigma_{f5} + 1)/\exp(C_5\sigma_{fs}), \qquad (12.30)$$

where $C_4 = (2\beta^2/Y^2) \exp(\alpha/\beta)$ and $C_5 = K_{Ic}/(\sigma_{Ic}\beta)$.

If a *power law* for crack velocity versus stress intensity is used [Evans (1974)] instead of Eq. (12.28).

$$V = AK_I^n \qquad (12.31)$$

than the time-to-failure is:

$$\log t_{fs} = -n \log \sigma_{fs} + \log C_6, \qquad (12.32)$$

where $C_6 = (2/AY^2(n - 2)) (K_{Ic}/\sigma_{Ic})^{2-n}$ which is the fracture mechanics analog to the Charles (1958b) Eq. (12.20) and $C_6 = C_1$.

The fracture mechanics descriptions of time-dependent fracture provide a more general predictive tool than Eqs. (12.20) and (12.24) because all of the constants can be established without doing environmental tests on a particular fiber of interest. For example, K_{Ic} is measured in bulk sample fracture mechanics tests and is a material constant (dependent on composition but independent of environment). A and n in Eq. (12.31) or α and β in Eq. (12.28) can be measured by either bulk sample fracture mechanics tests or static or dynamic fatigue tests and are representative of a given glass/environment system. Y is a calculated geometrical constant. Once these parameters are established one needs only to measure the inert strength, σ_{Ic}, to calculate t_{fs} as a function of σ_{fs}. Kalish and Tariyal (1977) have illustrated this approach for an ethylene–vinyl–acetate-coated optical fiber and their results are summarized in Fig. 12.14.

The more common method found in the literature for presenting static and dynamic fatigue results is in terms of the Charles' Eqs. (12.20) and (12.21), or the fracture mechanics analog, Eq. (12.30), with attention usually focused upon the value of the exponent "n" as a measure of susceptibility to static or dynamic fatigue. Charles (1958b) noted that "when n is infinitely large flaw growth would not occur until the critical conditions for spontaneous propagation are reached . . ." or, in fracture mechanics terminology, until K_{Ic} is reached. Conversely, ". . . the lower the value of n the greater will be the delayed characteristics of the breakage phenomena."

The data in Fig. 12.13, originally plotted semilogarithmically, were reanalyzed in terms of Eq. (12.20) to show that n for fused silica optical fibers in 97% humidity at 23°C is approximately 14 to 17 for various

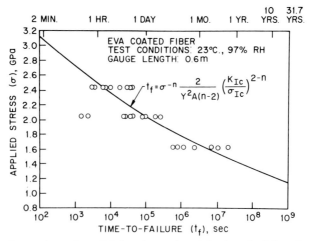

Fig. 12.14 Comparison of predicted and experimental values for failure in static fatigue of optical glass fiber at 97% RH and 23°C. Parameters for predicted curve based on dynamic fatigue results. Fiber coating: ethylene–vinyl–acetate. Gauge length: 0.6 m (Kalish and Tariyal, 1977).

coatings, drawing furnaces, and sources of glass. Comparable data (not shown here) for ambient humidity at 23°C give an n of about 21 to 24; this is in agreement with the data of Proctor *et al.* (1967) for uncoated fused silica tested in ambient humidity which give an n of 21.

12.4.5 Other Time-Dependent Strength Phenomena

12.4.5.1 Cyclic Fatigue. The term "fatigue" in engineering materials other than ceramics and glasses always refers to the degradation of strength caused by cyclic loading. Cyclic fatigue of metals or polymers is a consequence of inelastic deformation mechanisms that occur on an atomic scale such as dislocation motion in metals that lead to crack initiation, slow crack propagation and eventually to rapid fracture. The preponderance of evidence to date indicates that inelastic deformation of glass does not occur at temperatures well below the melting point. This should certainly be the case for fused silica at ambient temperature. Consequently, in glass there is no analogy to conventional cyclic fatigue. However, if a glass is subject to cyclic loading, where a portion or all of the cycle produces a tensile component of stress, then the glass can undergo dynamic fatigue during each cycle. Thus cyclic fatigue in glass should be a function of the total time a tensile stress is applied. The data of Gurney and Pearson (1948), Evans and Fuller (1974), and Evans and Linzer (1976) confirm that cyclic fatigue in glass is simply what Charles (1958c) calls dynamic fatigue.

12.4.5.2 Zero-Stress Aging. Mould (1960) showed that zero-stress aging of freshly abraded glass surfaces at room temperature could produce an increase in strength of up to about 30%, depending on the medium. Aging in water, solutions of various pH, and high humidity atmospheres increased the strength the most, while aging in low humidities increased the strength very little. These results supported some earlier studies on abraded glasses which are also discussed by Mould (1960, 1967) and it was proposed that crack tip rounding by the corrosive action of water led to the strength increases. The strength range wherein this phenomena was observed was below 100 Mpa (16 ksi).

In contrast Proctor *et al.* (1967) showed that high strength silica fibers stored in water or various humidities show a loss of strength but that the strength is completely recoverable with sufficient drying under vacuum. High strength polymer coated fused silica optical fibers also show a decrease in strength when aged in water or in high relative humidity (Wang and Zupko, 1978; Wang *et al.*, 1978; Fox *et al.*, 1977; Tariyal and Kalish, 1977b) but the results of these authors on recovery of strength after drying are contradictory. The mechanism for decrease in strength is probably absorption and adsorption of water molecules on the glass surface which then promote dynamic fatigue upon testing.

12.5 ENGINEERING DESIGN

The survival of long lengths of fibers exposed to stresses arising from the manufacture and installation of optical cables is crucial to an economically successful application of long-distance optical communication. From an engineering design point of view one needs to guarantee that long lengths of fibers have a certain minimum strength and a certain minimum static fatigue life under service conditions. It is very difficult to demonstrate compliance with these two requirements because of the size and time dependence of glass strength as discussed in Sections 12.3 and 12.4.

One way of establishing the long length strength is through a suitable statistical analysis of the short length data in order to obtain the pertinent statistical distribution parameters (Weibull, 1939; Kies, 1958; Phoenix, 1973, 1975; McClintock, 1974). These parameters can then be used to estimate long length strengths (Maurer, 1975); Kurkjian *et al.*, 1976; Tariyal and Kalish, 1977a). However, fiber strengths in general show multimodal strength distributions, arising from flaw populations of different physical origins. The predominant mode in any sample population is determined by factors such as gauge length, number of tests, and various fiber processing parameters (Tariyal and Kalish, 1977a; Kalish *et al.*, 1977) and this may not correspond to the flaw population in the entire sample which is controlling the strength in long lengths. Consequently, the estimation of the long length strength from short length data cannot be accomplished with any useful degree of reliability. In the absence of a reliable analytical estimation method, there are two possible alternatives for ensuring a minimum strength in long optical fibers, namely nondestructive inspection (NDI) and proof testing. At the present time, a suitable NDI technique to detect the appropriate flaw size range (0.02 to 2 μm) in fibers is not available and therefore, proof testing offers an attractive means of ensuring the mechanical reliability of glass fibers.

12.5.1 Theory of Proof Testing

In proof testing, an optical fiber is subjected to a tensile load greater than that expected in any subsequent step during cable manufacturing, installation, or service. The fibers which fail the proof test are rejected. The mathematical foundation for the selection of the proof test load and establishment of the proof test conditions for brittle materials have been developed by Wiederhorn (1973), Evans and Wiederhorn (1974), Evans and Fuller (1975), and Ritter and Meisel (1976). All specimens containing flaws larger than the critical flaw size will fail because the stress intensity factor at the crack tips will exceed the critical stress intensity factor K_{Ic}. For the fibers which survive the proof test, [see Eq. (12.11)],

$$K_{Ic} > K_{Ip} = \sigma_p Y(a_i)^{1/2}, \tag{12.33}$$

where K_{Ip} is the stress intensity factor developed during the proof test at the largest flaw in the surviving fiber, Y is a geometric constant, σ_p is the proof stress and a_i is the initial size of the largest flaw. Thus for any applied stress σ_a below the proof stress level, the maximum initial stress intensity factor K_{Ii} is

$$K_{Ii} < K_{Ic}(\sigma_a/\sigma_p). \tag{12.34}$$

Equation (12.34) describes the protection against short term overload on the fiber, i.e., a simple result of the fracture mechanics analysis is that the fiber will not break at instantaneous stresses below the proof test stress.

If the applied stress $\sigma_a < \sigma_p$ is maintained for a long time in an active environment then static fatigue can occur. However, since the fiber survived the proof test, a maximum possible size flaw has been guaranteed so that a a minimum guaranteed failure time due to static fatigue is also established. The development of the expression for the minimum time-to-failure follows the arguments presented in Section 12.4.4 except that Eq. (12.34) is used in place of Eq. (12.29). The result for the power law flaw growth model is:

$$t_{fs} = B\sigma_p^{n-2}\sigma_{fs}^{-n}, \tag{12.35}$$

where $B = 2/AY^2(n - 2)K_{Ic}^{(n-2)}$ and σ_{fs} is now the applied stress that leads to a static fatigue failure in time t_{fs} (min) or greater.

For a proof test to be useful in assuring a minimum strength and a minimum static fatigue life in service, it is neccessary to ensure that no fiber damage is introduced by the proof test itself (Tariyal et al., 1977; Tariyal and Kalish, 1978). Two possible types of additional damage ar the introduction of new flaws by abrasion of the fiber in the proof testing apparatus and the growth of the existing flaws due to dynamic fatigue. By measuring the strength distributions before and after proof testing it has been shown that mechanical damage does not occur in polymer-coated fibers run between rubber-faced capstans (Tariyal et al., 1977). However, the possibility of crack growth during proof testing has been confirmed by comparing the strength distribution of two groups of tensile tested fibers differing only in that one group had been prestressed in the tensile machine prior to testing to failure (Tariyal and Kalish, 1978).

The effect of crack growth during proof testing in a normal laboratory atmosphere can be calculated by using the model of Evans and Fuller (1975) which leads to:

$$\sigma_g = \sigma_p \left\{ 1 - \left[\frac{n-2}{n+1} \right] \left[\frac{AY^2 K_I^{(n-2)} \sigma_p^3}{2\dot{\sigma}_u} \right] \right\}^{1/n-2} \tag{12.36}$$

where σ_g is the guaranteed inert strength after proof testing and $\dot{\sigma}_u$ is the unloading rate during proof testing. It is seen that the guaranteed strength is equal to the proof stress for a proof test conducted in vacuum ($n = \infty$) or for a proof test where $\dot{\sigma}_u = \infty$. For a finite unloading rate and laboratory atmosphere, however, the guaranteed strength is lower than the proof stress. Some examples of the effect of n and $\dot{\sigma}_u$ on the guaranteed strength are shown in Fig. 12.15. The precipitous drop in σ_g at the various combinations of $\dot{\sigma}_u$ and n reflect the condition from Eq. (12.36) when,

$$\dot{\sigma}_u = \left[\frac{n-2}{n+1} \right] \left[\frac{AY^2 K_I^{(n-2)} \sigma_p^3}{2} \right], \tag{12.37}$$

i.e., when unloading is slow enough so that a flaw that was just subcritical at the onset of unloading can grow completely across the specimen during the unloading period. At a proof stress level of 240 MPa (35 Ksi) and an unloading rate of 200 MPa/sec, the crack growth due to dynamic fatigue is negligible and the proof stress can be used for design purposes. If the proof stress level were raised by an order of magnitude then the effect of dynamic fatigue would be significant and either the proof test conditions should be modified or the actual guaranteed strength should be calculated for design purposes.

The benefits of proof testing of long length optical fibers on the subsequent survival of these fibers in lightguide cables have been documented by Tariyal et al. (1977). The cable tests showed that cables containing fibers which had been proof tested at 207 MPa showed no fiber breaks

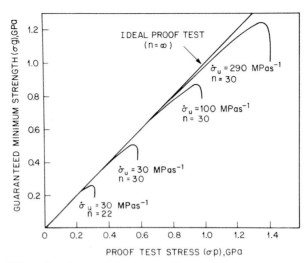

Fig. 12.15 Effect of crack growth during proof testing on guaranteed minimum strength (Tariyal and Kalish, 1978).

until the strain on the cable core (containing the fibers) was approximately equal to or greater than the prior proof test strain level. In contrast, a cable containing fibers that were not proof tested showed fiber breaks at cable core strains well below the equivalent proof test strain level.

12.5.2 Selection of a Suitable Proof Stress

Selection of a suitable proof stress level is extremely important in that it determines the reliability of the final cable as well as the economics of the cable manufacturing. It is obvious that the proof stress must be greater than the maximum instantaneous stress that a fiber will experience during the manufacturing and installation. This stress level must be determined by suitable analytical or experimental techniques. In addition, the maximum constant stress which a fiber could experience in service and the required service life (e.g., 40 yr) can be used in Eq. (12.35) to establish another criterion for the magnitude of the proof stress. Any constant long-term stress is likely to be far less in magnitude than the maximum instantaneous stress encountered in manufacturing. However, the long-term flaw growth characteristics could put a greater restriction on the maximum tolerable initial flaw size and in turn determine the minimum acceptable proof stress level. As the proof stress level is increased over the minimum acceptable level, a greater safety factor is built into the fiber performance but scrap rates in manufacturing might increase. The final selection of a proof stress level must balance these factors.

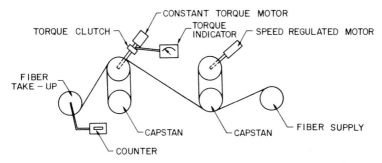

Fig. 12.16 Schematic of proof testing apparatus (Tariyal *et al.*, 1977).

12.5.3 Proof Test Apparatus

An apparatus for proof testing was constructed and successfully used by Tariyal *et al.* (1977). The apparatus, Fig. 12.16, consists of a double capstan arrangement at both the supply and take-up sides, which allows the fibers to start and finish on the reels with very little tension. The fiber is pulled from a supply reel mounted on ball-bearing centers, the drag being controlled by a hysteresis brake. The pulling force is developed at the supply capstan with a speed-regulated motor. A single wrap of fiber around two soft rubber wheels provides sufficient friction to prevent the transfer of the proof stress to the pay-off reel. The fiber then passes around the capstan at the take-up side and onto the take-up reel. A variable speed, constant torque dc motor coupled through a hysteresis clutch to the capstan provides the desired proof stress.

12.5.4 Life Prediction Diagrams

The relation between the time-to-failure and the applied stress for a proof tested fiber (Eq. (12.35)) can be used to provide a graphical aid for predicting service life. In Fig. 12.17, failure time is plotted as a function of applied stress for three different proof stress levels. Such a diagram is called a life prediction diagram (Evans and Wiederhorn, 1974).

Life prediction diagrams can be used in several ways to predict the lifetimes under certain residual loads, or to specify the maximum allowable residual stress in an installed cable for a specific service life. In addition, if the residual load is constrained at some minimum value then the life prediction diagrams can be used to determine the proof stress which will assure adequate service life under the given conditions. The reliability of all these predictions, however, depends on, (1) conducting a proper proof test and (2) obtaining accurate values of the material parameters to use in Eq. (12.35).

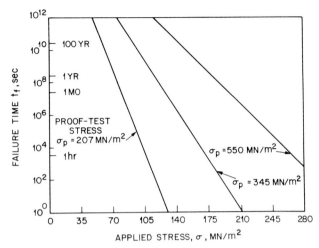

Fig. 12.17 Life prediction diagram for optical fibers based on proof test stress (Tariyal *et al.*, 1977).

REFERENCES

Adams, R., and McMillan, P. W. (1977). Review: Static fatigue in glass. *J. Mater. Sci.* **12**, 643.

Anderegg, F. O. (1939). Strength of glass fiber. *Ind. Eng. Chem.* **31**, 290.

Anderson, O. L. (1959). The Griffith criterion for glass fracture. *In* "Fracture" (B. L. Auerbach *et al.*, eds.), p. 331. Wiley, New York.

Bartenev, G. M. and Bovkunenko, A. N. (1956). Various factors which affect the strength of glass fibers. *Zh. Tekh. Fiz.* **26**, 2508.

Bartenev, G. M. and Izmailova, L. K. (1962). Flawless glass fibers. *Dokl. Akad. Nauk SSSR* **146**, 1136.

Bartenev, G. M., and Izmailova, L. K. (1964). Nature of the high strength and structure of glass fibers. *Sov. Phys.—Solid State (Engl. Transl.)* **6**, 920.

Boyle, W. S. (1977). Light-wave communications. *Sci. Am.*, **237**, 40.

Brenner, S. S. (1951). Properties of whiskers. *In* "Growth and Perfection of Crystals." (R. H. Doremus, B. W. Roberts, and D. Turnbull, eds.), p. 157. Wiley, New York.

Cameron, N. (1965). An investigation into the effect of environmental treatments on the strength of E-glass fibres. Ph.D. Thesis, Dept. of Chemical Engineering, University of Illinois, Urbana.

Charles, R. J. (1958a). Static fatigue of glass. I. *J. Appl. Phys.* **29**, 1549.

Charles, R. J. (1958b). Static fatigue of glass. II. *J. Appl. Phys.* **29**, 1554.

Charles, R. J. (1958c). Dynamic fatigue of glass. *J. Appl. Phys.* **29**, 1657.

Charles, R. J. (1975). Comment on 'Stress corrosion of a low temperature solder glass' by R. R. Tummula. *J. Non-Cryst. Solids* **19**, 273.

Charles, R. J., and Hillig, W. B. (1961). The kinetics of glass failure by stress corrosion. *In* "Symposium on the Mechanical Strength of Glass and Ways of Improving It," p. 511. Union Scientifique Continentale du Verre,

Davidge, R. W., McLaren, J. R., and Tappin, G. (1973). Strength-probability-time (SPT) relationship in ceramics. *J. Mater. Sci.* **8**, 1699.

DiMarcello, F. V., and Hart, A. C. (1978). Furnace-drawn silica fibers with tensile strengths >3.5 GPa (500 ksi) in 1 km lengths. *Electron. Lett.* **14**, 578.

Evans, A. G. (1974). Slow crack growth in brittle materials under dynamic loading conditions. *Int. J. Fract.* **10**, 251.

Evans, A. G., and Fuller, E. G. (1974). Crack propogation in ceramic materials under cyclic loading. *Metall. Trans.* **5**, 27.

Evans, A. G., and Fuller, E. R. (1975). Proof testing—the effect of slow crack growth. *Mater. Sci. Eng.* **19**, 69.

Evans, A. G., and Linzer, M. (1976). High frequency cyclic crack propogation in ceramic materials. *Int. J. Fract.* **12**, 217.

Evans, A. G., and Wiederhorn, S. M. (1974). Proof testing of ceramic materials—an analytical basis for failure prediction. *Int. J. Fract.* **10**, 379.

Fisher, R. A., and Tippett, L. H. C. (1928). Limiting forms of the frequency distribution of the largest or smallest members of a sample. *Proc. Cambridge Philos. Soc.* **24**, 180.

Fox, A., Fuchs, E. O., and Key, P. L. (1977). Strength and static fatigue of epoxy acrylate coated optical glass fibers. *Top. Meet. Op. Fiber Transm., 2nd, 1977* Paper TUA5.

Griffith, A. A. (1920). The phenomena of rupture and flow in solids. *Philos. Trans. R. Soc. London, Ser. A* **221**, 163.

Gurney, C., and Pearson, S. (1948). Fatigue of mineral glass under static and cyclic loading. *Proc. R. Soc. London, Ser. A* **192**, 537.

Hillig, W. B. (1961). Strength of bulk fused quartz. *J. Appl. Phys.* **32**, 741.

Hillig, W. B. (1962). Sources of weakness and the ultimate strength of brittle amorphous solids. *In* "Modern Aspects of the Vitreous State" (J. D. MacKenzie, ed.), Vol. 2, p. 152. Butterworth, London.

Hillig, W. B. and Charles, R. J. (1965). Surfaces, stress-dependent surface reactions, and strength. *In* "High Strength Materials" (V. F. Zackay, ed.), p. 682. New York.

Holloway, D. G. (1959). The strength of glass *Philos. Mag.* [7] 4, 1101.

Holloway, D. G. (1973). "Physical Properties of Glass." Springer-Verlag, Berlin and New York.

Inglis, C. E. (1913). Stresses in a plate due to cracks and sharp corners. *Trans. Inst. Nav. Architects* **55**, 219.

Irwin, G. R. (1958). Fracture. *Encycl. Phys.* **6**, 55.

Irwin, G. R. and Wells, A. A. (1965). A continuum mechanics view of crack propagation. *Met. Rev.* **10**, 223.

Justice, B. (1977). Strength consideration of optical waveguide fibers. *Fiber Integ. Opt.* **1**, 115.

Kalish, D., and Tariyal, B. K. (1977). Static and dynamic fatigue of fused silica optical fibers, *79th Annu. Am. Ceram. Soc., 1977* Paper 45-G-77.

Kalish, D., Tariyal, B. K., and Pickwick, R. O. (1977). Strength distributions and gage length extrapolations in optical fibers. *Am. Ceram. Soc. Bull.* **56**, 491.

Kao, C. K., Maklad, M., and Reed, T. (1976). Testing of tensile strength of optical fiber waveguides. *Program Int. Wire & Cable Symp., 25th, 1976* p. 223.

Kies, J. A. (1958). "The Strength of Glass," NRL Rep. 5093. Nav. Res. Lab., Washington, D.C.

Krause, J. T., and Kurkjian, C. R. (1977). Dynamic and static fatigue of high strength epoxy-acrylate coated fused silica fibers. *Top. Meet. Opt. Fiber Transm., 2nd, 1977* Paper TuA3.

Kurkjian, C. R. (1977). Tensile strength of polymer-coated fibers for use in optical communications. *Proc. Int. Congr. Glass, 11th, 1977* p. 469.

Kurkjian, C. R., Albarino, R. V., Krause, J. T., Vazirani, H. N., DiMarcello, F. V., Torza, S., and Schonhorn, H. (1976). Strength of 0.04-50 m lengths of coated fused silica fibers. *Appl. Phys. Lett.* **28**, 588.

Lawn, B. R., and Wilshaw, T., R. (1975). "Fracture of Brittle Solids." Cambridge Univ. Press, London and New York.

McClintock, F. A. (1974). Statistics of brittle fracture. *In* "Fracture Mechanics of Ceramics"

(R. C. Bradt, D. P. H. Hasselman, and F. F. Lange, eds.), Vol. 1, p. 93. Plenum, New York.

Macmillan, N. H. (1972). Review: The theoretical strength of solids. *J. Mater. Sci.* **7**, 239.

Mallinder F. P., and Proctor, B. A. (1964). Elastic constants of fused silica as a function of large tensile strain. *Phys. Chem. Glasses* **5**, 91.

Maurer, R. D. (1975). Strength of optical fibers. *Appl. Phys. Lett.* **27**, 220.

Maurer, R. D. (1977). Effect of dust on glass fiber strength. *J. Appl. Phys.* **30**, 82.

Maurer, R. D., Miller, R. A., Smith, D. D., and Trondsen, J. J. (1974). "Optimization of Optical Waveguides—Strength Studies," Final Report on Navy Contract No. 00014-73-c-0293 AD 777118 (unpublished).

Metcalfe, A. G., and Schmitz, G. K. (1964). Effect of length on the strength of glass fibers. *Proc. Am. Soc. Test. Mater.* **64**, 1075.

Morley, J. G., Andrews, P. A., and Whitney, I. (1964). Strength of fused silica. *Phys. Chem. Glasses* **5**, 1.

Mould, R. E. (1960). Strength and static fatigue of abraded glass under controlled ambient conditions. III. Aging of fresh abrasions. *J. Am. Ceram. Soc.* **43**, 160.

Mould, R. E. (1967). The strength of inorganic glasses. *In* "Fundamental Phenomena in the Materials Sciences" (L. J. Bonis, J. J. Duga, and J. J. Gilman, eds.), p. 119. Plenum, New York.

Novozhilov, V. V. (1953). "Foundations of the Nonlinear Theory of Elasticity." Graylock Press, Rochester, New York.

Otto, W. H. (1955). Relationship of tensile strength of glass fibers to diameter. *J. Am. Ceram. Soc.* **38**, 122.

Otto, W. H. (1965). "The Effect of Moisture on the Strength of Glass Fibers-A Literature Review," Defense Doc. Cent. Rep. AD-629370.

Pavelchek, E. K., and Doremus, R. H. (1976). Static fatigue in glass-A reappraisal. *J. Non-Cryst. Solids* **20**, 305.

Phillips, C. J. (1965). The strength and weakness of brittle materials. *Am. Sci.* **53**, 20.

Phoenix, S. L. (1973). Probablistic concepts in modeling the tensile strength behavior of fiber bundles and unidirectional fiber/matrix composites. *In* "Composite Materials Testing and Design," p. 130. Am. Soc. Test. Mater., STP 546, Philadelphia, Pennsylvania.

Phoenix, S. L. (1975). Statistical analysis of flaw strength spectra of high modulus fibers. *In* "Composite Reliability," p. 77. Am. Soc. Test. Mater., STP 580, Philadelphia, Pennsylvania.

Pierce, F. T. (1926). Tensile tests for cotton yarns—the weakest link. *J. Text. Inst.* **17**, 355.

Powell, B. E., and Skove, M. J. (1968). Measurement of higher order elastic constants using finite deformations. *Phys. Rev.* **174**, 977.

Proctor, B. A., Whitney, I., and Johnson, J. W. (1967). The strength of fused silica. *Proc. R. Soc. London Ser. A* **297**, 534.

Ritter, J. E., Jr., and Meisel, J. A. (1976). Strength and failure predictions for glass and ceramics. *J. Am. Ceram. Soc.* **59**, 478.

Runk, R. B. (1977). A zirconia induction furnace for drawing precision silica wave guides. *Top Meet. Opt. Fiber Transm., 2nd, 1977* Paper TuB5.

Schonhorn, H., Kurkjian, C. R., Jaeger, R. E., Vazirani, H. N., Albarino, R. V., and DiMarcello, F. V. (1976). Epoxy-acrylate-coated fused silica fibers with tensile strengths > 500 ksi (3.5 GN/m²) in 1-km gauge lengths. *Appl. Phys. Lett.* **29**, 712.

Shand, E. B. (1954). Experimental study of fracture of glass. II. Experimental data. *J. Am. Ceram. Soc.* **37**, 559.

Tanaka, S., Naruse, T., Osanai, H., Inada, K., and Akimoto, T. (1976). Properties of cabled low-loss silicone-clad optical fiber. *Proc. Eur. Conf. Opt. Fiber Commun., 2nd, 1976* p. 189.

Tariyal, B. K., and Kalish, D. (1977a). Application of Weibull-type analysis to the strength of optical fibers. *Mater. Sci. Eng.* **27,** 69.

Tariyal, B. K., and Kalish D. (1977b). Effect of aging on the strength of optical fibers. *Glass Div., Fall Meet. Am. Ceram. Soc., 1977* Paper 8-CG-77F.

Tariyal, B. K., and Kalish, D. (1978). Mechanical behavior of optical fibers. *In* "Fracture Mechanics of Ceramics" (R. C. Bradt, D. P. H. Hasselman, and F. F. Lange, eds.), Vol. 3, p. 161. Plenum, New York.

Tariyal, B. K., Kalish, D., and Santana, M. R. (1977). Proof testing of long length optical fibers for a communications cable. *Am. Ceram. Soc. Bull.* **56,** 204.

Thomas, W. B., and Brown, S. D. (1972). Nonlinear elasticity of fused silica and room temperature. *Phys. Chem. Glasses* **13,** 94.

Thomas, W. F. (1960). An investigation of the factors likely to affect the strength and properties of glass fibres. *Phys. Chem. Glasses* **1,** 4.

Wachtman, J. B. (1974). Highlights of progress in the science of fracture of ceramics and glass. *J. Am. Ceram. Soc.* **57,** 509.

Wang, T. T., and Zupko, H. M. (1978). Long-term mechanical behavior of optical fibers coated with a UV-curable epoxy acrylate. *J. Mater. Sci.* **13,** 2241.

Wang, T. T., Vazirani, H. N., Schonhorn, H., and Zupko, H. M. (1978). Effects of water and moisture on strengths of optical glass (silica) fibers coated with UV-cured VIF epoxy acrylate. *J. Appl. Polym. Sci.* (in press).

Weibull, W. (1939). A statistical theory of the strength of materials. *Proc. R. Swed. Inst. Eng. Res.* No. 151 (unpublished).

Wiederhorn, S. M. (1969). Fracture surface energy of glass. *J. Am. Ceram. Soc.* **52,** 99.

Wiederhorn, S. M. (1973). Prevention of failure in glass by proof-testing. *J. Am. Ceram. Soc.* **56,** 226.

Wiederhorn, S. M. (1975). Crack growth as an interpretation of static fatigue. *J. Non-Cryst. Solids* **19,** 169.

Wiederhorn, S. M. (1978). Mechanisms of subcritical crack growth in glass. *In* "Fracture Mechanics of Ceramics" (R. C. Bradt, D. P. H. Hasselman, and F. F. Lange, eds.), p. 549. Plenum, New York.

Wiederhorn, S. M., and Bolz, L. H. (1970). Stress corrosion and static fatigue of glass. *J. Am. Ceram. Soc.* **53,** 543.

Zhurkov, S. Effect of increased strength of thin filaments. *Zh. Tekh. Fiz.* **1,** 386.

Chapter 13

Optical Cable Design

MORTON I. SCHWARTZ
DETLEF GLOGE
RAYMOND A. KEMPF

13.1 INTRODUCTION

The large bandwidth, low-loss, and small physical size of optical fibers can be exploited by using individual fibers as isolated transmission media with separate sources and detectors. In this situation each broken fiber results in a lost channel, not just a reduction of signal-to-noise ratio, as would be the case if a bundle of fibers were dedicated to each channel. In addition, splicing requirements are considerably more stringent for the case where individual fibers represent separate communication channels.

Because of the great telecommunications potential of optical fiber systems which use individual fibers as separate transmission media, this chapter deals solely with this case. The design considerations for optical fiber bundles are covered elsewhere in the literature; see Tiedeken (1967), for example.

Most telecommunication applications of optical fibers require losses of less than 50 dB/km. Since only glass core fibers have achieved losses in this range, the cabling of glass optical fibers is considered herein.

13.2 DESIGN OBJECTIVES

This section examines optical cable performance objectives and provides a qualitative description of the manner in which they affect optical cable design.

435

13.2.1 Fiber Survival

One of the principal optical cable design objectives must be fiber survival. Because glass is a brittle material, the strength of which is determined by its largest surface flaw, the glass fiber must be protected with a coating in-line with the fiber draw process to avoid surface abrasion. The relationship between the fracture tensile strength σ and the depth of the maximum surface flaw a is given in Eq. (12.11). Thus, if a = 4 microns the breaking tensile stress is 322 MPa or 47 kpsi. Therefore, for this fiber to survive cable manufacture, the maximum tensile stress must be less than 47 kpsi. Since the economics of cable manufacture suggest that cables be made in long lengths, perhaps in the neighborhood of a kilometer or more, glass surface flaws larger than a few microns will have to be avoided over lengths of the order of a kilometer. This task is very difficult, but recent results (Schonhorn et al., 1976) suggest that very high fiber strengths are possible even over such long lengths.

After glass fibers are coated they must be handled with care in subsequent cable manufacturing steps. Whereas copper can be elongated more than 20% (plastically) without fracture, strong optical fibers may well break at 1% elongation. In addition, as described in Section 12.4, in the presence of moisture surface flaws will grow while under stress by a process called *static fatigue*. If the applied stress is below 20% of the short-term strength then static fatigue occurs very slowly; i.e., fracture would not occur for tens of years or more. However, at stresses above about 40% of the short-term strength, static fatigue is fairly rapid. Thus, if the short-term breaking elongation of a certain fiber is 4% (a very strong fiber) then the largest elongation one would wish to permit in cable manufacture is about 1%. More typically, long-length breaking elongations of good fibers are of the order of 0.5 to 1% so one would like to limit fiber elongations in cable manufacture to 0.12 to 0.25%.

The fragility and small elongation capability of glass fibers influence optical cable structures in many ways. Unlike multipair cable, where the wires themselves are the principal load-bearing members, it is desirable to avoid significant tensile loading of glass fibers. Also, because fibers do not deform plastically they have little capacity to absorb energy. Hence, the optical cable design must isolate the fibers from impact loads or they will fracture. In addition to these considerations, the small elongation capability of optical fibers makes it desirable to keep the fibers close to the neutral axis of the cable or to provide space for them to move to positions of reduced stress when the cable is bent.

All the foregoing considerations point to the fact that the outer sheath of an optical cable must be more than an envelope. It must isolate the fibers from impact loads, limit cable bending radii, and in some instances also be the principal tensile load bearing element in the cable.

13.2.2 Maintaining Fiber Transmission Characteristics

The transmission characteristics of optical fibers are affected by bending of the fiber axis. Bending of the axis can cause coupling of energy between guided modes in a multimode fiber and can cause radiation losses in both multimode and single mode fibers. Random bending of the fiber axis on a microscopic scale (Fig. 13.1) called microbending (Gardner, 1975; Gloge, 1975), results from fiber coating variations or other constraints imposed by the cable structure. If the fiber axis deflections have significant spatial frequency components in a critical range (Marcuse, 1973) the fiber transmission characteristics will be changed substantially. In addition to microbending, long period bending (in the range of a few centimeters or less for common fiber parameters) known as macrobending, results in added loss (see Chapter 3). Periodic bends of this type can readily occur due to stranding or other periodicities in the cable structure. The optical cable should be designed and manufactured in such a manner as to control macrobending and microbending losses in the packaging of fibers in cables. However, this is not the entire job; the fiber parameters and the cable structure and materials must be chosen such that the thermally and mechanically induced forces generated during the cable's intended life do not produce significant transmission degradation.

13.2.3 Organization of Fibers to Ease Mass Splicing and Handling

Because of the small size and fragility of optical fibers and because of the difficulty of splicing them it has been suggested that they be organized in regular geometric units (Schwartz, 1975). The resulting increase in the size and strength of the objects (fiber units rather than fibers) simplifies connectorization and mass splicing operations. For small optical cables, those with fewer than about 25 fibers, it may be practical, perhaps even economical, to field-splice individual fibers. However, for large cables and when space is at a premium, handling and splicing individual fibers appears to be a task manageable only by highly trained technicians. Requiring such

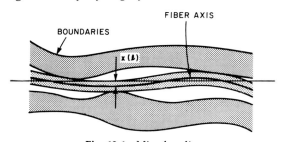

Fig. 13.1 Microbending

skill levels would almost surely relegate optical fiber transmission to special applications. Subsequently, the splicing advantages and disadvantages of different fiber units will be considered.

13.2.4 Space Efficiency and Weight

The small size of optical fibers makes them attractive from space and weight considerations. However, in order to compensate for the fragility of fibers, it is usually necessary to expend a larger portion of cable cross section on strength and support members than one would have to on conducting media. Fiber packing density is also constrained by the fact that the smaller the spacing between fibers is, the more difficult it is to mechanically isolate the fibers, and thereby avoid microbending. A qualitative discussion of the relative space efficiencies of several cable design approaches is given in Section 13.5.

13.2.5 Manufacturing and "Repairability"

Another very important cable design objective is manufacturability. The design must be capable of being fabricated with reasonable yields at acceptable production rates. In this case, yield means both fiber survival and a suitably narrow range of transmission characteristics for the cabled fibers.

The ability to repair a cable with broken fibers will depend on the cable design. In the case of large cables, a unit-based design which permits unit-to-unit splicing will be advantageous.

13.2.6 Design Directions

Section 13.2 has covered the cable design objectives of concern in this chapter (they are by no means exhaustive). Consideration of these design objectives has led to the following important design directions:

—Avoid appreciable loading of fibers
—Isolate fibers from other cable components (long mechanical coupling length)
—Keep fibers close to neutral axis or provide space for them to move
—Design outer sheath to protect against external environment
—Choose cable materials to minimize differential thermal expansion of cable components
—Unit-based cable design—this eases handling and splicing

Similarly, appropriate manufacturing directions can be identified:

—Only use fibers with a specified minimum strength
—Minimize fiber loads during cable manufacture

13.3 PHYSICAL PROTECTION

Section 13.2.1 pointed out a number of physical limitations of optical fibers which influence optical cable design. Consideration herein is given to methods, procedures, and materials properties which can be used to overcome these limitations.

13.3.1 Fiber Strength and Prooftesting

In Chapter 12 it was pointed out that the strength of optical fibers is inherently random. This is because fiber strength usually depends on the maximum surface flaw depth in the glass, which itself is a random variable. This phenomenon raises serious problems for the cable designer. If the designer chooses a conservative cable design in which the fibers experience negligible strain at full-rated cable load the cost of the cable will tend to be high (furthermore there remains the risk of some breakage in cable manufacture). If, on the other hand, he allows the fibers to bear appreciable load, he runs the risk of substantial fiber breakage.

A way out of this dilemma, which is described in Section 12.5, is to prooftest the fibers before they are incorporated into the cable structure. If the fiber has a weak point, corresponding to a strength below the prooftest value, the fiber will break and therefore it will not be put into a cable. By use of prooftesting the problems of design and manufacture of the random strength fiber medium can be converted to those corresponding to a known minimum strength medium. Proper choice of the fiber prooftest level involves economic trade-offs between fiber yield at a given prooftest level and the system cost of broken fibers in a cable. The latter costs may depend strongly on the system splicing strategy. For example, the approach of field-joining cable sections by way of prefabricated non-arrangeable connectors leads to high requirements on fiber survival. This is because when a number of cable sections are straight-through spliced, unbroken fibers in one section may be spliced to broken fibers in the next section, which can result in a proliferation of broken fiber transmission lines. Such a system splicing strategy leads to a relatively high prooftest level requirement. In the Bell System Chicago Lightwave Communication Project (Schwartz et al., 1977), a modest prooftest level of 35 kpsi was employed * and all ten connectorized optical cable sections were installed without fiber breakage. In the case where individual fiber splicing is performed, spare fibers can be used to compensate for fiber breakage, hence lower prooftest levels may be economically appropriate. The actual choice of prooftest level should be based on splicing strategy,

---

* Higher prooftest levels are deemed desirable for large scale manufacture of cables which use straight-through splicing.

and on the strength required to survive manufacture, installation, and in place use, with appropriate allowance for static fatigue.

13.3.2 Fibers as "Nonload Bearing" Cable Elements

Because of their low strain capability, care must be taken to avoid having the high elastic modulus fibers take up load before the intended load-bearing members. There are two approaches, or a combination thereof, which can be used to accomplish this. They are:

(1) incorporate load-bearing materials in the cable such that $\Sigma_i E_i A_i \gg N E_f A_f$; where E_i denotes the elastic modulus of the ith cable element and A_i its cross-sectional area, E_f and A_f are the modulus and cross-sectional area of each of the N fibers in the cable—in this case it is assumed fiber strain and cable strain are equal;

(2) place the fibers in the cable such that the length of the fibers is greater than the length of the cable; when the cable is loaded the fiber geometry is altered so that fiber strain is less than the cable strain.

Most cable designs used thus far, which avoid treating fibers as the principal load-bearing members, are based on the first approach. Examples are the BICC PSP cable (Dean and Slaughter, 1976), Corning's COREGUIDE® cable (R. A. Miller, 1975), "cable for public communication" (Murata et al., 1977), and Sumitomo's underground cable (Nakahara et al., 1975). In the case of Bell Laboratories Atlanta Experiment Cable, the 144 fibers carried about 18% of the tensile load (Buckler et al., 1977, 1978). The Chicago Lightguide cable, which was similar in structure except that it contained 24 fibers, had the fibers carry less than 4% of the tensile load.

An example of the second design approach is given by Beal (1977). He suggests stranding one or more fibers in a cable around a compressible cellular material. When a tensile load is applied to the cable, the fiber reduces its helix radius by causing diametral contraction of the compressible cellular material. This results in a fiber strain which is less than the cable strain. The same principle is employed in the cable structure shown in Fig. 13.2 in which fiber ribbons are helically wrapped around a cylindrical compressible core. Here too strain relief is accomplished principally by diametral contraction.

Many communication cable designs involve conductors which are not near the cable neutral axis. When such a cable is bent, the conductors distant from the neutral axis can be subject to significant bending strains. A standard approach for alleviating the problem is to "strand the conductors" such that they follow helical paths in the cable with a lay length comparable to the minimum cable bending radius. This approach is also applicable to most optical cable designs, but in some cases there will be

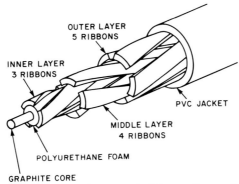

Fig. 13.2 Ribbon cable with strain relief due to diametral contraction.

constraints on helix lay length. A pertinent example is furnished by the ribbon-based cable core shown in Fig. 13.3. In this case 12 ribbons, each containing 12 optical fibers are stacked to form a rectangular array, with the cross section shown in the figure, and twisted to improve cable bending properties. With this geometry a short lay length causes high strains in the fibers. Hence, there is a trade-off between choosing the lay length short enough to provide protection against bending and long enough to avoid breakage in manufacture or due to static fatigue because of "built-in stresses."

This problem has been analyzed (Eichenbaum and Santana, 1977) under the following assumptions:

(i) With the cable axis straight, the individual fiber axes coincide with helices of appropriate diameter and pitch.

(ii) All the fibers within a given ribbon are completely coupled to each other.

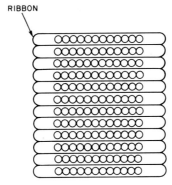

Fig. 13.3 Stranded rectangular array ribbon cable.

(iii) Induced stresses are supported entirely by the fibers, i.e., other ribbon materials are ignored.

(iv) The twisted unit maintains a rectangular cross section.

(v) The fibers are treated as filaments which follow the geometric axes of the real fibers.

(vi) The tensile and compressive moduli of the fibers are equal.

(vii) Torsionally induced shear strains, which may be significant, are neglected.

Under these assumptions the most severe stresses in this cable core are applied to the "center ribbon" and the model provides an upper bound of 3.6 kpsi tensile stress for the outside fiber in this ribbon and 2.3 kpsi compressive stress for the "inside fiber" for a 4-in. lay length. Since this analysis neglects torsionally induced shear strains, principal plane stresses can be significantly higher. In fact Santana* has shown that this is indeed the case. Nevertheless, the analysis performed indicated that cable lay length should be greater than 4 in.

13.3.3 Materials, Packaging, and Thermal Considerations

The materials used in any cable affect the cable's weight, strength, frictional characteristics, and thermal expansion. Because of the limited strain capability of optical fibers, and because of the microbending phenomenon, proper materials choices are crucial in the case of optical cables.

Materials such as KEVLAR®, steel, graphite, and glass fiber reinforced plastic have been used as the strength members in optical cables because of their high tensile moduli. If the strength member is "yarnlike in nature" it may not, unless it is well-coupled mechanically to the rest of the cable structure, provide much compressive strength. Compressive strength can be important in preventing polymer shrinkback in cable manufacture and in limiting compressive strain due to thermally induced compression. Shrinkback in manufacture can be reduced by utilizing noncrystalline polymers. An example of this approach is Corning's COREGUIDE™ (R. A. Miller, 1975) in which urethane buffering and sheathing materials have been used (Fig. 13.4). The urethane does not tend to shrink back; however, it is likely to result in a higher friction coefficient and is not as tough as polyethylene, for example. In the case of COREGUIDE™, it appears that the KEVLAR® is well-encapsulated, so it should be effective in limiting compressive strain.

Gloge (1975) has shown the desirability, with respect to microbending loss, of obtaining a large ratio of fiber coating longitudinal elastic modulus

---

* Unpublished work.

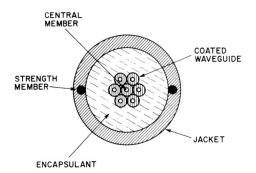

Fig. 13.4   Cross section of COREGUIDE® cable.

to transverse elastic modulus.* Jackson (Jackson *et al.*, 1976) has reported a technique of packaging optical fibers in loose fitting tubes of oriented polymer, thereby increasing the longitudinal modulus. By annealing, the tendency of the polymer to shrink back is reduced such that up to 80°C no discernible shrink back is noted in a 3-day period. One of the concerns with a loose-fitting fiber jacket is the possibility of the fiber buckling in the jacket as a result of jacket contraction (thermally or mechanically induced) with consequent increase in loss due to bending loss and microbending. However, the linear thermal coefficient of the oriented polymer is reduced by more than a factor of ten from that of the unoriented polymer and, as a result, Jackson reports that stable cable operation can be achieved.

Rokunohe (Rokunohe *et al.*, 1976), reported the results of extensive studies aimed at selecting a fiber jacketing material. They point out the importance of the jacket from the point of view of physical protection but especially with respect to the stability of transmission loss. Based on a study of a variety of materials, they found nylon 12 the preferred material. They then investigated a wide range of loose and tight-fitting nylon 12 jackets on fibers. Their results were significantly better for the case of the tight-fitting jackets, in that the loosely jacketed fibers exhibited increased transmission losses at low temperatures. These results, in conjunction with Jackson's, suggest that if loose-fitting polymer fiber jackets are used, they should be oriented polymers.

13.4 UNIT DESIGN

Before a fiber enters the final cabling process, it is subjected to a number of preparatory steps designed to provide the necessary protection and re-

---

* The dependence of microbending loss on the index difference delta and on other fiber parameters is discussed in Chapter 6 of this volume.

siliance which guarantee survival and unimpaired transmission. Often the preparation involves forming small subgroups or units of fibers in well-defined geometrical configurations. These subgroups are then maintained in the cable, thereby providing easy identification and facilitating group splicing methods (see Chapter 14). The fiber pair, the fiber array or ribbon, and the six-around-one unit are examples.

These units are often constructed and protected in a way similar to fiber arrangements used in semiprotected environments. Fiber cable for data bus applications inside a computer housing may, for example, closely resemble a ribbon or pair unit destined for underground-cable assembly. We therefore begin this section by examining unit designs for semiprotected environments, considering the single-fiber unit as the fundamental member of this group. We then proceed to multifiber units and finally discuss the variety of configurations which result from incorporating such units into a cable.

13.4.1 Single-Fiber Units

The application of a plastic coating simultaneously with the drawing process is essential for reasonable product life and has been discussed in Chapter 10. We consider this coating an intrinsic part of the fiber itself and not part of the cable or unit structure. It is therefore understood that a cross section labeled as "fiber" in the following figures is the cross section of the coated fiber. The simplest unit or cable usable in a protected environment is obtained by adding another thicker and more rigid coating to the first.

A configuration of this kind is shown in Fig. 13.5a. The outer coating or jacket may be made from nylon, polyethylene, or polypropylene (Tanaka et al., 1976; Foord and Lees, 1976). These materials typically have a Young's modulus 100 times smaller than that of glass, but constitute an important load-bearing member since the jacket cross section is typically 100 times larger than the glass cross section (the outer jacket diameter is in the range of 1 mm). Units of this kind survive the tension encountered in lightweight cable-making machinery and meet the requirements of certain permanent indoor installations. Of course, the protection that this jacket provides against harsh flexing or crushing, strong abrasion or other abuse is limited.

It is desirable that a soft plastic be interposed between the outer rigid jacket and the fibers (Gloge, 1975) so that the fiber does not conform to the unavoidable microscopic thickness variations of the jacket. Microbends so introduced into the fiber can increase transmission loss by tens of decibels (Foord and Lees, 1976). A soft intermediate layer allows the fiber to retain its natural straight condition. If the inner coating is not a soft one, it may

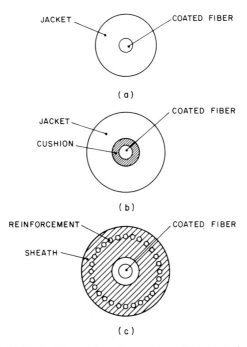

Fig. 13.5 Single-fiber unit configurations. (Murata *et al.*, 1975.)

be necessary to add a third layer made for example from urethane (see Fig. 13.5b).

In spite of the difficulties described in Section 13.3.3 the loose tube design proves useful for some applications, as, for example when heavy duty protection is required. A case in point are single-fiber jumper cables made for use as patch cords in cross-connect fields.* In this case, protection against flexing, crushing, twisting, or pulling must be substantially better than can be provided by a 1-mm nylon jacket. A tube wall of several millimeters thickness has been found useful, preferably incorporating a woven matrix of KEVLAR® or other strength members, as indicated in Fig. 13.5c.

13.4.2 Multifiber Units

This section examines the strategy of organizing groups of two or more fibers in special units either as a preparation for the final cabling process or for direct use in a semiprotected environment. In the simplest case, a unit may consist of a group of individually protected fibers wrapped or sheathed together to form a bundle (each fiber represents one com-

---

* Cross-connect fields are arrays of jacks which allow rearrangeable interconnection between system components.

munications channel). The sheath which holds the bundles together may also serve other purposes, i.e., providing a thermal barrier during the cabling process, providing protection against microbending in the cable, or contributing to the tensile strength required during cable assembly. The organization of fibers into groups simplifies the handling process, facilitates the provision of multifiber terminations and simplifies field-splicing. In special cases the unit may include strength members or electrical conductors.

Two unit configurations of interest are shown in Fig. 13.6. The "circularly symmetric" unit design shown in Fig. 13.6a was used by Furukawa Cable Co. (Murata *et al.*, 1975). Six nylon-coated fibers are stranded around a central strength member and enclosed in a sheath with or without a filler. Stranding improves the cable bending properties if the bending radius is greater than or comparable to the stranding pitch.

A unit design called COREGUIDE℗ (Miller and Pomeratz, 1974) is shown in Fig. 13.4. It is similar to the first one except that the central member is replaced by another fiber available as a transmission channel. All seven fibers are not stranded and are subject to the same tensile loads. They are embedded in a resilient urethane filler and placed in a sheath that includes two KEVLAR® strength members. This design protects the fibers quite efficiently against tensile and bending forces, but flexibility is limited to the direction transverse to the plane of the strength members.

The same design objective is evident in the cable shown in Fig. 13.6b which was constructed by British Insulated Callender's Cables in cooperation with Plessey (Slaughter *et al.*, 1975). The fibers are placed in the same plane with the strength members for better protection. Flexibility in the

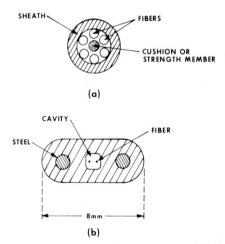

Fig. 13.6 Fiber units for small cables. (Slaughter *et al.*, 1975.)

transverse direction is enhanced by the flat cross section. Both COREGUIDE℗ and the BICC cable were actually not designed as units for inclusion in bigger cables; they have a heavy sheath which withstands tough handling and a harsh environment.

Both of the designs in Fig. 13.6 are well-suited for cables with a small number of fibers. Each of these designs can be used with or without individual factory applied fiber connectors. Individual fiber connectorization or field splicing of individual fibers is less desirable in cables with a large number of fibers. Large-fiber-count cables benefit from a unit structure which permits mass cable termination and simplified field splicing. A well-controlled *regular geometric unit* with geometry similar to that to be employed in the connector or splice is needed to accomplish this. An example of a unit which has these features is shown in Fig. 13.7a. This ribbon structure, which was proposed at Bell Laboratories (Standley, 1974), has high space efficiency and, because the fibers are positioned linearly in the ribbon with sufficient accuracy to provide preliminary alignment, simplifies mass termination and splicing (C. M. Miller, 1975; Schwartz, 1976). Ribbons made by sandwiching 12 coated fibers between two adhesive-backed polyester tapes (Saunders and Parham, 1977) measure only a fraction of 1 mm in thickness and a few millimeters in width. Ribbons of this type have been stacked to form the core of optical cables with as many as 144 fibers (see Fig. 13.3). Similar ribbons, perhaps with suitable reinforcement, have potential for use as circuit connections between computer frames or similar data processing units because the flat design is particularly well-matched to circuit board and circuit terminal design.

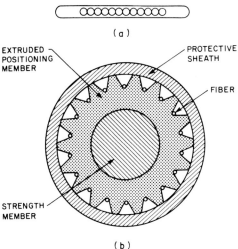

Fig. 13.7 Fiber units suited for mass termination. (Le Noane, 1976.)

Another regular geometric unit proposed by CNET (Le Noane, 1976) is shown in Fig. 13.7b. This unit, which employs circular symmetry, positions the fibers in much the same arrangement that they appear in CNET's splice connector (Le Noane, 1976).

13.5 CABLE STRUCTURES AND PERFORMANCE

The general principles that govern cable design for a hostile environment have been discussed in Section 13.3. This section reviews some of the cable designs that have been implemented and considers their performance characteristics. The cable designs considered are based on the multifiber units discussed in Section 13.4.1 and 13.4.2.

13.5.1 Specific Cable Designs and Their Characteristics

Cable designs may employ a central strength member as shown in Fig. 13.8a (Mizukami *et al.*, 1975) which takes most of the tensile load. A specially shaped plastic spacer which contains a stranded steel center is used. Individual coated fibers are loosely confined in four cavities formed by plastic tape wrapped around the spacer. This cable structure, which is 16 mm in diameter, provides good physical protection under tensile load and inherently low microbending loss but is neither space efficient nor inexpensive to manufacture.

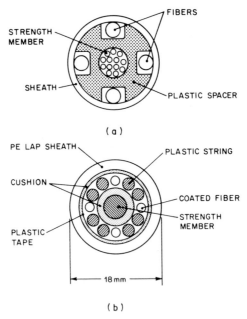

Fig. 13.8 Cables for underground application.

Figure 13.8b shows the cross section of a four-fiber optical cable designed for underground applications (Nakahara *et al.*, 1975). The cable has a central strength member, made of fiber reinforced plastic, over which a cushioning layer is placed. The four fibers are placed in a concentric arrangement around the cushion separated by plastic strings. An outer cushioning layer surrounds the fiber layer and an outer sheath of aluminum covered by polyethylene (PE LAP sheath) brings the cable o.d. to 18 mm. This structure provides very good mechanical protection for the fibers and also provides protection against moisture. Here again, the structure is not inexpensive. A similar cable design, with eight fibers and an o.d. of 14 mm has been described by (Tanaka *et al.*, 1976). Figure 13.9a illustrates the incorporation of multifiber units of the kind shown in Fig. 13.6a in a heavy-duty outdoor cable (Murata *et al.*, 1975). In this case four units and four plastic spacers are stranded around a central strength member thereby supplementing the load-bearing capability of the individual units. A plastic sheath surrounds the structure resulting in an 18 mm o.d. An eight-fiber cable (Murata *et al.*, 1977) is shown in Fig. 13.9b. The eight fibers are stranded around a cushioned central strength member to

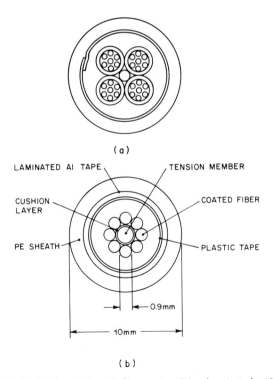

Fig. 13.9 Cables with multifiber units. (Mizukami *et al.*, 1975.)

form the cable core. The core is loosely contained in a sheath consisting of a laminated aluminum tape surrounded by a 10-mm o.d. polyethylene outer jacket. The microbending loss and the temperature stability of this design should be good if fiber coatings and jackets are properly selected (Section 13.3.3) because the fibers are free to move and thereby to relieve strains.

The three cable structures shown in Figure 13.10 (Rokunohe *et al.*, 1976) all have cylindrically symmetric strength members built into the cable sheath rather than a central strength member. The cable structure shown in Fig. 13.10c uses a nonmetallic strength member. Rokunohe shows that with properly jacketed fibers all three of the cable structures he considered exhibited loss variations of no more than 0.3 dB/km over a temperature range from −50 to 60°C.

Figure 13.11 shows a large fiber count cable based on the flat ribbon unit design (Schwartz *et al.*, 1976). Twelve ribbons, each containing 12 fibers are stacked, stranded and covered with paper insulation and a loose polyethylene tube. Polyolefin twine and load-bearing steel wires are applied helically over the tube before the outer sheath is extruded around the cable. The cable, which weighs about 920 N/km (63 lb/kft.), has a rated

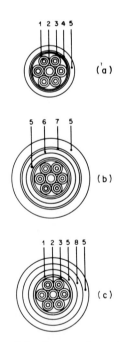

Fig. 13.10 Cables with outer reinforcing members. 1. Jacketed fiber. 2. Dummy fiber. 3. Core wrap. 4. Aluminium tape. 5. Polyethylene. 6. Corrugated steel. 7. Bituminous compound. 8. Glass reinforced nylon compound. (Rokunohe *et al.*, 1976.)

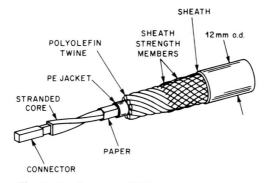

Fig. 13-11 Cutaway of BTL experimental cable.

tensile load of 1550 N (350 lb). This cable structure, which is 13 mm in o.d., is space-efficient, provides easy identification of individual fibers and greatly facilitates splicing. However, the twisted ribbon stack introduces torsional forces on the fibers, and in addition the fibers are relatively more confined, resulting in less opportunity to relieve strains due to manufacture, installation, and thermal variations. As a result, it is harder to achieve low microbending losses, good temperature stability, and low fiber breakage. By careful choices of materials, and by using an appropriate prooftest level, these problems can be mitigated.

An optical cable, based on the flat ribbon structure (Soyka, 1977), which was designed and built by the General Cable Corporation, is being used by the General Telephone and Electronics Corporation to provide commercial telephone service in Long Beach, California. The cable, shown in Fig. 13.12 has a single ribbon with six fibers which follows a helical path around a copper wire reinforced plastic member which has two helical grooves channeled in it. Three 22-gauge copper pairs are placed in the second helical groove. The pairs are used for powering, order wire and fault

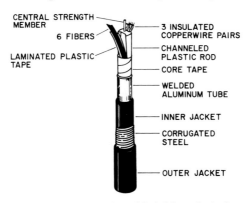

Fig. 13.12 Composite cable (ribbon design).

location. A laminated tape serves to confine the ribbon and pairs to their channels as well as to provide thermal isolation. A welded aluminum tube (55 mils) is applied over the tape followed by a polyethylene jacket. Corrugated steel armor, flooded with an asphaltic compound to provide a moisture barrier, surrounds the polyethylene jacket. Finally, an outer polyethylene jacket is applied bringing the cable o.d. to about 28 mm. The cable, which weights about 6500 N/km (450 lb/kft.), has a rated tensile load of 4000 N (900 lbs).

13.5.2 Cable Design Selection

Table 13.1 compares cable designs based on "circularly symmetric units" with designs based on flat ribbon units. Each type of cable design affords specific advantages and disadvantages. The proper choice of cable design depends strongly on the specific set of application requirements. However, some general guidelines are as follows:

Large bandwidth–small number of fibers—if a small number of fibers with very large bandwidth per fiber is required (of the order of 100 Mbit/sec or more), the circularly symmetric unit is probably the better choice. This is because it is easier to obtain and maintain low cabled fiber loss, because space efficiency is not important, and since more complex splicing is tolerable.

Long repeater spacing–ultralow loss fibers—for repeater spacings of the order of 15 km or more, the choice is also the circularly symmetric cable which can more readily maintain the low fiber losses. Individual fiber splicing is preferable to attain the lowest possible losses.

Large fiber count (>30 fibers)–bandwidths <20 Mbit/sec—the flat ribbon

TABLE 13.1
Comparison of Cable Design Approaches

Factor	Circularly symmetric unit	Flat ribbon unit
Microbending loss	Easier to control	Harder to control –but doable
Loss temperature stability	Easier to control	Harder to control –but doable
Space efficiency (fibers/area)	Lower	Higher
Ease of fiber handling & identification	More complex	Simpler
Suitability for mass splicing	Poorer	Superior
Suitability for repair splicing	Poorer	Superior
Ease of manufacture	Good	More difficult
Compatibility with conventional cable manufacture	Good	Poor

TABLE 13.2
Preferred Designs for Communication Applications

Application	Bandwidth (Mbit/sec)	Repeater spacing (km)	No. of fibers—M	Preferred cable structure
Point to point—either long	<100	10 or less	>10	FRU
haul or interoffice	>100		<10	CSU
		greater than 20		CSU
Large feeder—with branching capability			>100	FRU
Small feeder or			10 < M < 30	?
distribution			50<	FRU
	<20		30<	FRU
Drop cables			<4	CSU

FRU—flat ribbon unit.
CSU—circular symmetric unit.

cable is the better choice because of the difficulties and costs of handling and splicing a large number of individual fibers. Whenever the cost of splicing becomes significant on a per channel basis, a ribbon-based cable design, which affords the possibility of mass splicing, becomes attractive.

We conclude this chapter with our estimates of the preferred cable geometry for a number of telecommunication applications. Table 13.2 gives our current "guesstimates." The table is by no means exhaustive.

REFERENCES

Beal, R. E. (1977). Optical guides with compressible cellular material. U.S. Patent 4,037,923.

Buckler, M. J., Santana, M. R., and Shores, S. C. (1977). Design and performance of an optical cable. *Proc. Int. Wire & Cable Symp., 26th, 1977* pp. 276–280.

Buckler, M. J., Santana, M. R., and Saunders, M. J. (1978). Lightguide cable manufacture and performance. *Bell Syst. Tech. J.* **57**, No.6, Part 1, 1745–1757.

Dean, N. S., and Slaughter, R. J. (1976). Development of a robust optical fiber cable and experience to date with installation and jointing. *Proc. Int. Wire & Cable Symp., 25th, 1976* pp. 245–254.

Eichenbaum, B. R., and Santana, M. R. (1977). Analysis of longitudinal stress imparted to fibers in twisting an optical communication cable unit. *Bell Syst. Tech. J.* **56**, 1503–1512.

Foord, S. G. and Lees, J. (1976). Principles of fiber-optical cable design. *Proc. IEEE* **123**, 597–602.

Gardner, W. B. (1975). Microbending loss in optical fibers. *Bell Syst. Tech. J.* **54**, No. 2, 457–465.

Gloge, D. (1975). Optical-fiber packaging and its influence on fiber straightness and loss. *Bell Syst. Tech. J.* **54**, No. 2, 245–262.

Jackson, L. A., Reeve, M. H., and Dunn, A. G. (1976). Oriented polymer for optical fibre packaging. *Proc Eur. Conf. Opt. Fibre Commun., 2nd, 1976* p. 175–176.

Le Noane, G. (1976). Optical fibre cable and splicing techniques. *Proc. Eur. Conf. Opt. Fibre Commun., 2nd, 1976* p. 247–252.

Marcuse, D. (1973). Losses and impulse response of a parabolic index fiber with random bends. *Bell Syst. Tech. J.* **52**, No. 8, 1423–1437.

Miller, C. M. (1975). A fiber-optic cable connector. *Bell Syst. Tech. J.* **54**, 1547–1555.

Miller, R. A. (1975). Fiber cabling. *Tech. Dig., Top. Meet. Opt. Fiber Transm., 1975* paper WAI.

Miller, R. A., and Pomerantz, M. (1974). Tactical low loss optical fiber cable for army applications. *Proc. Int. Wire & Cable Symp., 23rd, 1974* pp. 266–271.

Mizukami, T., Hatta, T., Fukuda, K., Mikoshiba, K., Shimohori, Y. (1975). Spectral loss performances of optical fibre cables using plastic spacer and metal tube. *Proc. Eur. Conf. Opt. Fibre Commun., 1st, 1975* p. 191–193.

Murata, H., Inao, S., and Matsuda, S. (1975). Step-index type optical fibre cable. *Proc. Eur. Conf. Opt. Fibre Commun., 1st, 1975* p. 70–72.

Murata, H., Nakahara, T., and Tanaka, S. (1977). Recent development of optical fiber and cable for public communication. *Tech. Dig., Int. Conf. Integr. Opt. Opt. Fiber Commun., 1977* p. 281–284.

Nakahara, T., Hoshikawa, M., Suzuki, S., Shiraishi, S., Kurosaki, S., and Tanaka, G. (1975). Design and performance of optical fibre cables. *Proc. Eur. Conf. Opt. Fibre Commun., 1st, 1975* p. 81–83.

Rokunohe, M., Shintani, T., Yajima, M., and Utsumi, A. (1976). Stability of transmission properties of optical fibre cables. *Proc. Eur. Conf. Opt. Fibre Commun., 2nd, 1976* p. 183–188.

Saunders, M. J., and Parham, W. L. (1977). Adhesive sandwich optical fiber ribbons. *Bell Syst. Tech. J.* **56**, No. 6, 1013–1014.

Schonhorn, H., Kurkjian, C. R., Jaeger, R. E., Vazarani, H. N., Albarino, R. V., and Di Marcello, F. V. (1976). Epoxy-acrylate-coated fused silica fibers with tensile strengths >500 ksi ($3.5GN/M^2$) in 1-km gauge lengths. *Appl. Phys. Lett.* **29**, No. 11, 712–714.

Schwartz, M. I. (1975). Optical fiber cabling and splicing, *Top. Meet. Opt. Fiber Transm., 1st, 1975* Paper WA2.

Schwartz, M. I. (1976). Optical cable design associated with splicing requirements. *Proc. Eur. Conf. Opt. Commun., 2nd, 1976* pp. 325–329.

Schwartz, M. I., Reenstra, W. A., and Mullins, J. H. (1977). The Chicago lightwave communications project. *Post-Deadline Pap., Int. Conf. Integr. Opt. Opt. Fiber Commun., 1977* pp. 53–56.

Schwartz, M. I., Kempf, R. A., and Gardner, W. B. (1976). Design and characterization of an exploratory fiber optic cable. *Proc. Eur. Conf. Opt. Fibre Commun., 2nd, 1976* pp. 311–314.

Slaughter, R. J., Kent, A. H., and Callan, T. R. (1975). A duct installation of 2-fibre optical cable. *Proc. Eur. Conf. Opt. Fibre Commun., 1st, 1975* pp. 84–86.

Soyka, R. (1977). Optical fiber cable placing techniques-long section lengths. *Proc. Int. Wire & Cable Symp. 26th, 1977* pp. 281–286.

Standley, R. E. (1974). Fiber ribbon optical transmission lines. *Bell Syst. Tech. J.* **53**, No. 6, 1183–1185.

Tanaka, S., Naruse, T., Osanai, H., Inada, K., and Akimoto, T. (1976). Properties of cabled low-loss silicone clad optical fiber. *Proc. Eur. Conf. Opt. Fibre Commun., 2nd, 1976* pp. 189–192.

Tiedeken, R. (1967). "Fibre Optics and Its Application." Focal Press, New York.

Chapter 14

Fiber Splicing

DETLEF GLOGE
ALLEN H. CHERIN
CALVIN M. MILLER
PETER W. SMITH

14.1 INTRODUCTION

The loss exhibited by an optical fiber splice is strongly dependent upon the lateral alignment of the core diameters of the fibers being spliced. One of the principal advantages that a multimode fiber has compared to a single-mode fiber is its larger core diameter. But even for multimode fibers, lateral misalignment tolerances are only a fraction of the core diameter or typically a few micrometers. Many laboratories all over the world are heavily engaged in finding simple, rugged, and inexpensive means of connecting fiber ends in ways that cause the least amount of loss. Some less obvious and novel ways which have been suggested to do this include holographic methods or means of directional coupling, notably of the adiabatic type (Yamamoto *et al.*, 1976); but so far simple end-to-end butt joints seem to be unsurpassed in terms of their low loss and ruggedness. Thus, the various designs discussed in the following sections vary mainly in the ways the fiber ends are prepared, lined up, and secured in an end-to-end joining arrangement.

Different designs have evolved for different applications. Some splices, for example, are intended to stay permanently connected while others must offer an opportunity for demounting. Joints which are designed for frequent demounting and reconnection are termed "connectors." Design principles that apply to fiber connectors are covered in Chapter 15.

Another class of fiber optic connections, used primarily for joining a relatively large number of fibers in one operation, is designed for occasional demounting, for example, when connecting fiber cables. These designs are also called connectors although they, in general, do not allow for the same ease of connection or number of connect–disconnect cycles as a "connector" in the usual sense of the word.

The structure of a fiber splice also varies with the cable configuration for which it is designed. Cables comprising few fibers usually have these fibers arranged randomly and independently in the cable core (see Chapter 13). In this case, the fibers may be spliced independently using a so-called "single-fiber splice." For cables comprising a large number of fibers, it is more economical to arrange the fibers in structured arrays which can be spliced in groups without handling individual fibers. As low-cost mass production of fibers develops, the latter cable structures and multiple or array fiber splices should gain importance.

The next section describes the concepts and techniques of preparing the ends of fibers, fiber arrays, and fiber cables. A section on single-fiber splices and one on multiple fiber splices follow. The last two sections contain a discussion of the various parameters which influence the splice loss and the methods used to determine and analyze splice loss.

14.2 FIBER END PREPARATION

All of the fiber splicing techniques described in this chapter require that the ends of the fibers be prepared in such a way that, with suitable index matching, the propagating light beam will not be deflected or scattered at the fiber interfaces, and that the two ends to be spliced can be brought into close proximity. In practice this means that the ends must be flat, perpendicular to the fiber axis, and fairly smooth.

Although in principle a rough fiber surface would not cause appreciable scattering loss if an index-matching material were used which matched perfectly the refractive index of the core of the fiber, in practice it is found that smooth fiber surfaces are necessary to obtain low splice losses.

Several techniques have been proposed for fiber end preparation: sawing, grinding and polishing, and controlled fracture. The environment that a splice or connector will be fabricated in and the application that it will be used for are important factors in choosing a specific end preparation technique. It is generally agreed that for a field repair splice, the controlled fracturing or possibly the sawing technique are most appropriate. Cherin has proposed a field repair splicing method using an injection molded plastic splice where the fiber ends are prepared by a controlled fracturing technique (Cherin and Rich, 1976). Grinding and polishing of fiber ends is well-adapted to a controled factory environment. Miller has described a factory prepared fiber optic cable connector where a 144 fiber array is potted and the ends are all prepared simultaneously in approximately 15 min by a grinding and polishing technique (Miller, 1978). In a fiber optic system where different types of splices and connectors are used, it is likely that fiber ends will be prepared using both the controlled fracturing and grinding and polishing techniques. A brief description of both of these techniques will follow.

14.2.1 Controlled Fracture

Glass fibers fracture in such a way that the ends usually resemble one of the three types of fracture illustrated in Figure 14.1. The rough surface region in Fig. 14.1b is called the hackle zone. The smooth region adjacent to the fracture origin is called the mirror zone. It can be shown (Johnson *et*

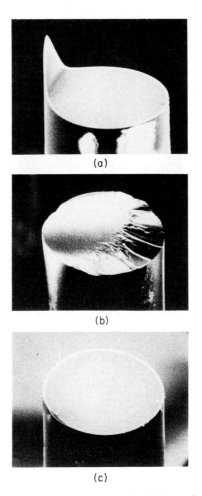

(a)

(b)

(c)

Fig. 14.1 Morphology of optical fiber ends.

al., 1966) that the boundary between the mirror and hackle zones is described by the equation

$$Z(r)^{1/2} = K, \tag{14.1}$$

where Z is the local stress at the point in question, r is the distance from the point of origin of the fracture to the mirror-hackle boundary, and K is a constant for a given material. For soda-lime-silicate glass $K = 6.1$ kg/mm$^{3/2}$ and for fused silica $K = 7.5$ kg/mm$^{3/2}$.

In order to break an optical fiber in such a way that the mirror zone extends across the entire fiber, it is necessary to have the stress at all points within the fiber low enough so that $Z(r)^{1/2} < K$ (Gloge *et al.*, 1973). The required value of Z at the origin of the fracture depends on the size of the crack or flaw from which the fracture originates. A second requirement is that Z is positive and essentially parallel to the fiber axis at any point across the fiber, or the crack will propagate in a direction which is not perpendicular to the axis of the fiber. Under these conditions, a lip is formed on one of the fiber ends, as shown in Figure 14.1a. We see, then, that in order to make a reliable clean mirror zone fracture, the stress distribution across the fiber must be suitably adjusted.

The simplest way to do this is to bend the fiber over a form of suitable curvature radius R while simultaneously applying tension. This method produces a uniform tensile stress T and, superimposed, a stress (E/R) $(b - x)$ where b is the fiber radius, E Young's modulus for silica and x a coordinate as indicated in Fig. 14.2. When a fracture is initiated by a surface flaw at $x = 0$, it advances from this point essentially at equal velocity

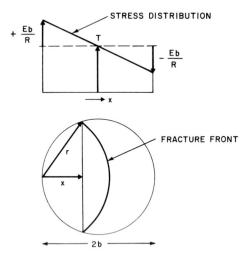

Fig. 14.2 Coordinate system for fiber stressed by tension T and bent over form of radius R (E-Young's modulus of glass).

in all directions. The stress present in the fracture front at any given instant is largest at the fiber circumference where

$$r = (2xb)^{1/2}. \tag{14.2}$$

After introducing (14.2) into (14.1) we find the limits

$$0 < Z(2xb)^{1/4} < K \qquad \text{for } x < b \tag{14.3}$$

as a necessary condition for avoiding both hackle and lip formation. Figure 14.3 is a plot of $Z(x)$ which shows the limits (14.3) cross-hatched for a fiber diameter $2b = 100$ μm.

The stress Z' needed to initiate a fracture depends on the scoring technique used and on the fiber material. On the average, $Z' = 25$ kg/mm² for fused silica scored with diamond or carbide tips (Gloge *et al.*, 1973). The variations in Z' from score to score can be significant. As indicated by the dashed line in Fig. 14.3, tensile stress without curvature (uniform stress for all x) produces hackle in fibers having a diameter larger than 75 μm if $Z' = 25$ kg/mm². Conversely, bending without tension produces lips since Z vanishes at $x = b$. The simultaneous application of tension and curvature offers an adequate margin against both limits. Consider, for example, an increase of the uniform tensile stress from 15 to 20 kg/mm²

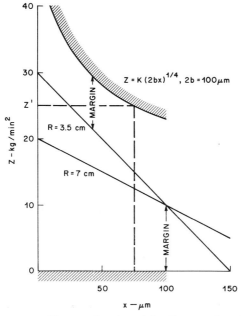

Fig. 14.3 Local stress Z in fiber as a function of the distance x from the origin of the fracture.

while the curvature radius R decreases from 7 to 3.5 cm. In this case, the stress distribution $Z(x)$ changes between the two straight lines in Fig. 14.3 ($E = 7000$ kg/mm² for silica). As Z at $x = 0$ reaches a magnitude sufficient for fracture initiation (typically between 20 and 30 kg/mm²), the break occurs at conditions limited by the two lines of Fig. 14.3 with a margin of 10 kg/mm² against lip formation and 8 kg/mm² against hackle.

For fibers with diameters $2b > 100$ μm, the hackle limit (upper cross hatched line in fig. 14.3) moves to lower stress values resulting in less margin for safe operating conditions. The limit of achieving mirror zone fracture is reached when it becomes impossible to draw a line through Z' that does not intersect the cross-hatched areas. The evaluation of this condition yields

$$b_{max} = 1.75(K/Z')^2 \qquad (14.4)$$

which is the radius of the largest fiber that can be fractured successfully with this method. for fused silica and $Z' = 25$ kg/mm², $b_{max} = 157$ μm, and $R_{max} \approx 10$ cm.

There are some fibers that have built-in stress due to the differential compression between the core and the cladding regions on cooling from the melt. Values up to 10 kg/mm² are possible (Paek and Kurkjian, 1975). When these stresses are of the order of, or larger than, the margin available for a safe fracture process, corrugated or even hackle surfaces result, as one must expect from (14.1) where Z is now the total stress produced by internal and external forces. Recent work (Albanese and Maggi, 1976) indicates that it is possible in many cases to relieve the internal stresses by heating the fiber prior to fracturing; in this way, mirror zone fractures can be obtained across the entire face of the fiber.

Fiber breaking machines based on these ideas have been described by several authors (Gloge et al., 1977; Hensel, 1975; Fulenwider and Dakss, 1977). These machines used diamond or tungsten carbide blades to initiate the score. It has also been shown that a spark erosion technique can be used to score fibers with azimuthal scratches so that a low value of Z' is obtained (Sklyarov, 1975); Caspers and Neumann, 1976; Hensel, 1977). Kohe and Kuyt have described a hot wire technique for scoring and creating a nonuniform stress distribution (Kohe and Kuyt, 1977).

It has been shown that under some conditions mirror zone fractures made with these fiber breaking machines are not perpendicular to the fiber axis, but may deviate from perpendicularity by a few degrees (Gordon et al., 1977a). Even for dry splices, however a deviation from perpendicularity of 4° results in a contribution to splice loss of only ~0.15 dB (Gordon et al., 1977b). A negligable contribution would be expected for index-matched splices with typical multimode fibers.

It seems clear that the controlled fracture technique is useful in the preparation of single fiber ends. The technique can easily be extended to linear arrays of fibers (Chinnock *et al.*, 1975; Cherin and Rich, 1975) and all-glass optical fiber tapes (Bisbee and Smith, 1975).

14.2.2 Grinding and Polishing

Grinding and polishing equipment and methods used for metallographic sample preparation often apply directly to fiber end preparation especially when the fiber is supported in a glasslike matrix. End preparation for the single-fiber connectors discussed in Chapter 15 sometimes involves polishing techniques; multifiber potted array connectors to be discussed later in this chapter rely solely on polishing methods for end preparation. Grinding and polishing have the advantage that the fiber ends can be prepared reproducibly perpendicular to the fiber axis and with the required degree of smoothness and that many fiber ends can be prepared in a short time without handling individual fibers.

Mechanical polishing principles are conceptually simple. (1) Abrasive action is applied by motion of the sample relative to an abrasive surface. (2) Successively finer abrasives are used until the scratches caused by the previous abrasive have been replaced by the finer scratches of the current abrasive. (3) This is repeated until the desired degree of polish is obtained. In metallographic sample preparation, many stages of grinding and polishing are usually required to obtain metal sample surfaces that exhibit the true metal microstructure. Grinding and polishing requirements for optical fiber end preparation are generally less stringent than that required for metallography so that fewer stages of polishing are needed, usually two or three.

There is an art to preparing high-quality hand-polished surfaces; however, the use of automated equipment that controls polishing pressure and time, greatly reduces the required operator skill level. At present, grinding and polishing as a method for fiber end preparation seems applicable only to a controlled environment (a lab or factory); however, advancements in equipment and methods may eventually allow polishing in a field environment.

14.3 SINGLE-FIBER SPLICES

The two fiber ends joined in a fiber splice can either be held in place by a surrounding structure, a capillary tube, a grooved substrate, or precision pins, or the two fibers can be permanently attached to each other by adhesives or welding. In the latter case, the temperature of the joint is raised to the softening temperature of the fiber material so that the core and cladding materials from either end fuse together. If the fibers consist of glasses

having a softening temperature of 700°C, the heat from a nichrome wire
loop surrounding the joint is sufficient for welding (Bisbee, 1971; Dyott *et
al.*, 1972).

Silica fiber joints require an electric arc (Kohanzadek, 1976; Bisbee,
1976) or a microtorch (Jocteur and Tardy, 1976) to generate the required
welding temperature of 1600°C. Figure 14.4 shows the laboratory apparatus used by Bisbee to prepare silica fiber splices. The arc electrodes (A)
are sharpened tungsten welding rods, 3.18 mm in diameter, mounted in a
piece of transite for insulation from heat and high voltage. The transite
holding the electrodes is mounted on a manipulator. Its shaft (B) is
motor-driven to carry the arc into the slot of a vacuum chuck (C) (Benson
and Mackenzie, 1975). The vacuum chuck holds and aligns the fibers (D)
while they are being spliced.

The fiber ends are prepared by scoring, stressing, and bending as discussed in Section 14.2. Two fiber ends so prepared are placed in the top
and bottom of the vacuum chuck, also shown in Fig. 14.4, so their end
faces touch in the middle of the slot. After adjusting the position of the
electrodes vertically so the arc will pass across the junction of the fibers,
the arc is moved across, fusing the fibers together.

Good results were obtained with an ac arc current of 22 mA, an arc gap
of 1.3 mm, and a manipulator speed of 0.6 mm/sec. At that speed, the arc
is in contact with the fibers for about 2 sec. Figure 14.5 is a microphotograph of a typical splice. The joint leaves no discernable deformation.

Figure 14.6 is a histogram of the loss per splice, and Fig. 14.7 is a graph
of the cumulative distribution of loss obtained for 45 splices. The lowest
loss is 0.03 dB, the highest 0.55 dB, and the average 0.14 dB. The higher
loss values are probably caused by contamination. The fiber ends tend to

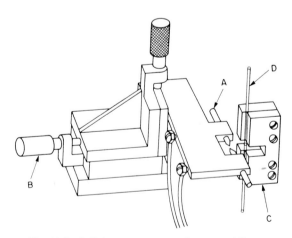

Fig. 14.4 Splicing arrangement for arc welding.

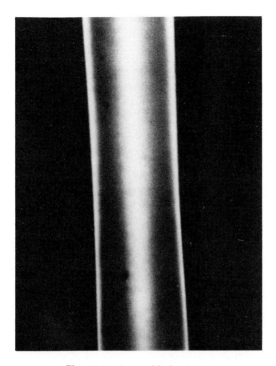

Fig. 14.5   Arc-welded splice.

develop a static charge that attracts dust particles. Defective fiber end surfaces can be conveniently heat polished prior to fusion by applying the electric arc to the surfaces while the fibers are still several micrometers apart (Hirai and Naoya, 1977).

Surface tension present during the fusion process generates forces in the molten silica which tend to correct a transverse misalignment of the two

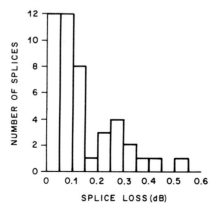

Fig. 14.6   Histogram of loss in arc-welded splices.

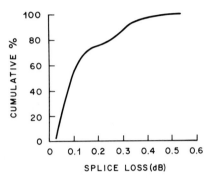

Fig. 14.7 Cumulative percentage of splices having loss less than abscissa for arc-welded joints.

fiber ends (Hatakeyama and Tsuchiya, 1978a). A threefold reduction of the loss resulting from transverse or angular misalignment of multimode fibers has been achieved. The effect is especially helpful in single-mode fiber splices where minute displacements cause noticeable loss. By maximizing the effect of surface tension, Hatakeyama and Tsuchiya achieved average splice losses of 0.1 dB in single-mode fibers having a core diameter of 10 μm and an index difference of 0.19% (Hatakeyama and Tsuchiya, 1978b). They found that a heating time of 3 sec at a discharge power of 8.5 W (1-mm arc length) established thermal equilibrium without causing detrimental overheating. The fusion temperature at this optimal condition was 2000°C. Alignment improved fourfold. As a result, alignment loss decreased almost sixteen-fold as expected from the square-law relationship between the loss c_s and the offset s in (3.107).

Even though the molten core ends usually fuse together with the initial misalignment and subsequently suffer a distortion while the fiber claddings align under surface tension, the fundamental mode traverses the fusion region essentially on a straight-line path determined by the core ahead of the fusion zone and aligned with the core behind it. Residual splice loss depends to a large extent on the effectiveness of the surface forces and on any core eccentricity which remains of course uncorrected.

A mechanical evaluation on 20 splices showed that spliced fiber strength may vary anywhere from original fiber strength to one-third strength, with the average strength of spliced fibers equal to 61% of the average original fiber strength (Kohanzadeh, 1976). It is likely that handling of uncoated fiber ends during splicing causes surface flaws, thus reducing fiber strength, whereas the splice itself is possibly as strong as the fiber.

Several investigators have fabricated single-fiber splices using snug-fitting capillary tubes into which the fiber ends are pushed from opposite sides (Derosier and Stone, 1973; Pinnow, 1974). Thermal shrinking plastic tubes and glass sleeves are being used. The glass sleeves may be fabricated with a center hole for insertion of an adhesive (Murata *et al.*, 1975).

If the glass sleeve is to support the fiber with the required alignment accuracy, its bore must be less than typically 0.0001 in. larger than the fiber. Both fiber and sleeve must be highly circular, and the fiber diameter must be controlled to at least the same tolerances. The insertion of the fibers into the tube is difficult unless the tube ends are flared. Contaminants that may be scraped off the inside wall of snug-fitting sleeves during fiber insertion are trapped between the fiber ends where the effect of contamination is worst. These dificulties have discouraged efforts in using snug-fitting tubes.

Better results have been obtained with "loose" metal tubes which are crimped to form a three-point contact with the fiber (Dalgleish *et al.*, 1975). Losses with this technique are reported to be 0.25 to 0.35 dB with index-matching material.

A loose tube with a square cross section (see Fig. 14.8) has the advantage that the fiber ends can be pushed to one corner of the square cross section by bending the fiber outside the tube for better alignment (Miller, 1975a). The tube has nearly flat interior walls and a small radius in the interior corners, as shown in the cross section in Fig. 14.9. Epoxy is forced into the square tube prior to insertion of the fibers and serves as an index-matching adhesive. The flow of epoxy around the fiber ends during insertion of the fibers also helps in removing contaminants from the critical junction area.

Assembly of a splice involves inserting two fibers with good ends approximately halfway into each end of a square cross section tube filled with uncured epoxy. No particular orientation of the square tube cross section is required. The fibers are placed on a flat surface and bent in a curved pattern. This causes forces to be generated that rotate the tube so

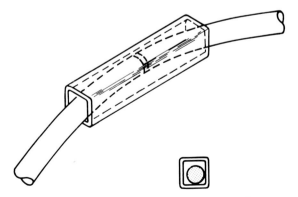

Fig. 14.8 Loose-tube splice using square capillary.

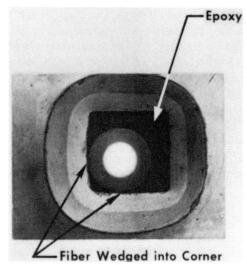

Fig. 14.9 Micrograph of splice cross section.

that a diagonal of the square cross section is in the same plane as the bent fibers and the fibers line up in the same corner of the tube. The bent fibers are then pushed into the tube until they touch each other. Figure 14.9 is a cross-sectional photograph of a splice showing a fiber in a corner of the square tube. Figure 14.10 shows a magnified view of one splice and part of a coin for dimensional comparison. Eight epoxied square-tube splices fabricated in series produced a total splice loss of 0.58 dB or 0.073 dB per splice. The splices were spaced by about 1.5 m of fiber.

Inexpensive precision steel pins accurate to ± 0.5 μm are commercially available from roller-bearing manufacturers. These pins have been clustered around fibers for alignment and held in place by heat shrinkable tubes or metal clips (O'Hara, 1975; Nippon Telephone and Telegraph Co., 1975). Figure 14.11 shows a sectional view of a splice of this kind. The splice is made by inserting the optical fiber between the three guides for

Fig. 14.10 Loose-tube splice 1.5 cm long, 0.3 mm wide.

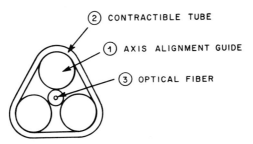

② CONTRACTIBLE TUBE

① AXIS ALIGNMENT GUIDE

③ OPTICAL FIBER

Fig. 14.11 Cross-sectional view of splice using steel pin alignment. (E.C.L. Technical Publication #131, 1975.)

alignment and by shrinking a contractible tube around this assembly. Index-matched epoxy is applied to prepare a permanent splice. Figure 14.12 is a splice loss histogram obtained by splicing multimode fibers with this technique which had a core diameter of 80 μm.

14.4 ARRAY SPLICES

The method of aligning fibers in a small v-groove, which was initially used to make single-fiber splices in the laboratory (Someda, 1973), turned out to be most conveniently adaptable to linear array splices. For this purpose, the fibers are arranged in linear arrays or ribbons in the cable (Standley, 1974). (See Chapter 13 for cable fabrication.)

Grooved substrates have been prepared by embossing or molding plastic materials (Cherin and Rich, 1975), by stamping metal (Miller, 1975b), and by preferentially etching silicon wafers (Schroeder, 1978). The best control of groove dimensions is achieved by the preferential etching of silicon. In this case, groove dimensions can be held within a tolerance of less than 1 μm (Schroeder, 1978). The transverse groove dimensions in injection molded plastic parts have been held to within 2 μm of the dimensions of the metal master used in the mold (Cherin and Rich, 1975). Both the sili-

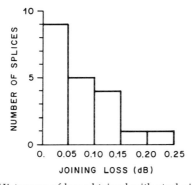

Fig. 14.12 Histogram of loss obtained with steel pin alignment.

con wafer and injection molded plastic technologies are well adapted for large-scale manufacturing processes.

Groove structures have been used for a number of different splicing and connector applications. Descriptions of repair splicing techniques and of a remountable factory-prepared cable connector will follow.

14.4.1 Array Repair Splices

To prepare a ribbon end for repair splicing, the plastic cover is removed from around the fibers so that the bare fibers protrude from the ribbon end in a linear array (see Fig. 14.13). The fiber ends are prepared using the controlled fracturing technique described in Section 14.2 (see also Chinnock *et al.*, 1975) and placed into a grooved substrate. The grooves have about the same spacing as the fibers of the array so that the fibers slide into the grooves without additional guidance. The two ribbon ends to be spliced are placed in the grooves from opposite sides. The fiber ends are pushed together and pressed down into the grooves by a cover plate. An index-matched epoxy is added to fabricate a permanent assembly.

Variations of this technique differ in the way the substrate is prepared and in the tools used to close the splice. Figure 14.14 illustrates the tool used to fabricate a splice with an embossed grooved plastic substrate (Cherin and Rich, 1975). A substrate of plastic was placed into the tool and was cold embossed by applying pressure to the positive embossing head. Using slotted ramps in the tool fiber arrays are guided into the grooves in the substrate to form a butt joint. The permanent splice is then completed by bonding a cover plate to the plastic substrate using a holding fixture fastened to the tool.

Figure 14.15 shows how this technique can be expanded to splice 12 ribbons side by side in one operation. In this case, the substrate was an injection molded plastic part that contained 12 sections each equipped

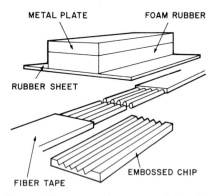

Fig. 14.13   Groove alignment technique used for ribbon splicing.

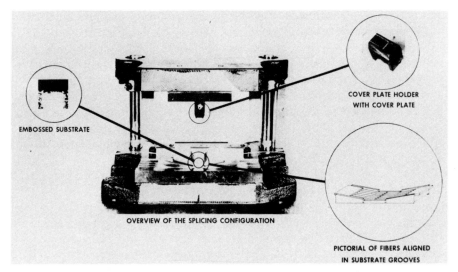

COVER PLATE HOLDER
WITH COVER PLATE

EMBOSSED SUBSTRATE

OVERVIEW OF THE SPLICING CONFIGURATION

PICTORIAL OF FIBERS ALIGNED
IN SUBSTRATE GROOVES

Fig. 14.14 Ribbon splicing technique adaptable to field splicing.

with a prealignment slot and 12 grooves 90 μm apart. The splicing tool placed the ribbons into the prealignment slots and guided the fibers into their respective grooves in the substrate (Cherin and Rich, 1976). A cover plate was then attached to the substrate and matching material injected through a slot in the cover plate to complete the splice. The completed splice can join two optical cables, each consisting of 12 ribbons that house 90 μm o.d. fibers. This type of molded splice yielded an average splice loss of 0.2 dB for the 425 splice joints measured. Fifty percent of the losses measured were less than 0.1 dB and 95% of the splice joints had losses less than 0.8 dB.

Contamination of the fiber ends as they slide along the prealignment slots and the grooves may be a problem encountered in a field environment. This problem can be avoided by a modification of the groove align-

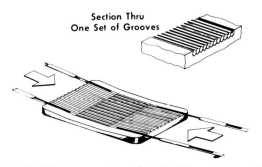

Section Thru
One Set of Grooves

Fig. 14.15 Substrate used to splice 12 ribbons side by side.

ment technique (Chinnock *et al.*, 1975). This method employs a substrate with two sets of grooves which line up with respect to each other when the substrate is folded about a central groove (see Fig. 14.16). Each ribbon end is attached to one set of grooves from opposite sides. Only then are the fiber ends prepared by the method described in Section 14.2. Immediately after the end preparation, the substrate is folded and the fiber ends come to line up in the channel formed by two grooves. An index-matched epoxy is used for a permanent splice assembly. Laboratory tests produced a splice loss distribution with 50% of the splices having a loss below 0.16 dB and 95% below 0.4 dB.

Another ribbon splice that could be used is illustrated in Figure 14.17 (Smith *et al.*, 1975). The splice is cast around the fibers in a casting form which contains a ridge having small indentations. The fibers rest in these indentations and are aligned by them. After the form is removed from the casting, the fibers are exposed where they rested on the ridge; they can be scored and fractured in these places in the way explained in Section 14.2. A brittle casting material is used which fractures along the same plane as the fibers. Two connectors so prepared are inserted into a sleeve or channel using the narrow sides of the connectors for alignment. A cumulative loss distribution for this type of splice is shown in Figure 14.18. It

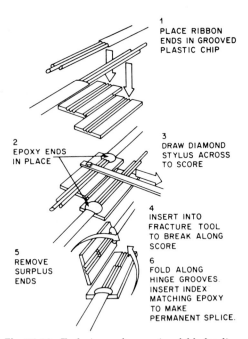

Fig. 14.16 Technique of preparing folded splice.

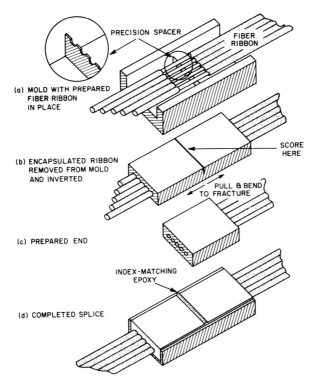

Fig. 14.17 Technique of preparing cast ribbon connector.

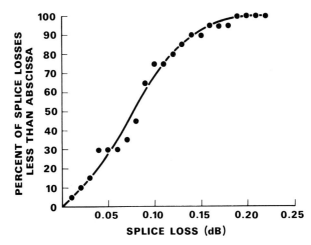

Fig. 14.18 Cumulative percentage of splices having loss less than abscissa for cast ribbon connector.

shows that 50% of the splices had losses of less than 0.08 dB, and 95% had losses less than 0.2 dB.

14.4.2 Array Connectors

A fiber optic array connector based on the groove alignment technique can be used for joining cable sections in the field (Miller, 1976). As presently conceived, fabrication of an array connector involves the following operations:

(i) aligning all fibers of one end of a fiber-optic cable into a uniform matrix,
(ii) potting the structure to retain the geometry,
(iii) grinding and polishing the ends of the potted array,
(iv) joining two cable ends prepared by the previous three operations.

Several techniques have been studied to align fibers in a two-dimensional array. Threading fibers through holes as opposed to laying fibers in grooves is in general a more difficult and less accurate method of fiber alignment. The following paragraphs elaborate on a technique that involves a thin chip design which is grooved on both sides (Miller, 1977).

These chips provide the primary alignment mechanism for assembly of fibers into a uniform rectangular array. These chips are fabricated by preferential etching of silicon wafers (Schroeder, 1978). Ribbon ends are prepared by removing the supporting material and are then interleaved between chips to form a stacked rectangular array (see Fig. 14.19).

A stacked array, held by a vise, can be potted by allowing epoxy to seep through the array. Approximately 15 min is required for this operation

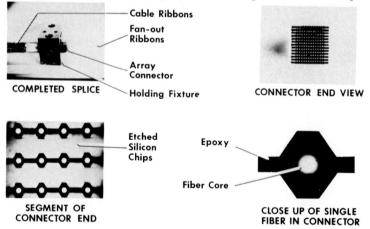

Fig. 14.19 Details of ribbon cable connector; fiber alignment between silicon chips (bottom); side and front view of connector (top).

after which the epoxy cures. Fast curing of the epoxy is possible at elevated temperatures; however, at room temperature, several hours are required for the epoxy to completely cure. Chips with ridges that mate with the unoccupied grooves of the stacked array will be referred to as negative chips. These negative chips are used to align the top and bottom chips of the stacked array while in the vise.

Grinding and polishing has been found to be the best end preparation technique for this array connector. Sawing is possible, but this method is more time consuming and does not yield fiber ends as smooth as grinding and polishing. A polishing fixture is used for supporting the array and aligning the axis of the array during the polishing operation. Using a grit sequence of 800x, 9.5 and 0.3 μm, an array connector containing 144 fibers can be polished in approximately 15 min.

Since the chips are grooved on both sides, the top and bottom chips of the stacked array have unoccupied grooves which can be used as references for alignment during polishing and subsequently to align the two rectangular arrays in forming the butt joint. Plexiglass and steel fixtures have been used for final alignment in conjunction with grooved negative chips. These negative chips are pressed against the top and bottom of the two connectors and then placed in a final alignment fixture which further aligns the two connectors and holds them in place. This arrangement is shown in Fig. 14.19 along with magnified views of a connector end.

Splice losses in array connectors with identical fibers average approximately 0.2 dB with array uniformity averaging 2–4 μm.

14.5 SPLICE LOSS PARAMETERS

We distinguish two sources of splice loss: (1) imperfections in the splice preparation, and (2) differences in the characteristics of the two fibers to be spliced. Only the former can be influenced by the splice design. The critical parameters in this case are fiber alignment and placement, the preparation of good end faces, and a contamination-free, transparent, index-matching material that fills the gap between the fiber ends. The preparation and quality of end faces are discussed in Section 14.2. Some precautions against contamination are mentioned in the previous sections. Theoretical aspects of fiber misalignment and dimensional variations are treated in Section 3.9. It remains here to derive from this information the critical aspects of splice design and, where possible, to attempt a prediction of splice losses to be expected in a practical situation.

The elimination of reflection loss in a dry joint requires almost perfect surface contact (separation not more than nanometers). Attempting to achieve this consistently with fiber end surfaces seems hopeless indeed. If the gap is larger than a wavelength, the mean total reflection loss at the two end faces is

$$c_r = 2\left[\frac{n_1 - n_{gap}}{n_1 + n_{gap}}\right]^2,\qquad(14.5)$$

where n_1 is the refractive index of the core and n_{gap} is the refractive index of the gap. In the absence of an index matching material, $n_{gap} = 1$ and $c_r = 8\%$ or 0.4 dB.

Even if core and gap index are matched on axis, the gap lacks the lateral index variation necessary for guidance causing a loss which increases with the length of the gap. If both fiber surfaces are plane and perpendicular to the fiber axes, a ray analysis applicable to a uniformly excited step-index multimode fiber yields the estimate (Nippon Telephone and Telegraph Co., 1975).

$$c_z = (z/4a)(2\Delta)^{1/2},\qquad(14.6)$$

where z is the gap length. A nonuniform mode power distribution results in a smaller loss than indicated by (14.6), but estimates for this case as well as for graded-index fibers are not available. A comparison of (14.6) with the offset loss (3.110) shows that the tolerance for transverse displacements is by a factor $8/\pi(2\Delta)^{1/2}$ more stringent than the tolerance for end separation.

If the fiber ends are good and the gap is small, the index-match need not be very accurate for the reflection loss to be small. This fact is evident from (14.5) and supported by measurements. For fiber end faces that are chipped or not perpendicular to the fiber axis, a splice loss results, which is critically dependent on the index mismatch and the gap width. Slight end face imperfections as they occur with the fracture technique described in Section 14.2 are usually not serious enough to cause a significant increase in splice loss when an index-matching material is used.

Adequate fiber alignment was the overriding consideration in the designs discussed in Sections 14.3 and 14.4. The theoretical treatment in Section 3.9 distinguishes between three possibilities: (1) transverse misalignment (offset), (2) angular misalignment (tilt), and (3) end separation (gap). For typical graded-index multimode fibers and a nonuniform (steady-state) power distribution in the modes, (3.117) predicts that an offset of 10% of the core radius *or* a tilt of 7% of the numerical aperture, each in the absence of the other, cause 0.1 dB of splice loss. This estimate includes excess loss behind the splice.

The values correspond to 2.5 μm offset and 1° tilt for a typical multimode fiber having a core radius of 25 μm and a numerical aperture of 0.25. The use of grooves, pins, or tubes typically provides an angular alignment accuracy of this magnitude or better; the offset considered above is more difficult to achieve.

If we assume perfect angular alignment, an rms offset of 10% of the core radius, and a normal distribution of offset errors, curve (1) of Fig. 14.20 represents the cumulative loss distribution to be expected on the basis of the splice loss formula derived for c_σ in Section 3.9. The loss of 90% of all splices is less than 0.25 dB. If, in addition, an rms tilt of 7% of the numerical aperture exists and the distribution of tilt errors is also normal, curve (2) applies. The 90% point is at 0.45 dB. Practical splices meet narrower angular tolerances; their loss distribution can be found between curves (1) and (2). Figure 14.20 applies only to graded-index multimode fibers having a steady-state power distribution, but should serve as a good first estimate in all situations where the splice is not closer than a few meters from the source.

The loss distribution from alignment errors in a single-mode fiber splice is given by (3.107) and (3.108). As an example, consider a step-index single-mode fiber having a core radius of 5 μm and a relative index difference of 0.2% of the core index. If the offset is 1 μm or 20% of the core radius, the loss is 0.17 dB in the absence of any other imperfection. Similarly, a 0.2° tilt causes a loss of 0.17 dB in the absence of any other imper-

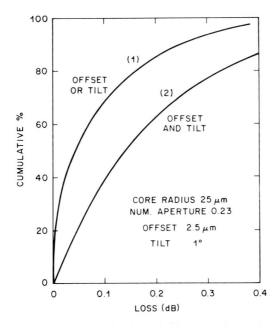

Fig. 14.20 Cumulative percentage of multimode fiber joints having loss less than abscissa in the case of offsets and/or angular misalignment.

fection. If both alignment errors are present, have a normal probability distribution, and rms values of 1 μm and 0.2°, respectively, Figure 14.21 applies. The loss of 10% of all splices is expected to be larger than 0.8 dB.

Differences in the two fibers to be spliced lead to loss no matter what the quality of the splice. The difference may be in the diameters of the cores or the claddings, in the index profiles, or it may be a possible asymmetry of the fiber cross sections. Some of these imperfections, as, for example, a difference in the outer fiber diameters or eccentricities of the cores, may indirectly lead to transverse offset of the kind discussed earlier. However, if the core or the index difference of the receiving fiber is smaller than that of the transmitting fiber, a loss results even with perfect alignment and symmetry.

The theoretical aspects of this situation are discussed in Section 3.9. As in the case of misalignment losses, the power distribution in the modes affects the loss encountered in splices of unequal multimode fibers. Figure 14.22 shows typical cumulative loss distributions for the case of core radius differences with normal probability distribution and a standard deviation of 4.2% of the core radius. The assumption of a uniform mode power distribution (curve 1) leads to significantly higher losses than those obtained for an equilibrium distribution (curve 2). Most practical cases are likely to fall between the two distributions shown, but losses even lower than those expected for the equilibrium are possible if the higher modes are strongly attenuated.

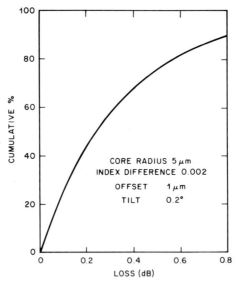

Fig. 14.21 Cumulative percentage of single-mode fiber joints having loss less than abscissa in the case of offsets and angular misalignment.

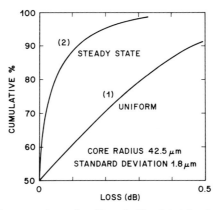

Fig. 14.22 Cumulative percentage of multimode fiber joints having loss less than abscissa in the case of diameter variations.

In practice, diameter and profile differences must be expected at the same time. Figure 14.23 illustrates the case of a 1.8-μm-diameter variation in the presence of differences in the numerical aperture and the profile exponent (Thiel and Davis, 1976). The standard deviation of the numerical aperture variation is 0.009, that of the profile exponent 0.16. The plot is the result of 50,000 trial computations obtained with the assumption of a uniform mode power distribution. As mentioned earlier, this assumption is likely to lead to a highly conservative result which must be considered more as an upper bound than as the description of a practical situation.

Nevertheless, even if losses of this kind turn out to be smaller than indicated in Fig. 14.23, they should receive serious attention. Tighter fiber tolerances require intensive efforts in improving fiber manufacturing pro-

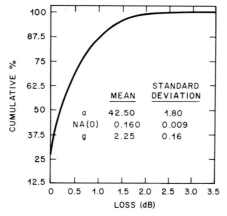

Fig. 14.23 Cumulative percentage of multimode fiber joints having loss less than abscissa in case of diameter, index and profile variations. (Thiel and Davis, 1976.)

cesses and are therefore likely to be more expensive than the development of more sophisticated splicing techniques. Hence, the splice loss caused by variations in the fiber characteristics may ultimately prove to be a more significant barrier to splice loss reduction than alignment accuracy.

14.6 MEASUREMENT OF SPLICING EFFECTS

Figure 14.24 shows the basic setup for splice loss measurements. Power is coupled into the transmitting fiber and propagates through the splice to a receiver arrangement. The measurement is performed by comparing the loss of a spliced transmission line to a reference line without splice. Since the splice loss is usually much smaller than the line loss or the launching loss, care must be taken that the spliced line and the reference line are identical in every respect except for the splice. This is most easily guaranteed if the total transmission loss can be measured before the line is broken and reconnected at the splice point. In this case, the launching assembly need not be changed between the two measurements.

If this procedure is not convenient or possible, the establishment of identical launching conditions is an essential consideration. The optimization of the launching conditions for maximum output (Cherin and Rich, 1975) is convenient, but fails to guarantee reproducible launching conditions in some graded-index fibers. In this case, a telescope can be incorporated in the launching arrangement which permits a visual inspection of the fiber front face during the adjustment of the source beam (Smith *et al.*, 1975).

In multimode fibers, splice loss depends strongly on the mode power distributions (Section 3.9). A measurement must attempt to establish definitive mode conditions and results should be quoted together with relevant information on what these conditions were or how they were achieved. For example, if a splice loss measurement under conditions of uniform mode power distribution is intended, it seems natural to use an incoherent source which fills the near- and far-field distribution of the

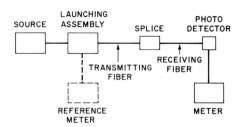

Fig. 14.24 Principal arrangement for the purpose of measuring splice loss.

fiber modes and to take care that the launching assembly does not obstruct the source beam in a way that would jeopardize that objective.

Even with these precautions, a uniform mode power distribution may be difficult to achieve, as indicated by Fig. 14.25. Figure 14.25 shows measurements of offset and tilt loss in a splice made with graded-index multimode fibers (Chu and McCormick, 1978) together with the theoretical expectations for a uniform and an equilibrium mode power distribution. Even though a uniform mode excitation was attempted and probably achieved at the input, the fiber between splices seems to have produced a sufficient amount of mode coupling and high-order mode loss that the uniform distribution was significantly modified at the splice point. Under such circumstances, it may become somewhat irrelevant what source is being used at the input. In fact, a coherent source (He–Ne laser) can give results very similar to those depicted in Fig. 14.25 if care is taken that the Gaussian laser beam fills the solid angle of acceptance of the fiber (Gloge, 1976).

If the intent is to measure splice loss under equilibrium conditions, the transmission lengths on both sides of the splice should be chosen long enough that the equilibrium mode power distribution can establish itself before and after the splice (see Section 3.11). The length required depends on the fiber used and can be as short as 10 m or as long as 1 km or more. As Fig. 14.26 shows (Miller, 1976; Gloge, 1976), offset loss measurements performed under such conditions have produced results that are in good agreement with computations on the basis of (3.117).

For some routine measurements, an articifical generation of equilibrium conditions in the splice is necessary or convenient. To do this, a strong

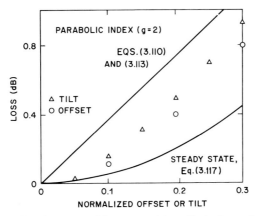

Fig. 14.25 Expected and measured loss caused by offset of angular misalignment of graded-index multimode fiber joints; measurement was aimed at achieving uniform mode power distribution.

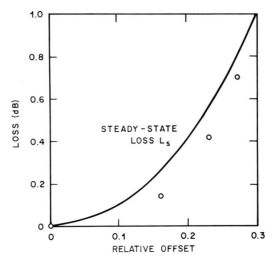

Fig. 14.26 Expected and measured loss caused by offsets in graded-index multimode fiber joints; equilibrium achieved with the help of ½-km fiber on both sides of joint.

mode power exchange can be produced in the fiber by pressing sections of the transmitting or receiving fiber between corrugated surfaces. Emery paper has been used. The mode power distribution established in this way is reproducible and usually close to the equilibrium distribution. Whatever method is used, it is recommendable that the source beam fill the solid angle of acceptance at the fiber input.

Apart from being less dependent on the launching conditions, splice measurements under equilibrium conditions are believed to yield results that are more definitive and more representative of transmission line conditions than measurements designed to achieve a uniform mode power distribution.

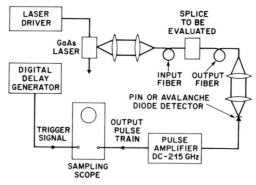

Fig. 14.27 Arrangement to measure pulse distortion caused by fiber splice.

Everything said here about the dependence of splice loss on the mode power distribution is even more pertinent to the effect that the splice has on the impulse response. The most relevant measurement seems again to be the one which allows an equilibrium distribution to establish itself between source, splice, and receiver. In this case, the splice may be considered as a discrete source of mode mixing (Ikeda *et al.*, 1977) and high-order mode loss both of which add randomly to transmission line effects. The result is a net shortening of the impulse response. Such predictions are supported in a qualitative way by available measured results (Cherin and Rich, 1977). The measurement system that was used is sketched in Fig. 14.27. The measurement procedure is in principle the same used for fiber dispersion measurements (Section 11.4).

REFERENCES

Albanese, A., and Maggi, L. (1977). New fiber breaking tool. *Appl. Opt.* **16**, 2604.

Benson, W. W., and Mackenzie, D. R. (1975). Optical fiber vacuum chuck. *Appl. Opt.* **14**, 816.

Bisbee, D. L. (1971). Optical fiber joining technique. *Bell Syst. Tech. J.* **50**, 3153.

Bisbee, D. L. (1976). Splicing silica fibers with an electric arc. *Appl. Opt.* **15**, 796.

Bisbee, D. L., and Smith, P. W. (1975). All-glass optical fiber tapes. *Bell Syst. Tech. J.* **54**, 479.

Caspers, F. R., and Neumann, E. G. (1976). Optical-fibre end preparation by spark erosion. *Electron. Lett.* **12**, 443.

Cherin, A. H., and Rich, P. J. (1975). Multigroove embossed-plastic splice connector for joining groups of optical fibers. *Appl. Opt.* **14**, 3026.

Cherin, A. H., and Rich, P. J. (1976). An injection molded plastic connector for splicing optical cables. *Bell Syst. Tech. J.* **55**, 1057.

Cherin, A. H., and Rich, P. J. (1977). Delay distortion characteristics of optical fiber splices. *Appl. Opt.* **16**, 497.

Chinnock, E. L., Gloge, D., Bisbee, D. L., and Smith, P. W. (1975). Preparation of optical fiber ends for low-loss tape splices. *Bell Syst. Tech. J.* **54**, 471–477.

Chu, T. C., and McCormick, A. R. (1978). Measurements of loss due to offset, end separation and angular misalignment in graded index fiber excited by an incoherent source. *Bell Syst. Tech. J.* **57**, 595–602.

Dalgleish, J. R., Lukas, H. H., and Lee, J. D. (1975). Splicing of optical fibers. *Proc. Eur. Conf. Opt. Fibre Commun.*, *1st, 1975* pp. 87–89.

Derosier, R. M., and Stone, J. (1973). Low-loss splices in optical fibers. *Bell Syst. Tech. J.* **52**, 1229.

Dyott, R. B., Stern, J. R., and Steward, J. H., (1972). Fusion junction for glass-fibre waveguides. *Electron. Lett.* **8**, 290.

Fulenwider, J. E., and Dakss, M. L. (1977). Hand-held tool for optical-fibre end preparation. *Electron. Lett.* **13**, 578.

Gloge, D. (1976). Offset and tilt loss in optical fiber joints. *Bell Syst. Tech. J.* **55**, 905–916.

Gloge, D., Smith, P. W., Bisbee, D. L., and Chinnock, E. L. (1973). Optical fiber end preparation for low-loss splices. *Bell Syst. Tech. J.* 1579–1588.

Gloge, D. C., Smith, P. W., and Chinnock, E. L. (1977). Apparatus for breaking brittle rods or fibers. U.S. Patent 4,027,817.

Gordon, K. S., Rawson, E. G., and Nafarrate, A. B. (1977a). Fiber-break testing by interferometry: A comparison of two breaking methods. *Appl. Opt.* **16**, 818.

Gordon, K. S., Rawson, E. G., and Norton, R. E. (1977b). Splice losses in step-index fibers: Dependency on fiber-break angle. *Appl. Opt.* **16**, 2372.

Hatakeyama, I., and Tsuchiya, H. (1978a). Fusion splices for optical fibers by discharge heating. *Appl. Opt.* **17**, 1959–1964.

Hatakeyama, I., and Tsuchiya, H. (1978b). Fusion splices for single-mode optical fibers. *J. Quantum Electron.* **QE-14**, 614–619.

Hensel, P. (1975). Simplified optical-fibre breaking machine. *Electron. Lett.* **11**, 581.

Hensel, P. (1977). Spark-induced fracture of optical fibres. *Electron. Lett.* **13**, 603.

Hirai, M., and Naoya, w. (1977). Melt splice of multimode optical fiber with an electric arc. *Electron. Lett.* **13**, 123–125.

Ikeda, M., Murakami, Y., and Kitayama, K. (1977). Mode scrambler for optical fibers. *Appl. Opt.* **16**, 1045.

Jocteur, R., and Tardy, A. (1976). Optical-fiber splicing with plasma torch and oxyhydric microburner. *Proc. Eur. Conf. Opt. Fiber Commun., 2nd, 1976* pp. 261–266.

Johnson, J. W., and Holloway, D. G. (1966). On the shape and size of the fracture surfaces. *Philos. Magn* [8] **14**, 731–743.

Khoe, G. D., and Kuyt, G. (1977). Cutting optical fibres with a hot wire. *Electron. Lett.* **13**, 147.

Kohanzadeh, Y. (1976). Hot splices of optical wavbeguide fibers. *Appl. Opt.* **15**, 793.

Miller, C. M. (1975a). Loose tube splices for optical fibers. *Bell Syst. Tech. J.* **54**, 1215.

Miller, C. M. (1975b). A fiber-optic-cable connector. *Bell Syst. Tech. J.* **54**, 1547.

Miller, C. M. (1976). Transmission vs. transverse offset for parabolic fiber splices with unequal core diameters. *Bell Syst. Tech. J.* **55**, 917–927.

Miller, C. M. (1978). Fiber optic array splicing with etched silicon chip. *Bell Syst. Tech. J.* **57**, 75–90.

Murata, H., Inao, S., and Matsuda, Y. (1975). Connection of optical fiber cable. *Top. Meet. Opt. Fiber Transm., 1st, 1975* p. WA5.

Nippon Telephone and Telegraph co. (1975). *Elect. Commun. Lab. Tech. J. Publ.* **131**.

O'Hara, S. (1975). Status of fiber transmission system research in Japan. *Top. Meet. Opt. Fiber transm., 1st, 1975* p. ThA2.

Paek, V. C., and Kurkjian, C. R. (1975). Calculation of cooling rate and induced stresses in drawing of optical fibers. *J. Am. Ceram. Soc.* **58**, 330.

Pinnow, D. A. (1974). U.S. Patent 3,810,802.

Schroeder, C. M. (1978). Accurate silicon spacer chips for an optical-fiber cable connector. *Bell Syst. Tech. J.* **57**.

Sklyarov, O. K. (1975). An electric-spark method of treating the ends of an optical fiber. *Sov. J. Opt. Technol. (Engl. Transl.)* **42**, 606.

Smith, P. W., Bisbee, D. L., Gloge, D., and Chinnock, E. L. (1975). A molded-plastic technique for connecting and splicing optical-fiber tapes and cables. *Bell Syst. Tech. J.* **54**, 971.

Someda, C. G. (1973). Simple low-loss joints between single-mode optical fibers. *Bell Syst. Tech. J.* **52**, 583.

Standley, R. D. (1974). Fiber ribbon optical transmission lines. *Bell Syst. Tech. J.* **53**, 1183–1185.

Thiel, F. L., and Davis, D. H. (1976). Contributions of optical waveguide manufacturing variations to joint loss. *Electron. Lett.* **13**, 340.

Yamamoto, Y., Naruse, Y., Kamiya, T., and Yanai, H. (1976). A large-tolerance single mode optical fiber coupler with a tapered structure. *Proc. IEEE* **64**, 1013.

Chapter 15

Optical Fiber Connectors

JACK COOK
PETER K. RUNGE

15.1 INTRODUCTION

Fiber connectors are considered here to be distinguished from fiber splices, in that connectors are demountable. They can be connected and disconnected any reasonable number of times. In this chapter we shall discuss problems and alternatives unique to fiber connectors.

Much of the work at Bell Laboratories has been oriented toward the use of optical fibers in telephone trunk systems. For many reasons graded-index, multimode fibers are favorable for that application, and our connector work has concentrated on connecting graded-index fibers. Although this chapter deals with that specific problem, the reader will find the information and approach useful in dealing with the general problem of connecting any kind of fibers end-to-end.

15.2 THE ROLE OF CONNECTORS

Anyone who tried to assemble a fiber communication system during the early exploratory years quickly recognized the need for connectors. Reasonably good fibers, sources, and detectors became available for experimentation before connectors. The result was a communication technique, incredibly simple in concept, but dedicated to perversity in the hands of a novice.

That was only the beginning. As more exploratory systems have been assembled the importance of simple, reliable, and, above all, interchangeable connectors has become apparent. It is in the very nature of manufacture, assembly, testing, installation, and maintenance of a communication system that the transmitter, the medium, and the receiver be

483

treated as separate, interchangeable entities. Furthermore, the administration of a telephone trunk system is made easier if three connectors lie between the fundamental medium and the terminal devices. This allows for simple patch cords to interconnect any of a group of terminals and any of a number of fiber transmission lines that appear on a common patch panel, in short, a fiberguide distributing frame.

There is yet a third need, not so apparent. It is the development process itself. In designing the molded connector, which will be discussed in this chapter, and providing the needs for the exploratory development and implementation of two major experimental systems and several lesser ones, Bell Laboratories molded over 10,000 fiber connector plugs. This was before any formal development of a fiber system for WE manufacture had begun. The tyranny of numbers works in other ways, too. It is clear that low-loss, low-cost, and reliable fiber connectors will be important to communications by optical fiber.

Reliability of connection carries perhaps even more importance for fibers than for copper systems. The connector resides "in line" with the communication signal. In today's systems, the signal is usually tested by breaking the line at a connector and remaking it after test. One needs the assurance of a quality connection to be sure that the signal that was measured is the signal that will be seen on line. Note, here, too, the importance of interchangeability.

15.3 OPTICAL MEASUREMENT PROBLEMS

Measurement of multimode optical connectors is complicated by the fact that no univeral semantics have been established. A connector alters not only the total optical energy passing along the fiber but its distribution among the modes. Ideally, one would like to be able to attach a loss figure on each fiber plug that would characterize its contribution to system loss. In fact, we have not figured out how to do that.

Connectors are even more difficult to characterize than splices in this regard. The majority of splices are flanked by a few hundred meters of cabled fiber which establish some equilibrium distribution of energy among the modes entering the splice, and again at the next splice or the receiving point a few hundred meters further along the cable. Connectors are typically located near the ends of a communication link.

Connectors and splices alike produce results that are a complex composite of the effects of both the mating plugs and fibers. There may be a time when production control on either fibers or connectors will be such that the other can be evaluated separately, but it has not been true so far and may not be for some time to come. Control of both fiber configuration and connector accuracy seem to be progressing together.

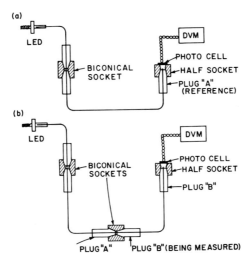

Fig. 15.1   Connector loss measuring arrangement (a) reference (b) measurement.

In some sense one can idealize a connector measurement by breaking a good fiber carefully, connectorizing the broken ends and rejoining them without rotating the fibers about their axes. We have chosen not to do this, because though it might represent the state of the "connector art," it does so outside the bounds of the optical fiber systems art.

A more difficult choice comes in the question of how to excite the connector for measurement. At one end of the extremes is excitation by a laser with a narrow angle beam. At the other end is excited by an LED. We have chosen to use wide-aperture LED excitation with a mode stripper on each side of the connector, but no equilibrator (no artificial means for establishing power equilibrium among the guided modes). Our choice is driven not only by the need to be conservative, but because it provides the most measurement sensitivity. The connector loss measurement arrangement is shown in Fig. 15.1; (a) shows the reference, (b) the measurement.

15.4 LATERAL, LONGITUDINAL, AND ANGULAR DISPLACEMENT

A number of theoretical and experimental studies have been made to determine loss due to offsets and end separation in fiber butt joints (Young, 1973; Bisbee, 1971; Chu and McCormick, 1978). McCormick at Bell Laboratories made the careful measurements recorded here using LED excitation of a graded-index fiber arranged as shown in Fig. 15.2.

The Burrus LED had a fiber (NA of 0.63 and nominal core diameter of 55 μm) connected permanently to it. The light output from that fiber was collimated by a microscope objective (25×) with a numerical aperture of 0.5.

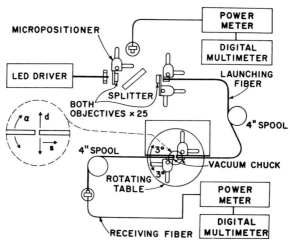

Fig. 15.2 Sketch of measurement setup. (Chu and McCormick, 1978.) Reprinted with permission from *Bell Syst. Tech. J.* Copyright 1978, the American Telephone and Telegraph Company.

The collimated beam was focused into the launching fiber of another objective of the same kind. The graded-index fibers had o.d. of 110 μm, core diameter of 55 μm, and nominal NA of 0.2. The index of refraction of the core center and cladding were 1.472 and 1.458, respectively. The launching and receiving fibers were each approximately 10 m long and nylon-coated which provided the stripping function. Glycerol was used to provide an optical match between fiber ends.

Results of the measurements are shown in Figs. 15.3 through 15.8. S, the end separation of the fibers, and d, the axial displacement of the fiber ends, are normalized with respect to R, the core radius. α is simply the axial angular displacement in degrees.

Various checks by axial rotation of the fibers and the use of other fibers show that the core of the fiber on which these measurements were made

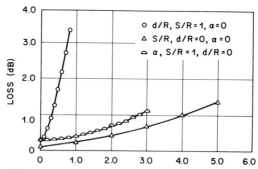

Fig. 15.3 Loss in decibels versus normalized separation, offset, and angular misalignment. (Chu and McCormick, 1978.) Reprinted with permission from *Bell Syst. Tech. J.* Copyright 1978, the American Telephone and Telegraph Company.

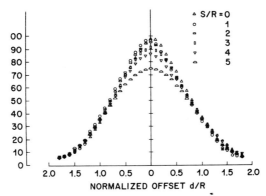

Fig. 15.4 Percent transmission versus normalized offset with normalized end separation as a parameter. (Chu and McCormick, 1978.) Reprinted with permission from *Bell Syst. Tech. J.* Copyright 1978, the American Telephone and Telegraph Company.

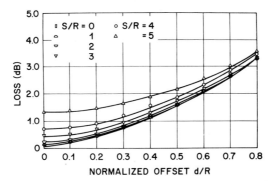

Fig. 15.5 Loss in decibels versus normalized offset for the same normalized separations as in Fig. 15.4. (Chu and McCormick, 1978.) Reprinted with permission from *Bell Syst. Tech. J.* Copyright 1978, the American Telephone and Telegraph Company.

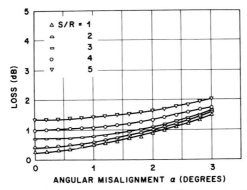

Fig. 15.6 Graph of loss in decibels versus angular misalignment for the same normalized separations as in Fig. 15.4. (Chu and McCormick, 1978.) Reprinted with permission from *Bell Syst. Tech. J.* Copyright 1978, the American Telephone and Telegraph Company.

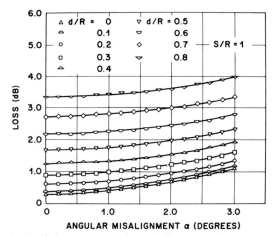

Fig. 15.7 Loss in decibels versus angular misalignment with normalized displacement and normalized separation as a parameter. (Chu and McCormick, 1978.) Reprinted with permission from *Bell Syst. Tech. J.* Copyright 1978, the American Telephone and Telegraph Company.

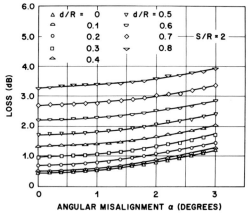

Fig. 15.8 Same as Fig. 15.7, but with a normalized separation of 2. (Chu and McCormick, 1978.) Reprinted with permission from *Bell Syst. Tech. J.* Copyright 1978, the American Telephone and Telegraph Company.

was not perfectly symmetrical about its axis, but that it was sufficiently good to provide useful results.

As others have found, the lateral displacement of the fiber is most critical, and end separation the least. Section 3.9 of Chapter 3 gives theoretical relations relevant to losses at fiber joints.

15.5 CONNECTOR ALIGNMENT TECHNIQUES

The goal in any good connector design is to provide maximum transfer of optical energy from one fiber to any other fiber in the same system. (It has already been pointed out in Chapters 3 and 14 that reestablishment of the equilibrium mode pattern is part of this requirement.) To achieve that, many parameters must be matched. One of the most demanding is the alignment of the core axis, as stated above.

All connectors that we have studied can be classified as belonging to one of two classes: channel-centered or cone-centered. We have studied both, conceptually and experimentally, and prefer the latter, but each has its advantages.

The channel-centered connector usually comprises precision ferrules or clustered precision rods with the fibers to be joined locked "on-axis" by one of several possible techniques; a common sleeve, rod cluster, or collet that aligns the ferrules (Suzuki et al., 1977). This is a conceptually attractive arrangement because it lends itself to simple, well-established precision machining processes. Furthermore, the ferrules, or plugs, can be inserted to the point where they touch, making index-matching with transparent plastic particularly simple as discussed below.

It is the absolute precision required that makes the channel alignment technique difficult. Typically multimode fiber cores are about 50 μm in diameter. A few micrometers displacement of their axis can significantly reduce the coupled power (Fig. 15.3). The need for close mechanical confinement of the connectors in their coupled state is perpetually in conflict with the need to permit relatively easy, wear-free insertion of the plug into the channel. Perhaps the double-collet channel, which snugs down when the plug retainer is screwed tight, is a viable remedy for this problem.

The cone-centered connector comprises two conical plugs with the fibers centered therein and a biconical sleeve that receives and aligns the plugs (Cook and Runge, 1976; Runge et al., 1977). As we shall explain below, conical plugs can be molded, an inherently inexpensive production process, and when inserted into the biconical sleeve no wearing contact is made until the "moment of truth." A possible disadvantage is that the fiber end separation must be well controlled; the plugs "bottom" on the biconic sleeve rather than the plug ends. This may make provision of index continuity through the connector somewhat more difficult as discussed in the next section. A cone-centered connector is shown in Fig. 15.9.

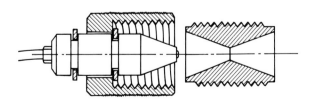

Fig. 15.9   A cone-centered plug and sleeve.

TABLE 15.1
Some Developers and Manufacturers of Fiber Optic Connectors

AMP	Hewlett Packard
Augat	Hughes
Bell Northern Research	ITT Cannon
Bell Telephone Laboratories	Meret
British Post Office Research	NEC
Bunker Ramo/Amphenol	NTT
Cables Cortaillod	Opto Micron Industries
Cablewave Systems	Plessey
Centronic	Radiation Devices
CNET	Sealectro
Corning	Siemens/SiCor
Cinch-TRW	Spectronics
Deutsch	Socapex
Electro-Fiberoptics	Sumitomo
Fibre Link	Telefunken
Fujitsu	Thomas & Betts/Ansley
Furokawa Electric	Thomson CSF
Galileo	Valtec
Harris	

In recent years the fiber optic technology has developed at a rapid pace. At the publishing time of this book a large number of fiber optic connectors, differing significantly in designs, are either on the commercial market or have been published in the literature. Table 15.1 attempts to list the names of developers or manufactures of fiber connectors known to the authors with no claim of completeness made.

A quantitative comparison of these connectors was not attempted, because of the lack of suitable standardization in connector measurements. To begin with, there is no standard on the loss measurement, e.g., on the kind of excitation, coherent or noncoherent and on the length of fiber behind the connection; yet significantly different "loss figures" can be obtained for the same connector.

Second, in many cases, the connector insertion loss means the added insertion loss due to the connector on one and the same fiber. For the systems designer, however, this figure alone is insufficient, since it does not reveal the sensitivity of the connector design to fiber diameter variations.

15.6 INDEX MATCHING

The typical graded-index fiber core has an optical index of about 1.45 to 1.47. At the interface between the core and air about $3\frac{1}{2}\%$ of the light is reflected. If in going from one fiber to the next the light enters the air, it is reflected at both interfaces causing, reflection, on average, of about 7%. This is a loss of about 0.3 dB. If the medium between fibers nearly matches the fiber index, and optically contacts the fiber faces, that reflection and the attendant loss can be eliminated.

As corrosion and organic film have always been a problem for electrical contacts, so foreign matter will eventually interfere with optical connectors. It will cause plugs to be forced out of alignment on the one hand, and simply block the passage of light between fibers on the other. So the use of an index-matching material can possibly serve two purposes—increase light transmission and keep dirt from between the fibers.

The need to keep the plug and its guide clean strongly suggests that fluid-matching media not be used. What then? The optical matching problem can be solved by depositing appropriate dielectric layers on each fiber end. That eliminates the reflection at each fiber end independently of the other. If used with a cone-centered connector, the match coatings will not abrade. On the othe hand, the open space between fibers could permit the intrusion of foreign matter.

Another alternative is the use of a compliant plastic material either attached to one or both plugs, or held as a septum in the connector sleeve. This can work with either channel- or cone-centered connectors.

15.7 THE MOLDED CONE CONNECTOR

Molding is one of the cheapest means for producing parts in quantity. On the other hand molding has not been generally considered a good way to make precision piece parts. There are two problems. It is almost impossible to control molding materials and conditions sufficiently to maintain absolute dimensions throughout a continuous production process, and dimensions of individual piece parts can change with time.

In the molded cone connector these problems are relieved primarily because the spacing between fibers, as indicated in the previous section, is not nearly as critical as displacement. While batch-to-batch control of molding material is difficult, material homogeneity can be good, and it is circular symmetry that must be held to control lateral displacement.

A second factor is that the plug length is determined after the plug has been molded and cured. Lapping and polishing of the fiber-containing plug end can be done in a hardened reference fixture. This determines the penetration of the finished plug into the biconic sleeve.

Finally, there is a tiny amount of surface plasticity of the plugs and sleeves by which they accommodate to variations of the cone angles of the molded plugs and sleeves by as much as the achievable 5' total angular variation.

The material used in these connectors is a heavily silica-filled epoxy. Its

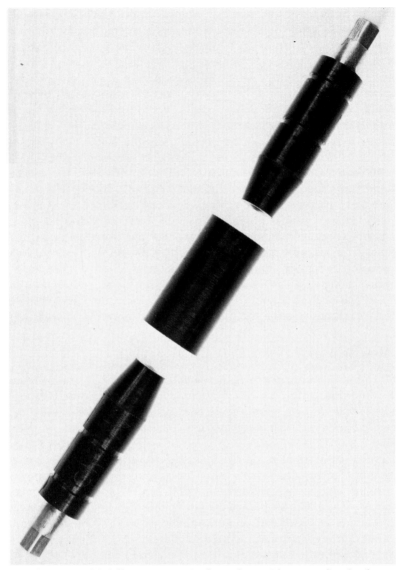

Fig. 15.10 A molded fiber connector is shown here without coupling hardware.

dimensional stability with time and humidity appears to be good, although extensive testing has yet to be done.

Figure 15.10 shows a photograph of a molded fiber connector. The connector consists of two transfer-molded plugs with precision alignment tapers and a transfer-molded biconical sleeve. The hardware to assemble the individual parts is omitted in the figure.

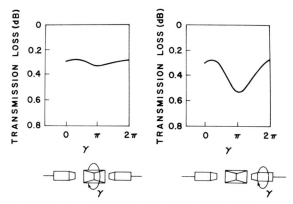

Fig. 15.11 The effect on loss of rotation of socket and plug. (Runge *et al.*, 1977.)

Figure 15.11 demonstrates the precision alignment of the optical fibers that was accomplished with this connector. The left-hand graph shows the measured transmission loss through a connector as a function of the angle of rotation of the biconical socket. Both connector plugs, in this case, were held stationary.

The variation in transmission loss is in the order of ± 0.03 dB corresponding to a variation in mechanical alignment of about ± 1 μm. (All transmission loss data were obtained with fiber excitation by a light emitting diode, a nominal 25-μm end separation between fibers, and glycerin as the index-matching medium.) The observed 1-μm eccentricity of the sleeve was confirmed by mechanical measurements.

The right-hand graph of Fig. 15.11 demonstrates the variation in transmission loss due to rotation of one connector plug. The loss varies typically by ± 0.1 dB with respect to the average loss, which in this case was 0.4 dB. A distribution of the measured average transmission loss of 51 connectors is plotted in Fig. 15.12. The mean average loss is about 0.38 dB.

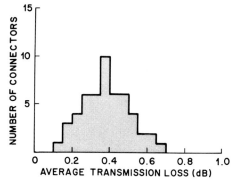

Fig. 15.12 Loss histogram of connectors molded for the Chicago experimental optical fiber installation.

TABLE 15.2
Precision Transfer Molded Single-Fiber Connector

Material:
   Epoxy, heavily silica filled
   High-dimensional stability, excellent abrasion resistance
Fiber:
   Ge-doped, graded index—55-μm core,
   110-μm cladding diameter protected in jumper cable sheath
Transmission loss: (LED excited)

Loss due to encapsulation, crimps, Fiber mismatch, angular misalignment	0.2 dB
Loss due to fiber misalignment	0.2 dB $\pm$ 0.1 dB
Total loss with index match	0.4 dB $\pm$ 0.1 dB
Total loss without index match	0.8 dB $\pm$ 0.2 dB

Four consecutive jumper cables:
   Total— $\left.\begin{array}{l}1.18 \text{ dB MIN.} \\ 1.8 \text{ dB MAX.}\end{array}\right\}$ 0.37 dB mean average per connection

When two connector plugs are aligned with respect to each other, without a socket, by an x–y–z manipulator, the offset error of those plugs can be tuned out. The remaining transmission loss is then due to the combined effects of microbending introduced by the epoxy shrinking onto the fiber, the crimp providing the strain relief for the cable sheath, the angular misalignment of the fibers, and the mismatch of fiber core parameters. This latent loss was measured to be in the range from 0.1 to 0.2 dB. The additional loss due to the lateral offset of the fiber cores within the plugs is then 0.2 $\pm$ 0.1 dB, which corresponds to a maximum lateral misalignment

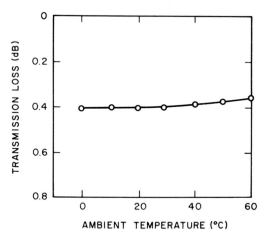

Fig. 15.13 Transmission loss through a typical molded connector versus ambient temperature. (Runge *et al.*, 1977.)

error of ± 3 μm. That error is about equally shared between fiber and plug eccentricities.

Table 15.2 summarizes the results of the loss measurements. Noteworthy is a loss measurement made on four consecutive 1-m-long jumper cables. All four connectors tuned for maximum loss resulted in 1.8 dB total loss. Tuned for minimum loss, the four connectors had 1.18 dB total loss yielding a mean average transmission loss of 0.37 dB. This is consistent with losses measured individually as previously described.

Figure 15.13 shows a plot of the temperature dependence of the transmission loss. The variation in loss is negligible for the temperature range from 0 to 60°C indicating almost-perfect match in thermal expansion coefficients of the connector parts. Several cycles traced the same curve with no observable hysteresis.

15.8 A CHANNEL-CENTERED CONNECTOR

Hardened, ground, and lapped stainless steel rollers with diameter tolerance of ± 1 μm can be purchased for less than 3¢ apiece in lots of a few thousand. Closer tolerance rollers can also be purchased at higher cost. If three identical rods are clustered they form an orifice that will accept a cylindrical fiber of diameter $(2 - 3^{1/2})/3^{1/2}$, or 0.1547 times that of the rods. To first order, the accepting orifice is an equilateral triangle, and by geometrical construction it can be shown that if the rods vary in diameter by $\pm \epsilon$, and the mean rod diameter is chosen to allow no interference over the full range of $\pm \epsilon$, then the maximum fiber center displacement for a zero-tolerance fiber will be about $2(2 - 3^{1/2})\epsilon/3^{1/2}$. A similar calculation for maximum displacement of the rod cluster center within a zero-tolerance guiding hole shows that to be about $2(2 + 3^{1/2})\epsilon/3^{1/2}$. So given a perfect noninterfering fiber and perfect noninterfering guiding hole for a three-rod cluster, where the rods have ± 1 μm tolerance range, the minimum fiber center displacement will be about 0.3 μm, and the maximum displacement about 4.3 μm. If the diameter tolerance on the guiding hole is ± 2 μm, the displacement can exceed 8 μm. Fiber tolerance of 1 μm diameter would permit total displacement of more than 10 μm. Since that is the possible displacement of each of two matching plugs, the worst possible plug center displacement relative to another plug would be over 20 μm. Of course this is the worst possible addition of many tolerance variations and its realization is indeed unlikely. Somewhat better, perhaps, than a precision hole to center the plugs is a large cluster of precision rods. If, for example, six rods are clustered to channel two three-rod plugs together, and the rods have diameter tolerance of ± 1 μm, the maximum plug-center displacement due to that tolerance would be about 1.3 μm. Furthermore, it is very unlikely that all rods would be the maximum extreme of the tolerance

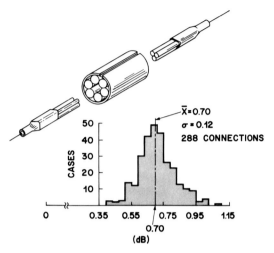

Fig. 15.14 A six-rod channel-centered connector.

range. For zero-tolerance channel rods, ± 1 μm plug rods would permit maximum plug center displacement of the 2.5 μm inside the channel cluster. Tolerance of ± 1 μm on the fiber still permits 2 μm fiber-center displacement in the three-rod plug, but not in the direction of maximum displacement of the plug in the channel. With the above tolerances of ± 1 μm on the fiber and all rods, maximum possible displacement of each fiber-center in the channel is about 5 μm.

If the 3-rod plugs are constrained to enter the hexagonal channel so that the plug rods abutt rather than being displaced 60°, the maximum relative fiber center displacement will be a little less than 10 μm.

Four different versions of three rod plug connectors have been designed and tested at Bell Laboratories. One of these was discussed at CLEA in 1977 (Warner, 1977). Another, a six-rod-channel-centered connector, was designed and built by C. R. Sandahl. The latter connector and measurement results are shown in Fig. 15.14. In this instance the channel rods were chosen to provide an interference fit with the three-rod plug. The outer rods were retained in a stainless steel split sleeve. This eliminates displacement due to looseness of the plugs in the channel except for that due to differences in the dimensions of the two mated plugs. It should be noted, however, that unless the retaining sleeve is clamped tight after insertion of the three-rod plugs, lateral force on the back of the plug, caused even by the weight of the fiber cable, can cause a shift in the fiber center and significant change in the connector loss.

ACKNOWLEDGMENTS

Many people have contributed to the understanding and to the specific connector work discussed in this chapter. In addition to the contributions of C. R. Sandahl and A. R. McCormick mentioned in the text, we would like to acknowledge most particularly the work of W. C. Young, L. Curtis, T. C. Chu, R. E. Spicer, and F. L. Porth.

REFERENCES

Bisbee, D. L. (1971). Measurements of loss due to offsets and end separation of optical fibers. *Bell Syst. Tech. J.* **49,** 3159–3168.

Chu, T. C., and McCormick, A. R. (1978). Measurements of loss due to offset, end separation, and angular misalignment in graded index fibers excited by an incoherent source. *Bell Syst. Tech. J.* **57,** 595–602.

Cook, J. S., and Runge, P. K. (1976). An exploratory fiberguide interconnection system. *Proc. Eur. Conf. Opt. Fibre Commun., 2nd, 1976* pp. 253–256.

Runge, P. K., Curtis, L., and Young, W. C. (1977). Precision transfer molded single fiber optic connector and encapsulated connectorized devices. *Tech. Dig., Opt. Fiber Transm., 2nd., 1977* WA4.

Suzuki, N., Koyana, M., Kurochi, N., Koyama, Y., Furuta, H., and Oguro, S. (1977). A new demountable connector developed for a trial optical transmission system. *Tech. Dig., Int. Conf. Integr. Opt. Opt. Fiber Commun., 1977* p. 351.

Warner, A. W., Jr. (1977). An optical fiber connector. *Dig. Tech. Pap., IEEE Conf. Laser Eng. Appl., 1977* p. 20.

Young, M. (1973). Geometrical theory of multimode optical fiber to fiber connectors. *Opt. Commun.* **7,** 253–255.

Chapter 16

Optical Sources

CHARLES A. BURRUS
H. CRAIG CASEY, JR.
TINGYE LI

16.1 INTRODUCTION

Solid-state optical sources for use in communications are required to have characteristics quite different from those of optical sources designed for other applications. Most commonly, the aim for noncommunications applications has been to achieve very high power in the infrared or near-infrared, or to produce maximum visibility with minimum power consumption for visual display. In the field of semiconductor injection devices, this emphasis has led to pulsed multimode lasers operating at high peak currents and to large-area low-radiance light-emitting diodes (LEDs), both of which are largely unsuitable for all but the most elementary communications applications.

In contrast, emitters to serve as sources in sophisticated communications systems do not require high power per se, but it is essential that they be capable of stable, continuous (CW) operation at room temperature for many years; that they be of size and configuration compatible with the transmission line, generally taken to be an optical fiber; and that they be capable of signal modulation at rates which might be as low as audio frequencies but more usefully should extend well beyond a gigahertz.

To be maximally useful, the communications sources should have output at wavelengths for which the transmission fiber has both low loss and low dispersion. (Figure 16.1 shows loss characteristics which now can be achieved in experimental fibers and indicates the wavelengths at which devices of the more promising materials systems can serve as sources; Fig. 16.2 shows the dispersion vs wavelength behavior of various fiber

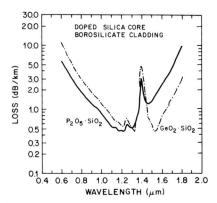

Fig. 16.1 Transmission loss vs wavelength for state-of-the-art optical fibers (Osanai *et al.*, 1976). Older fibers exhibited a loss peak near 0.9-μm wavelength due to OH⁻ absorption.

glasses.) The sources also should operate with reasonable efficiency and at wavelengths where detectors are efficient. Further, while output power is not the most important consideration, the sources must be capable of coupling at least microwatts, and preferably a milliwatt or more, of optical power into the transmission fiber.

In general, lasers offer the benefits of narrow spectral bandwidth (20 Å or less, useful in minimizing the effects of fiber dispersion), modulation capabilities extending to perhaps hundreds of megahertz, relatively directional output (permitting reasonable efficiency in coupling power into a fiber), and coherence (orginally considered essential to allow heterodyne detection in high capacity systems, but now considered primarily of use in single-mode systems). Incoherent LEDs offer advantages of inherent simplicity of construction and operation, and thus the expectation of ex-

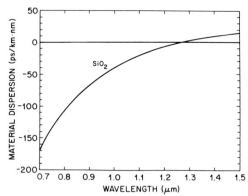

Fig. 16.2 Dispersion of SiO_2 vs wavelength (Malitson, 1965). Addition of dopants (to raise the refractive index for use in fiber cores) displaces this curve only slightly (Fleming, 1976).

tended trouble-free operational life. It is to be noted especially that, for long repeater spans, wideband incoherent sources must be made to operate at 1.2- to 1.4-μm wavelength where fiber dispersion is low (Fig. 16.2), but that the choice of wavelength for narrow-band lasers is much less critical.

Although injection devices so far have been the simplest and thus the most obvious choice of sources for communications, the use of other solid-state lasers is not precluded—provided, of course, the necessary pumping can be accomplished with another simple, long-life, solid-state device! Optically pumped lasers with neodymium as an active element seem particularly attractive in this area, since the common Nd-doped materials are stable, oscillate at 1.06 or 1.3 μm wavelength where the transmission fibers have attractive characteristics (Figs. 16.1, 16.2), and have a strong pump band at a wavelength (0.805 μm) conveniently produced by well-known III–V compound semiconductor devices.

It is highly unlikely that any single device will be a panacea as a source for communications use. The injection laser, the injection LED, the solid-state laser pumped by an injection device, and probably other devices to come, all in time undoubtedly will serve in different applications to which their various properties can be fitted most economically. In the simplest terms, lasers now seem best suited for applications requiring long-distance transmission with large modulation bandwidths, where the cost and complexity of their use can be tolerated; incoherent LEDs probably will be more economical for use in simpler, shorter distance, lower capacity systems operating at modest modulation rates.

In this chapter, we shall review the present status and prospects of lasers and LEDs as they pertain to use in communications systems employing low-loss single-fiber transmission lines. We first shall consider the semiconductor materials which are proving to be useful in the fabrication of light-emitting devices and then look at the devices themselves (noting, as an interesting commentary on technological progress, that research on semiconductor optical sources preceded the development of the optical fiber, and that it was carried out without regard to the requirements of lightwave communication). The discussion will, of necessity, center on devices prepared for laboratory use since fully proved "commercial" devices as yet do not exist, and it will take the viewpoint of device use rather than device physics.

16.2 SEMICONDUCTOR MATERIALS

16.2.1 Historical Background

In the late 1950s and early 1960s, considerable effort was devoted to materials preparation and studies of radiative recombination mechanisms for light-emitting diodes. These studies were concerned with visible-wavelength emitters made from the binary III–V compound GaP and the ternary solid solution $GaAs_xP_{1-x}$, as well as the binary III–V compound GaAs which emits in the infrared near 0.9-μm wavelength (Casey and Trumbore, 1970).

In 1962, the first GaAs semiconductor lasers operating at low temperature were reported (Hall *et al.*, 1962; Nathan *et al.*, 1962). These devices are based on p–n junctions of a single semiconductor, and therefore were called homojunction lasers. Considerable effort to improve these lasers was undertaken immediately with emphasis on GaAs, a direct energy-gap semiconductor suitable for such use (Dumke, 1962) and also the III–V compound in the most advanced technological state at the time. It was suggested early (Kroemer, 1963; Alferov and Kazarinov, 1963) that improved semiconductor lasers could be prepared with heterojunctions, which are junctions between two semiconductors with different energy gaps and refractive indices. Heterojunctions could confine the injected carriers and provide waveguiding more effectively than homojunctions and thus their use could drastically reduce the device current. The disclosure (Woodall *et al.*, 1967; Rupprecht *et al.*, 1967) that high quality $Al_xGa_{1-x}As$ could be grown on GaAs by liquid-phase epitaxy (LPE) led rapidly to the preparation of $GaAs$-$Al_xGa_{1-x}As$ heterostructure lasers. Reduced-threshold $GaAs$-$Al_xGa_{1-x}As$ single-heterostructure (SH) lasers (Hayashi *et al.*, 1969; Kressel and Nelson, 1969) and then $Al_xGa_{1-x}As$-$GaAs$-$Al_xGa_{1-x}As$ double-heterostructure (DH) devices that operated continuously at room temperature soon were reported (Hayashi *et al.*, 1970; Alferov *et al.*, 1971). With the achievement of CW room temperature operation, the impetus for the practical development of high-radiance semiconductor LEDs and injection lasers was established.

16.2.2 Materials for Semiconductor Light Sources

The most intensively studied and thoroughly documented materials for optical-fiber light sources are GaAs and $Al_xGa_{1-x}As$. The variation of the energy gap E_g and the refractive index $\bar{n}$ with AlAs mole fraction x are shown in Fig. 16.3. In the direct-energy gap material, an electron can make a transition from the valence to the conduction band (or vice vesa) directly by the absorption (or emission) of a photon. For the indirect energy gap, the minimum energy transition also must involve the emission or absorption of a phonon and thus the radiative efficiency is reduced. As x increases, E_g becomes larger and $\bar{n}$ is decreased. A relatively lager E_g at a heterojunction interface is required for carrier confinement, while a reduced refractive index in $Al_xGa_{1-x}As$ is required for waveguiding. With these restrictions, the structure of a DH laser in this system then is

$$N\text{-}Al_xGa_{1-x}As|n\text{- or } p\text{-}GaAs|P\text{-}Al_xGa_{1-x}As.$$

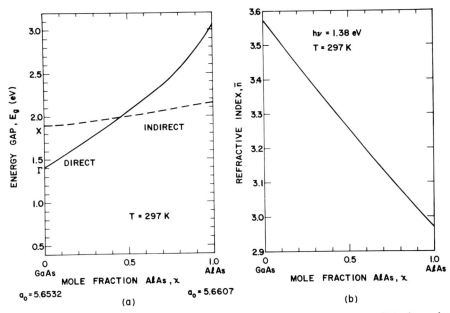

Fig. 16.3 (a) Compositional dependence of the Al$_x$Ga$_{1-x}$As energy gap (Dingle *et al.*, private communication). (b) Compositional dependence of the Al$_x$Ga$_{1-x}$As refractive index at 1.38 eV (Casey *et al.*, 1974a).

The conductivity type of the wide energy gap material is designated by N or P, and the conductivity type of the narrow energy gap material is designated by n or p. The stimulated emission occurs in the active layer, which is the central n- or p-region, with wavelength characteristic of the energy gap of the active-layer material. The wider energy gap N- and P-regions provide confinement of carriers to the active region (increasing the injected carrier concentration for a given current density) and wave confinement which leads to guiding in the higher index active region.

The lattice constant a_0 also is noted on Fig. 16.3a to emphasize that a_0 for GaAs and AlAs differs, but by an amount less than 0.14%. The lattice constant varies linearly with composition (Vegard's law), and therefore the lattice mismatch is considerably less than 0.14% for Al$_x$Ga$_{1-x}$As. Lattice matching is very important for several reasons: (1) a close lattice match is necessary in order to obtain high-quality crystal layers by epitaxial growth, (2) excess lattice mismatch between the heterostructure layers results in crystalline imperfections which lead to nonradiative recombination and thus prevents lasing, and (3) lattice mismatch causes degradation in devices during operation.

Early fibers with a large OH-ion loss peak near 0.9 μm wavelength (a problem now largely remedied) exhibited lowest transmission loss near 0.85 μm. Since emission in this region could be obtained from injection devices with an active layer of $Al_{0.08}Ga_{0.92}As$, device emphasis was largely focused on this system. However, minimum loss for fibers without OH-ion absorption occurs in the wavelength region 1.2–1.6 μm (Fig. 16.1) and thus there is strong motivation to develop optical sources in this region. Since $Ga_xIn_{1-x}P_yAs_{1-y}$ at selected values of x and y lattice-matches perfectly to InP and has an emission wavelength in this region, considerable effort is being devoted to this system. Recent work (Shen et al., 1977; Yamamoto et al., 1978) has resulted in long-life CW room temperature operation of DH lasers of

$$\text{N-InP|undoped } Ga_{0.12}In_{0.88}P_{0.77}As_{0.23}\text{|P-InP}$$

that were grown on n-type InP substrates by LPE. Discussion of this system provides a good example of a binary-to-quaternary III–V lattice matched system for heterostructure lasers.

Since the ratio of the total number of group III atoms to the total number of group V atoms is unity, the composition of this quaternary is uniquely represented by the two parameters x and y. The composition of this solid solution may be represented by a square in the x–y plane with the four binary compounds at the corners. With the composition represented by the x–y plane, the three-dimensional compositional dependence of the energy gap for this system has been drawn in Fig. 16.4. Both the direct- and indirect energy gap surfaces are shown, and their intersec-

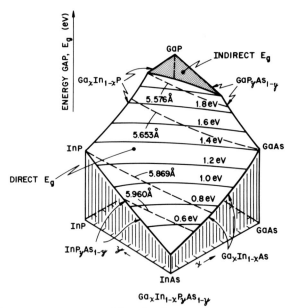

Fig. 16.4 Energy gap vs composition for $Ga_xIn_{1-x}P_yAs_{1-y}$. Nearly horizontal solid curves are iso-energy gaps. Dashed curves are iso-lattice constants (Nuese, 1977).

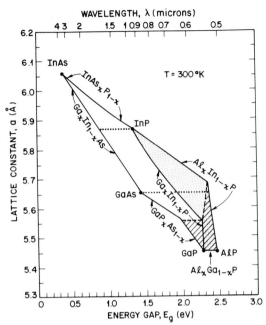

Fig. 16.5 Energy gap and lattice constant for $Ga_xIn_{1-x}P_yAs_{1-y}$ (clear) and $(Al_xGa_{1-x})_yIn_{1-y}P$ (shaded). Dashed lines separate direct and indirect energy gap regions, and the cross-hatching designates the indirect energy gap. Dotted lines show lattice match to binary compounds (Casey and Panish, 1978).

tion demonstrates that most of this system is in the direct energy gap region. The compositions that lattice-match to InP are of primary interest. Constant values of E_g have been drawn in the direct energy gap surface together with the InP lattice constant. Emission at 1.2 μm is possible with InP-$Ga_{0.22}In_{0.78}P_{0.53}As_{0.47}$ DH lasers. Efficient broad-area LEDs (Pearsall *et al.*, 1976) as well as small-area high-radiance LEDs compatible with multimode fibers (Dentai *et al.*, 1977; Oe *et al.*, 1977b) also have been prepared with $Ga_xIn_{1-x}P_yAs_{1-y}$. The exceptional lattice match possible with this system, its capability of operation at optimum fiber transmission wavelengths, and the absence of a readily oxidized component all suggest that this quaternary semiconductor system will be highly favored in the investigation of future optical sources. Considerable research is now being devoted to these materials for the fabrication of both DH lasers and LEDs.

There has been substantial experience in the growth of compositionally graded III–V ternary compounds by chemical-vapor deposition (CVD). This experience has been achieved primarily with the growth of GaP_xAs_{1-x} on GaAs for the production of red light-emitting diodes. Compositional grading also has been used for III–V solid solutions grown by LPE. For example, DH injection lasers have been made both in stepwise

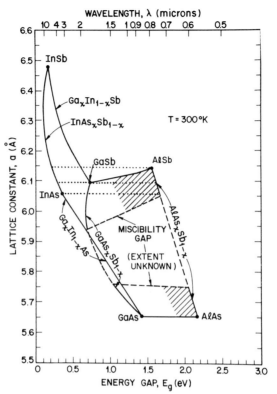

Fig. 16.6 Energy gap and lattice constant for $Ga_xIn_{1-x}As_ySb_{1-y}$ and $Al_xGa_{1-x}As_ySb_{1-y}$. The dashed lines show the edge of a misability gap of uncertain extent in the $Al_xGa_{1-x}As_ySb_{1-y}$ system. The misability gap along $GaAs_xSb_{1-x}$ must extend into the $Ga_xIn_{1-x}As_ySb_{1-y}$ system, but is not shown in the figure because of its uncertain extent. Cross-hatching designates indirect energy gap regions. Dotted lines show lattice match to binary compounds (Casey and Panish, 1978).

compositionally graded $GaAs_xSb_{1-x}$ material, LPE-grown on GaAs substrates (Nahory *et al.*, 1976) and in $Ga_xIn_{1-x}As$ lattice-matched to continuously compositionally graded $Ga_xIn_{1-x}P$ material, CVD-grown on GaAs substrates (Nuese *et al.*, 1976). CW operation at room temperature was obtained in both cases with emission in the 1-μm wavelength region.

The variation of energy gap (emission wavelength) with lattice constant is summarized in Figs. 16.5 and 16.6 for several ternary and quaternay III–V solid solutions. For conversion between photon energy E and wavelength λ,

$$E(eV) = 1.2398/\lambda \ (\mu m). \tag{16.1}$$

The boundaries joining the binary compounds give the ternary energy gap and lattice constant. The crosshatching indicates indirect-energy-gap material. Note that both direct and indirect-energy-gap material may be useful for the construction of heterostructure lasers: the active region requires an energy gap that is direct, but the surrounding wider energy gap material may be direct or indirect. The dotted lines in these figures give the lattice match to the indicated binary. Further description of semiconductor materials for heterostructures is given in Casey and Panish (1978).

16.2.3 Heteroepitaxy

Preparation of multilayered structures has required the development of heteroepitaxial growth processes to provide very thin grown multilayers of III–V semiconductor materials. This growth of multilayered structures generally is accomplished by liquid-phase epitaxy (LPE). For GaAs–$Al_xGa_{1-x}As$ DH lasers, a carefully polished single-crystal substrate of n-type GaAs (0.05 cm thick, 1.5 cm long, and 1 cm wide) is brought successively into contact with metallic solutions. The solutions, commonly held in a graphite boat, are principally Ga, but they also may contain small amounts of Al, As, and dopant elements such as Te, Sn, Ge, Si, and Zn. The compositions of these solutions are selected carefully so that, from each one, the proper $Al_xGa_{1-x}As$ alloy composition is grown with the proper amount of impurity. Growth takes place when the solution is cooled while in contact with the substrate as illustrated in Fig. 16.7. The solutions rest on a graphite slider that also holds the substrate wafer.

The growth process starts with the apparatus in a pure hydrogen atmosphere at a temperature near 800°C. An expendable seed (precursor seed) precedes the growth seed to help relieve supersaturation in some of the solutions. The slider is moved so that the precursor seed contacts solution 1 (step 1). A cooling rate of typically 0.1°C/min is initiated, and the slider is moved again (step 2) so that the precursor seed is under solution 2 while the growth seed is under solution 1. After cooling, about 2 μm of n-type $Al_xGa_{1-x}As$ has grown on the substrate. The slide again is moved so that the precursor seed is under solution 3, and the growth seed under solution 2 (step 3). After 20 sec about 0.2 μm of p-type GaAs has grown, and the slide is moved again (step 4). This process is repeated for four (or more) layers. After removal from the furnace, the multilayered wafer is processed into individual (discrete) devices. Further description of LPE is given by Casey and Panish (1978).

There are four different CVD techniques used for III-V compounds. The so-called water-vapor transport process (Frosch, 1964) is based on the reversible reaction of water vapor with the III-V compound to form the group III element oxide and the group V element vapor. This process is now used primarily for the preparation of high purity GaP and has not been used for preparation of heterostructures. The next two CVD processes are based on the same chemistry. In the process called the halide

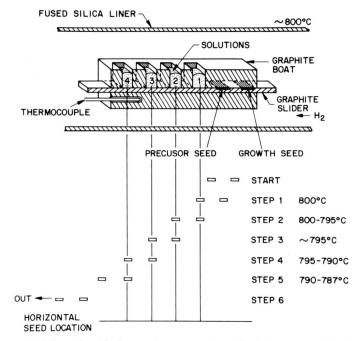

Fig. 16.7 Multilayer liquid-phase epitaxy apparatus. The bottom part of the figure illustrates the temperature-position sequence for the growth of the four-layer $Al_xGa_{1-x}As$ DH laser (Casey and Panish, 1975).

(arsenic or phosphorus trichloride) process (Finch and Mehal, 1964), $AsCl_3$ (and/or PCl_3), H_2, and Ga (and/or In) are the initial reactants to grow binary or more complex III-V compounds. In the hydride (arsine and phosphine) process, AsH_3 and PH_3 are used as the sources of arsenic and phosphorus (Tietjien and Amick, 1966). Gallium (and/or In) is transported via volatile gallium chloride(s) which are produced by passing HCl over the heated Ga. The arsine and phosphine process will be discussed for the growth of $Ga_xIn_{1-x}P_yAs_{1-y}$ on InP. In the fourth CVD technique, the decomposition of organometallic compounds is used as a source of group III elements and arsine or other group V hydrides as the source of the group V element (Dupuis and Dapkus, 1978).

Solid solutions of $Ga_xIn_{1-x}P_yAs_{1-y}$ have been grown by the arsine and phosphine process. The CVD system is shown schematically in Fig. 16.8 (Olsen and Ettenberg, 1977). Both $GaAs$-$Ga_{0.51}In_{0.49}P$ and $Ga_{0.32}In_{0.68}P$-$Ga_{0.84}In_{0.16}As$ DH lasers can be prepared with this system. It also may be used to grow $Ga_xIn_{1-x}P_yAs_{1-y}$ on InP. The growth process begins with the substrate in a forechamber. The system entry valve is closed, the forechamber is flushed with H_2, flows of HCl are established over the Ga and In, and the AsH_3 and PH_3 gases are introduced. An n-type dopant (sulfur)

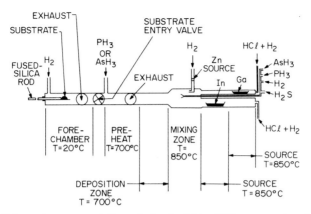

Fig. 16.8 Schematic representation of the hydride CVD system for the growth of $Ga_xIn_{1-x}P_yAs_{1-y}$ (Olsen and Ettenberg, 1977).

is supplied by H_2S while the p-type dopant is obtained by flowing H_2 over Zn. Composition is determined by the flow rates of HCl or the group V gases. When the flows have been established, the entry valve is opened and the substrate is placed in the preheat zone. A flow of either AsH_3 or PH_3 is passed over the substrate to prevent decomposition. The sample is than moved to the deposition zone and the desired layers are grown by controlling the gas flows.

One of the more exciting recent developments in the growth of heteroepitaxial wafers is the success of the organometallics. There are several alkyl compounds of most group III elements that are readily available. The compounds containing organic radicals that have only one or two carbon atoms are usually moderately volatile liquids at room temperature and decompose at several hundred degrees Celsius. The growth of GaAs by the decomposition of trimethyl gallium typifies the process. The trimethyl gallium is usually transported into the reaction chamber as a dilute vapor by bubbling H_2 through it at 0°C. Arsenic is transported as AsH_3. The reactor is usually a glass chamber with water-cooled walls. The substrate is located on an inductively heated susceptor. For GaAs, epitaxial growth is usually obtained with the substrate in the temperature range 550–700°C. Epitaxial crystal growth takes place as the result of the decomposition of both the organometallic compound and the hydride. The net reaction is primarily

$$Ga(CH_3)_3 + AsH_3 \xrightarrow{H_2} GaAs + 3CH_4. \quad (16.2)$$

Details of the mechanism by which this reaction takes place are not known; however, high quality GaAs and InP epitaxial layers are readily achieved. Doping is generally done by incorporating either organometallics [e.g., $Zn(C_2H_5)_2$] or hydrides (e.g., H_2S) into the reactant gas stream. Since there are no heated walls in the apparatus, the reactants can be present at high supersaturation at the growing surface and the degree of nonequilibrium is probably more representative of the molecular-beam epitaxy described below than of conventional CVD techniques.

Initial studies with broad area GaAs-$Al_xGa_{1-x}As$ DH lasers prepared by the organometallic process (Dupuis and Dapkus, 1978a) showed that very low thresholds (<1000 A/cm²) can be achieved. In particular, they reported a threshold current density of 700 A/cm² for a DH laser with $d = 0.18$ μm and $x = 0.52$. Single-longitudinal-mode operation of CW room temperature $Al_xGa_{1-x}As$-GaAs DH lasers has also been demonstrated (Dupuis and Dapkus, 1978b). If a reasonably long life can be demonstrated for lasers prepared by the CVD-organometallic process, then this technique of heteroepitaxial wafer preparation will prove an attractive alternative to liquid-phase epitaxy.

The chief advantage of LPE is the high quality of the resulting semiconductor layers. However, there is another growth technique, called molecular-beam epitaxy (MBE), that permits better precision in layer thickness control. Molecular-beam epitaxy is a relatively new thin-film growth technique (Arthur, 1968), but for some time (Cho and Casey, 1974) it has been possible to produce device-quality MBE layers. However, although GaAs-$Al_xGa_{1-x}As$ DH lasers prepared by MBE have operated CW at room temperature (Cho et al., 1976) the threshold current densities are still about a factor of two greater than for LPE devices of comparable geometry.

In MBE, beams of atoms and molecules from small ovens in an ultrahigh vacuum impinge on a heated GaAs substrate, as illustrated schematically in Fig. 16.9. Since each Ga or Al atom that strikes the surface condenses and binds one As atom while excess As is reflected from the surface, growth of GaAs or $Al_xGa_{1-x}As$ may occur. Typical impurity elements to provide n-type conductivity are Sn, Si, or Ge; and to give p-type conductivity are Be, Mn or Mg (elements with very high vapor pressures, such as Zn or Te, are more difficult to use). Composition of the layers is adjusted by regulating the temperature of the Ga and Al ovens and by opening and closing appropriate shutters. A more detailed discussion of MBE may be found in the review paper by Cho and Arthur (1975).

16.3 LIGHT-EMITTING DIODES

16.3.1 General Considerations

Under proper but easy-to-achieve conditions, forward-biased p–n junctions of many semiconductors, notably those composed of elements

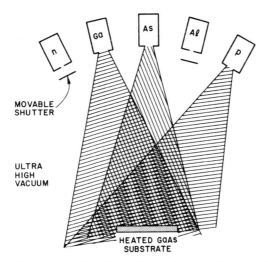

Fig. 16.9 Schematic representation of a molecular-beam epitaxy growth apparatus. The heated ovens for $Al_xGa_{1-x}As$ are represented by Ga, As, and Al. The donor and acceptor impurity ovens are represented by n and p, respectively. Instaneous composition changes may be achieved by opening and closing the movable shutters.

from group III and group V of the periodic table, can emit external spontaneous radiation in the visible or infrared regions of the spectrum. Such devices are called light-emitting diodes or LEDs. In operation, the normally empty conduction band of the semiconductor is populated with electrons injected into it by a forward current through the junction, and light is generated when these electrons recombine with holes in the valence band and emit a photon. The energy of the emitted photon is approximately that of the energy gap between the conduction and valence bands of the particular semiconductor. This spontaneous emission, which usually occurs physically in the p-layer close to the p–n junction, is referred to as recombination radiation. Unfortunately, the recombination of the injected electrons also can occur by processes which do not emit a photon, the so-called nonradiative processes, and thus the internal quantum efficiency of an LED is not 100%. One aim in the fabrication of LEDs is to maximize this internal conversion, i.e., minimize the crystalline imperfections and impurities which lead to nonradiative traps for the injected electrons. In practice the internal quantum efficiency can be quite high, certainly exceeding 50% in simple homostructure LEDs (Hill, 1965; Archer and Kreps, 1967). In the double-heterostructure (DH) LEDs to be the principal topic of discussion here, measurements of recombination lifetime suggest that internal quantum efficiencies of 60–80% are being achieved (Lee and Dentai, 1978), with the losses primarily ascribed to nonradiative recombination at semiconductor surfaces and interfaces. Another aim is to produce the radiation in a geometry from which it can be collected and thus to maximize the useful external power efficiency. A third aim is to produce diodes in which the light output can be directly current-modulated at high rates with information-carrying signals. Still another aim is to produce geometries from which heat can be extracted efficiently, since the output of an LED drops by 2–3 dB if the junction temperature rises 100°C.

It was recognized early that, if significant power were to be coupled from an incoherent LED into a small fiber, the source would have to exhibit very high radiance. For light-emitting diodes this meant that it would be necessary to use direct energy gap semiconductors and that structures which could be driven at high current densities would have to be devised. It also was recognized that such a diode probably would take the geometrical form of either a very small-area emitter (Kibler *et al.*, 1964) in which the light from the surface of a small junction would be collected perpendicular to the junction plane through a thin or transparent layer of semiconductor above the junction, or of an edge-emitter (Zargar'yants *et al.*, 1971) in which the light would be emitted directly from the exposed edge of the junction. Both configurations have been made in planar form and applied to optical fiber uses.

16.3.2 Surface Emitters

A homojunction surface emitter is illustrated in five slightly different forms in Fig. 16.10. The philosophy of this design (Burrus, 1969; Burrus and Dawson, 1970; Goodfellow and Mabbitt, 1976) is that the small-area p–n junction will benefit from rapid spreading of heat into the large heat sink and thus can be driven at relatively high current densities before overheating. In these designs, the emitting area of the junction is confined in various ways to a small dot, usually 15 to 100 μm diameter, and the semiconductor through which the emission must be collected is made very thin, 10–15 μm, to minimize absorption and allow the end of the fiber to be very close to the emitting surface. They may be operated at current densities of a few kiloamperes per square centimeter for the larger sizes to several tens of kiloamperes per square centimeter for the smallest.

Considerable advantage can accrue from the use of somewhat more complicated semiconductor structures combined with any of the junction confinement geometries of Fig. 16.10, and an example is illustrated in Fig. 16.11. These structures (Burrus and Miller, 1971; King and SpringThorpe, 1975), employing the heterostructures described before, have several advantages compared to the single-material or homostructure devices. These advantages are (1) increased efficiency resulting from the electron confinement (mentioned in the previous section) provided by the layers of higher energy gap semiconductor surrounding the recombination regions near

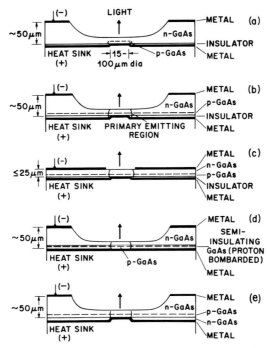

Fig. 16.10 Several of the many possible configurations for small-area, high-radiance surface-emitter LEDs. Although equally applicable to DH, SH and homostructures, they are shown for simplicity as homostructures. (a) Small-area junction diffused or grown through window in insulator (Burrus, 1969); (b) and (c), large-area junctions with small-area contacts to confine current flow, with and without etched "well" (Burrus and Dawson, 1970; C. A. Burrus, unpublished, 1974); (d) and (e) designs provide current confinement without an insulator layer and thus have improved thermal conduction to the heat sink [(d) after Dyment et al. (1977); (e) after T. P. Lee, unpublished 1976). Structures shown have been realized in practice; drawings approximately to scale.

the junction; (2) increased transmission of the emitted radiation to the outside, since the higher energy gap confining layers do not absorb radiation from the lower energy gap emitting regions; and (3) not limited to heterostructures but especially easy to obtain with them, the emitted wavelength may be varied readily by controlling the semiconductor composition of the layers containing the junction. The disadvantages are that (1) heat conduction is poorer in the currently-used configurations involving Al-containing semiconductors to provide the higher energy gap layers, and (2) increased degradation is observed in devices made from the many-layered structures, probably the result of imperfect crystalline lattice-matching. A somewhat simpler intermediate structure, comprised of a single heterostructure with a higher energy gap layer on only one side of the emitting junction, also has been made in high-radiance form

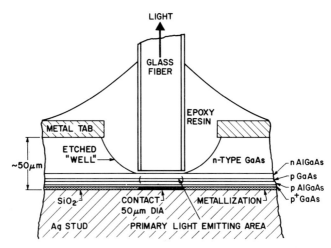

Fig. 16.11 Detailed cross section of a particular small-area, high-radiance AlGaAs DH surface-emitter LED with attached fiber, not to scale. The top n-AlGaAs layer is transparent to the radiation from the GaAs active layer (Burrus and Miller, 1971).

(Burrus and Ulmer, 1971; Yamakoshi *et al.*, 1977). This configuration demonstrates output power and heat conduction properties intermediate to those of the homostructure and double heterostructure.

The output of these devices is approximately Lambertian, that is, the surface radiance is constant in all directions, but maximum emission is perpendicular to the junction and falls off to the sides in proportion to the cosine of the viewing angle because the apparent area varies with this angle. The expected coupling of this output to the fiber depends upon the exact geometry and conditions, such as the LED emission pattern and radiance, fiber size and refractive index grading, effective NA of the fiber, relative area of the fiber core and LED emitting area, distance and alignment between the two, the medium between them, etc. Thus coupling in an individual case is complex (Colvin, 1974; Yang and Kingsley, 1975; DiVita and Vannucci, 1975) and probably can be determined accurately only by measurement. However, it can be estimated for step-index fiber by $P_{coupled} = RA\Omega$, where R is the radiance into free space of the source, Ω is the solid acceptance angle of the fiber, and A is the smaller of the fiber core cross-sectional area or the emitter area. Since the core area and the acceptance angle of the fiber are dictated largely by the desired transmission characteristics, radiance of the source is of prime importance. If the fiber core can be significantly larger than the emitter, a lens may be used between the two to decrease the angular divergence of the source, and practical gains of 3 to 6 dB have been realized by use of very small lenses in the diode surface (King and SpringThorpe, 1975) or at the ends of fibers (Kato, 1973, for example), and much greater benefits have been calculated

(Abram *et al.*, 1975). For a graded-core fiber, optimum direct coupling requires that the source diameter be about one-half the fiber core diameter and, again, a lens offers some advantages for small-source, large-fiber combinations (Marcuse, 1975) at the expense of added complexity. Use of a lens cannot increase the power which can be coupled from a large-area incoherent source into a small-area fiber.

A factor which further complicates the accurate prediction of the results to be expected in LED-fiber coupling is the transmission characteristics of the so-called leaky modes or large-angle off-axis rays (Chapter 3). Much of the light from an incoherent source is initially coupled into these large-angle rays, which fall within the acceptance angle of the fiber but which have much higher attenuation than do the meridional or on-axis rays; energy from these rays tunnels or "leaks" into the cladding and is lost. The result is that much of the light coupled into a multimode fiber from an LED is lost within meters to hundreds of meters, and the power apparently coupled into a shorter fiber significantly exceeds that coupled effectively into a longer fiber. A concise review is available in the literature (Snyder, 1974).

By designing a low-loss large-core (115 μm dia.) fiber for use with a particular surface-emitter LED (O'Connor *et al.*, 1977), it has been possible to couple about 1 mW of optical power at 0.8 μm wavelength into graded-core fiber of 0.36 NA butted directly against the diode surface. The DH LED had a contact area of 75 μm diameter and without antireflection coatings delivered 15 mW of optical power into free space in the 130° cone defined by the sunken-well structure of Fig. 16.11. The diode was operated at 3.4 kA/cm², about one-half its saturation value.

16.3.3 Edge Emitters

An efficient edge emitter LED, as illustrated in Fig. 16.12 (Kressel and Ettenberg, 1975; Horikoshi *et al.*, 1976; Ettenberg *et al.*, 1976), emits part of its radiation in a relatively directed beam and thus has the advantage of improved efficiency in coupling light into a fiber. This is particularly important in coupling to a fiber with a small acceptance angle. The decrease in emission angle for this configuration is in the plane perpendicular to the junction, and it results from waveguiding effects of the heterostructure (Dumke, 1975; Wittke, 1975; Seki, 1976). In the structures of Fig. 16.12 the active layer is extremely thin, $\sim$500 Å, so that light generated there is not totally contained but leaks into the surrounding waveguide layers and is coupled strongly to the lowest-order guided mode perpendicular to the junction. Although absorption in the active layer itself is high, the absorption in the surrounding Al-containing waveguide layers is low, so that most of the light propagating in this mode is transmitted to the end faces and emitted with a beamwidth determined by the waveguide parameters.

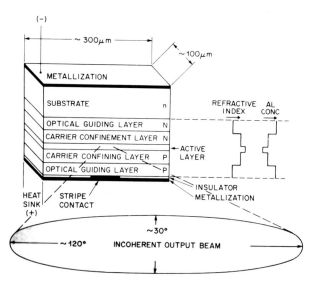

Fig. 16.12 AlGaAs edge-emitter DH LED with separate optical and carrier confinement layers to provide semidirectional output. A schematic representation of the refractive index and Al concentration of the various layers is shown at the right. For best directivity, thickness of the active layer must be very small, ~0.05 μm, and that of the guiding layers thicker, ~2–3 μm (Ettenberg *et al.*, 1976). For improved heat sinking, the bottom guiding layer may be omitted (Seki, 1976). Not to scale.

A reflector at one endface and an antireflection coating at the other assures that most of the propagating light is emitted at one endface. The emitted beam then is Lambertian with a half-power width of 120° in the plane of the junction where there is no waveguide effect, but has been made as small as 25–35° in the plane perpendicular to the junction (Ettenberg *et al.*, 1976) by properly proportioning the waveguide.

A second advantage of this structure is that, as a result of the channeling of light to a very small endface by the waveguide, the effective radiance at this face can be very high. The value achieved to date is 1000—1500 W/cm²-sr in an emitting area of 2–4 $\times$ 10^{-6} cm² (Horikoshi *et al.*, 1976; Ettenberg *et al.*, 1976). This value is several times the 300 W/cm²-sr achieved in air without antireflection coatings with a 15-μm diameter DH surface emitter (C. A. Burrus, unpublished, 1974) and an order of magnitude greater than that reported for 50-μm diameter DH devices operated at one-half to two-thirds saturation (Burrus and Miller, 1971). With the aid of a lens at the end of the fiber (effective because the emitter area was smaller than the fiber cross section), 0.8 mW of optical power has been coupled into a 0.14-NA, 90-μm core fiber from one of these edge-emitters operating near diode saturation at 5.1 kA/cm² (Ettenberg *et al.*, 1976). This is a coupling improvement of 7 dB compared to that expected from a purely Lambertian source.

On the basis of idealized assumptions, recent analysis (Marcuse, 1977) has shown that an edge-emitter LED with a guiding region should be capable of coupling 7.5 times more power into a fiber than can a surface emitter of equal intrinsic radiance and active layer thickness. When more realistic assumptions are made, including, for example, the more difficult heat-sinking geometry of the edge emitter, this same analysis shows that the expected advantage drops toward a factor of two for practical devices. However, for a variety of practical manufacturing and handling reasons, present day usage seems to favor surface emitters as incoherent sources for prototype systems.

16.3.4 Output Spectrum

The spectral width of the output of an LED operating at room temperature in the 0.8- to 0.9-μm region usually is 250–400 Å at the 3-dB points, and 500–1000 Å for materials with smaller energy gaps operating in the 1- to 1.3-μm wavelength region. The spectral width is a consequence of the fact that the carrier distributions in the conduction and valance bands between which the radiative transitions occur are given by the product of the band density of states and the Fermi distribution function. Since these carrier distributions spread at high temperatures, the output spectrum broadens as the temperature rises. This effect can be very pronounced in diodes heated up as a result of the power dissipated in the device resistance by the driving current. The peak wavelength also is shifted 3 to 4 Å/°C temperature rise in the junction. The spectrum may be broadened further by band tails in the density of states due to heavy doping or carrier injection. These effects are noticeable above concentrations of about 10^{18} cm^{-3}. The spectral width of the LEDs, narrow compared to a blackbody radiator but broad compared to even a multimode laser, is an inherent property of the materials, and it is greater for lower energy (longer wavelength) emission. It can be decreased only by cooling to reduce the thermal distributions (impractical in most everyday uses) or by filtering the output (a lossy process which lowers the output efficiency).

The peak wavelength of the emission from a surface emitter and from an edge emitter made from the same semiconductor can differ significantly, sometimes by 200–300 Å. The apparent peak wavelength is dependent upon both the internal radiative recombination spectrum and the absorption characteristics of the semiconductor window through which the radiation must pass or the waveguide and/or gain region in which it must propagate.

16.3.5 Power Efficiency

In considering the operating efficiency of an LED, the total external quantum efficiency and similar measures of junction performance often are of little practical use. It is more convenient to think of a "useful power efficiency" as the ratio of the optical power which can be collected in a useful way to the electrical power applied at the diode terminals. The latter is easy to determine unambiguously, but the collectible optical power depends upon the application. It may range from the total forward emission collected by a large detector near the top of the well of the device of Fig. 16.11, which emits into an unobstructed forward angle of about 130°, to that collected by a 0.14-NA fiber with an acceptance angle of only 16°. In the first case, power from a 75-μm diameter DH LED of the type illustrated in Fig. 16.11 at 150 mA (one-half diode saturation) and 2 V can be 15–18 mW for air interfaces and no antireflection coatings (22–24 mW at saturation) and results in a useful external power efficiency of 6%. The simple addition of a drop of epoxy resin in the well to decrease the index mismatch and thus increase the internal emission cone at the semiconductor surface can increase this number to 8–10%. This is not the total optical power emitted by the junction, but the total useful power for a particular application. Each application then leads to a different value.

16.3.6 Efficiency and Modulation Bandwidth

The efficiency and power of light generation in these devices (Lee and Dentai, 1978) is related directly to the effective electron density in the active region. This density in turn is a function of the active layer thickness, the carrier diffusion length and the surface recombination velocity, as well as self-absorption in the active layer which is itself a function of the doping density. The speed at which the junction can be directly current modulated with an information-carrying signal (Chapter 17) is fundamentally limited by the recombination lifetime of the carriers. The carrier recombination lifetime is primarily a function of the doping concentration, the number of electrons injected into the active region, the surface recombination velocity (the rate at which electrons recombine without radiation at crystalline defects at nearby semiconductor interfaces), and the physical thickness of the active layer. All of these parameters controlling the efficiency and the modulation bandwidth, many of them interdependent, are adjustable, within limits, in present-day technology. In practice, then, the device fabricator seeks to optimize both the radiative output and the modulation capabilities. The desirable characteristics for the materials are large injection rates, proper (not necessarily highest) doping levels, very thin ($\ll 1$ μm) active layers, and highly perfect crystalline layers and interfaces. In the devices themselves, it is important to minimize parasitic capacitance, which can range from 150–200 pF at zero bias for the structure of Fig. 16.10(b or c) to perhaps only 10–20 pF for that of Fig. 16.10(d or e). The main effect of parasitic capacitance is introduction of a delay between the driving signal and the response of the emitting junction (Lee, 1975), but it may be difficult in an LED to distinguish between a delay and a degraded risetime.

The state-of-the-art results for $Al_xGa_{1-x}As$ DH surface emitters are shown in Fig. 16.13, where the dc output near saturation is plotted vs device modulation bandwidth measured at a lower operating point. The plot shows that large modulation bandwidth and large output power have not yet been achieved simultaneously. The modulation bandwidth (see Chapter 17) is defined here in electrical terms, i.e., the 3-dB bandwidth of the detected electrical power due to the modulated portion of the optical signal, the form in which the information is useful in systems design. Sometimes the 3-dB bandwidth of the modulated optical power is quoted as the modulation bandwidth of LEDs, a practice which appears to inflate the "modulation bandwidth" by the factor $3^{1/2}$. The high-frequency devices depicted in Fig. 16.13 were achieved with highly doped, moderately thin (~ 0.5 μm) active layers and this heavy doping caused the increase in modulation bandwidth shown. However, it has been demonstrated in edge-emitters that modulation bandwidths to 145 MHz in electrical terms can be achieved with moderate doping if extremely thin active regions can be made (Lockwood et al., 1976; Ettenberg et al., 1976), and thus heavy doping is not always necessary for high speed. Measured modulation bandwidths to about 600 MHz in electrical terms have been reported for

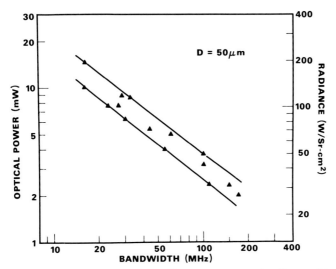

Fig. 16.13 Experimental variation of optical power output and radiance with modulation bandwidth. High output devices can be directly modulated only at low rates; if high modulation rates are required, output is sacrificed (Lee and Dentai, 1978).

high-radiance surface emitting LEDs (Goodfellow and Mabbit, 1976; Heinen et al., 1976).

The speed at which any LED can be modulated may be greatly extended at reduced efficiency if additional driving power is available (Dawson and Burrus, 1971), or the response time can be made arbitrarily small at the expense of output power (Harth et al., 1976). Surface emitters have been driven in a digital pseudorandom word configuration up to 250 Mbit/sec (Dawson, 1975), but the required driving power increased at about the 6-dB/octave rate predicted for the 40-MHz bandwidth LEDs (Lee et al., 1975).

16.3.7 Output Power Limitations

The output power of LEDs as a function of current is observed to increase approximately linearly at first, then to saturate gradually so that further increases of current lead to diminished light output. At the point of maximum output the diode is said to be saturated. It is of interest to see what limits the power output of these devices.

First, it is an accepted but not completely understood fact that the internal quantum efficiency of p–n junction light-emitting devices decreases exponentially with increasing temperature. Second, the thermal conductivity of the III–V semiconductor compounds is an exponentially decreasing function of increasing temperature (Carlson et al., 1965). Thus the junction temperature at which maximum power output occurs (the diode saturates) can be calculated for a given semiconductor, and it is found to be about 360°K for $Al_xGa_{1-x}As$ and slightly higher, about 400°K, for GaAs (Lee and Dentai, 1978). The maximum output is set by the junction temperature regardless of the size of the junction, but the temperature rise due to resistive heating from the driving current is very much dependent upon the structure.

Since in the structures of Fig. 16.10 the oxide layer, used to define the contact and confine the current flow, has very low heat conductivity, the heat flow in the emitting region of these structures can be approximated by a simple model assuming several stacked layers and one-dimensional heat flow. With such a model, and separation of the heat conduction into temperature-dependent (semiconductor) and temperature-independent (metallic) parts, the temperature rise expected from a bias current applied to several emitter diameters, semiconductor compositions and layer configurations has been calculated (Lee and Dentai, 1978). The results show that, since power saturation occurs at a constant junction temperature and smaller devices run cooler than larger ones for a given current density, small-area units can operate at relatively higher current densities (and hence higher radiance) than can larger area diodes. Similarly, smaller devices can be driven at higher current densities before the critical junction temperature is reached, so that again the smaller devices provide higher radiance. The results of this work provide a remarkably close fit to experimental observations on the small-area surface emitters.

16.3.8 Operating Lifetime

The operating lifetime of high-radiance LEDs has not yet been characterized fully. However, indications from accelerated aging tests suggest that the operating life can be made very long, probably of the order of 10^6 to 10^7 hr (100 to 1000 yr) at a fixed dc current (King *et al.*, 1975; Horikoshi *et al.*, 1976; Yamakoshi *et al.*, 1977; Gibbons, 1977). Lifetime is defined as operation to one-half the initial output. Since the LED is a simple, compact structure which operates without a threshold, it seems reasonable to expect that, with fabrication by comparable technology, its operational life always will exceed that of an injection laser dependent upon a long, narrow resonator cavity for its operation.

16.3.9 Superluminescent LED

Still a third device geometry which may eventually offer advantages of (1) increased power, (2) a well-directed output beam, and (3) reduced spectral bandwidth is an elongated version of the stripe-geometry edge emitter. It has been called a superluminescent diode, or SLD, and one form of its construction is illustrated in Fig. 16.14 (Lee *et al.*, 1973). While optimization for each application may lead to some variations in the configuration of the semiconductor layers, the edge emitter, the SLD and the laser (next section) are strikingly similar in construction.

In the SLD structure, which requires a p–n junction in the form of a

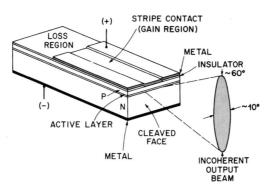

Fig. 16.14 Single-pass gain "superluminescent" LED. The region without a contact provides loss to suppress lasing; otherwise structure is that of a long ($\sim 1000\ \mu$m) laser. Shown as a homostructure for simplicity, but realized as an AlGaAs double heterostructure (Lee *et al.*, 1973; Iwamoto *et al.*, 1976).

long rectangular stripe, any one of the current-confinement arrangements of Fig. 16.11 may be used. One end of the stripe is made optically lossy to prevent reflections and thus suppress lasing, and the output is from the opposite end. In operation, the injection current is increased until stimulated emission (the first step in the onset of lasing, next section) and amplification occurs, but because of high loss at one end of the stripe, no feedback exists and no oscillation builds up. Therefore, in the current region of stimulated emission, there is gain and the output increases rapidly with current due to the single-pass amplification. At the same time, the spectral width of the output decreases to less than 100 Å. Devices have been made and operated in a pulsed mode to provide a peak output of 60 mW in an optical bandwidth of 60–80 Å at a center wavelength near 0.87 μm (Lee et al., 1973), and CW devices have delivered 25 mW (Iwamoto et al., 1976).

The disadvantage of these devices is that, to produce power comparable to that of a laser, the required current density is about three times that of a laser and the absolute current is high due to the long length (large area). However, a long-life CW version, if achievable, would seem to offer several much-sought advantages as an optical source: high output power, relatively narrow beamwidth and usefully small optical bandwidth, without the limitations of a threshold device.

16.3.10 Other Materials

Illustrations have been confined to the GaAs–$Al_xGa_{1-x}As$ systems due to space limitations and because these have, so far, been the most-studied systems. However, various other materials systems are potentially attractive, particularly as sources to operate further into the infrared where fiber transmission characteristics are optimum. Of special interest for LEDs is the wavelength region near 1.2–1.4 μm, where material dispersion of fibers can be zero or very small (Fig. 16.2). For this longer wavelength region, LEDs have been made with the geometries of Fig. 16.10 from $Ga_xIn_{1-x}As$ at 1.06 μm (Mabbit and Goodfellow, 1975), $Al_xGa_{1-x}As_ySb_{1-y}$ at 1 μm (Cho et al., 1977), and $Ga_xIn_{1-x}P_yAs_{1-y}$ at 1.3 μm (Dentai et al., 1977; Oe et al., 1977b). The solid solution $Ga_xIn_{1-x}P_yAs_{1-y}$ on InP would seem to hold particular promise for the longer wavelengths because of the extremely close lattice match which is possible (Antypas and Moon, 1973), and it is receiving increasing attention. In addition to LEDs, this heterostructure combination has resulted in successful fabrication of long-life, CW, room-temperature DH lasers at 1.1–1.3 μm (Hsieh, 1976; Oe et al., 1977a; Abbott et al., 1978). Although the difficulties of preparing semiconductiors rise rapidly as the number of components increase, preparation of the quaternary $Ga_xIn_{1-x}P_yAs_{1-y}$ seems to have proceeded with remarkable ease compared with the many difficulties encountered in the ternary $Al_xGa_{1-x}As$. The "perfect" lattice-match possible with $Ga_xIn_{1-x}P_yAs_{1-y}$ on InP certainly is a significant advantage; but perhaps the greatest advantage arises from the fact it contains no Al, an element which forms an extremely stable oxide with notorious ease.

While it is expected that efforts to improve devices in the 1.1- to 1.6-μm wavelength region will accelerate, the semiconductor systems now widely used for visible light emitters, GaP and GaP_yAs_{1-y}, probably will have little import for use with fibers because of the relatively poor fiber transmission characteristics at the shorter wavelengths.

16.3.11 Applications

Since LEDs, whether at emission wavelengths of 0.8–0.9 or 1.1–1.6 μm, are the simplest solid-state sources, they can be expected to find a place in a number of communications applications where optical fibers serve as transmission lines. They promise useful output power, moderate optical bandwidth, convenient direct modulation, and long operational life. The limitations are dictated primarily by the fiber dispersion characteristics rather than by fiber losses. The inherent spectral bandwidth and modulation capabilities suggest that (1) available diodes emitting at wavelengths of 0.8–0.9 μm can serve as low-bit-rate sources (a few tens of megabits) with relatively long (several kilometer) transmission lines; (2) essentially the same devices can be used at distances up to a few hundred meters at rates of a few hundred megabits; and (3) longer wavelength LEDs can be used at both moderately high bit rates and long distances. The modulation-bandwidth fiber-length product in the vicinity of 1.2–1.4 μm could be above 500 Mbit-km which suggests, for example, the use of 50-Mbit rates at distances greater than 10 km (Muska *et al.*, 1977). Certainly the future of communications-type LEDs would seem to lie in the direction of the longer wavelengths, in spite of the difficulties of working with the less-studied and (perhaps) harder-to-handle quaternary semiconductors. Whatever the difficulties with forward-biased devices of these semiconductors, however, they probably will be easier to fabricate than reverse-biased detectors of these same materials—other devices badly needed at the longer wavelengths.

16.4 INJECTION LASERS

16.4.1 General Description

By adding a cavity to provide feedback, the LED becomes a laser at high current densities. The requirements for stimulated emission are presented later in this section. For high-quality $GaAs$-$Al_xGa_{1-x}As$ DH lasers, typical threshold current densities J_{th} for broad-area lasers are between 1 and 2 kA/cm², while J_{th} is about a factor of two larger for stripe-geometry lasers with 10- to 15-μm-wide stripes. A typical stripe-geometry DH laser is illustrated in Fig. 16.15. The parallel reflecting surfaces that form a Fabry–Perot interferometer are obtained by cleaving. The laser dimensions shown are typical of GaAs-Al$_x$Ga$_{1-x}$As DH lasers that may be used as a source in lightwave communications. Proton bombardment has been used to convert regions of the semiconductor to high resistivity so that current flow is confined to the 13-μm-wide stripe. Layers of SiO₂ are an alternative technique to confine the current to the stripe area. The stripe-geometry not only reduces the device area so that the operating current can be low, but it also gives improved operating life by removing most of the junction perimeter from the surface. The side with the epitaxial layers is mounted onto the Cu heat sink. For a current of 150–200 mA at 1.6 V (about ⅓ the power necessary to operate an ordinary flashlight), the output power is approximately 5 mW from one face.

The circular inset in Fig. 16.15, a scanning-electron photomicrograph, illustrates the important details of the CW heterostructure laser. The light generation region, which is called the active layer, is the p-Al$_{0.05}$Ga$_{0.95}$As layer. The layer shown is 0.1 μm thick, and generally it would be in the range 0.1 to 0.2 μm. The adjacent N- and P- regions are Al$_x$Ga$_{1-x}$As to confine the injected carriers and provide the waveguide. The p⁺-GaAs layer aids in making the ohmic contact. This structure currently is most often grown by LPE, but both CVD and MBE also are used as discussed previously. The small size of the laser chip is demonstrated by the fact that

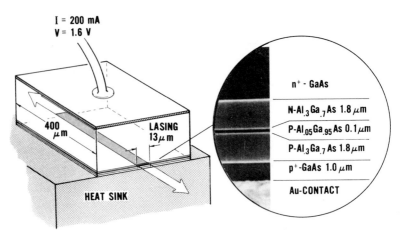

I = 200 mA
V = 1.6 V

400 μm   LASING 13μm

HEAT SINK

n⁺ - GaAs
N-Al$_{.3}$Ga$_{.7}$As 1.8 μm
P-Al$_{.05}$Ga$_{.95}$As 0.1μm
P-Al$_{.3}$Ga$_{.7}$As 1.8 μm
p⁺-GaAs 1.0 μm
Au-CONTACT

Fig. 16.15 Schematic representation of a stripe-geometry DH laser for lightwave communications. The circular insert is a scanning electron photomicrograph of the heteroepitaxial layers (Casey and Panish, 1975).

100,000 laser chips can be prepared from a 1-gm wafer. Continuous operation at room temperature and up to ~100° C is possible.

The emission properties of the stripe geometry DH laser are illustrated in Fig. 16.16, where the upper part of the figure shows the far-field emission pattern. Because the emission is from such a small area, it is not as well collimated as for most other types of lasers, but diverges strongly. Typical emission beam divergences at one-half the peak intensity are 45° perpendicular to the junction plane and 9° parallel to the junction plane, and this beam divergence must be considered when coupling the laser emission to an optical fiber. The lower part of Fig. 16.16 illustrates that the injection laser emission is not a single-wavelength line, but is a family of

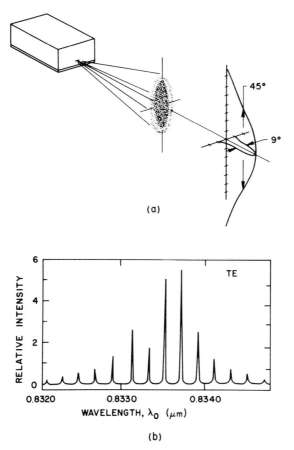

Fig. 16.16 (a) Schematic representation of the far-field emission of a stripe-geometry GaAs–Al$_x$Ga$_{1-x}$As DH laser. Typical values of the full angle at half power are illustrated perpendicular to and along the junction plane (Casey and Panish, 1978). (b) Emission spectra for a stripe-geometry DH laser operating CW at 1.1 J_{th} (Paoli, 1975).

longitudinal modes corresponding to an integral number of cavity lengths. The spacing depends on the cavity length. Certain types of stripe-geometry lasers have single-longitudinal mode emission. The polarization of the emission is along the junction plane and in the direction perpendicular to the cavity length, which is the transverse electric (TE) polarization.

16.4.2 Waveguiding in Injection Lasers

Many of the properties of injection lasers are related to the optical field distribution within the semiconductor device. Consideration of both the threshold current density and the emission-beam divergence requires a detailed description of the propagating optical field distribution (mode). The variation of the energy gap, the refractive index and the optical field for the GaAs-Al$_x$Ga$_{1-x}$As DH laser are illustrated in Fig. 16.17. The N- and P-Al$_x$Ga$_{1-x}$As lasers are made sufficiently thick to prevent the outer n$^+$- and p$^+$-GaAs layers from influencing the field distribution. The optical field distribution then may be obtained as the solution of the reduced wave equation (Casey et al., 1973).

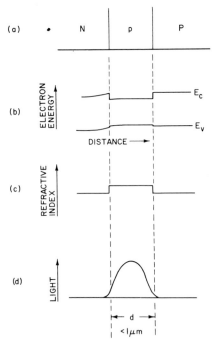

Fig. 16.17 (a) Schematic representation of a N–p–P GaAs–Al$_{0.3}$Ga$_{0.7}$As DH laser (b) energy-band diagram at high forward bias, (c) refractive-index profile, and (d) optical-field distribution.

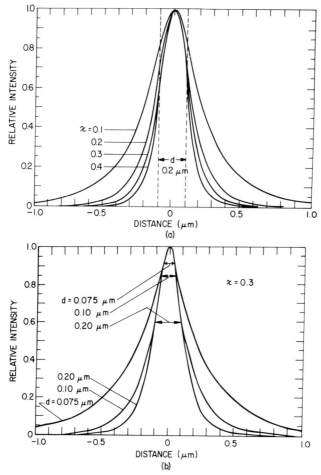

Fig. 16.18 The optical field intensity as a function of position within a GaAs–Al$_x$Ga$_{1-x}$As DH laser. The distance is measured from the center of the active region normal to the plane of the junction. (a) Variation of Al concentration x for constant width of active region $d = 0.2$ μm. (b) Variation of active region width for constant Al concentration $x = 0.3$. The arrows indicate d for each curve (Casey and Panish, 1978).

The influence of the active layer thickness d and the AlAs mole fraction x on the confinement of light within the active layer is illustrated by the calculated light intensity as shown in Fig. 16.17. The variation of the refractive index with x was illustrated in Fig. 16.3 (Section 16.2). In Fig. 16.18a, the active layer thickness of 0.2 μm is held constant while the Al composition is varied. It is readily seen that a significant reduction in confinement occurs when x is reduced from 0.2 to 0.1. This spreading of the intensity distribution with decreasing Al composition results in an increased threshold current density and a smaller beam divergence. In Fig. 16.18b, the Al composition is held constant at $x = 0.3$ while the active layer width is varied from $d = 0.075$ to $0.2 \ \mu m$. For $d = 0.1 \ \mu m$, very little light intensity is within the active layer. Clearly, the intensity distribution and the related beam divergence are dramatically influenced by both the active layer thickness and the refractive index step.

The extent of light confinement with the active layer may be represented by a confinement factor Γ, defined as the ratio of the light energy within the active layer to the sum of light energy both within and outside the active layer. The confinement factor is therefore the fraction of the energy of the propagating mode within the active layer. The variation of Γ with d and x is shown in Fig. 16.19. This factor Γ summarizes the spreading of the light intensity with d and x as illustrated in Fig. 16.18.

Figures 16.18 and 16.19 apply only to the lowest-order TE mode. Higher order modes are possible for larger values of d. For a given width d, the modes designated by N which can propagate in a symmetrical DH are determined by the condition (Marcuse, 1972)

$$N < 2|\bar{n}_1^2 - \bar{n}_2^2|^{1/2} \, d/\lambda_0, \tag{16.3}$$

where $\bar{n}_1$ is the refractive index of the active layer, $\bar{n}_2$ is the refractive index of the $Al_xGa_{1-x}As$ bounding layers, and λ_0 is the emission wavelength. Even at large d where higher order modes are possible, the larger Γ and smaller mirror reflection loss for the lowest order mode results in its domination. For example, with $x = 0.3$, the lowest-order mode ($N = 0$) is still observed for all $d < 0.76 \ \mu m$, although the first odd-order mode ($N = 1$) is possible at $d = 0.38 \ \mu m$. The $N = 2$ mode is not observed until $d > 0.76$. The $N = 3$ mode is observed for $d > 1.14 \ \mu m$.

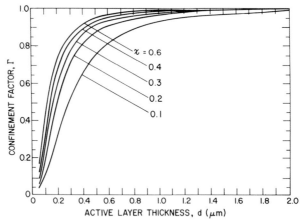

Fig. 16.19 Fraction of the propagating mode within the active layer Γ as a function of active layer thickness d and Al composition x for DH lasers (Casey and Panish, 1978).

The far-field pattern of the injection laser may be calculated from the optical field distribution within the optical cavity of the laser (Casey and Panish, 1978). The beam divergence, taken as the full beamwidth at the half-power (3 dB) points perpendicular to the junction plane, is shown in Fig. 16.20 to vary with d and x. As was shown in Fig. 16.16, it is $\sim 9°$ along the junction plane. This beam divergence influences the coupling of the DH laser to the optical fiber.

There are several other configurations of heterostructure lasers. The single-heterostructure (SH) laser has the p–n junction within the narrow energy gap semiconductor and is bounded on the p-side by a wide energy gap semiconductor to provide electron and light confinement. Because there is only one heterojunction, neither the optical field nor the holes are as well confined as in the DH configuration. The SH laser is represented in Fig. 16.21a. Note the asymmetry of the optical field. Minimum room temperature threshold current densities J_{th} for SH lasers are $\sim 10^4$ A/cm^2 (Hayashi et al., 1969). The DH laser that was illustrated in Fig. 16.17 provides greatly reduced threshold current density due to its better carrier and light confinement, and threshold current densities as low as 475 A/cm^2 have been obtained (Ettenberg, 1975). In this case, the compositional discontinuity at the heterojunction was 0.45–0.68 rather than the

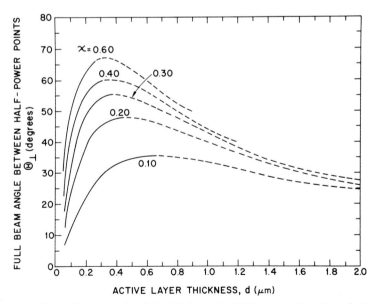

Fig. 16.20 Beam divergence for GaAs-Al$_x$Ga$_{1-x}$As DH laser as a function of active layer thickness d and composition x for the lowest order TE mode. The dashed portion of the curve represents active layer thicknesses where higher order modes are possible (Casey and Panish, 1978).

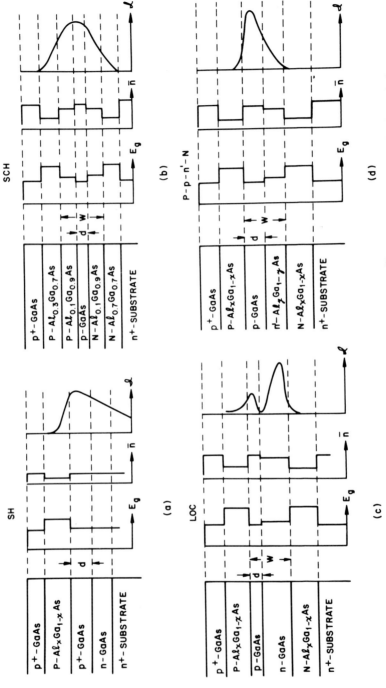

Fig. 16.21 Schematic representation of the energy gap E_g, index of refraction $\bar{n}$, and optical intensity $\mathscr{I}$. (a) Single heterostructure (b) separate confinement, (c) large optical cavity, and (d) Ppn'N with p- and n-layer thicknesses for fundamental mode operation. The active layer thickness is d and the optical cavity thickness is w.

usual ~ 0.3. Low J_{th} also has been obtained with the separate confinement heterostructure (SCH) (Casey et al., 1974b). The SCH illustrated in Fig. 16.21b also is called the localized-gain heterostructure laser (Thompson and Kirby, 1973). Values of J_{th} have been obtained as low as 575 A/cm^2 with such a structure (Thompson and Kirby, 1973). In the SCH shown in Fig. 16.21b, the step in E_g between GaAs and Al$_{0.1}$Ga$_{0.9}$As is sufficient to confine the carriers within the GaAs layer, but the step in $\bar{n}$ does not confine the light. However, the larger step in the refractive index between Al$_{0.3}$Ga$_{0.7}$As and Al$_{0.1}$Ga$_{0.9}$As serves to confine the light and thereby provides the optical waveguide of width w.

In order to obtain greater power output than is available from DH lasers, heterostructure lasers have been prepared with a GaAs p–n junction between the P-Al$_x$Ga$_{1-x}$As and N-Al$_x$Ga$_{1-x}$As layers (Lockwood et al., 1970). The structure is illustrated in Fig. 16.21c and was designated a large optical cavity (LOC) heterostructure. The carrier confinement is similar to that of the more common configurations, but the optical waveguide cross section is relatively large so that the possibility of facet damage is reduced. As can be seen in this figure, the GaAs waveguide is divided into a thin p-type active recombination region and a wide passive region. The advantage of this structure is that increasing the (LOC) waveguide thickness does not result in as large an increase in J_{th} as for the DH laser. A four-layered heterostructure also has been described (Paoli et al., 1973) in which fundamental transverse mode operation was observed with optical waveguides 2 μm thick. This structure was designated the P–p–n'–N heterostructure where N and P represent the P- and N-type Al$_x$Ga$_{1-x}$As layers, p represents the p-type GaAs active layer, and n' represents an Al$_x$Ga$_{1-x}$As layer with $x \approx 0.01$ to 0.02. It has been shown (Krupka, 1975) that mode discrimination occurs for the wide cavity, and that the mode-selection behavior is governed largely by the thickness of the p-layer and the step in $\bar{n}$ between the p- and n'-layer. Although there are other possible combinations of heterojunctions to form additional structures, these examples in Fig. 16.21, together with the DH laser, illustrate the most common types of heterostructure lasers.

16.4.3 Threshold Current Density

The necessary condition for stimulated emission in a semiconductor may be expressed very simply (Bernard and Duraffourg, 1961), and it has been shown that the separation of the quasi-Fermi levels within the conduction and valence bands must exceed the photon emission energy in order for the stimulated emission rate to exceed the absorption rate. For a laser formed by parallel reflecting surfaces bounding a gain medium in the form of a Fabry–Perot interferometer, the condition for lasing is (Casey and Panish, 1978, Chapter 3)

$$g\Gamma = \alpha_1 + (1/L)\ln(1/R), \qquad (16.4)$$

where g is the gain coefficient, Γ is the optical confinement coefficient, α_1 is the internal loss, L is the cavity length, and R is the mirror reflectivity. The internal losses include free-carrier absorption and scattering of radiation due to irregularities at the heterostructure interfaces or within the waveguide region.

The mirror reflectivity enters the threshold relations as $(1/L)\ln(1/R)$. It has been shown (Ikegami, 1972) that the TE polarization selection is due to the difference in mirror reflectivity for TE and TM waves. Also, it is both the confinement factor and the mirror reflectivity which determine when the higher order modes appear as the active layer thickness is increased. For low-threshold lasers, $(1/L)\ln(1/R)$ is ~ 30 cm^{-1} where α_1 is ~ 10 to 15 cm^{-1}.

From fundamental principles, an expression may be obtained for the threshold current density of a DH laser. Calculations of the gain coefficient (Stern, 1976) permit writing the current dependence of gain as

$$g = 5.0 \times 10^{-2} |(J\eta/d) - 4.5 \times 10^3|, \qquad (16.5)$$

where J is the current density, η is the internal quantum efficiency, and d is the active layer thickness in microns. Combining Eqs. (16.4 and 16.5) gives the threshold current density as

$$J_{th}(A/cm^2) = \frac{4.5 \times 10^3 d}{\eta} + \frac{20}{\eta\Gamma}\left[\alpha_1 + \frac{1}{L}\ln\left(\frac{1}{R}\right)\right]. \qquad (16.6)$$

The confinement factor Γ was given in Fig. 16.19. A comparison of experimental J_{th} vs d data and J_{th} calculated from Eq. (16.6) are shown in Fig. 16.22. The fit for $x = 0.6$ is very good, while for $x = 0.3$ the calculated and experimental J_{th} differ by 400 A/cm^2 at $d = 0.1$ μm. The reason for this disagreement is not presently understood.

16.4.4 Heterojunctions

As mentioned previously, a heterojunction is a junction between two semiconductors with different energy gaps. A model has been developed for heterojunctions (Anderson, 1962) in which part of the energy gap difference is assigned to the conduction band and the remainder to the valence band. The heterojunction energy band diagrams are the same as for homojunctions except for these conduction and valence band discontinuities. A detailed analysis of heterojunctions for heterostructure lasers has been given (Casey and Panish, 1978, Chapter 4). Complete reviews of the

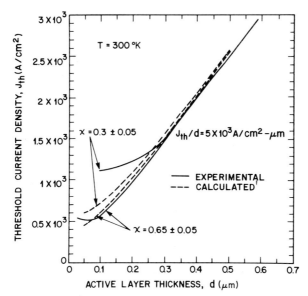

Fig. 16.22 Comparison of the experimental J_{th} as a function of d with J_{th} calculated from Eq. (16.5). The value of Γ is from Fig. 16.19 and η is taken as 1.0 (Casey and Panish, 1978).

heterojunction literature are given in the books on heterojunctions (Milnes and Feucht, 1972; Sharma and Purohit, 1974).

The energy band diagrams for DH lasers are shown in Fig. 16.23 with an applied voltage near the value necessary to obtain sufficient current for stimulated emission. Part (a) is for a p-type active layer, while part (b) is for a n-type active layer. Similarity of the energy band diagrams illustrates that either n- or p-type active layers may be used. For the N–p–P structure, the wider energy gap N-region injects electrons into the p-layer. The step in the conduction band at the p–P interface confines these injected carriers to the active layer. A carrier density of about 1–2 × 10¹⁸ cm⁻³ is necessary for stimulated emission. The bandbending and step in the valence band prevents holes from leaking into the N-layer. For the N–n–P structure, the P-layer injects holes into the n-layer. This feature of injection and carrier confinement is one of the primary reasons for using heterojunctions in semiconductor lasers.

16.4.5 Operating Properties of Heterostructure Lasers

A mounted DH laser was represented in Fig. 16.15, and the emission properties were given in Fig. 16.16. The beam divergence in the direction perpendicular to the junction plane was summarized in Fig. 16.20. The light output from one mirror as a function of input current for a low threshold stripe-geometry $Ga_xAl_{1-x}As$ DH laser is shown in Fig. 16.24.

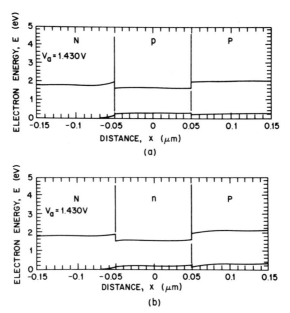

Fig. 16.23 Energy band diagram for GaAs–Al$_{0.3}$Ga$_{0.7}$As double heterostructure at 1.430 V forward bias at room temperature. (a) N–p–P DH and (b) N–n–P DH (Casey and Panish, 1978).

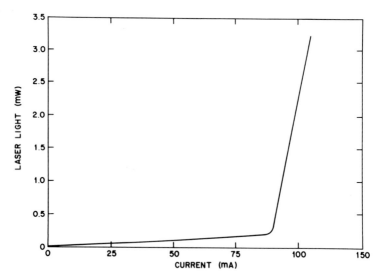

Fig. 16.24 Light output vs dc current for a stripe-geometry Al$_x$Ga$_{1-x}$As DH laser (D. D. Sell, private communication 1977).

The laser heat-sink is at room temperature. The sharp break in the curve of the laser light is the threshold for lasing. The temperature rise at the junction is approximately 5°C. Since this device was intended to have an emission wavelength at 0.85 μm in order to avoid the loss in many fibers near 0.9 μm due to OH, about 0.08 AlAs has been added to the active layer.

One of the most important properties of a source for lightwave communications is the operating life. In the initial DH lasers operated CW at room temperature, laser operation ceased within minutes or a few hours. This rapid degradation was shown to be due to the appearance of areas of reduced luminescence that penetrated the active layers. These defects were designated dark line defects (DLD) (DeLoach et al., 1973). The DLD has been identified (Petroff and Hartman, 1974) as a three-dimensional dislocation network that grows during excitation. For the 3-mW output shown in Fig. 16.24, catastrophic mirror damage is not important. However, for high power devices, the critical optical power density at which mirror facet damage occurs is 5–10 $\times$ 10^6 W/cm^2 (Hakki and Nash, 1974).

Data for accelerated aging tests on CW stripe-geometry DH GaAs-Al$_x$Ga$_{1-x}$As lasers suggest that the mean operating lifetime at room temperature, with power outputs exceeding 1 mW per laser face, can be expected to be in excess of 10^6 hr (Hartman et al., 1977). These results indicate both the feasibility of long operating life and that no fundamental problem exists; however, they do not indicate that any given device will operate for long periods.

Future-generation optical-fiber systems may use wavelength multiplexing as a means for increasing information-carrying capacity. In such cases, the wavelength stability of the emission of laser sources would be important. The wavelength shift of the peak emission of GaAs lasers with cleaved mirrors is observed to be 3.8 A/°C, as expected from the temperature dependence of the energy gap. However, distributed-feedback (DFB) lasers using periodic structures such as gratings instead of mirrors can greatly reduce the sensitivity to temperature of the emission wavelength (Casey et al., 1975). A temperature sensitivity of only 0.5 A/°C has been attained in a DFB laser operated CW at room temperature (Nakamura et al., 1975).

16.4.6 Other Stripe Geometry Laser Configurations

Numerous stripe-geometry DH laser configurations have been investigated in an attempt to improve the device operating behavior. For structures such as the stripe-contact or the proton-bombarded stripe-geometry laser, there often are nonlinearities in the optical-power-output versus current (L–I) characteristic as shown in Fig. 16.25. These L–I nonlinearities, generally called "kinks," are not only undesirable, but in some

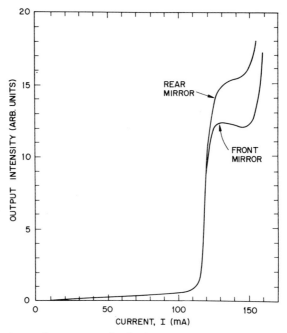

Fig. 16.25 Emission characteristics for the front and rear mirrors of a stripe-geometry DH laser operating CW at room temperature. Full scale corresponds to ~3 mW (Paoli, 1976a).

applications can make the device unacceptable. They can reduce the available output power and make stable operating intensities difficult to achieve.

Two other adverse operating characteristics are related to the dynamic properties. When a fast-rise-time current pulse is applied to a semiconductor laser, the light output power is delayed a few nanoseconds and then is characterized by a damped oscillation as shown in Fig. 16.26. The damped oscillation for this particular case has a frequency of ~1 GHz and a decay constant of a few nanoseconds in the light output. This behavior is called a relaxation oscillation. When the current pulse is applied, the carrier buildup initially is delayed by the carrier lifetime. The carrier concentration then builds up to levels that exceed the value necessary to reach threshold, and the resulting high optical fields deplete the carrier concentration; this in turn reduces the optical field so that the carrier density can build up again. This process continues, but each successive cycle is diminished. Thus the relaxation oscillations are related to the interactions between the carriers and the photons.

Another type of pulsation that has been observed is characterized by a lack of damping and is called self-pulsation. These pulsations, illustrated in Fig. 16.27, were found to occur during device aging in devices that initially exhibited no self-pulsations (Paoli, 1977). For these lasers, the self-

DC BIAS = 0

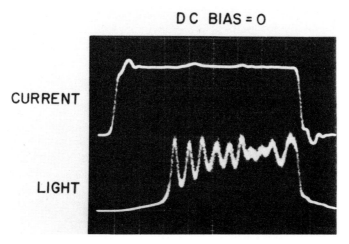

CURRENT

LIGHT

Fig. 16.26 Input excitation current and the resulting output light intensity for a stripe-geometry DH laser. The time scale is 2 nsec/div (Paoli, 1976b).

induced pulsations occur at frequencies ranging from 300 to 600 MHz, which is significantly less than the relaxation oscillation frequency of these same lasers. The mechanisms by which self-induced pulsations occur are not presently understood.

Although there are more than 20 different types of stripe geometry laser configurations, only a few will be described as illustrative examples.

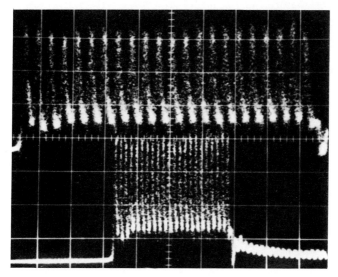

Fig. 16.27 Time variation of the output intensity from a DH laser operated pulsed with a dc bias current. The sustained pulsations occur after continuous operation at 70°C for 50 hr. The current is 13% above the threshold and the horizontal scale is 2 nsec/div for the top trace and 5 nsec/div for the bottom trace (Paoli, 1977).

These structures are intended to eliminate kinks and relaxation oscillations, and some of them are also relatively free of self-pulsations. A more complete listing of stripe-geometry laser configurations has been summarized elsewhere (Casey and Panish, 1978).

When the Zn diffusion for the planar-stripe laser is extended into the active layer, as shown in Fig. 16.28a, both kink-free and single-longitudinal mode operation was observed (Yonezu et al., 1977). It was found that planar-stripe lasers grown on channeled substrates, as shown in Fig. 16.28b, also eliminated kinks (Aiki et al., 1978). This structure has many good operating characteristics, such as single-longitudinal mode emission and absence from relaxation oscillations. A Zn-diffused structure called the transverse-junction-stripe (TJS) geometry laser is shown in Fig. 16.29a (Namizaki et al., 1974). Since the injection in this case is from the n-GaAs layer into the Zn-diffused p-GaAs, this device actually is a very thin homojunction laser. Both single-longitudinal-mode operation and threshold currents lower than 50 mA have been obtained with the TJS laser.

For the buried-heterostructure (BH) laser shown in Fig. 16.29b, the active region is completely surrounded by $Al_xGa_{1-x}As$ (Tsukada, 1974). The BH laser is characterized by an active region as small as 1 μm square in cross section. With such small active regions, the threshold current may be as low as 15 mA, and a nearly symmetrical far-field pattern is obtained. The active layer width and thickness must be $\leqslant 1$ μm to prevent higher order modes. It was found that neither kinks nor self-pulsations are observed in the majority of the lasers after long-term aging (Kajimura et al.,

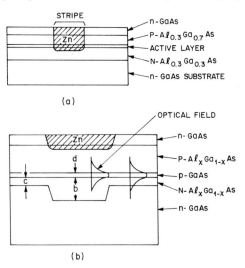

Fig. 16.28 (a) Schematic representation of cross section for planar-stripe laser (Yonezu et al., 1977). (b) Channeled substrate planar-stripe geometry laser. The active-layer thickness $d \approx 0.1$ μm, while $c \approx 0.4$ μm, and $b \approx 1.4$ μm (Aiki et al., 1977).

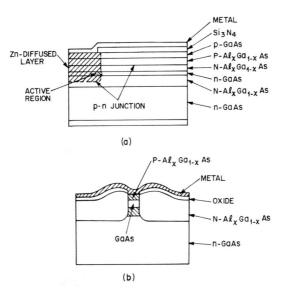

Fig. 16.29 (a) Transverse stripe-geometry (TJS) laser (Namizaki *et al.*, 1974). (b) Buried-heterostructure laser. The GaAs active region is completely surrounded by $Al_xGa_{1-x}As$ (Tsukada, 1974).

1978). In the strip-buried heterostructure (SBH), the introduction of a $N-Al_{0.1}Ga_{0.9}As$ layer, as shown in Fig. 16.30, converts the channel waveguide in a BH laser to a strip-loaded waveguide (Tsang *et al.*, 1978). The optical mode is no longer confined to the thin p-GaAs active region, but spreads mostly into the $N-Al_{0.1}Ga_{0.9}As$ low-loss waveguide and thereby results in a reduction of the refractive index along the junction plane. Therefore, the SBH laser retains the two-dimensional carrier confinement as in the BH laser, while permitting the light to spread out along the junction plane. The reduced light confinement permits stable fundamental-mode operation for a strip width as wide as 5 μm, and there are no light-current nonlinearities. Relaxation oscillations and self-pulsations also are absent (Tsang *et al.*, 1979). This recent work on stripe-geometry lasers demonstrates that the $L-I$ nonlinearities, relaxation oscillations, and self-pulsations can be eliminated. Although these structures are based on GaAs-$Al_xGa_{1-x}As$, future work can be expected to use systems such as InP-$Ga_xIn_{1-x}P_yAs_{1-y}$ for emission near 1.3 μm.

6.5 NEODYMIUM LASERS

16.5.1 General Considerations

Solid-state neodymium (Nd) lasers are attractive sources for optical-fiber transmission systems for several reasons. Their important emission

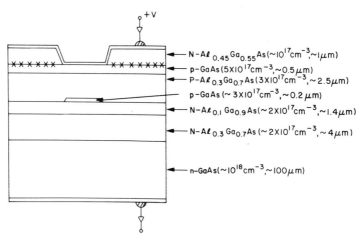

Fig. 16.30 Schematic representation of the strip buried heterostructure laser. The active region is the p-GaAs between the P-Al$_{0.3}$Ga$_{0.7}$As and N-Al$_{0.1}$Ga$_{0.9}$As layers. Lateral current confinement is provided by the reversed-biased p-N heterojunction marked by x–x (Tsang *et al.*, 1978).

wavelengths lie in the region 1.05–1.35 μm, where low-loss, doped silica fibers of today exhibit very low attenuation and vanishingly small material dispersion. The output linewidths of the Nd ion in lightly doped, crystalline hosts are both very narrow and homogeneously broadened, permitting single-frequency single-mode operation, which is a desirable condition for maximizing the transmission bandwidth of a fiber. The fact that Nd lasers can be pumped by long-lived Al$_x$Ga$_{1-x}$As LEDs makes them especially attractive for high-capacity, long-haul applications where reliability, modal purity, and narrow spectral width are important considerations.

A shortcoming of the Nd laser is that its upper laser level has a long fluorescence lifetime, on the order of 10^{-4} sec (compared to 10^{-9} sec for the recombination lifetime of the carriers in the Al$_x$Ga$_{1-x}$As injection laser), so that direct modulation of the Nd laser at megahertz rates is impossible. An external optical modulator is therefore an essential companion for the Nd laser in communications applications.

There are numerous host materials for Nd, some crystalline and others amorphous. The Nd ion can be either a dopant in some host material in low concentrations, or a stoichiometric constituent of a compound in high concentrations. Lasing action has been observed in many of these materials. Detailed discussions of Nd lasers can be found in several review articles (Geusic *et al.*, 1970; Findlay and Goodwin, 1970; Chesler and Geusic, 1972; Weber, 1975; Danielmeyer, 1975; 1976; Chinn *et al.*, 1976a).

16.5.2 Neodymium-Doped Yttrium–Aluminum–Garnet (Nd:YAG) Lasers

The crystalline material yttrium aluminum garnet ($Y_3Al_5O_{12}$) is a particularly good laser host for the Nd^{3+} ion because it has relatively low photoelastic constants and good optical, mechanical and thermal properties. The Nd^{3+} ion is incorporated substitutionally for yttrium. As shown in Fig. 16.31, the Nd:YAG laser material is a four-level system with several pumping bands and fluorescent transitions. The strongest pumping bands are at 0.81 and 0.75 μm and the most prominent laser transitions are at 1.064 and near 1.32 μm (Geusic et al., 1970). Since $Al_xGa_{1-x}As$ LEDs can be made to emit with sufficiently high radiance in these pump bands, they are well-suited for pumping the Nd laser.

LED pumping configurations have been studied both theoretically and experimentally (Barnes, 1973; Chesler and Singh 1973; Farmer and Kiang, 1974; Ostermayer, 1977). Figure 16.32 shows the side- and end-pumping schemes. Because the fluorescence efficiency of Nd:YAG at high doping levels is reduced by crystal lattice distortions, concentration quenching, cross relaxation, inhomogeneities and lowered heat conductivity, the op-

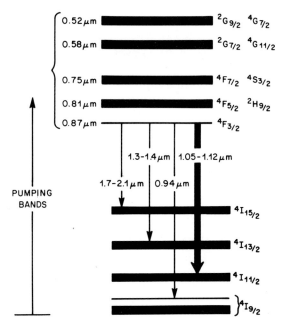

Fig. 16.31 Main pump bands and fluorescent transitions in Nd:YAG. The laser transitions near 1.06 and 1.3 μm are of greatest interest for fiber transmission; the pump bands near 0.81 and 0.87 μm can be reached with AlGaAs LEDs and lasers.

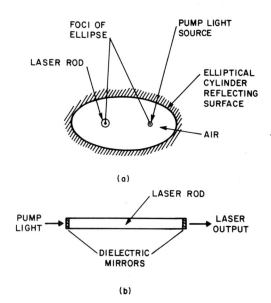

Fig. 16.32 (a) Typical side-pumping configuration for Nd:YAG laser. (b) End-pumping configuration.

timum Nd conncentration for efficient CW operation is about one atomic percent, i.e., 1% replacement of Y by Nd (Geusic, 1965; Danielmeyer and Blätte, 1973; Danielmeyer, 1976). This low concentration results in a relatively long pump absorption length (~1 cm) for the strongest pumping band at 0.805 μm (Chesler and Singh, 1973). The diameter of the laser rod in a side-pumping configuration therefore should be a few millimeters for efficient pumping. Reducing the side-pumped rod diameter to make the laser compatible with a fiber of similar size would drastically decrease the fraction of pump light absorbed and would result in a device of very low efficiency.

The situation is quite different for end-pumping, since the constraint is now on the length of the laser rod and not on its diameter. If a high-radiance type LED is used, the diameter of the laser can be the same as that of the fiber and the optimal length for efficient pumping is about 5 mm (Chesler and Singh, 1973).

At threshold, the roundtrip laser gain must equal the roundtrip loss within the resonator. The roundtrip gain g, in the small-gain approximation, is given by (Chesler and Singh, 1973)

$$g = 2P_p\eta\sigma\tau_f F/h\nu_p A, \tag{16.7}$$

where P_p is the input pump power, η is the fraction of the input pump power that is absorbed by the laser medium, σ is the cross section of laser transition ($\sigma = 4.6 \times 10^{-19}$ cm^2 for the $\lambda = 1.06$ μm transition) (Singh *et al.*, 1974), τ_f is the fluoresence lifetime of the upper laser level ($\tau_f = 230$ μsec) (Kushida *et al.*, 1968), $h\nu_p$ is the pump quantum energy, A is the cross sectional area of the laser rod, and F is the fraction of the inverted population in the upper laser level of the $4F_{3/2}$ manifold ($F \simeq 0.4$). The round-trip loss consists of (i) the mirror losses, (ii) the nonresonant losses in the laser material, and (iii) resonant absorption of the laser transition due to the small but finite population of the terminal laser level at room temperature. The resonant loss is equal to $2L\sigma N\beta/Z$, where L is the length of the laser medium, N is the density of lasing ions, β is the Boltzmann factor for the terminal laser level, and Z is the partition function. Equating the roundtrip gain with the total roundtrip loss $\mathscr{L}$ gives the threshold pump power, P_p^{th}:

$$P_p^{th} = h\nu_p A\mathscr{L}/2\eta\sigma\tau_f F. \tag{16.8}$$

For Nd:YAG material with 1 at.% doping, the resonant absorption loss is about 0.1% per cm and the absorbed pump power density required to reach threshold for $\mathscr{L} \simeq 0.2\%$ is about 10 W/cm^2 (Chesler and Singh, 1973; Singh *et al.*, 1975).

The first Nd:YAG laser that operated continuously at room temperature utilized a tungsten–halogen lamp in an elliptical cylinder for side-pumping (Gausic *et al.*, 1964). This combination has been highly developed as a research instrument and is now used in industry as a manufacturing tool. Multimode output power of hundreds of watts at $\lambda = 1.06$ μm can be obtained with overall efficiencies of a few percent (Geusic *et al.*, 1970; Findlay and Goodwin, 1970). Bulkiness, large power consumption, poor lamp life, and the necessity for forced cooling render this form of the Nd:YAG laser unsuitable for fiber system use.

Pumping with LEDs results in a more compact laser consuming much less power and requiring no forced cooling. The first LED-pumped, 1.06 μm, Nd:YAG laser operating continuously at room temperature had an array of 64 GaP$_y$As$_{1-y}$ diodes in a side-pumped configuration (Ostermayer *et al.*, 1971). Other experiments using arrays of Al$_x$Ga$_{1-x}$As diodes in similar pumping configurations but with the laser rods cooled below room temperature have been reported (Barnes, 1973; Farmer and Kiang, 1974).

Further reduction in size, weight, and power consumption has been obtained by end-pumping with a single LED. The feasibility of a miniature end-pumped laser was first demonstrated with a 5 × 0.45-mm Nd:YAG rod pumped by a GaP$_y$As$_{1-y}$ diode (Chesler and Draegert, 1973; Draegert, 1973). Laser threshold was achieved at room temperature in a pulsed mode of operation. When a hemispherical reflector was used to focus the pump radiation from a 0.46-mm-diameter, domed Al$_x$Ga$_{1-x}$As LED into a 5 × 0.46-mm laser rod, an output power of 0.25 mW was obtained in a quasi-CW mode of operation at room temperature (Ostermayer, 1977). A superluminescent diode (with a spectral width of 56 Å) also has been used, in conjunction with a graded-index (Selfoc®) lens for focusing, to end-pump a 5.4 × 3-mm laser rod (Washio et al., 1976). Continuous operation with 1.5-mW output power was obtained at room temperature.

The single-crystal fiber laser end-pumped by a high-radiance LED (Stone et al., 1976) is the first truly fiber-compatible Nd:YAG laser. The single-crystal fiber is grown from a 1% Nd-doped YAG rod heated by the focused beam of a CO_2 laser (Burrus and Stone, 1975). Successive regrowths can produce fibers with diameters as small as 50 μm. Measurements indicate that the properties of the small regrown crystals are identical with those of the larger original rods. Room temperature, CW operation at 1.06 μm and at 1.3 μm has been obtained in a single-crystal fiber laser, 5-mm long and 60 μm in diameter, end-pumped by a double-heterostructure $Al_xGa_{1-x}As$ high-radiance LED (Stone et al., 1976; Burrus

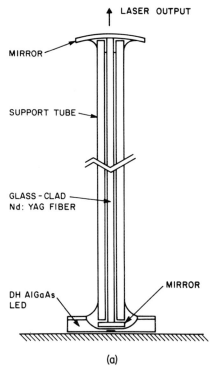

(a)

Fig. 16.33 Cross section (a) and photograph (b) of an early embodyment of a small Nd: YAG fiber laser pumped by a small-area high-radiance LED. Diameter of the as-grown Nd:YAG fiber is 50 μm; o.d. of the support tube is 150 μm; laser length is 0.5 cm. (Stone and Burrus, 1979).

(b)

Fig. 16.33 (*Continued*)

et al., 1976). The epitaxial growth and fabrication parameters of the diode were adjusted to provide, as nearly as possible, maximum output, spectral match to the 0.805-μm pumping band, and an emitter area equal to that of the fiber cross section. The diode-pumped Nd:YAG single-crystal fiber laser (an early embodiment is pictured in Fig. 16.33) has provided about 1 mW of output power at 1.06 μm with about 225 mA drive current to the LED (Stone and Burrus, 1979). Further optimization of the mirrors and fiber-diode coupling could lead to a 1-mW output at a diode current not over 75–100 mA, and comparable results (modified by the wavelength factor) would be attainable at 1.3 μm with suitable mirrors. (Schuöcker and Schiffner, 1978; Stone and Burrus, 1979).

16.5.3 High-Concentration Neodymium-Compound Lasers

A recently discovered group of crystalline materials that show promise as laser hosts consists of compounds in which Nd appears as a stoichiometric constituent rather than a dopant. Examples are neodymium pentaphosphate (NdP_5O_{14}), lithium-, potassium-, and sodium-neodymium tetraphosphate ($LiNdP_4O_{12}$, $KNdP_4O_{12}$, $NaNdP_4O_{12}$), neodymium-aluminum borate ($NdAl_3(BO_3)_4$) and others (Möckel, 1978). The Nd con-

TABLE 16.1
Laser Spectroscopic Parameters for Nd:YAG and Nd Pentaphosphates[a]

Material	Laser wavelength λ (μm)	Pump wavelength λ_p (μm)	LED pump absorption coefficient $\tilde{\alpha}_p$ (cm^{-1})	Laser transition cross section σ (cm^2)	Flourescent life-time τ_f (μs)	Concentration N (cm^{-3})	Fraction of ions in upper laser level F	Boltzmann factor of terminal laser level β	Partition function Z
$Y_{2.97}Nd_{0.03}Al_5O_{12}$	1.064	0.808	1.1	4.6×10^{-19}	230	1.39×10^{20}	0.4	4×10^{-5}	2.15
$La_{0.75}Nd_{0.25}P_5O_{14}$	1.051	0.798	8.1	1.8×10^{-19}	238	9.91×10^{20}	0.63	8.5×10^{-5}	2.55
NdP_5O_{14}	1.051	0.789	32.4	1.8×10^{-19}	120	3.96×10^{21}	0.63	8.5×10^{-5}	2.55

[a] Singh et al., 1975.

centration in these stoichiometric materials can be more than one order of magnitude larger than in optimally doped Nd:YAG without having the deleterious effects of concentration quenching appreciably degrade their fluorescence properties (Danielmeyer, 1975; Singh *et al.*, 1975; Weber, 1975). The high concentration of Nd ions results in strong absorption in the pump band near 0.8 μm and in high gain at the laser transitions, desirable properties for side pumping with LEDs and for miniaturization. The same high concentration also leads to proportionately higher resonant loss in the material. For end-pumping, high concentration materials give no advantage at 1.06 μm (Singh *et al.*, 1975). However, an interesting conclusion can be drawn for operation at 1.3 μm with side-pumping. Because resonant loss is negligibly small at 1.3 μm, the required threshold pump power density can be appreciably less than that for operation at 1.06 μm (Burrus *et al.*, 1976). This conclusion has been verified experimentally in argon-laser-pumped pentaphosphate lasers (Otsuka *et al.*, 1977).

A theoretical analysis that compares the threshold power of a Nd:YAG laser with that of a concentrated and a diluted Nd pentaphosphate laser has been made for LED pumping (Singh *et al.*, 1975). The results, based on parameters given in Table 16.1 and assuming an active-medium length of 1 cm, are given in Fig. 16.34. It is seen that the side-pumped $La_{0.75}Nd_{0.25}P_5O_{14}$ laser, the best performing of all three, shows only a slight advantage over the Nd:YAG laser when the total nonresonant loss is small. Nevertheless, the possibility of CW room-temperature operation with just a few hundred microwatts of pump power from high-radiance

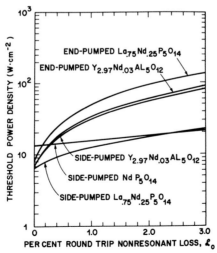

Fig. 16.34 Required threshold pump power density as a function of loss for several Nd-laser materials (Singh *et al.*, 1975).

LEDs and calculated optical gains up to 10 dB per wavelength at saturation make stoichiometric materials very interesting for miniature-laser applications (Möckel, 1978; Budin et al., 1978).

Fluorescence in NdP_5O_{14} was first reported in 1972 (Danielmeyer and Weber, 1972). Some of the measured spectroscopic parameters of NdP_5O_{14} and of a diluted material $LaNdP_5O_{14}$, pertinent to laser operation, are given in Table 16.1 (Weber et al., 1974; Singh et al., 1975). Laser action at $\lambda = 1.051$ μm was first observed in a 35-μm platelet pumped by a pulsed dye laser at $\lambda_p = 0.58$ μm (Weber et al., 1973). Subsequently, CW or quasi-CW laser action at room temperature has been achieved in NdP_5O_{14}, $LaNdP_5O_{14}$, and $ScNdP_5O_{14}$ (Damen et al., 1973; Danielmeyer et al., 1973; Weber et al., 1975; Chinn et al., 1976b). Of particular interest are a miniature LaNd pentaphosphate laser of acicular geometry (12 × 12 × 680 μm) (Weber et al., 1975) and a Nd pentaphosphate laser pumped by an $Al_xGa_{1-x}As$ injection laser at 0.8 μm with a threshold pump power of ~7 mW (Chinn et al., 1976b).

Fluorescence in $LiNdP_4O_{12}$ was first reported in 1974 (Yamada et al., 1974). The measured laser spectroscopic properties ($\tau_f = 120$ μs, $\sigma = 3.2 \times 10^{-19}$ cm^2) are similar to those of NdP_5O_{14} (Otsuka et al., 1975). Continuous-wave or quasi-continuous laser action at room temperature has been achieved in $LiNdP_4O_{12}$ ($\lambda = 1.048$ μm), $KNdP_4O_{12}$ ($\lambda = 1.052$ μm), and $NaNdP_4O_{12}$ ($\lambda = 1.051$ μm) (Otsuka and Yamada, 1975; Chinn and Hong, 1975a,b; Nakano et al., 1976). $Al_xGa_{1-x}As$ LEDs and lasers have been used to pump $LiNdP_4O_{12}$ lasers. Threshold pump power densities of 7.2 and 14.9 W/cm^2 have been obtained for a laser cooled to $-30°$C pumped from the side and from the end, respectively (Saruwatari and Kimura, 1976). Continuous operation at room temperature has been achieved in a laser consisting of $LiNdP_4O_{12}$ crystal 0.15-mm thick and a graded-index fiber 1.33-mm long sandwiched between a pair of flat mirrors and pumped by an $Al_xGa_{1-x}As$ injection laser (Saruwatari et al., 1976).

Continuous-wave laser action at room temperature also has been reported in $NdAl_3(BO_3)_4$, an acentric crystal which has a small electrooptic coefficient (Chinn and Hong, 1975b; Winzer et al., 1978). The measured threshold pump power density of 12 kW/cm^3, however, is about four times as large as those for $LiNdP_4O_{12}$ and NdP_5O_{14} lasers.

Research on high-concentration neodymium compounds, and on lasers made from them, is being pursued actively. Although the present state of the art of these lasers is not yet as advanced as that of the Nd:YAG laser, a miniature device side-pumped with a yet-to-be-developed LED could be an attractive source for use with fiber systems (Kubodera and Otsuka, 1977).

16.6 OTHER LASERS

Only solid-state lasers are considered to be serious candidates for operating use in communications systems with optical fiber transmission lines, for rather obvious reasons. However, many other laser systems in which the high operating voltages, large size, or other problems preclude use in unattended manholes are extremely useful in laboratory evaluations of materials and components. The many gas systems, He–Ne, Ar, Kr, etc., provide a wide range of wavelengths for spectral characterization of fibers, for example, and dye lasers can provide extremely high power, tunable signals for the same uses. Lasers which operate at wavelengths highly absorbed by fiber materials, the 10.6-μm CO_2 laser for example, have proved to be extremely useful as controlled sources of heat in the manufacture of both glass-transmission and single-crystal-laser fibers.

REFERENCES

Abbott, S. M., Muska, W. M., Lee, T. P., Dentai, A. G., and Burrus, C. A. (1978). 1.1-Gb/s pseudo-random pulse-code modulation of 1.27 μm wavelength CW InGaAsP/InP DH lasers. *Electron. Lett.* **14,** 349.

Abram, R. A., Allen, R. W., and Goodfellow, R. C. (1975). The coupling of light-emitting diodes to optical fibers using sphere lenses. *J. Appl. Phys.* **46,** 3468.

Aiki, K., Nakamura, M., Kuroda, T., and Umeda, J. (1977). Channeled-substrate planar structure AlGaAs injection lasers. *Appl. Phys. Lett.* **30,** 649.

Aiki, K., Nakamura, M., Kuroda, T., Umeda, J., Ito, R., Chinnone, N., and Maeda, M. (1978). Transverse mode stabilized $Al_xGa_{1-x}As$ injection laser with channeled-substrate-planar structures. *IEEE J. Quantum Electron.* **QE-14,** 89.

Alferov, Zh, I., and Kazarinov, R. F. (1963). Author's Certificate No. 181737, Claim No. 950840 of March 30.

Alferov, Zh. I., Andreev, V. M., Garbuzov, Yu. V., Zhilyaev, E. P., Morozov, E. P., Portnoi , E. L., and Triofim, V. G. (1971). Investigation of the influence of the AlAs-GaAs heterostructure parameters in the laser threshold current and realization of continuous emission at room temperature. *Sov. Phys.—Semicond.* (Engl. Transl.) **4,** 1573 (1971); *Fiz Tekh. Poluprovod.* **4,** 1826 (1970).

Anderson, R. L. (1962). Experiments on Ge-GaAs heterojunctions. *Solid-State Electron.* **5,** 341.

Antypas, G. A., and Moon, R. L. (1973). Growth and characterization of InP-InGaAsP lattice-matched heterojunctions. *J. Electrochem. Soc.* **120,** 1574.

Archer, R. J., and Kreps, D. (1967). The quantum efficiency of electro-luminescence in gallium arsenide diodes. *Gallium Arsenide Symposium, 1966,* Inst. Phys. and Phys. Soc. (1967), p. 103.

Arthur, J. R. (1968). Interaction of Ga and As_2 molecular beams with GaAs surfaces. *J. Appl. Phys.* **39,** 4032.

Barnes, N. P. (1973). Diode-pumped solid-state lasers. *J. Appl. Phys.* **44,** 230.

Bernard, M. G. A., and Duraffourg, G. (1961). Laser conditions in semiconductors. *Phys. Status Solidi* **1,** 699.

Budin, J. P., Neubauer, M., and Rondot, M. (1978). On the design of neodymium miniature lasers. *IEEE J. Quantum Electron* **QE-14,** 831.

Burrus, C. A. (1969). "Small-area electroluminescent diodes coupled to optical-fiber transmission lines: A report," Tech. Memo. Bell Laboratories (unpublished).

Burrus, C. A., and Dawson, R. W. (1970). Small-area high-current-density GaAs electroluminescent diodes and a method of operation for improved degradation characteristics. *Appl. Phys. Lett.* **17**, 97.

Burrus, C. A., and Miller, B. I. (1971). Small-area, double-heterostructure aluminum-gallium arsenide electroluminescent diode sources for optical-fiber transmission lines. *Opt. Commun.* **4**, 307.

Burrus, C. A., and Stone, J. (1975). Single-crystal fiber optical devices: A Nd:YAG fiber laser. *Appl. Phys. Lett.* **26**, 318.

Burrus, C. A., and Ulmer, E. A. (1971). Efficient small-area GaAs-$Ga_{1-x}Al_x$As heterostructure electroluminescent diodes coupled to optical fibers. *Proc. IEEE* **59**, 1263.

Burrus, C. A., Stone, J., and Dentai, A. G. (1976). Room-temperature 1.3-μm CW operation of a glass-clad Nd:YAG single-crystal fibre laser end-pumped with a single LED. *Electron. Lett.* **12**, 600.

Carlson, R. O., Slack, G. A., and Silverman, S. T. (1965). Thermal conductivity of GaAs and GaAsP laser semiconductors *J. Appl. Phys.* **36**, 505.

Casey, H. C., Jr., and Panish, M. B. (1975). Injection lasers and integrated optics with heteroepitaxial III-V compounds. *Ind. Res.* Sept., 57.

Casey, H. C., Jr., and Panish, M. B. (1978). "Heterostructure Lasers." Academic Press, New York.

Casey, H. C., Jr., and Trumbore, F. A. (1970). Single crystal electroluminescent materials. *Mater. Sci. Eng.* **6**, 69 (a review of single crystal electroluminescent semiconductor materials).

Casey, H. C., Panish, M. B., and Merz, J. L. (1973). Beam divergence of the emission from double-heterostructure injection lasers. *J. Appl. Phys.* **44**, 5470.

Casey, H. C., Sell, D. D., and Panish, M. B. (1974a). Refractive index of Al_xGa_{1-x}As between 1.2 and 1.8 eV. *Appl. Phys. Lett.* **24**, 63.

Casey, H. C., Jr., Panish, M. B., Schlosser, W. O., and Paoli, T. L. (1974b). GaAs-Al_xGa_{1-x}As heterostructure laser with separate optical and carrier confinement. *J. Appl. Phys.* **45**, 322.

Casey, H. C., Jr., Somekh, S., and Ilegems, M. (1975). Room-temperature operation of low-threshold separate-confinement injection laser with distributed feedback. *Appl. Phys. Lett.* **27**, 142.

Chesler, R. B., and Draegert, D. A. (1973). Miniature diode-pumped Nd:YAlG lasers. *Appl. Phys. Lett.* **23**, 235.

Chesler, R. B., and Geusic, J. E. (1972). Solid-state ionic lasers. *In* "Laser Handbook" (F. T. Arecchi and E. O. Schultz-Dubois, eds.), Vol. I, p. 325. North-Holland Publ., Amsterdam.

Chesler, R. B., and Singh, S. (1973). Performance model for end-pumped miniature Nd:YAlG lasers. *J. Appl. Phys.* **44**, 5441.

Chinn, S. R., and Hong, H. Y.-P. (1975a). Low-threshold CW $LiNdP_4O_{12}$ laser. *Appl. Phys. Lett.* **26**, 649.

Chinn, S. R., and Hong, H. Y.-P. (1975b). CW laser action in acentric $NdAl_3(BO_3)_4$ and $KNdP_4O_{12}$ *Opt. Commun.* **15**, 345.

Chinn, S. R., Hong, H. Y.-P., and Pierce, J. W. (1976a). Minilasers of neodymium compounds. *Laser Focus* **12**, 64.

Chinn, S. R., Pierce, J. W., and Heckscher, H. (1976b). Low-threshold transversely excited NdP_5O_{14} laser. *Appl. Opt.* **15**, 1444.

Cho, A. Y., and Arthur, J. R. (1975). *Prog. Solid State Chem.* **10**, 157.

Cho, A. Y., and Casey, H. C., Jr. (1974). Properties of Schottky barriers and p-n junctions prepared with GaAs and $Al_xGa_{1-x}As$ molecular beam epitaxial layers. *J. Appl. Phys.* **45,** 1258.

Cho, A. Y., Dixon, R. W., Casey, H. C., Jr., and Hartman, R. L. (1976). Continuous operation of GaAs-$Al_xGa_{1-x}As$ double heterostructure lasers prepared by molecular-beam epitaxy. *Appl. Phys. Lett.* **28,** 501.

Cho, A. Y., Casey, H. C., Jr., and Foy, P. W. (1977). Back-surface emitting $GaAs_xSb_{1-x}$ LEDs ($\lambda = 1$ μm) prepared by molecular-beam epitaxy. *Appl. Phys. Lett.* **30,** 397.

Colvin, J. (1974). Coupling (launching) efficiency for a light-emitting diode, optical fiber termination. *Opto-electronics* **6,** 387.

Damen, T. C., Weber, H. P., and Tofield, B. C. (1973). NdLa pentaphosphate laser performance *Appl. Phys. Lett.* **23,** 519.

Danielmeyer, H. G. (1975). Stoichiometric laser materials. *Adv. Solid State Phys.* **15,** 253.

Danielmeyer, H. G. (1976). Progress in Nd:YAG lasers. *In* "Lasers, A Series of Advances" (A. K. Levine and A. J. DeMaria, eds.), Vol. IV, p. 1. Dekker, New York.

Danielmeyer, H. G., and Blätte, M. (1973). Fluorescence quenching in Nd:YAG. *Appl. Phys.* **1,** 269.

Danielmeyer, H. G., and Weber, H. P. (1972). Fluorescence in neodymium ultraphosphate. *IEEE J. Quantum Electron.* **QE-8,** 805.

Danielmeyer, H. G., Buber, G., Krühler, W. W., and Jeser, J. P. (1973). Continuous oscillation of a (Sc,Nd) pentaphosphate laser with 4 milliwatts pump threshold. *Appl. Phys.* **2,** 335.

Dawson, R. W. (1975). Pseudorandom 250 Mb/s modulation of GaAs LEDs. *Electron. Lett.* **11,** 144.

Dawson, R. W., and Burrus, C. A. (1971). Pulse behavior of high-radiance small-area electroluminescent diodes. *Appl. Opt.* **10,** 2367.

DeLoach, B. C., Jr., Hakki, B. W., Hartman, R. L., and D'Asaro, L. A. (1973). Degradation of CW GaAs double-heterostructure lasers at 300°K. *Proc. IEEE* **61,** 1042.

Dentai, A. G., Lee, T. P., Burrus, C. A., and Buehler, E. (1977). High-radiance InGaAsP LEDs emitting at 1.2-1.3 μm. *Electron. Lett.* **13,** 484.

Dingle, R., Logan, R. A., and Arthur, J. R. Private communication.

DiVita, P., and Vannucci, R. (1975). Geometrical theory of coupling errors in dielectric optical waveguides. *Opt. Commun.* **14,** 139.

Draegert, D. A. (1973). Single-diode end-pumped Nd:YAG laser *IEEE J. Quantum Electron.* **QE-9,** 1146.

Dumke, W. P. (1962). Interband transitions and laser-action. *Phys. Rev.* **127,** 1559.

Dumke, W. P. (1975). The angular beam divergence in double-heterojunction lasers with very thin active regions. *IEEE J. Quantum Electron.* **QE-11,** 400.

Dupuis, R. D., and Dapkus, P. D. (1978). Very low threshold $Ga_{1-x}Al_xAs$-GaAs double-heterostructure lasers grown by metalorganic chemical vapor deposition. *Appl. Phys. Lett.* **32,** 473.

Dyment, J. C., SpringThorpe, A. J., King, F. D., and Straus, J. (1977). Proton bombarded double-heterostructure LEDs. *J. Electron. Mater.* **6,** 173.

Ettenberg, M. (1975). Very low-threshold double-heterostructure $Al_xGa_{1-x}As$ injection laser. *Appl. Phys. Lett.* **27,** 652.

Ettenberg, M., Kressel, H., and Wittke, J. P. (1976). Very high radiance edge-emitting LED. *IEEE J. Quantum Electron.* **QE-12,** 360.

Farmer, G. I., and Kiang, Y. C. (1974). Low-current-density LED-pumped Nd:YAG laser using a solid cylindrical reflector. *J. Appl. Phys.* **45,** 1356.

Finch, W. F., and Mehal, E. W. (1964). Properties of $GaAs_xP_{1-x}$ by vapor phase reaction. *J. Electrochem. Soc.* **111,** 814.

Findlay, D., and Goodwin, D. W. (1970). The neodymium in YAG laser. In Adv. Quantum Electron. **1**, page 77.

Fleming, J. W. (1976). Material and mode dispersion on $GeO_2{:}B_2O_2{:}SiO_2$ glasses. J. Am. Ceram. Soc. **59**, 503.

Frosch, C. J. (1964). The epitaxial growth of GaP by a Ga_2O vapor transport mechanism. J. Electrochem. Soc. **111**, 180.

Geusic, J. E. (1965). The YAlG:Nd laser. In "Solid-State Maser Research (Optical)," Final Report, Contract DA-36-039-AMC-02333E. U.S. Army Electron. Mater. Agency.

Geusic, J. E. Marcos, H. M., and Van Uitert, L. G. (1964). Laser oscillation in Nd-doped yttrium aluminum, yttrium gallium, and gadolinium garnets Appl. Phys. Lett. **4**, 182.

Geusic, J. E., Bridges, W. B., and Pankove, J. I. (1970). Coherent optical sources for communications. Proc. IEEE **58**, 1419.

Gibbons, G. (1977). Progress in LEDs as signal sources. Top. Meet. Opt. Fiber Transm., 2nd, 1977 Paper WB-1.

Goodfellow, R. C., and Mabbitt, A. W. (1976). Wide-bandwidth high-radiance gallium-arsenide light-emitting diodes for fibre-optic communication. Electron. Lett. **12**, 50.

Hakki, B. W., and Nash, F. R. (1974). Catastrophic failure in GaAs double-heterostructure injection lasers. J. Appl. Phys. **45**, 3907.

Hall, R. N., Fenner, G. E., Kingsley, J. D., Soltys, T. J., and Carlson, R. O. (1962). Coherent light emission from GaAs junctions. Phys. Rev. Lett. **9**, 366.

Harth, W., Huber, W., and Heinen, J. (1976). Frequency response of GaAlAs light-emitting diodes. IEEE Trans. Electron devices **ED-23**, 478.

Hartman, R. L., Schumacher, N. E., and Dixon, R. W. (1977). Continuously operated AlGaAs DH lasers with 70° lifetimes as long as two years. Appl. Phys. Lett. **31**, 756.

Hayashi, I., Panish, M. B., and Foy, P. W. (1969). A low-threshold room temperature injection laser. IEEE J. Quantum Electron. **QE-5**, 211.

Hayashi, I., Panish, M. B., Foy, P. W., and Sumski, S. (1970). Junction lasers which operate continuously at room temperature. Appl. Phys. Lett. **17**, 109.

Heinen, J., Huber, W., and Harth, W. (1976). Light-emitting diodes with a modulation bandwidth of more than 1 GHz Electron. Lett. **12**, 533.

Hill, D. E. (1965). Internal quantum efficiency of GaAs electroluminescent diodes. J. Appl. Phys. **36**, 3405.

Horikoshi, Y., Takanashi, Y., and Iwane, G. (1976). High-radiance light-emitting iodes. Jpn. J. Appl. Phys. **15**, 485 (1976).

Hsieh, J. J. (1976). Room-temperature operation of GaInAsP/InP double-heterostructure diode lasers emitting at 1.1 μm, J. Appl. Phys. **28**, 283.

Ikegami, T. (1972). Reflectivity of mode at facet and oscillation mode in double-heterostructure injection lasers. IEEE J. Quantum Electron. **QE-8**, 470.

Iwamoto, K., Hino, T., Matsumoto, S., and Inoue, K. (1976). Room temperature CW operated superluminescent diodes for optical pumping of Nd:YAG laser. Jpn. J. Appl. Phys. **15**, 2191.

Kajimura, T., Saito, K., Shige, N., and Ito, R. (1978). Stable operation of buried-heterostructure $Ga_{1-x}Al_xAs$ lasers during accelerated aging. Appl. Phys. Lett. **33**, 626.

Kato, D. (1973). Light coupling from a stripe-geometry GaAs diode laser into an optical fiber with spherical end. J. Appl. Phys. **44**, 2756.

Kibler, L. U., Burrus, C. A., and Trambarulo, R. (1964). Light-emitting, formed-point-contact gallium arsenide and gallium arsenide-phosphide diodes. Proc. IEEE **52**, 1260.

King, F. D., and SpringThorpe, A. J. (1975). The integral lens coupled LED. J. Electron. Mater. **4**, 243.

King, F. D., SpringThorpe, A. J., and Szentesi, O. I. (1975). High-power long-lived double heterostructure LEDs for optical communications. Int. Electron Devices Meet. 1975 p. 480.

Kressel, H., and Ettenberg, M. (1975). A new edge-emitting (AlGa)As heterojunction LED for fiber-optic communications. *Proc. IEEE* **63**, 1360.

Kressel, H., and Nelson, H. (1969). Close confinement gallium arsenide p-n junction lasers with reduced optical loss at room temperature. *RCA Rev.* **30**, 106.

Kroemer, H. (1963). A proposed class of heterojunction injection lasers. *Proc. IEEE* **51**, 1783.

Krupka, D. C. (1975). Selection of modes perpendicular to the junction plane in GaAs large-cavity double-heterostructure lasers. *IEEE J. Quantum Electron.* **QE-11**, 390.

Kubodera, K., and Otsuka, K. (1977). Diode-pumped miniature solid-state laser; design considerations. *Appl. Opt.* **16**, 2747.

Kushida, T., Marcos, H., and Geusic, J. E. (1968). Laser transition cross section and fluorescence branching ratio for Nd^{3+} in yttrium aluminum garnet. *Phys. Rev.* **167**, 289.

Lee, T. P. (1975). Effect of junction capacitance on the rise time of LEDs and on the turn-on delay of injection lasers. *Bell Syst. Tech. J.* **54**, 53.

Lee, T. P., and Dentai, A. G. (1978). Power and modulation bandwidth of GaAs-AlGaAs high radiance LEDs for optical communication systems. *IEEE J. Quantum Electron.* **QE-14**, 150.

Lee, T. P., Burrus, C. A., and Miller, B. I. (1973). A stripe-geometry double-heterostructure amplified-spontaneous-emission (superluminescent) diode. *IEEE J. Quantum Electron* **QE-9**, 820.

Lee, T. P., Burrus, C. A., and Holden, W. S. (1975). Direct modulation efficiency of LEDS for optical fiber transmission applications. *Proc. IEEE* **63**, 318.

Lockwood, H. F., Kressel, H., Sommers, H. S., Jr., and Hawrylow, F. Z. (1970). Low-threshold LOC GaAs injection lasers. *Appl. Phys. Lett.* **17**, 499.

Lockwood, H. F., Wittke, J. P., and Ettenberg, M. (1976). LED for high data rate, optical communications. *Opt. Commun.* **16**, 193.

Mabbitt, A. W., and Goodfellow, R. C. (1975). High-radiance small-area gallium-indium arsenide 1.06 μm light emitting diodes. *Electron. Lett.* **11**, 274.

Malitson, I. H. (1965). Interspecimen comparison of the refractive index of fused silica. *J. Opt. Soc. Am.* **55**, 1205.

Marcuse, D. (1972). "Light Transmission Optics." Van Nostrand-Reinhold, Princeton, New Jersey.

Marcuse, D. (1975). Excitation of parabolic-index fibers with incoherent sources. *Bell Syst. Tech. J.* **54**, 1507.

Marcuse, D. (1977). LED fundamentals: Comparison of front and edge emitting diodes. *IEEE J. Quantum Electron.* **QE-13**, 819.

Milnes, A. G., and Feucht, D. L. (1972). "Heterojunctions and Metal-Semiconductor Junctions." Academic Press, New York.

Möckel, P. (1978). Optically pumped miniature solid-state lasers from stoichiometric neodymium compounds. *Frequenz* **32**, 85.

Muska, W. M., Li, Tingye, Lee, T. P., and Dentai, A. G., (1977). Material-dispersion-limited operation of high-bit-rate optical-fibre data links using l.e.d.s. *Elect. Lett.* **13**, 605.

Nahory, R. E., Pollack, M. A., Beebe, E. D., DeWinter, J. C., and Dixon, R. W. (1976). Continuous operation of 1.0 μm wavelength $GaAs_{1-x}Sb_x/Al_yGa_{1-y}As_{1-x}Sb_x$ double-heterostructure injection lasers at room temperature. *Appl. Phys. Lett.* **28**, 19.

Nakamura, M., Aiki, K., Umeda, J., and Yariv, A. (1975). CW operation of distributed-feedback GaAs-GaAlAs diode lasers at temperatures up to 300K. *Appl. Phys. Lett.* **27**, 403.

Nakano, J., Otsuka, K., and Yamada, T. (1976). Fluorescence and laser emission cross-sections in $NaNdP_4O_{12}$. *J. Appl. Phys.* **47**, 2749.

Namizaki, H., Kan, H., Ishii, M., and Ito, A. (1974). Transverse-junction-stripe-geometry double-heterostructure lasers with very low threshold current. *J. Appl. Phys.* **45**, 2785.

Nathan, M. I., Dumke, W. P., Burns, G., Dill, F. H., Jr., and Lasher, G. (1972). Stimulated emission of radiation from GaAs p-n junctions. *Appl. Phys. Lett.* **1,** 62.

Nuese, C. J. (1977). III-V alloys for optoelectronic applications. *J. Electron. Mat.* **6,** 253.

Neuse, C. J., Olsen, G. H., Ettenberg, M., Cannon, J. J., and Zamerowski, T. J. (1976). CW room temperature $In_xGa_{1-x}As/In_yGa_{1-y}P$ 1.06-μm lasers. *Appl. Phys. Lett.* **29,** 807.

O'Connor, P. B., MacChesney, J. B., and Melliar-Smith, C. M. (1977). Large-core high-NA fibres for data link applications. *Electron. Lett.* **13,** 170.

Oe, K., Ando, S., and Sugiyama, K. (1977a). 1.3 μm CW operation of GaInAsP/InP DH diode lasers at room temperature. *Jpn. J. Appl. Phys.* **16,** 1273.

Oe, K., Ando, S., and Sugiyama, K. (1977b). Surface emitting LEDs for the 1.2-1.3 μm wavelength with GaInAsP/InP double heterostructures. *Jpn. J. Appl. Phys.* **16,** 1693.

Olsen, G. H., and Ettenberg, M. (1978). *In* "Crystal Growth: Theory and Techniques" (C. Goodman, ed.), Vol. II. Plenum, (in press).

Osanai, H., Shioda, T., Moriyama, T., Araki, S., Horiguchi, M., Isawa, T., and Takata, H. (1976). Effect of dopants on transmission loss of low-OH-content optical fibres. *Electron. Lett.* **12,** 549.

Ostermayer, F. W., Jr. (1977). LED end-pumped Nd:YAG lasers. *IEEE J. Quantum Electron.* **QE-13,** 1.

Ostermayer, F. W., Jr., Allen, R. B., and Dierschke, E. G. (1971). Room-temperature CW operation of a $GaAs_{1-x}P_x$ diode-pumped YAG:Nd laser. *Appl. Phys. Lett.* **19,** 289.

Otsuka, K., and Yamada, T. (1975). Continuous oscillation of a lithium neodymium tetraphosphate laser with 200 μW pump threshold. *IEEE J. Quantum Electron.* **QE-11,** 845.

Otsuka, K., Yamada, T., Saruwatari, M., and Kimura, T. (1975). Spectroscopy and laser oscillation properties of lithium neodymium tetraphosphate. *IEEE J. Quantum Electron.* **QE-11,** 330.

Otsuka, K., Miyazawa, S., Yamada, T., Iwasaki, H., and Nakano, J. (1977). CW laser oscillators in $MeNdP_4O_{12}$ (Me = Li, Na, K) at 1.32 μm. *J. Appl. Phys.* **48,** 2099.

Panish, M. B. (1976). Heterostructure injection lasers. *Proc. IEEE* **64,** 1512.

Paoli, T. L. (1975). Depolarization of the lasing emission from CW double-heterostructure junction lasers. *IEEE J. Quantum Electron.* **qe-11,** 489.

Paoli, T. L. (1976a). Nonlinearities in the emission characteristics of stripe-geometry (AlGa)As double-heterostructure junction lasers. *IEEE J. Quantum Electron.* **QE-12,** 770.

Paoli, T. L. (1976b). Modulation characteristics of CW laser diodes. *Techn. Digest 1976 Int. Electron DEvices Meeting,* p. 136.

Paoli, T. L. (1977). Changes in the optical properties of CW (AlGa)As junction lasers during accelerated aging. *IEEE J. Quantum Electron.* **QE-13,** 351.

Paoli, T. L., Hakki, B. W., and Miller, B. I. (1973). Zero-order transverse mode operation of GaAs double-heterostructure lasers with thick waveguides. *J. Appl. Phys.* **44,** 1276.

Pearsall, T. P., Miller, B. I., Capik, R. J., and Bachman, K. J. (1976). Efficient lattice-matched double-heterostructure LEDs at 1.1μm from $Ga_xIn_{1-x}As_yP_{1-y}$. *Appl. Phys. Lett.* **28,** 499.

Petroff, P., and Hartman, R. P. (1974). Rapid degradation phenomenon in heterojunction GaAlAs-GaAs lasers. *J. Appl. Phys.* **45,** 3899.

Reinhart, F. K., Logan, R. A., and Shank, C. V. (1975). $GaAs-Al_xGa_{1-s}As$ injection lasers with distributed Bragg reflectors. *Appl. Phys. Lett.* **27,** 45.

Rupprecht, H., Woodall, J. M., and Pettit, D. G. (1967). Efficient visible electroluminescence at 300°K from $Ga_{1-x}Al_xAs$ p-n junctions grown by liquid phase epitaxy. *Appl. Phys. Lett.* **11,** 81.

Saruwatari, M., and Kimura, T. (1976). LED pumped lithium neodymium tetraphosphate lasers. *IEEE J. Quantum Electron.* **QE-12,** 584.

Saruwartari, M., Kimura, T., and Otsuka, K. (1976). Miniaturized CW $LiNdP_4O_{12}$ laser pumped with a semiconductor laser. *Appl. Phys. Lett.* **29,** 291.

Schuöcker, D., and Schiffner, G. (1978). Optimierung des diodengepumpten YAG:Nd^{3+}-faserlaser. *Arch. Elek. Übertrag.* **32**, 171.

Seki, Y. (1976). Light extraction efficiency of the LED with guide layers. *Jpn. J. Appl. Phys.* **15**, 327.

Sharma, B. L., and Purohit, R. K. (1974). "Semiconductor Heterojunctions." Pergamon, Oxford.

Shen, C. C., Hsieh, J. J., and Lind, T. A. (1977). 1500-hr continuous CW operation of double-heterostructure GaInAsP/InP lasers. *Appl. Phys. Lett.* **30**, 353.

Singh, S., Smith, R. G., and Van Uitert, L. G. (1974). Stimulated-emission cross section and fluorescent quantum efficiency of Nd^{3+} in yttrium aluminum garnet at room temperature. *Phys. Rev. B* **10**, 2566.

Singh, S., Miller, D. C., Potopowicz, J. R., and Shick, L. K. (1975). Emission cross section and fluorescence quenching of Nd^{3+} lanthanum pentaphosphate. *J. Appl. Phys.* **46**, 1191.

Snyder, A. W. (1974). Leaky-ray theory of optical waveguides of circular cross section. *Appl. Phys.* **4**, 273.

Stern, F. (1976). Calculated spectral dependence of gain in excited GaAs. *J. Appl. Phys.* **47**, 5382.

Stone, J., and Burrus, C. A. (1979). Self-contained LED-pumped single crystal Nd:YAG fiber laser. *Fiber Integr. Opt.* **2**, 19.

Stone, J., Burrus, C. A., Dentai, A. G., and Miller, B. I. (1976). Nd:YAG single-crystal fiber laser: Room-temperature CW operation using a single LED as an end pump. *Appl. Phys. Lett.* **29**, 37.

Thompson, G. H. B., and Kirby, P. A. (1973). Low threshold current density in 5-layer-heterostructure (GaAl)As/GaAs localized-gain-region injection lasers. *Electron. Lett.* **9**, 295.

Tietjen, J. J., and Amick, J. A. (1966). The preparation and properties of vapor-deposited epitaxial GaAs$_{1-x}$P$_x$ using arsine and phosphine. *J. Electrochem. Soc.* **113**, 724.

Tsang, W. T., Logan, R. A., and Ilegems, M. (1978). High-power fundamental-transverse-mode strip buried heterostructure lasers with linear light-current characteristics. *Appl. Phys. Lett.* **32**, 311.

Tsang, W. T., Logan, R. A., and van der Ziel, J. P. (1979). Low current-threshold strip buried heterostructure lasers with self-aligned current injection stripes. *Appl. Phys. Lett.* (To be published).

Tsukada, T. (1974). GaAs-Ga$_{1-x}$Al$_x$As buried-heterostructure injection lasers. *J. Appl. Phys.* **45**, 4899.

Washio, K., Iwamoto, K., Inoue, K., Hino, I., Matsumoto, S., and Saito, F. (1976). Room-temperature CW operation of an efficient miniaturized Nd:YAG laser end-pumped by a superluminescent diode. *Appl. Phys. Lett.* **29**, 720.

Weber, H. P. (1975). Nd-pentaphosphate lasers. *Opt. Quantum Electron.* **7**, 431.

Weber, H. P., Damen, T. C., Danielmeyer, H. G., and Tofield, B. C. (1973). Nd-ultraphosphate laser. *Appl. Phys. Lett.* **22**, 534.

Weber, H. P., Liao, P. F., and Tofield, B. C. (1974). Emission cross-section and fluorescent efficiency of Nd-pentaphosphate. *IEEE J. Quantum Electron.* **QE-10**, 563.

Weber, H. P., Liao, P. F. Tofield, B. C., and Bridenbaugh, P. M. (1975). CW fiber laser of NdLa-pentaphosphate. *Appl. Phys. Lett.* **26**, 692.

Winzer, G., Möckel, P. G., and Krühler, W. W. (1978). Laser emission from miniaturized NdAl$_3$(BO$_3$)$_4$ crystals with directly applied mirrors. *IEEE J. Quantum Electron.* **QE-14**, 840.

Wittke, J. P. (1975). Spontaneous-emission-rate alteration by dielectric and other wave-guiding structures. *RCA Rev.* **36**, 655.

Woodall, J. M., Rupprecht, H., and Pettit, G. D. (1967). Efficient electroluminescence from epitaxially grown Ga$_{1-x}$Al$_x$As p-n junctions. *Solid State Device Res. Conf., 1967.*

Yamada, T., Otsuka, K., and Nakano, J. (1974). Fluorescence in lithium neodymium ultraphosphate single crystal. *J. Appl. Phys.* **45,** 5096.

Yamakoshi, S., Hasegawa, O., Hamaguchi, H., Abe, M., Yamaoka, T., Komatsu, Y., and Isozumi, S. (1977). Degradation of high-radiance $Ga_{1-x}Al_xAs$ LEDs. *Appl. Phys. Lett.* **31,** 627.

Yamamoto, T., Sakai, K., Akiba, S., and Suemastsu, Y. (1978). $In_{1-x}Ga_xAs_yP_{1-y}/InP$ DH laser fabricated on InP (100) substrates. *IEEE J. Quantum Electron.* **QE-14,** 95.

Yang, K. H., and Kingsley, J. D. (1975). Calculation of coupling losses between light emitting diodes and low-loss optical fibers. *Appl. Opt.* **14,** 288.

Yonezu, H., Matsumoto, Y., Shinohara, T., Sakuma, I., Suzuki, T., Kobayashi, K., Lang, R., Nannichi, Y., and Hayashi, I. (1977). New stripe geometry laser with high quality lasing characteristics by horizontal transverse mode stabilization—a refractive index guiding with Zn doping. *Jpn. J. Appl. Phys.* **16,** 209.

Zargar'yants, M. N., Mezin, Yu. S., and Kolonenkova, S. I. (1971). Electroluminescent diode with a flat surface emitting continuously 25 W/cm^2-sr at 300°K. *Sov. Phys. —Semicond. (Engl. Transl.)* **4,** 1371.

Chapter 17

Modulation Techniques

IVAN P. KAMINOW

TINGYE LI

17.1 INTRODUCTION

In order to convey information on an optical wave, it is necessary to modulate a property of the wave in accordance with the information signal. The wave property may be its intensity, phase, frequency, polarization, or direction; the modulation format may be analog or digital. The choices are dictated by the characteristics of the transmission medium, the available carrier sources and detectors, and by systems considerations. The optical fiber, whether single-mode or multimode has loss which attenuates the optical signal and dispersion which imposes limitations on the modulation bandwidth. In general, the polarization state of the optical wave will not be maintained along a fiber of circular cross section because of coupling between orthogonally polarized modes due to uncontrolled birefringence and dimensional variations. Thus, any signal-processing method that requires a stable and predictable polarization state at the receiver will be difficult to implement. A case in point is optical heterodyne detection which is necessary for the demodulation of phase- or frequency-modulated waves. The stringent requirements for stable, single-frequency, single-mode sources as carrier generator and local oscillator make heterodyne detection much more difficult to implement than direct detection (photon counting) even if the polarization problem were solved. Thus, intensity modulation with direct detection appears to be the most practical approach at present.

Both analog and digital formats are suitable for modulating the optical carrier in a fiber system. Analog modulation has the appeal of simplicity, but the large signal-to-noise ratios required limit its use to relatively narrow-bandwidth, short-distance applications. Digital modulation pro-

557

vides noise immunity at the expense of large bandwidth; it is therefore ideally suited to fiber transmission where the available bandwidth is large. Most potential applications in medium-to-long-distance fiber transmission systems involve the digital modulation format.

Intensity modulation, simple in concept, is also simple to implement, especially with presently available optical carrier sources such as the electroluminescent light-emitting diodes (LEDs) and injection lasers. Because of the short lifetimes of their radiative levels, these sources can be modulated directly by variation of their driving currents at rates up to hundreds of megahertz. However, future systems may utilize still broader bandwidths or employ other types of carrier sources (such as the neodymium laser) that cannot be directly modulated at high frequency; these systems will require separate external optical modulators. External modulators can also be used for time-multiplexing and switching of optical signals. Hence it is worthwhile at this early stage of optical communications research to consider both direct modulation of sources and external modulation.

The optical wave property that is most amenable to external modulation by a number of physical effects is the phase or frequency of the wave. Once phase modulation of the wave is realized, intensity, polarization, or direction variations can be achieved with passive components. The physical effects that have been commonly exploited in external modulators are the electrooptic, magnetooptic, and acoustooptic effects whereby the index n of a transparent medium is changed by Δn in proportion to an applied electric field E, magnetization M, or strain S, respectively. Each of these reactive effects serves to change the phase of the optical carrier passing through a length L of the material by $2\pi\Delta nL/\lambda$, where λ is the optical wavelength. Intensity modulation is obtained by interfering a phase-modulated wave with an unmodulated wave or by combining two waves modulated in phase opposition. Alternatively, phase gratings created acoustooptically or electrooptically may be used to deflect an optical beam and thereby produce intensity modulation. Still another physical effect that provides intensity modulation directly is electroabsorption in which a strong electric field shifts the band edge of the modulating medium.

Recently, considerable effort has been devoted to developing modulators in planar, single-mode waveguides rather than in bulk configurations. The waveguide devices have the potential advantages of ease of fabrication using photolithographic techniques and compatibility with single-mode fibers. In addition, because diffraction effects do not limit the modulator dimensions, their modulating power efficiencies can be orders of magnitude better than bulk devices. In the future, these waveguide devices may be fabricated in arrays on a single substrate to serve as compact switching networks or multiplexers. However, applications of these waveguide devices in fiber systems will require the development of fibers that can maintain a single-polarization state over long distances (Steinberg and Giallorenzi, 1976; Ramaswamy et al., 1978b; Kaminow and Ramaswamy, 1979), or alternatively the development of polarization-independent devices (Burns et al., 1978).

Direct modulation of semiconductor light-emitting diodes and lasers, and external bulk and waveguide modulators are described in the following sections.

17.2 DIRECT MODULATION OF ELECTROLUMINESCENT DEVICES—LIGHT-EMITTING DIODES AND LASERS

17.2.1 Light-Emitting Diodes (LEDs)

Since minority carriers injected into a forward-biased p–n junction recombine spontaneously, it is possible to modulate the output of a LED by varying the injected current, provided the rate of the variation is slower than the reciprocal of the carrier lifetime τ. Junction and parasitic capacitances may further increase the time required to alter the minority carrier density injected into the recombination region. However, since both the diode conductance and diffusion capacitance (which is due to τ) increase with the drive current, the effect of the space-charge and parasitic capacitances (which are nearly invariant with current) becomes small when a constant dc bias is maintained; the modulation bandwidth of the diode is then limited by τ (Lee, 1975).

The output characteristic of a typical AlGaAs LED illustrated in Fig. 17.1 shows that the output light intensity is proportional to the input current over a wide range of the current. If the drive current consists of a small ac signal of frequency $f = \omega/2\pi$ superimposed on a dc bias, the output light intensity contains a modulated part $I(\omega)$ superimposed on a quiescent part I_0. The amplitude of the modulated intensity, $|I(\omega)|$, which represents the response of the diode to the ac drive current, is given approximately by (Namizaki et al., 1974b; Liu and Smith, 1975)

$$|I(\omega)| = |I(0)|/(1 + (\omega\tau)^2)^{1/2}, \tag{17.1}$$

where $I(0)$ is the modulated intensity extrapolated to zero frequency. The dc bias is assumed to be sufficiently large that the effects of space-charge and parasitic capacitances can be neglected. Since the detected electrical signal power $p(\omega)$ is proportional to $|I(\omega)|^2$, it is useful and appropriate to define the (3-dB) modulation bandwidth $\Delta\omega$ as the frequency band over which $p(\omega) \geq \frac{1}{2}p(0)$, or $|I(\omega)|^2 \geq \frac{1}{2}|I(0)|^2$; that is

$$\Delta\omega = 1/\tau. \tag{17.2}$$

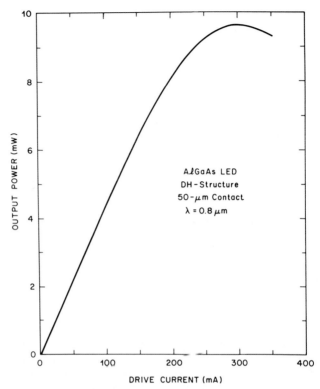

Fig. 17.1 Output characteristic of a typical Burrus-type AlGaAs double-heterostructure LED (C. A. Burrus, private communication).

When both radiative and nonradiative processes are present, the carrier lifetime τ is given by

$$1/\tau = (1/\tau_r) + (1/\tau_{nr}), \qquad (17.3)$$

where τ_r is the radiative lifetime and τ_{nr} is the nonradiative lifetime. The internal quantum efficiency η_i is τ/τ_r. High efficiency requires the band-to-band radiative recombination process to dominate over the nonradiative process occuring at impurity and defect centers in the bulk and at the heterointerfaces (i.e., $\tau_r \ll \tau_{nr}$). For doping concentrations less than the injected carrier density, τ_r is proportional to $(d/J)^{1/2}$ where d is the thickness of the active region and J is the injected current density (Namizaki *et al.*, 1974a). At higher doping concentrations, τ_r is inversely proportional to the doping level (Hall, 1960). For heterojunction LEDs at lower doping levels ($< 10^{18}$) the nonradiative lifetime is dominated by surface recombination at the heterointerfaces and τ_{nr} is proportional to d/s, where s is the surface recombination velocity (Ettenberg and Kressel, 1976). There-

fore it is possible to increase the modulation bandwidth either by reducing the thickness d or by increasing the doping level. Since a small d
gives a small τ_{nr} and heavily doped material containing a large number of
nonradiative recombination centers also gives a small τ_{nr}, both effects lead
to lower quantum efficiency. Thus there is a trade-off between modulation
bandwidth and modulated output intensity.

Among dopant materials for AlGaAs, germanium is most often used for
making fast diodes because it can be incorporated into AlGaAs to concentration levels greater than 10^{19} cm^{-3}, yielding carrier lifetimes of 1–2 nsec
(Ettenberg et al., 1973; Casey et al., 1973; Acket et al., 1974; Zucker et al.,
1976). Other dopants include zinc (for GaAs homostructure LEDs) and silicon. Silicon-doped AlGaAs is especially interesting because it produces
diodes of very high efficiency (20–30%) but with much reduced bandwidth ($\tau \sim 500$ nsec) (Dawson, 1977).

Homostructure, single-heterostructure (SH), and double-heterostructure (DH), high-radiance, Burrus-type LEDs of AlGaAs-GaAs material have been investigated with respect to their output, modulation,
and spectral characteristics (King et al., 1975; Harth and Heinen, 1975;
Goodfellow and Mabbitt, 1976; Harth et al., 1976; Lee and Dentai, 1978).
Figure 17.2 shows how the measured bandwidth and the carrier lifetime
calculated from Eq. (17.2) vary with the doping level for typical high-
radiance LEDs fabricated with Ge-doped DH AlGaAs material (Lee and
Dentai, 1978). Figure 16.13 in Chapter 16 gives the measured dc output intensity near saturation of these diodes as a function of their modulation

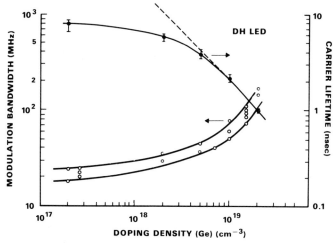

Fig. 17.2 Experimentally determined modulation bandwidth and carrier lifetime
versus doping level for typical high-radiance, Burrus-type, Ge-doped, DH AlGaAs LEDs
(Lee and Dentai, 1978).

bandwidth measured at an operating point lower than saturation. This curve shows that the maximum output intensity falls off at higher bandwidths because the injected carrier density decreases with the reduced carrier lifetime.

Homostructure GaAs LEDs heavily doped with Zn tend to have larger bandwidths than DH devices; a modulation bandwidth of 350 MHz with an output radiance of 11 W/sr-cm^2 at 300 mA drive has been achieved in a Burrus-type diode with a Zn doping level of 4×10^{19} cm^{-3} (Goodfellow and Mabbitt, 1976). However, LPE-grown DH materials cannot be heavily doped with Zn because of its high diffusion rate. AlGaAs DH-structure LEDs more lightly doped with Ge have smaller bandwidth (~ 200 MHz) but more output (~ 30 W/sr-cm^3 at 300 mA drive with a Ge-doping level of 2×10^{19} cm^{-3}) (Lee and Dentai, 1978).

The variation of modulation bandwidth with active-layer thickness has been investigated in homostructure and SH-structure surface emitters (Harth et al., 1975) and in DH-structure edge emitters (Lockwood et al., 1976; Ettenberg and Kressel, 1976). Narrow-active-layer, stripe-geometry, edge-emitting LEDs are especially interesting because of their high coupling efficiency into fibers of low numerical aperture. Edge-emitting stripe diodes having very thin active layers (≤ 0.1 μm) have been produced with output radiances ~ 1000 W/sr-cm^2 and modulation bandwidths ~ 140 MHz (Ettenberg et al., 1976).

One of the LED characteristics that influences its usefulness as a modulated source is its large optical spectral width $\Delta\lambda$. Material dispersion in the fiber limits its information transmission capacity for $\Delta\lambda > 0$. The spectral width of a LED at a given temperature is determined mainly by the influence of the doping concentration on the density of states. Increasing the doping level results in a greater concentration of bandtail states within the bandgap. These states then give rise to a larger spectral width (Casey and Stern, 1976). The spectral widths of typical Ge-doped LEDs range from 250 Å for lightly-doped materials to over 500 Å for heavily-doped materials (Lee and Dentai, 1978).

Because material dispersion in high-silica fibers is minimum in the wavelength range 1.25 to 1.35 μm, there is considerable interest in LEDs operating in this region. Double-heterostructure InGaAsP/InP high-radiance Burrus-type LEDs emitting in the above wavelength region have been fabricated (Dentai et al., 1977). Even though the spectral widths of these LEDs are about three times as wide as those of AlGaAs LEDs, the minimal material dispersion enables a fiber data link using InGaAsP LEDs to operate with a potential bandwidth at least several times that of a link using AlGaAs LEDs (Muska et al., 1977; Machida et al., 1979).

The linearity between output light intensity and drive current is important if the modulation format is analog. The output characteristics in Fig.

17.1 shows that, depending upon operating conditions, nonlinear distortion in the output of a directly-modulated LED can be appreciable. Nonlinear distortion is also dependent upon diode structure, active-layer thickness and carrier concentration, and operating frequency (Straus and Szentesi, 1975; King et al., 1976; Lee, 1977; Asatani and Kimura, 1978a; Dawson, 1978; Straus, 1978b). The limit on the tolerable level of nonlinear distortion in an analog system is determined by the particular application and may be exceedingly stringent when cross-talk among frequency-multiplexed signals is involved (Szentesi and Szanto, 1976). Measurements of nonlinear distortion in AlGaAs LEDs have shown that the total harmonic distortion is in the range of 30–40 dB below fundamental for modulation depth of about 0.5 (Straus and Szentesi, 1975; Ozeki and Hara, 1976; King et al., 1976; Dawson, 1978). Compensation methods used to reduce nonlinear distortion to still lower levels include complementary distortion (or predistortion), negative feedback, phase shift modulation (using a matched pair of LEDs), and feedforward schemes (Straus et al. 1977; Straus and Szentesi, 1977a,b; Umebu et al., 1977; Asatani and Kimura, 1978b). An improvement of 30–40 dB (i.e., total harmonic distortion level ~ −70 to −80 dB) has been achieved using a quasi-feedforward technique (Straus and Szentesi, 1977a; Straus, 1978a).

Nonlinearity is much less important in digital systems. In fact, its importance diminshes as the number of levels decreases in an intensity-modulated multilevel digital signal and vanishes entirely when the number of levels is two (binary signal format). When binary signals are used, the LED is driven between the *off* state of zero output, and the *on* state near saturation. In order to operate at higher bit rates, a small dc bias current is necessary to keep the effects of junction capacitance small (Lee, 1975). Since heating effects will limit the peak *on*-state current, the presence of the bias will degrade the extinction ratio and thereby impair system performance. The reduced extinction ratio is manifested as a loss of receiver sensitivity, which is not serious when the extinction ratio is greater than 10. For example, the loss of sensitivity is less than 2 dB for a receiver with an avalanche detector of optimal gain and is less than 1 dB for a receiver with a PIN detector (no gain) when the extinction ratio is 10 (Personick, 1973).

17.2.2 Injection Lasers

As in the case of light-emitting diodes, semiconductor injection lasers can be modulated directly by varying the driving current. Both analog and digital modulation can be used because the laser output intensity is proportional to changes in the injected current over a broad current range above the threshold level, as shown in Fig. 17.3 for a typical stripe-

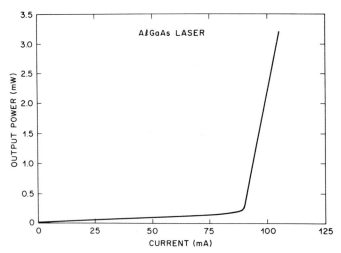

Fig. 17.3 Output characteristic of a typical stripe-geometry DH AlGaAs laser (B. C. De-Loach, private communication).

geometry AlGaAs double-heterostructure laser. The modulation bandwidth of the laser, however, is significantly larger than that of the LED because the recombination lifetime of the carriers ($\tau \lesssim 10^{-8}$ sec) is shortened appreciably by the action of stimulated emission; the lifetime of photons in the cavity ($\tau_p \lesssim 10^{-11}$ sec) is insignificant by comparison. Thus, ideally, injection lasers are expected to have modulation bandwidths up to a few gigahertz.

The transient behavior or the modulation response of the injection laser is strongly influenced by a resonance phenomenon found in solid-state lasers. This resonance effect may manifest itself as a peaked high-frequency noise (i.e., "spiking" oscillations) in the light output (Kurnosov et al., 1966), as damped relaxation oscillations when the laser is suddenly switched on (Roldan, 1967), or as regular intensity pulsations at some fixed repetition rate (D'Asaro et al., 1968). The phenomenon is attributable to the nonlinear interaction of the injected carrier density and the optical field in the cavity (McCumber, 1966; Haug and Haken, 1967; Ikegami and Suematsu, 1968; Basov et al., 1969; Paoli and Ripper, 1970; Ikegami et al., 1970; Adams, 1973; Harth, 1975; Boers et al., 1975; Schicketanz, 1975; Paoli, 1975; Daikoku, 1977).

When the drive current is switched abruptly from zero to a value above threshold, there is a delay of a few nanoseconds as the carrier density inversion builds up before the laser output occurs. Moreover, the delay and the light pulse amplitude are influenced by the presence or absence of preceding drive current pulses (Ozeki and Ito, 1973a,b; Lee and Derosier, 1974). This pulse-pattern-dependent effect can be reduced by operating with a dc bias current above the threshold (Basov *et al.*, 1967a; Dyment *et al.*, 1972) or by using compensating pulses (Ozeki and Ito, 1973b).

As the laser output builds up, a ringing (or relaxation oscillation) is observed on the leading edge of the pulse as shown in Fig. 17.4a. What happens is that the inversion increases well beyond threshold before lasing action takes place; the high lasing field produced by the excess gain then rapidly depletes the inversion, reducing the optical field in the cavity, thereby allowing the inversion to build up again; then the process is repeated. Frequently the ringing is damped out in a few cycles after which the laser output is constant. However, some lasers exhibit a repetitive pulsing behavior as shown in Fig. 17.4b. With aging, lasers may develop

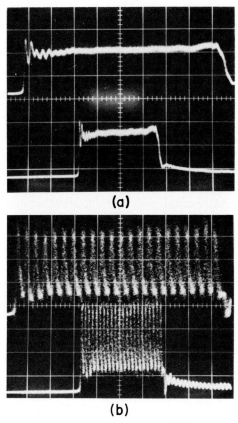

(a)

(b)

Fig. 17.4 The variation of the output intensity from a DH laser operated pulsed with a dc bias current. Part (a) shows the stable behavior of the device prior to aging and part (b) shows the sustained pulsations occurring after continuous operation at 70°C for 50 hr. For both (a) and (b) the peak current is 13% above threshold and the horizontal scale is 2 nsec/div, for the top trace and 5 nsec/div for the bottom trace (Paoli, 1977a).

pulsating behavior and the pulsating rate may decrease with time (Yang *et al.*, 1974; Paoli, 1977a). At present, it is not clearly understood why some lasers pulsate spontaneously while others do not, nor·how the pulsing behavior correlates with the drive current and with aging.

A linearized, small-signal analysis gives a first-order approximation to the resonance or pulsing frequency ν_r of a single-mode laser (Ikegami and Suematsu, 1968; Paoli and Ripper, 1970; Adams, 1973):

$$\nu_r \simeq \frac{1}{2\pi} \left(\frac{1}{\tau \tau_p} \left(\frac{J}{J_{th}} - 1 \right) \right)^{1/2}, \tag{17.4}$$

where τ is the carrier recombination lifetime (which contains both radiative and nonradiative components), τ_p is the photon lifetime in the lasing mode (which is related to cavity losses), J is the drive current density, and J_{th} is the threshold current density. The resonance frequency ν_r can vary from a few hundred megahertz to several gigahertz under typical operating conditions depending on the particular diode being examined. Aging can reduce ν_r with time (Paoli, 1977a). Experimentally ν_r is found to be proportional to $[(J/J_{th}) - 1]^n$ where n is close to $\frac{1}{2}$ for operation near threshold where the self-induced amplitude fluctuations are weakly modulated, but may deviate from $\frac{1}{2}$ in either direction when the small-signal approximation no longer holds (Lee and Serra, 1976; Paoli, 1977a).

The large-signal modulation behavior of injection lasers has been studied analytically and experimentally (Harth, 1975; Schicketanz, 1975); it is found that the resonance frequency and the modulation efficiency decrease with increasing modulation current, and that hysteresis effects may lead to serious signal distortion.

The small-signal theory yields a modulation response exhibiting a peak at the resonance frequency ν_r, as illustrated in Fig. 17.5. A well-behaved laser can be modulated by a wideband signal without distortion, so long as the bandwidth is substantially less than ν_r. When the modulating frequency approaches ν_r, the highly peaked response will cause distortion of the output signal and, in addition, relaxation oscillations may be excited.

The troublesome effects of relaxation oscillation and pulsation can be reduced by biasing the laser slightly above threshold so that the carrier density and the optical field do not undergo large excursions. A further advantage of such prebias is that the laser can be made to operate in a single longitudinal mode, thus ensuring very narrow spectral width $\Delta\lambda$ (Selway and Goodwin, 1976). The penalty that one must pay is the degraded extinction ratio as discussed in the preceding subsection (17.2.1).

Another way to overcome the difficulties of relaxation oscillation and large spectral width is by means of light injection into the laser (Russer, 1975; Hillbrand and Russer, 1975; Lang and Kobayashi, 1976; Arnold *et*

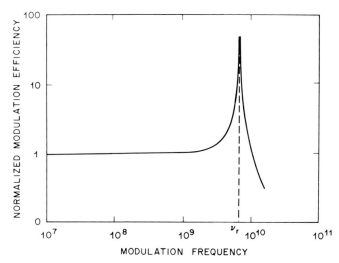

Fig. 17.5 Normalized modulation efficiency of a GaAs laser vs frequency showing the relaxation frequency peak at ν_r.

al., 1978). The injected coherent radiation acts as a damping mechanism in suppressing the relaxation oscillations and increases the stimulated emission into the irradiated mode at the expense of other lasing modes (Russer, 1975). The drawbacks, of course, are the requirement for an auxiliary laser and the added complications of biasing, coupling, etc.

Transverse carrier diffusion also plays a role in damping relaxation oscillations by stabilizing the inversion level. It has been shown in theory and in experiment that the resonance peak at ν_r and the associated relaxation oscillations are much reduced when the stripe width for light and current confinement is comparable with the carrier diffusion length (Furuya *et al.*, 1978).

In spite of the difficulties of relaxation oscillations and self-induced pulsations, selected lasers have been modulated at high rates in the 1–3 Gbit/sec range. A variety of devices have been used as high-speed drivers; among these are bipolar transistors (Chown *et al.*, 1973), GaAs field-effect transistors (Ostoich *et al.*, 1975; Baack *et al.*, 1978), Gunn-effect diodes (Thim *et al.*, 1973), trappatt diodes (Carroll and Farrington, 1973), and step-recovery diodes (Russer and Schultz, 1973). Furthermore, as discussed in Sections 17.3 and 17.4 many of the difficulties associated with direct modulation of the semiconductor laser can be avoided by operating the laser as a CW source and providing a separate external modulator.

Strong nonlinearities, or "kinks" are often observed in the light-output-versus-current characteristics of injection lasers (Paoli, 1976, 1977a). This phenomenon is usually accompanied by other undesirable anomalous behavior such as enhanced intensity fluctuations, deterioration of modulation capability and beam shift. The "kink" behavior is thought to be the result of transverse-mode instability with current. Since lateral confinement in stripe-geometry lasers is provided by weak gain-induced guidance (Paoli, 1977b), small mutual interactions among gain, current, and mode profiles can lead to transverse-mode instabilities, resulting in kinks (Chinone, 1977). Techniques for creating stronger lateral confinement with passive waveguiding mechanisms have led to kink-free lasers; thus single-transverse-mode, kink-free operation has been obtained in buried heterostructure lasers (Tsukada, 1974, Kajimura et al., 1978), lateral-injection lasers (Namizaki, 1975; Susaki et al., 1977), channeled-substrate planar lasers (Aiki et al., 1978), strip-buried heterostructure lasers (Tsang and Logan, 1978), and curved stripe lasers (Scifres et al., 1978). There is also evidence that pulsation and relaxation–oscillation effects are reduced with single-transverse-mode operation. Single-longitudinal-mode operation has been observed in lateral-injection and channel-substrate-planar lasers (Susaki et al., 1977; Aiki et al., 1978), although the reason that single-longitudinal-mode operation is obtained in these laser but not in others is not well understood. Stable operation of buried-heterostructure lasers during accelerated aging has been reported recently (Kajimura et al., 1978). It was observed that neither kinks nor self-pulsations developed in the majority of these lasers after long-term aging at 70°C, in contrast to conventional stripe-geometry lasers.

As with LEDs, research efforts on lasers are being directed toward operation in the 1.1 to 1.6-μm spectral region. Because lasers have much narrower spectral widths than LEDs, the advantage of minimal material dispersion is less important than is the lower fiber loss in this region. Double-heterostructure InGaAsP lasers have been operated continuously at room temperature with potentially good reliability (Shen et al., 1977; Yamamoto et al., 1978). Modulation bandwidths up to 2.5 GHz have been observed (Akiba et al., 1978), and repeater experiments have been performed with these lasers operating at wavelengths near 1.3 μm and at data rates up to 1.2 Gbit/sec (Abbott et al., 1978; Yamada et al., 1978).

Because laser threshold depends critically on temperature and somewhat on age (Hartman and Dixon, 1975; Goodwin et al., 1975; Thompson et al., 1976), it is important that the laser output be stabilized against variations due to these causes. The stabilization feedback circuitry is thus an essential part of the laser driver-bias subsystem. Mode instabilities associated with kink phenomenon tend to complicate feedback stabilization (Chen et al., 1978). However, a stabilized AlGaAs laser source has been successfully implemented in a 45-Mbit/sec repeater for an optical-fiber transmission experiment (Schumate et al., 1978).

A pulse-train consisting of pulses at a fixed repetition rate may be required for some applications. A source of short, high-repetition-rate pulses is the double-diode structure in which two sections of the laser are electrically isolated but optically coupled in tandem [Basov *et al.*, 1967b; Lee and Roldan, 1970). The current in one section of the diode is maintained at a level high enough to provide gain while the current in the other section is adjusted to give saturable loss with a suitable relaxation time. Under the appropriate conditions, predictable by theory, the laser produces sustained self-induced pulsations of about 100-psec duration at a repetition rate between about 10^2 and 10^3 MHz, dependent on the driving current and the ratio of the lengths of the two sections. Sustained, self-induced pulsations of several hundred megahertz to a gigahertz have also been observed in the output of conventional single-section stripe-geometry injection lasers (Paoli and Ripper, 1969; Ripper and Paoli, 1971). The repetition rate of these self-pulsing lasers can be synchronized to a stable microwave source by current injection (Ripper and Paoli, 1969); these diodes, when combined with external modulators, are suitable for application in a high-bit-rate time-multiplexed PCM system. If the external microwave source is angle-modulated, the output pulses become modulated in position (PPM) (Ripper and Paoli, 1969). Such a technique for deriving PPM optical pulses from an angle-modulated signal can provide appreciable bandwidth with very little modulating power.

17.3 BULK MODULATORS

17.3.1 General Considerations

Many design parameters and performance criteria must be considered when evaluating modulators for optical-fiber systems. They include bandwidth, modulating power, optical insertion loss, modal restrictions, extinction ratio, temperature sensitivity, and coupling efficiency to source and fiber. Since the required modulating power P is generally proportional to the bandwidth Δf for external reactive modulators, a common and useful figure of merit is the power per unit bandwidth ($P/\Delta f$) required to achieve a nominal degree of modulation (typically, one radian for phase modulation or 70–100% for intensity modulation) at some particular wavelength. For the best bulk modulators, this figure of merit is in the neighborhood of 1 mW/MHz-rad^2 at $\lambda = 0.63$ μm. Since the microscopic physical effects that produce the modulation are fast, properly designed modulators can operate well beyond the gigahertz range. Many electrooptic and acoustooptic materials are transparent over the wavelength range of interest (0.7–1.6 μm) and have negligible insertion loss. Bulk modulators are frequently designed as monochromatic, single-mode, single-polarization devices and their performance characteristics (such as figure of merit and extinction ratio) degrade when they are used with wideband or multimode lasers. Material effects, such as crystal imperfection, temperature sensitivities, and birefringence can also degrade extinction ratio and affect performance. So far, very little attention has been paid to the problem of efficient coupling of modulators to sources and fibers.

17.3.2 Acoustooptic Modulators

The periodic density changes produced in any transparent medium by an acoustic wave give rise to periodic refractive index changes due to the photoelastic effect. The medium then serves as a three-dimensional phase grating. As illustrated in Fig. 17.6a, the incident beam will be partially diffracted from the zero-order into a multiplicity of higher orders when the grating length L is sufficiently small—this is the Raman–Nath regime that applies when $Q \equiv 2\pi\lambda L/n\Lambda^2 < 1$. If L is sufficiently large, the zero-order beam will be partially deflected into only one order—this is the Bragg regime that applies when $Q > 10$. In the definition of Q, L is the width of the acoustic wave of wavelength Λ in a medium of index n at optical wavelength λ. In either of the above cases, the device will modulate the intensity of either the zero-order beam or one of the higher order beams as the intensity of the acoustic subcarrier is modulated.

The bandwidth of the piezoelectric transducer that excites the acoustic wave is generally limited to about 25% of the subcarrier frequency. Thus a subcarrier oscillator at a frequency approximately $4\Delta f$ is required for an acoustooptic modulator, where Δf is the modulator bandwidth. Moreover, acoustic losses and transducer design problems set a practical limit for the subcarrier frequency of several hundred megahertz, so that Δf is usually limited to about 200 MHz.

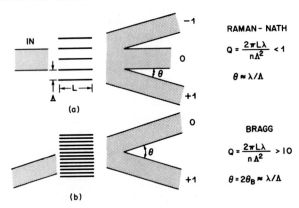

Fig. 17.6 (a) Raman–Nath and (b) Bragg deflection by thin ($Q < 1$) and thick ($Q > 10$) gratings, respectively (Kaminow, 1975).

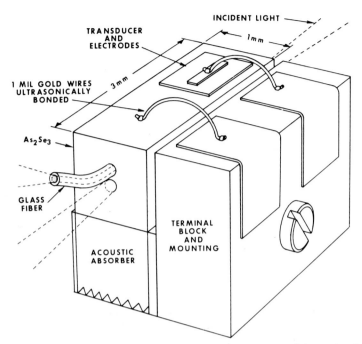

INCIDENT LIGHT

1 mm

TRANSDUCER
AND
ELECTRODES

1 MIL GOLD WIRES
ULTRASONICALLY
BONDED

3 mm

As$_2$Se$_3$

GLASS
FIBER

TERMINAL
BLOCK
AND
MOUNTING

ACOUSTIC
ABSORBER

Fig. 17.7 Sketch of the 50 Mbit/sec As$_2$Se$_3$ acoustooptic modulator developed for optical-fiber application (Warner and Pinnow, 1973).

The performance of bulk acoustooptic modulators is illustrated by the Bragg deflector shown in Fig. 17.7 which was designed for optical-fiber systems at 1.06 μm (Warner and Pinnow, 1973). It employs As$_2$Se$_3$ glass as the acoustooptic medium, has an acoustic subcarrier frequency of 200 MHz and bandwidth of 50 MHz. The electrical modulating power required for 70% deflection into the first-order beam is 1 mW/MHz. Since the deflected order, rather than the zero-order, beam is transmitted, the on–off ratio is 100%.

Further information on acoustooptic devices can be found in several reviews (Sittig, 1972; Uchida and Niizeki, 1973; Chang, 1976). Bulk acoustooptic modulators are commercially available from several suppliers.

17.3.3 Electrooptic Modulators

Certain classes of crystals exhibit a linear electrooptic (Pockels) effect whereby an applied electric field E produces a proportional change in refractive index Δn. Since Δn and E are related by a tensor, the field must be applied in special crystal directions for maximum effect and the induced index change depends upon the polarization of the optical beam. For a rod cut from a birefringent crystal as shown in Fig. 17.8, there are two orth-

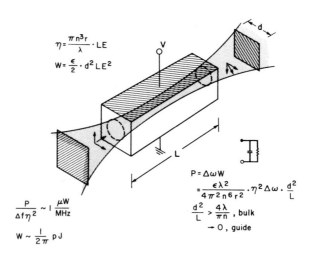

Fig. 17.8 Focused Gaussian beam passing through a bulk electrooptic modulator rod (Kaminow, 1975).

ogonal polarization directions that correspond to normal modes of propagation through the crystal. Light incident with one of these polarizations, denoted by subscript 1 or 2, is phase modulated with modulation index $\eta_{1,2}$ as the applied field E is varied

$$\eta_{1,2} = 2\pi L \,\Delta n_{1,2}/\lambda, \qquad (17.5)$$

where L is the rod length and Δn_1 and Δn_2 are in general different. If the incident beam of intensity I_0 is divided between these normal modes by a polarizer oriented at 45° to the principal polarization directions and then, after experiencing a differential phase retardation Γ in the crystal, these modes are recombined in an output polarizer at ±45° to the principal directions, interference between the modes will cause the output intensity to vary as a function of Γ; that is (Kaminow, 1974)

$$I/I_0 = \sin^2 \Gamma/2, \qquad (17.6)$$

where

$$\Gamma = 2\pi L(n_1 - n_2)/\lambda \qquad (17.7)$$

and

$$n_{1,2} = n_{1,2}(0) + \Delta n_{1,2}(E). \qquad (17.8)$$

We see from Eq. (17.6) that a nearly linear intensity modulation results if $\Gamma = \Gamma_{\text{bias}} + \Gamma(t)$, with the retardation bias $\Gamma_{\text{bias}} = q \cdot (\pi/2)$ (q an odd integer) and the electrooptically determined time-dependent retardation $\Gamma(t)$ limited to a range much less than $\pm \pi/2$ radian (approximately ± 1 radian gives 90% intensity modulation). Since the crystal is usually birefringent,

$n_1(0) \neq n_2(0)$, a portion of Γ_{bias} is introduced by the crystal itself; the remaining portion of the bias must be provided by a separate adjustable compensator. Thus, focusing lenses, polarizers, and a compensator are required in addition to the crystal. Since small changes in temperature ($\sim 1°C$) can vary the birefringence of the crystal substantially, it is best to use a crystal with small natural birefringence. Otherwise, it is necessary to control the temperature and/or to use two crystals in tandem to cancel the natural birefringence (Chen and Benson, 1974). Small natural birefringence has the further advantage of relaxing the requirement for parallelism of the end faces of the rod to ensure uniform Γ_{bias} over the cross section.

The electrooptic effect operates at frequencies up to approximately 10^3 GHz, where lattice resonances of the crystal occur. In practice the modulator bandwidth is determined by the external circuit employed to supply the modulating voltage across the modulator capacitance. The peak phase modulation index η for either of the normal modes is proportional to the field-length product EL while the energy W stored in the capacitance is proportional to E^2 times the crystal volume d^2L. Hence, the modulating power P per unit bandwidth for unit η^2 is given by

$$P/\eta^2 \Delta f = W/\eta^2 = Kd^2/L, \qquad (17.9)$$

where K is a proportionality constant and d^2 is the rod cross section. According to Eq. (17.9), the modulating power is proportional to the geometric factor d^2/L. But diffraction of the focused beam passing through the rod requires that d^2/L must be larger than $4\lambda/\pi n$ for the fundamental Gaussian mode. Hence, the minimum P in bulk devices is diffraction limited. Since d^2/L may be reduced without limit in an optical waveguide, the guided-wave devices to be discussed later, are intrinsically capable of much higher efficiency than bulk devices. The constant K in Eq. (17-9) is given by

$$K = (\epsilon_0/2\pi) \cdot \lambda^2 \cdot (\epsilon/n^6 r^2), \qquad (17.10)$$

where ϵ_0 is the vacuum permittivity, ϵ is the relative dielectric constant, and r is the electrooptic coefficient (Kaminow, 1974). Note that P increases as λ^3 for the diffraction-limited case.

Among the many electrooptic materials that have been investigated for bulk modulators, lithium niobate ($LiNbO_3$) and lithium tantalate ($LiTaO_3$) have emerged as most practical and useful in the 0.4 to 4-μm-wavelength range. Because of its smaller natural birefringence and greater resistance to optical damage (Lines and Glass, 1977), $LiTaO_3$ is often preferred. Lumped-element modulators of very wide bandwidth have been constructed and tested (at visible wavelengths) using these materials. A $LiTaO_3$ baseband modulator required 1.1 mW/MHz of modulating power for $\sim 100\%$ modulation at $\lambda = 0.633$ μm over a bandwidth of 1.3 GHz (Denton et al., 1967). A LiNbO$_3$ microwave modulator has been reported (Chow and Leonard, 1970) to require 4.5 mW/MHz for 50% modulation at $\lambda = 0.633$ μm over a band extending from 1.0 to 2.1 GHz. A 0–1.5 GHz baseband LiTaO$_3$ modulator in a matched 50-Ω microstrip transmission line (White and Chin, 1972) required 0.6 mW/MHz for 30% modulation at $\lambda = 0.496$ μm. A traveling-wave modulator consisting of two LiTaO$_3$ crystals fed by a zigzag microstrip transmission line, in which the optical path length is adjusted to eliminate transit-time effects, had a bandwidth of 3 GHz (An et al., 1976).

A baseband intensity modulator using LiNbO$_3$ was developed (Chen and Benson, 1974) specifically for optical-fiber applications at the wavelength of 1.06 μm with emphasis placed on low cost, compactness and simplicity. The modulator has been operated at 70 Mbit/sec with extinction ratio better than 40 to 1, 1-dB optical insertion loss and 20 mW/MHz drive power requirement for $\sim 100\%$ modulation.

Perhaps the most thoroughly engineered wideband modulator is the 1 Gbit/sec LiTaO$_3$ device designed for a space communication system (Ross et al., 1978). It operates at 0.532 μm (the second harmonic of the Nd:YAG laser); its operating point is stabilized by means of feedback, temperature control, and compensating crystal orientations; and it requires 50 W of electrical power.

More complete reviews of bulk electro-, acousto-, and magnetooptic modulators are given elsewhere (Kaminow and Turner, 1966; Chen, 1970; Kaminow, 1974). General-purpose electrooptic modulators are commercially available from several suppliers.

17.4 OPTICAL WAVEGUIDE DEVICES

17.4.1 Optical Waveguides

Planar waveguides varying in thickness from ~ 0.5–100 μm have been produced in acousto- and electrooptic materials by epitaxial growth and by diffusion. Photolithographic methods can be employed to form epitaxial or diffused strip guides that confine the optical wave in both transverse dimensions, in contrast to the planar guide which permits diffraction of the beam within the plane. Thus, strip guides allow the geometrical factor d^2/L discussed in Section 17.3 to be reduced below the bulk diffraction limit in order to reduce modulating power. In addition, the strip guides can be designed to carry the optical wave along paths that include bends in the plane of the substrate.

For semiconductor electrooptic materials such as GaAs, liquid-phase epitaxial (LPE) (Logan and Reinhart, 1975), molecular beam epitaxial

(MBE) (Merz and Cho, 1976) and vapor phase epitaxial (VPE) (Stillman *et al.*, 1976) methods of waveguide fabrication have been studied. These growth methods yield planar waveguides that can be converted to strip guides by ion milling or chemical etching to produce narrow ridge guides. For LiNbO$_3$ and LiTaO$_3$, diffusion of metal into the surface of the crystal has provided a simple means for realizing both planar and strip guides (Hammer and Phillips, 1974; Schmidt and Kaminow, 1974). For the latter, the metal pattern is simply defined by lithography before diffusion (Standley and Ramaswamy, 1974). Typical guides may have a refractive index about 0.01 greater than the substrate, cross-sectional dimensions of a few microns and length of a few centimeters. These guides can be made to support one mode or many.

For experimental studies, laser light can be coupled into a guide by a prism coupler set on its surface or light can be focused into the end of the guide made accessible by polishing or cleaving the edge of the substrate (Hsu and Chang, 1977). For single-mode guides, the fiber-to-guide coupling problem involves the design of a suitable transformer to match the field distribution of the mode of an optical fiber to that of the strip guide. For example, if we wish to couple a circular fiber of core diameter $a \approx 10$ μm to a rectangular guide of cross section $b \times c \approx 3 \times 1$ μm^2, the transformer must change both the size and symmetry of the mode (Dalgoutte *et al.*, 1975).

The theory and fabrication of planar optical waveguides is reviewed in more detail elsewhere (Tamir, 1975; Tien, 1976). Optical waveguide modulators, including magnetooptic devices not mentioned here, have also been reviewed previously (Kaminow, 1975; Hammer, 1975; Tien, 1976).

17.4.2 Bragg Deflection Modulators

Since LiNbO$_3$ is a piezoelectric material, a surface interdigital electrode with period Λ can generate an acoustic surface wave at the resonant acoustic wavelength Λ, whose energy will be confined to within a depth on the order of Λ (at 300 MHz, $\Lambda \approx 10$ μm). If, in addition, a planar optical waveguide is formed by diffusion of a thin layer of titanium into the surface (Schmidt and Kaminow, 1974) then an optical wave will be confined by the waveguide to within a few μm of the surface. A high-frequency acoustooptic interaction will then take place within a thin surface layer (Schmidt, 1976) and will be much more efficient than a bulk interaction which must be distributed over a larger volume because of diffraction of the unguided acoustic and optic beams. Figure 17.9a shows a Bragg surface-wave acoustooptic deflector employing a titanium diffused LiNbO$_3$ planar waveguide (Schmidt and Kaminow, 1975). Even though LiNbO$_3$ does not have as large an acoustooptic effect as say As$_2$Se$_3$, the

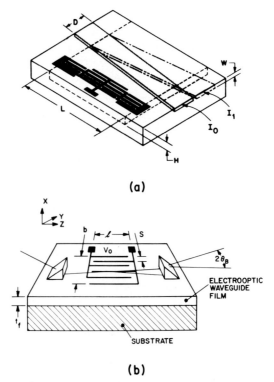

(a)

(b)

Fig. 17.9 (a) Schematic of a surface-wave acoustooptic Bragg deflector illustrating optical beam of width D and surface acoustic wave of depth H and width L. Series-parallel connected transducer has center bonding pad tied to ground through inductor and outer two bonding pads wired together and connected to the oscillator (Schmidt and Kaminow, 1975). (b) Schematic of an electrooptic grating modulator (Hammer, 1975).

electrical modulating power required is only 1.4 mW/MHz for 70% deflection at $\lambda = 0.633$ μm with subcarrier frequency 175 MHz and transducer-limited bandwidth of 35 MHz.

Multiple, staggered, surface-wave acoustic transducers on $LiNbO_3$ have been used to achieve Bragg deflectors capable of large bandwidths (250 MHz) and multiport beam switching (Kim and Tsai, 1976; Nguyen and Tsai, 1977). High-acoustic-frequency operation has been realized by propagating the acoustic beam normal to the optical waveguide plane (Brandt et al., 1978).

A Bragg deflector can also be produced electrooptically as shown in Fig. 17.9b (Hammer and Phillips, 1974; Hammer, 1975). Here a planar waveguide is formed by diffusing niobium into $LiTaO_3$ (Phillips and Hammer, 1975) and the optical beam passes under interdigital electrodes on the surface. The fringing electric fields from the electrodes produce a phase grating that deflects the beam when the electrode voltage is switched on. At $\lambda = 0.633$ μm, 80% deflection has been achieved for 6.6 V applied to the electrodes, which have a capacitance of 20 pF corresponding to a power requirement of 1.4 mW/MHz (Hammer, 1975). No acoustic wave is generated as long as the electrode period does not equal the acoustic wavelength at the modulating frequency. Important advantages of the electrooptic deflector over the acoustooptic deflector are that no subcarrier is required and the bandwidth is not limited by the resonance of the acoustic transducer.

17.4.3 Junction Modulators

Binary semiconductors such as GaAs and CdS exhibit a linear electrooptic effect. The resistivities of these crystals are often too low to permit the effective application of modulating voltages. To avoid this problem one can employ a back-biased pn-junction to provide the modulating field in the junction. The depletion region of the junction serves as a weak optical waveguide to confine the light to the neighborhood of the junction plane. The efficiency of the electrooptic interaction can be enhanced by stronger guidance provided by a heterojunction (i.e., a thin layer of high index material at the junction sandwiched between lower index material). A back-biased Schottky barrier can also be used as a junction modulator (Reinhart et al., 1976).

The $Al_xGa_{1-x}As$ system with variable x allows waveguiding layers to be grown by LPE (liquid phase epitaxy) or MBE (molecular beam epitaxy). The GaAs waveguiding layer, sandwiched between lower-index $Al_xGa_{1-x}As$ layers, is typically 0.5 μm thick with the depletion layer in the junction of similar thickness. Thus, extremely high fields can be realized in the junction plane with small applied voltages. In an intensity modulator (Reinhart, 1974), for which 1.06-μm light polarized at 45° to the junction plane is focused into the end of the guide, a modulating voltage of 4 V and power of 0.1 mW/MHz produced 90% modulation over a baseband of ~ 1 GHz. With a different crystal orientation, the incident beam may be polarized in or normal to the junction, permitting the incorporation of integrated planar polarizers. In this device the power required is 0.15 mW/MHz (McKenna and Reinhart, 1976).

Alternatively if λ is close to the bandedge of GaAs, the modulating field will shift the edge and vary the absorption (Franz–Keldysh effect) for a wave polarized either in or normal to the junction plane (Reinhart, 1973; Dyment and Kapron, 1976). Such a device operating at $\lambda = 0.9\,\mu$m with 4V and $\Delta f \approx 1$ GHz yielded 0.2 mW/MHz for 90% modulation (Reinhart, 1974). Another device consisting of a buried heterojunction in AlGaAsSb/GaAsSb operated at $\lambda = 1.08$ μm with an extinction ratio exceeding 25 dB and 6.5-dB loss (Campbell *et al.*, 1978). Although electroabsorption modulators do not require external polarizers, they can only operate at wavelengths near the absorption edge and hence can be quite lossy (~5 dB) and are restricted in operating wavelength.

As epitaxial growth techniques become more sophisticated, one may look forward to integrated devices containing semiconductor laser, modulator, and detector (for monitoring) on the same substrate. A step in that direction consists of a multilayer AlGaAs chip with cleaved ends to form the resonator of a laser contained in a particular waveguiding layer. Laser gain is provided by an amplifier section taper-coupled into that guide and the optical length of the resonator is varied by a phase modulator section in the guide. Thus, the laser can be frequency modulated or, with a suitable discriminator such as an external spectrometer or an internal periodic grating, it can be intensity modulated (Reinhart and Logan, 1975). Another AlGaAs integrated device consists of a laser section with etched mirrors coupled by a passive waveguide to a detector section (Merz and Logan, 1977).

17.4.4 Strip Waveguide Modulators

The planar waveguide devices described in the two preceding sections require external focusing to collimate the beam within the waveguide plane. Hence they do not lend themselves to coupling to optical fibers; nor can such a planar device approach the efficiency of a strip waveguide device, for which the ratio d^2/L is not limited by diffraction effects.

As an example, a strip waveguide modulator was fabricated by diffusing a photolithographically defined titanium metal strip into $LiNbO_3$, and applying coplanar metal electrodes on either side of the guide to provide the modulating field in the guide as indicated in Fig. 17.10a (Kaminow *et al.*, 1975). The electrodes are 3 cm long and spaced 9 μm apart. The guide is 5 μm wide and about 1 μm deep. Light can be injected by a prism coupler on top of the guide or by focusing into the squared and polished end of the guide. Both TE and TM waves (polarized in and normal to the plane, respectively) can be guided and individually phase modulated. For the TE (extraordinary) wave at 0.633 μm a phase modulation index η of 1 rad is produced by 0.3 V with a modulating power of 1.7 μW/MHz over a calculated bandwidth $\Delta f = 530$ MHz for a 50-Ω load resistor in parallel with the electrodes. Using external polarizer and compensator, intensity modulation with 10 to 1 extinction ratio was also achieved in this device by injecting a beam polarized at 45° to the plane with a voltage swing of ± 0.5 V (I. P. Kaminow, unpublished, 1976). Two such Ti-diffused $LiNbO_3$ waveguides in tandem separated by a miniature half-wave plate gave a temperature-stabilized intensity modulator at 1.06 μm with a bandwidth of 850 MHz and extinction ratio of 13 dB (Kubota *et al.*,

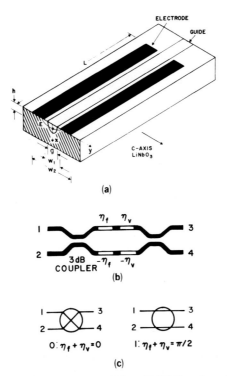

Fig. 17.10   (a) Schematic drawing of a Ti-diffused LiNbO$_3$ strip guide modulator. (b) Proposed balanced bridge waveguide switch and (c) schematic representation of the two states (0,1) of the switch (Kaminow *et al.*, 1975).

1978). This modulator was employed in a broadband optical-fiber transmission experiment (Kimura *et al.*, 1978). A very broadband intensity modulator makes use of a novel microstrip traveling-wave transmission line and Ti-diffused strip-guide (Izutsu *et al.*, 1978); bandwidths up to 10 GHz were reported.

In order to realize an intensity modulator or a four-pole switch without external polarizers, one can incorporate the phase modulator into a waveguide balanced-bridge as in Fig. 17.10b. The bridge is analogous to the conventional bulk intensity modulator of Fig. 17.8; but here, 3-dB waveguide couplers replace the 45° polarizers and the two strip waveguide arms of the bridge replace the two normal modes of polarization in the crystal. With equal and opposite phase shifts in each arm consisting of a phase bias $\eta_f = \pm \pi/4$ and a phase shift η_v varying between $\pm \pi/4$, the output would switch between the ports as indicated in Fig. 17.10c. For the phase modulator described above (which is 3 cm long), the switching voltage would be ± 0.24 V, the drive power 2 μW/MHz and the switching energy only 2 pJ to go from state 0 to state 1. The preceding discussion assumes single-mode waveguides guiding polarized light for proper operation of the 3-dB couplers and phase shifters. Various attempts have been made to realize such a bridge modulator on a single substrate. The difficult problem is to fabricate couplers that are precisely 3 dB in order to obtain a large extinction ratio. For this purpose, the coupler dimensions must be accurately controlled. Alternatively, the couplers can be tuned electrooptically to obtain a 3-dB operation (Ramaswamy *et al.*, 1978a).

While adequate dimensional control to obtain passive 3-dB couplers could be achieved with sufficient effort, another approach to realizing a four-pole switch is simply to make a single electrooptically controlled directional coupler as illustrated in Fig. 17.11a. If the propagation constants $\beta_{1,2}$ in both guides are identical, then for a certain coupling length L all the power will couple from guide 1 to guide 2. Electrodes placed over the two guides produce fields normal to the surface that are positive for one guide and negative for the other. Such a field can create a phase retardation $(\beta_1 - \beta_2)L$ which when equal to π will reduce the output coupling to zero. The electrooptic coupler switch has been realized in Ti diffused LiNbO$_3$: for 2-μm wide guides separated by 3-μm over a 3-mm coupling length, the switching voltage was 6 V at $\lambda = 0.5145$ μm (Papuchon *et al.*, 1975).

An electrooptic directional coupler switch has been reported at $\lambda = 1.06$ μm in expitaxial GaAs. For 6-μm wide guides separated by 7 μm and a

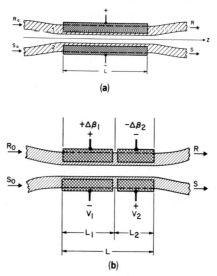

(a)

(b)

Fig. 17.11 (a) Electrooptically switched directional coupler consisting of optical strip waveguides with an interaction length L. The electrodes are in the "COBRA" configuration (Papouchon *et al.*, 1975). (b) The split electrode configuration with stepped $\Delta\beta$ reversal (Kogelnik and Schmidt, 1976).

length of 8 mm, switching occurs for 35 V applied to the electrodes, which form a part of a reverse-biased Schottky barrier. A drive power of 180 μW/MHz, bandwidth of 100 MHz and extinction ratio of 13 dB were reported (Campbell *et al.*, 1975). A similar device using rib guides had a 30-dB extinction ratio and potential 2-GHz bandwidth (Kawaguchi, 1978).

The chief difficulty with the directional coupler switch is that its dimensions must also be carefully controlled in order that complete coupling from guide 1 to guide 2 take place for $V = 0$; otherwise 100% switching cannot be realized. This difficulty has been overcome in the so-called "$\Delta\beta$-switch" by splitting the electrode pattern in two as in Fig. 17.11b in order to introduce an additional degree of freedom, i.e., one can apply independent voltages to each electrode pair (Schmidt and Kogelnik, 1976). Analysis of this structure shows that complete switching is possible for $\Delta\beta L = \pi$, as before, but without the critical dimensional requirements. A split electrode switch was fabricated by diffusing Ti into LiNbO$_3$ to form a coupler with 3-μm-wide guides spaced by 3 μm over a 3-mm coupling length. At $\lambda = 0.633$ μm, switching voltages on the order of 26 V gave an extinction ratio of 26 dB. Improved performance, i.e., 3-V drive, 13-dB extinction ratio, 100-Mbit/sec data rate and 1-mW/Mbit/sec figure of merit, was obtained in a six-section $\Delta\beta$-switch at 0.633 μm (Schmidt and Cross, 1978). The bandwidth was extended to 1 Gbit/sec in a later device (Cross and Schmidt, 1979).

A four-pole switch of the type we have been discussing and which is represented schematically in Fig. 17.10c is a well-known element in electrical switching arrays. One such array that is compatible with planar optical waveguide technology [i.e., does not require waveguide crossovers (Taylor, 1974)] is illustrated in Fig. 17.12a. For an $M \times M$ nonblocking switch, $N = \frac{1}{2}M(M - 1)$ elements are required. A 4×4 switch using five of the split electrode switches (not nonblocking) of Fig. 17.11b on a single Ti-diffused LiNbO$_3$ substrate has been demonstrated recently (Schmidt and Buhl, 1976). Similar $\Delta\beta$-switches (Leonberger and Bozler, 1977) and switching arrays (Shelton *et al.*, 1978) operating at 1.06 μm in AlGaAs have been reported.

Still another type of array that can be used for binary addressing is illustrated in Fig. 17.12b. In this case, for $q = 2^r$ addresses, we require r ranks of elements with all the elements in a given rank in the same state (0 or 1), for a total of $p = q - 1$ elements.

Insertion loss is an important consideration for any device application. In a waveguide switch these losses include waveguide attenuation in straight and curved sections, Fresnel reflections, and coupling loss due to misalignment and mismatch between input (fiber) and waveguide mode patterns. Attenuation in straight, single-mode Ti:LiNbO$_3$ waveguides is as low as 1.0 dB/cm at 0.63 μm (Noda *et al.*, 1978; Kaminow and Stulz, 1978; Fukuma *et al.*, 1978) and 0.3 dB/cm at 1.06 μm (Kubota *et al.*, 1978).

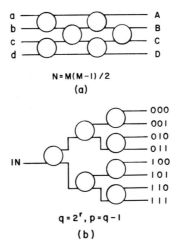

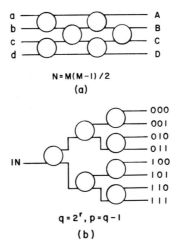

Fig. 17.12 (a) An array of switching elements that yields a 4 × 4 nonblocking switch without cross-overs of connecting waveguides. N is the number of elements required for a $M \times M$ switch. (b) A binary addressing array or multiplexer made up of p elements in r ranks to provide q addresses (I. P. Kaminow, unpublished, 1976).

In AlGaAs the attenuation is 5.6 dB/cm at 1.06 μm (Shelton et al., 1978), and in GaAs rib guides the attenuation is 2.2 dB/cm at 1.15 μm (Kawaguchi, 1978). The mismatch loss for circular-Gaussian to Ti-diffused-strip-guide mode is theoretically 0.8 dB (Burns and Hocker, 1977) as has been observed experimentally using lens coupling (Kaminow and Stulz, 1978). Experimental fiber-to-strip guide coupling losses as low as 1.5 dB have been reported (Hsu et al., 1978; Noda et al., 1978). Proper mode transformers and improved waveguide processing should further reduce these insertion losses.

17.4.5 Multimode Waveguide Modulators or Switches

The strip waveguide modulators or switches discussed in the preceding section are all single-mode, single-polarization devices. As such, they pose difficult coupling problems, especially if the laser source is not mode stabilized. Multimode waveguide devices should ease the coupling problem but require much greater drive voltage. Switching action has been demonstrated in an electrooptic field-induced multimode waveguide in LiNbO$_3$ with an applied voltage of several hundred volts (Soref et al., 1976). An electrooptic multimode waveguide switch that utilizes the concept of imaging in multimode waveguides has been shown to switch a few low-order modes simultaneously with relatively low drive voltage (Campbell and Li, 1978). Switching action was obtained electrooptically by nonuniform modulation of the index profile of a Ti-diffused LiNbO$_3$ waveguide, and extinction ratios of 13–20 dB were observed with 20–80 V drive. Important aspects of this device are that electrical compensation of small fabrication errors is possible and that it is realizable in a variety of sizes and geometries to allow easy and efficient coupling to lasers, fibers, and other components. An electrooptic waveguide device that simultaneously modulates both modes of polarization equally has been demonstrated (Burns et al., 1978) for potential use with "single-mode" fibers that do not maintain a fixed state of polarization.

17.5 CONCLUSIONS

A number of modulation methods suitable for optical-fiber transmission systems have been described. The most highly developed from the systems standpoint are the directly modulated LEDs and injection lasers.

Material dispersion effects in fibers will probably limit bandwidths to about 70 MHz (or a bit rate of about 140 Mbit/sec) for AlGaAs–LED systems with a repeater spacing of 1 km. However, LEDs of other material systems such as InGaAsP that operate near 1.3 μm where material dispersion falls to zero offer the possibility of extending the bandwidth beyond 400 Mb · km/sec (Muska et al., 1977). Since the fabrication and reliability of LEDs is satisfactorily understood, LED systems pose no fundamental problems for applications at moderate bandwidths.

Directly-modulated injection-laser systems have been tested at data rates up to 1.2 GHz; theoretical and experimental performances are found to be in good agreement (Li, 1978). At present, the chief uncertainties lie with the reliability, the reproducibility and the uncontrolled pulsation of the lasers. Satisfactory operation at bandwidths of a few hundred megabits per second is presently feasible, and, if the poorly understood pulsing behavior can be overcome, practical operation at bandwidths over 1 gigabits per second may be possible. A number of added difficulties arises in a high-bit-rate single-mode fiber system: the spectral width of present diodes may increase with aging thus introducing a material dispersion limit on bandwidth; a small instability of the transverse mode pattern may cause a large variation of coupling into the single-mode fiber; the influence of previous pulses on the amplitude of a given pulse (pattern-effect) will be more pronounced for the closely-spaced pulses.

Bulk acoustooptic modulators operate satisfactorily for bandwidths less than 200 Mbit/sec. A shortcoming of these devices is the requirement of a subcarrier oscillator at a frequency about four times the bandwidth.

Bulk electrooptic modulators have been operated at bandwidths above 1 GHz. System studies have been reported on two separate 1-Gbit/sec systems (Ward and Peterson, 1973; Abernathy et al., 1973; Ross et al.,

1978). These modulators require focusing lenses, polarizers and means for controlling the operating retardation bias. Their use is justified for broadband systems and/or systems employing non-semiconductor sources.

Optical waveguide modulators are at a relatively early stage of development. They offer the prospects of batch fabrication and low modulation power. They are also compatible with single-mode optical fibers and, in addition, may be incorporated in integrated optical circuits. For the most part, integrated-optical devices operate only with polarized light. Thus, operation of these devices in the receiver of an optical-fiber system (say, as switches or demultiplexers) would require the development of optical fibers that can maintain the state of polarization over long lengths. Alternatively, polarization-insensitive devices must be developed.

REFERENCES

Abbott, S. M., Muska, W. M., Lee, T. P., Dentai, A. G., and Burrus, C. A. (1978). 1.1-Gb/s pseudo-random pulse-code-modulation of 1.27 μm wavelength of CW InGaAsP/InP DH lasers. *Electron. Lett.* **14**, 348.

Abernathy, J. L., Brand, J. P., Carson, L. M., Chenoweth, A. J., Dreesewerd, D. W., Federhofer, J. F., and Rice, R. R. (1973). A 1-Gb/s mode-locked frequency-double Nd:YAG laser communications laboratory system. *IEEE J. Quantum Electron* **qe-9**, 691.

Acket, G. A., Nijman, W., and Lam, H. 't. (1974). Electron lifetime and diffusion constant in germanium-doped gallium arsenide. *J. Appl. Phys.* **45**, 3033.

Adams, M. J. (1973). Rate equations and transient phenomena in semiconductor lasers. *Opto-electronics* **5**, 201.

Aiki, K., Nakamura, M., Kuroda, T., Umeda, J., Ito, R., Chinone, N., and Maeda, M. (1978). Transverse mode stabilized $Al_xGa_{1-x}As$ injection lasers with channeled substrate planar structure. *IEEE J. Quantum Electron.* **qe-14**, 89.

Akiba, S., Sakai, K., and Yamamoto, T. (1978). Direct modulation of InGaAsP/InP double heterostructure lasers. *Electron. Lett.* **14**, 197.

An, J. C., Cho, Y., and Matsuo, Y. (1976). A cascaded traveling-wave electrooptic modulator. *Opt. Commun.* **18**, 597.

Arnold, G., Peterman, K., Russer, P., and Berlec, F. (1978). Modulation behavior of double heterostructure injection lasers with coherent light injection. *Arch. Elektr. Ubertrag.* **32**, 129.

Asatani, K., and Kimura, T. (1978a). Analyses of LED nonlinear distortions. *IEEE Trans. Electron Devices* **ED-25**, 199.

Asatani, K., and Kimura, T. (1978b). Linearization of LED nonlinearity by predistortions. *IEEE Trans. Electron Devices* **ED-25**, 207.

Baack, C., Elze, G., Enning, B., Heydt, G., Knupke, H., Löffler, R., and Walf, G. (1978). 1.12 Gbit/s regeneration experiment for an optical transmission system. *Frequenz* **32**, 151.

Basov, N. G., Drozhbin, Yu. A., Zahkarov, Yu. P., Nikitin, V. V., Semenov, A. S., Stepanov, B. M., Tolmachev, A. M., and Yakovlev, V. A. (1967a). Effect of injected current on the time dependence of the emission from GaAs lasers. *Sov. Phys.—Solid State (Engl. Transl.)* **8**, 2254.

Basov, N. G., Morozov, V. N., Nikitin, V. V., and Semenov, A. S. (1967b). Investigation of GaAs laser radiation pulsations. *Fiz. Tekh. Poluprovod.* **1**, 1570; *Sov. Phys.—Semicond. (Engl. Transl.)* **1**, 1305 (1968).

Basov, N. G., Nikitin, V. V., and Semenov, A. S. (1969). Dynamics of semi-conductor injection lasers. *Usp. Fiziol. Nauk* **97**, 561; *Sov. Phys.—Usp.* (*Engl. Trans.*) **12**, 219 (1969).

Boers, P. M., Vlaardingerbroek, M. T., and Danielson, M. (1975). Dynamic behavior of semiconductor lasers. *Electron. Lett.* **11**, 206.

Brandt, G. B., Gottlieb, M., and Weinert, R. W. (1978). Gigahertz modulators using bulk acoustooptic interactions in thin-film waveguides. *Fiber and Integrated Optics* **1**, 417.

Burns, W. K., and Hocker, G. B. (1977). Endfire coupling between optical fibers and diffused channel waveguides. *Appl. Opt.* **16**, 2048.

Burns, W. K., Giallorenzi, T. G., Moeller, R. P., and West, E. J. (1978). Interferometric waveguide modulator with polarization independent operation. *Appl. Phys. Lett.* **33**, 944.

Campbell, J. C., and Li, T. (1978). Electro-optic multimode waveguide switch. *Appl. Phys. Left,* **33**, 710.

Campbell, J. C., Blum, F. A., Shaw, D. W., and Lawley, K. L. (1975). GaAs electro-optic directional-coupler switch. *Appl. Phys. Lett.* **27**, 202.

Campbell, J. C., DeWinter, J. C., Pollack, M. A., and Nahory, R. E. (1978). Buried heterojunction electroabsorption modulator. *Appl. Phys. Lett.* **32**, 471.

Carroll, J. E., and Farrington, J. G. (1973). Short-pulse modulation of gallium arsenide lasers with trappatt diodes. *Electron. Lett.* **9**, 166.

Casey, H. C., Jr., and Stern, F. (1976). Concentration dependent absorption and spontaneous emission of heavily doped GaAs. *J. Appl. Phys.* **47**, 631.

Casey, H. C., Jr., Miller, B. I., and Pinkas, E. (1973). Variation of minority-carrier diffusion length with carrier concentration in GaAs liquid-phase epitaxial layers. *J. Appl. Phys.* **44**, 1281.

Chang, I. C. (1976). Acoustooptic devices and applications. *IEEE Trans. Sonics Ultrason.* **SU-23**, 2.

Chen, F. S. (1970). Modulators for optical communications. *Proc. IEEE* **58**, 1440.

Chen, F. S., and Benson, W. W. (1974). Lithium niobate light modulator for optical communications. *Proc. IEEE* **62**, 133.

Chen, F. S., Karr, M. A., and Shumate, P. W. (1978). Effects of beam displacement and front-back mistracking of junction laser on lightwave trasmitter output stability. *Appl. Opt.* **17**, 2219.

Chinone, N. (1977). Nonlinearity in power-output-current characteristics of stripe-geometry injection lasers. *J. Appl. Phys.* **48**, 3237.

Chow, K. K., and Leonard, W. B. (1970). Efficient octave-bandwidth microwave light modulators. *IEEE J. Quantum Electron.* **QE-6**, 789.

Chown, M., Goodwin, A. R., Lovelace, D. F., Thompson, G. H. G., and Selway, P. R. (1973). Direct modulation of double-heterostructure lasers as rates up to 1 Gbit/s. *Electron. Lett.* **9**, 34.

Cross, P. S., and Schmidt, R. V. (1979). A 1-Gbit/sec integrated optical modulator. To be published.

Daikoku, K. (1977). Direct modulation characteristics of semiconductor laser diodes. *Jpn. J. Appl. Phys.* **16**, 117.

Dalgoutte, D. G., Mitchell, G. L., Matsumoto, R. L. K., and Scott, W. D. (1975). Transition waveguides for coupling fibers to semiconductor lasers. *Appl. Phys. Lett.* **27**, 125.

D'Asaro, L. A., Cherlow, J. M., and Paoli, T. L. (1968). Continuous microwave oscillations in GaAs junction lasers. *IEEE J. Quantum Electron.* **QE-4**, 164.

Dawson, L. R. (1977). High-efficiency graded-band-Gap $Ga_{1-x}Al_xAs$ light-emitting diodes. *J. Appl. Phys.* **48**, 2485.

Dawson, R. W. (1978). Frequency and bias dependence of video distortion in Burrus-type homostructure and heterostructure LEDs. *IEEE Trans. Electron Devices* **ed-25**, 550.

Dentai, A. G., Lee, T. P., and Burrus, C. A. (1977). Small-area, high-radiance C. W. InGaAsP L.E.D.s emitting at 1.2 to 1.3μm. *Electron. Lett.* **13**, 484.

Denton, R. T., Chen, F. S., and Ballman, A. A. (1967). Lithium tantalate light modulators. *J. Appl. Phys.* **38,** 1611.

Dyment, J. C., and Kapron, F. P. (1976). Extinction ratio limitations in GaAlAs electroabsorption light modulators. *J. Appl. Phys.* **47,** 1523.

Dyment, J. C., Ripper, J. E., and Lee, T. P. (1972). Measurement and interpretation of long spontaneous lifetimes in double heterostructure lasers. *J. Appl. Phys.* **43,** 452.

Ettenberg, M., and Kressel, H. (1976). Interfacial recombination at (AlGaAs)As/GaAs heterojunction structures. *J. Appl. Phys.* **47,** 1538.

Ettenberg, M., Kressel, H., and Gilbert, S. L. (1973). Minority carrier diffusion length and recombination lifetime in GaAs: Ge prepared by liquid-phase epitaxy. *J. Appl. Phys.* **44,** 827.

Ettenberg, M., Kressel, H., and Wittke, J. P. (1976). Very high radiance edge-emitting LED. *IEEE J. Quantum Electron.* **QE-12,** 360.

Fukuma, M., Noda, J., and Iwasaki, H. (1978). Optical properties in titanium-diffused LiNbO$_3$ strip waveguides. *J. Appl. Phys,* **49,** 3693.

Furuya, K., Suemastu, Y., and Hong, T. (1978). Reduction of reasonance-like peak in direct modulation due to carrier diffusion in injection laser. *Appl. Opt.* **17,** 1949.

Goodfellow, R. C., and Mabbitt, A. W. (1976). Wide-bandwidth high-radiance galliumarsenide light-emitting diodes for fibre-optic communication. *Electron. Lett.* **12,** 50.

Goodwin, A. R., Peters, J. R., Pion, M., Thompson, G. H. B., and Whiteaway, J. E. A. (1975). Threshold temperature characteristics of double heterostructure Ga$_{1-x}$Al$_x$As lasers. *J. Appl. Phys.* **46,** 3126.

Hall, R. N. (1960). Recombination processes in semiconductors. *Proc. IEE* **106,** Suppl. 17, Part B, 923.

Hammer, J. M. (1975). Modulation and switching of light in dielectric waveguides. *In* "Integrated Optics" (T. Tamir, ed.), p. 139. Springer-Verlag, Berlin and New York.

Hammer, J. M., and Phillips, W. (1974). Low-loss single-mode optical waveguides and efficient high-speed modulators of LiNb$_x$Ta$_{1-x}$O$_3$ on LiTaO$_3$. *Appl. Phys. Lett.* **24,** 545.

Harth, W. (1975). Properties of injection lasers at large-signal modulation. *Arch. Elektr. Ubertrag.* **29,** 149.

Harth, W., and Heinen, J. (1975). Investigation of GaAs : GaAlAs single-heterostructure light emitting diodes for optical communication systems. *Arch. Elektr. Ubertrag.* **29,** 489.

Harth, W., Heinen, J., and Huber, W. (1975). Influence of active-layer width on the performance of homojunction and single-heterojunction GaAs light-emitting diodes. *Electron. Lett.* **11,** 23.

Harth, W., Huber, W., and Heinen, J. (1976). Frequency response of GaAlAs light-emitting diodes. *IEEE Trans. Electron Devices* **ED-23,** 478.

Hartman, R. L., and Dixon, R. W. (1975). Reliability of DH GaAs lasers at elevated temperatures. *Appl. Phys. Lett.* **26,** 239.

Haug, H., and Haken, H. (1967). Theory of noise in semiconductor laser emission. *Z. Phys.* **204,** 262.

Hillbrand, H., and Russer, P. (1975). Large signal PCM behaviour of injection lasers with coherent irradiation into one of its oscillating modes. *Electron. Lett.* **11,** 372.

Hsu, H. P., and Chang, W. S. C. (1977). Coupling methods in prospective single-mode fiber integrated optics systems: A progress report. *Fiber Integr. Opt.* **1,** 153.

Hsu, H. P., Milton, A. F., Burns, W. K., and Sheem, S. K. (1978). Multiport coupling between single mode fibers and indiffused channel waveguides. *Tech. Dig., Top. Meet. Integr. Fiber Opt., 1978* WD6.

Ikegami, T., and Suematsu, Y. (1968). Direct modulation of semiconductor junction lasers. *Electron. Commun. (Jpn).* **51-B,** 51 (1968).

Ikegami, T., Kobayashi, K., and Suematsu, Y. (1970). Transient behavior of semiconductor injection lasers. *Electron. Commun. (Jpn.)* **53-B,** 82 (1970).

Izutsu, M., Itoh, T., and Sueta, T. (1978). 10 GHz bandwidth travelling-wave LiNbO$_3$ optical waveguide modulator. *IEEE J. Quantum Electron.* **QE-14**, 394.

Kajimura, T., Saito, K., Shige, N., and Ito, R. (1978). Stable operation of buried-heterostructure Ga$_{1-x}$Al$_x$As lasers during accelerated aging. *Appl. Phys. Lett.* **33**, 626.

Kaminow, I. P. (1974). "An Introduction to Electrooptic Devices." Academic Press, New York.

Kaminow, I. P. (1975). Optical waveguide modulators. *IEEE Trans. Microwave Theory Tech.* **23**, 57.

Kaminow, I. P., and Ramaswamy, V. (1979). Single polarization optical fibers: Slab model. *Appl. Phys. Lett.* **34**, 62.

Kaminow, I. P., and Stulz, L. W. (1978). Loss measurements in diffused LiNbO$_3$ waveguides. *Appl. Phys. Lett.* **33**, 62.

Kaminow, I. P., and Turner, E. H. (1966). Electrooptic light modulators. *Proc. IEEE* **54**, 1374.

Kaminow, I. P., Stulz, L. W., and Turner, E. H. (1975). Efficient strip-wave-guide modulator. *Appl. Phys. Lett.* **27**, 555.

Kawaguchi, H. (1978). GaAs rib-waveguide directional coupler switch with Schottky barriers. *Electron. Lett.* **14**, 387.

Kim, B., and Tsai, C. S. (1976). High performance guided wave acoustooptic scanning devices using multiple surface acoustic waves. *Proc. IEEE* **64**, 788.

Kimura, T., Saruwatari, M., Yamada, J., Uehara, S., and Miyashita, T. (1978). Optical fiber (800-Mbit/sec) transmission experiment at 1.05 μm. *Appl. Opt.* **17**, 2420.

King, F. D., Springthorpe, A. J., and Szentesi, O. I. (1975). High-power long-lived double heterostructure LEDs for optical communications. *Tech. Digest, Int. Electron Devices Meet. 1975* p. 480.

King, F. D., Straus, J., Szentesi, O. I., and Springthorpe, A. J. (1976). High-radiance long-lived L.E.D.s for analogue signalling. *Proc. Inst. Electr. Eng.* **123**, 619.

Kogelnik, H., and Schmidt, R. V. (1976). Switched directional couplers with alternating Δβ. *IEEE J. Quantum Electron.* **QE-12**, 396.

Kubota, K., Minakata, M., Saito, S., and Uehara, S. (1978). Temperature stabilized optical waveguide modulator. *Opt. Quantum Electron.* **10**, 205.

Kurnosov, V. D., Magalyas, V. I., Pleshkov, A. A., Rivlin, L. A., Trukhan, V. G., and Tsvetkov, V. V. (1966). Self modulation of emission from an injection semiconductor laser *JETP Lett. (Engl. Transl.)* **4**, 303.

Lang, R., and Kobayashi, K. (1976). Suppression of the relaxation oscillation in the modulated output of semiconductor lasers. *IEEE J. Quantum. Electron.* **QE-12**, 194.

Lee, T. P. (1975). Effect of junction capacitance on the rise time of LEDs and on the turn-on delay of injected lasers. *Bell Syst. Tech. J.* **54**, 53.

Lee, T. P. (1977). The nonlinearity of double-heterostructure LED's for optical communications. *Proc. IEEE* **65**, 1408.

Lee, T. P., and Dentai, A. G. (1978). Power and modulation bandwidth of GaAs-AlGaAs high-radiance LEDs for optical communication systems. *IEEE J. Quantum Electron.* **QE-14**, 150.

Lee, T. P., and Derosier, R. M. (1974). Charge sotrage in injection lasers and its effects on high-speed pulse modulation of laser diode. *Proc. IEEE* **62**, 1175.

Lee, T. P., and Roldan, R. H. R. (1970). Repetively Q-switching light pulses from GaAs injection lasers with tandem double-section stripe geometry. *IEEE J. Quantum Electron.* **QE-6**, 339.

Lee, T. P., and Serra, T. J. B. (1976). Characteristics of injection locking of self-pulsing in an AlGaAs DH junction laser. *IEEE J. Quantum Electron.* **qe-12**, 368.

Leonberger, F. J., and Bozler, C. O. (1977). GaAs directional-coupler switch with stepped Δβ reversal. *Appl. Phys. Lett.* **31**, 223.

Li, T. (1978). Optical fiber communication—the state of the art". *IEEE Trans. Commun.* **COM-26,** 946.

Lines, M. E., and Glass, A. M. (1977). "Principles and Applications of Ferroelectrics and Related Materials," p. 459. Oxford Univ. Press (Clarendon), London and New York.

Liu, Y. S., and Smith, D. A. (1975). The frequency response of an amplitude-modulated GaAs luminescence diode. *Proc. IEEE* **63,** 542.

Lockwood, H. F., Wittke, J. P., and Ettenberg, M. (1976). LED for high data rate optical communications. *Opt. Commun.* **16,** 193.

Logan, R. A., and Reinhart, F. K. (1975). Integrated GaAs-Al$_x$Ga$_{1-x}$As double-heterostructure laser with independently controlled optical output divergence. *IEEE J. Quantum Electron.* **QE-11,** 46.

McCumber, D. E. (1966). Intensity fluctuations in the output of CW laser oscillators. *I. Phys. Rev.* **141,** 306.

McKenna, J., and Reinhart, F. C. (1976). Double-heterostructure GaAs-Al$_x$GA$_{1-x}$As [110] P-N junction diode modulator. *J. Appl. Phys.* **47,** 2069.

Machida, S., Nagai, H., and Kimura, T. (1979). Modulation characteristics of InGaAsP/InP l.e.d.s at 1.5 μm wavelength. *Elect. Lett.* **15,** 175.

Merz, J. L., and Cho, A. Y. (1976). Low-loss Al$_x$Ga$_{1-x}$As waveguides grown by molecular beam epitaxy. *Appl. Phys. Lett.* **28,** 456.

Merz, J. L., and Logan, R. A. (1977). Integrated GaAs-Al$_x$Ga$_{1-x}$As injection lasers and detectors with etched reflectors. *Appl. Phys. Lett.* **30,** 530.

Muska, W. M., Li, T., Lee, T. P., and Dentai, A. G. (1977). Material-dispersion-limited operation of high-bit-rate optical-fibre data links using L.E.D.s. *Electron. Lett.* **13,** 605.

Namizaki, H. (1975). Transverse-junction-stripe lasers with a GaAs p-n homojunction. *IEEE J. Quantum Electron.,* **QE-11,** 427.

Namizaki, H., Kau, H., Ishii, M., and Ito, A. (1974a). Current dependence of spontaneous carrier lifetimes in GaAs-Ga$_{1-x}$Al$_x$As double heterostructure lasers. *Appl. Phys. Lett.* **24,** 486.

Namizaki, H., Nagano, M., and Nakahara, S. (1974b). Frequency response of Ga$_{1-x}$Al$_x$As light-emitting diodes. *IEEE Trans. Electron Devices* **ED-21,** 688.

Nguyen, L. T., and Tsai, C. S. (1977). Efficient wideband guided-wave acoustooptic bragg diffraction using phased surface acoustic wave arran in LiNbO$_3$ waveguides. *Appl. Opt.* **16,** 1297.

Noda, J., Mikami, O., Minikata, M., and Fukuma, M. (1978). Single mode optical waveguide-fiber coupler. *Appl. Opt.* **17,** 2092.

Ostoich, V., Jeppesen, P., and Slaymaker, N. (1975). Direct modulation of DH GaAlAs lasers with GaAs M.E.S.F.E.T.S. *Electron. Lett.* **11,** 515.

Ozeki, T., and Hara, E. H. (1976). Measurement of nonlinear distortion in light-emitting diodes. *Electron. Lett.* **12,** 78.

Ozeki, T., and Ito, T. (1973a). Pulse modulation of DH (GaAl)As lasers. *IEEE J. Quantum Electron.* **QE-9,** 388.

Ozeki, T., and Ito, T. (1973b). A new method for reducing pattern effect in PCM current modulation of DH-GaAlAs lasers. *IEEE J. Quantum Electron.* **QE-9,** 1098.

Paoli, T. L. (1975). Noise characteristics of stripe-geometry double-heterostructure junction lasers operating continuously—I. Intensity noise at room temperature. *IEEE J. Quantum Electron.* **QE-11,** 276.

Paoli, T. L. (1976). Nonlinearities in the emission characteristics of stripe-geometry (AlGa)As double-heterostructure junction lasers. *IEEE J. Quantum Electron.* **QE-12,** 770.

Paoli, T. L. (1977a). Changes in the optical properties of CW (AlGa)As junction laser during accelerating aging. *IEEE J. Quantum Electron.* **QE-13,** 351.

Paoli, T. L. (1977b). Waveguiding in stripe-geometry junction lasers. *IEEE J. Quantum Electron.* **QE-13**, 662.

Paoli, T. L., and Ripper, J. E. (1969). Optical pulses from CW GaAs injection lasers. *Appl. Phys. Lett.* **15**, 105.

Paoli, T. L., and Ripper, J. E. (1970). Direct modulation of semiconductor lasers. *Proc. IEEE* **58**, 1457.

Papuchon, M., Combemale, Y., Mathieu, X., Ostrowsky, D. B., Reiber, L., Roy, A. M., Sejourne, B., and Werner, M. (1975). Electrically switched optical directional coupler: COBRA. *Appl. Phys. Lett.* **27**, 289.

Personick, S. D. (1973). Receiver design for digital fiber-optic communication systems. II. *Bell Syst. Tech. J.* **52**, 875.

Phillips, W., and Hammer, J. M. (1975). Formation of lithium niobate-tantalate waveguides. *J. Electron. Mater.*, **4**, 549.

Ramaswamy, V., Divino, M. D., and Standley, R. D. (1978a). Balanced bridge modulator switch using Ti diffused $LiNbO_3$ strip waveguides. *Appl. Phys. Lett.* **32**, 644.

Ramaswamy, V., Kaminow, I. P., Kaiser, P., and French, W. G. (1978b). Single polarization optical fibers: Exposed cladding technique. *Appl. Phys. Lett.* **33**, 814.

Reinhart, F. K. (1973). Electroabsorption in $Al_yGa_{1-y}As-Al_xGa_{1-x}As$ double heterostructures. *Appl. Phys. Lett.* **22**, 372.

Reinhart, F. K. (1974). Phase and intensity modulation properties of $Al_yGa_{1-y}As-Al_xGa_{1-x}As$ double heterostructure P-N junction waveguides. *Tech. Dig., Top. Meet. Integr. Opt.*, 1974 WA6.

Reinhart, F. K., and Logan, R. A. (1975). Integrated electro-optic injection laser. *Appl. Phys. Lett.* **27**, 532.

Reinhart, F. K., Sinclair, W. R., and Logan, R. A. (1976). Single heterostructure $Al_xGa_{1-x}As$ phase modulator with SnO_2-dopted In_2O_3 classing layer. *Appl. Phys. Lett.* **29**, 21.

Ripper, J. E., and Paoli, T. L. (1969). Frequency pulling and pulse position modulation of pulsing CW GaAs injection lasers. *Appl. Phys. Lett.* **15**, 203.

Ripper, J. E., and Paoli, T. L. (1971). Optical self-pulsing of junction lasers operating continuously at room temperature. *Appl. Phys. Lett.* **18**, 466.

Roldan, R. (1967). Spikes in the light output of room temperature GaAs junction lasers. *Appl. Phys. Lett.* **11**, 346.

Ross, M., Freedman, P., Abernathy, J., Matassov, G., Wolf, J., and Barry, J. (1978). Space optical communications with the Nd:YAG laser. *Proc. IEEE* **66**, 319.

Russer, P. (1975). Modulation behavior of injection lasers with coherent irradiation into their oscillating mode. *Arch. Elektr. Ubertrag.* **29**, 231.

Russer, P., and Schultz, S. (1973). Direkte Modulation eines Doppelheterostrukterlasers mit einer Bitrate von 2-3 Gbit/s. *Arch. Elektr. Ubertrag.* **27**, 193.

Schicketanz, D. (1975). Large-signal modulation of GaAs lasers diodes. *Siemens Forsch.-Entwicklungs per.* **4**, 325.

Schmidt, R. V. (1976a). Acoustooptic interactions between guided optical waves and acoustic surface waves. *IEEE Trans. Sonics Ultrason.* **SU-23**, 22.

Schmidt, R. V., and Buhl, L. L. (1976). An experimental integrated optical 4×4 switching network. *Electron. Lett.* **12**, 575.

Schmidt, R. V., and Cross, P. S. (1978). Efficient optical waveguide switch/amplitude modulator. *Opt. Lett.* **2**, 45.

Schmidt, R. V., and Kaminow, I. P. (1974). Metal-diffused optical waveguides in $LiNbO_3$. *Appl. Phys. Lett.* **25**, 458.

Schmidt, R. V., and Kaminow, I. P. (1975). Acoustooptic Bragg deflection in $LiNbO_3$ Ti diffused waveguides. *IEEE J. Quantum Electron.* **QE-11**, 57.

Schmidt, R. V., and Kogelnik, H. (1976). Electrooptically switched coupler with stepped $\Delta\beta$ reversal using Ti-diffused $LiNbO_3$ waveguides. *Appl. Phys. Lett.* **28**, 503.

Scifres, D. R., Streifer, W., and Burnham, R. D. (1978). Curved stripe GaAs:GaAlAs diode lasers ant waveguides. *Appl. Phys. Lett.* **32,** 231.

Selway, P. R., and Goodwin, A. R. (1976). Effect of D. C. bias level on the spectrum of GaAs lasers operated with short pulses. *Electron. Lett.* **12,** 25.

Shelton, J. C., Reinhart, F. K., and Logan, R. A. (1978). "GaAs-Al$_x$Ga$_{1-x}$As rib waveguide switches with MOS electrooptic control for monolithic integrated optics. *Appl. Opt.* **17,** 2548.

Shen, C. C., Hsieh, J. J., and Lind, T. A. (1977). 1500-h continuous CW operation of double-heterostructure GaInAsP/InP lasers. *Appl. Phys. Lett.* **30,** 353.

Shumate, P. W., Chen, F. S., and Dorman, P. W. (1978). GaAlAs laser transmitter for lightwave transmission systems. *Bell Syst. Tech. J.* **57,** 1823.

Sittig, E. K. (1972). Elastooptic light modulation and deflection. *Prog. Opt.* **10,** 231.

Soref, R. A., McMahon, D. H., and Nelson, A. R. (1976). Multimode achromatic electro-optic waveguide switch for fiber-optic communications. *Appl. Phys. Lett.* **28,** 716.

Standley, R. D., and Ramaswamy, V. (1974). Nb diffused LiTaO$_3$ optical waveguides: Planar and embedded strip guides. *Appl. Phys. Lett.* **25,** 711.

Steinberg, R. A., and Giallorenzi, T. G. (1976). performance limitations imposed on optical waveguide switches and modulators by polarization. *Appl. Opt.* **15,** 2440.

Stillman, G. E., Wolfe, C. M., Rossi, J. A., and Hechscher, H. (1976). Low-loss high-purity GaAs waveguides for monolithic integrated optical circuits at GaAs laser wavelengths. *Appl. Phys. Lett.* **28,** 197.

Straus, J. (1978a). Linearized transmitters for analog fiber optical systems. *Tech. Dig., Conf. Laser Electroopt. Syst., 1978* p. 18.

Straus, J. (1978b). The nonlinearity of high-radiance light-emitting diodes. *IEEE J. Quantum Electron.* **QE-14,** 813.

Straus, J., and Szentesi, O. I. (1975). Linearity of high power, high radiance Ga$_x$Al$_{1-x}$As Ge double heterostructure LED's. *Tech. Dig., 1975 Int. Electron Devices Meet., 1975* p. 484.

Straus, J., and Szentesi, O. I. (1977a). Linearization of optical transmitters by a quasifeed-forward compensation technique. *Electron. Lett.* **13,** 158.

Straus, J., and Szentesi, O. I. (1977b). Linearized transmitters for optical communications. *Proc. IEEE Int. Symp. Circuits Syst., 1977* p. 288.

Straus, J., Springthorpe, A. J., and Szentesi, O. I. (1977). Phase-shift modulation technique for the linearization of analogue optical transmitters. *Electron. Lett.* **13,** 149.

Susaki, W., Tanaka, T., Kan, H., and Ishii, M. (1977). New structures of GaAlAs lateral-injection laser for low-threshold and single-mode operation. *IEEE J. Quantum. Electron.* **QE-13,** 587.

Szentesi, O. I., and Szanto, A. J. (1976). Fiber optics video transmission. *Proc. SPIE/SPSE Tech. Symp. East.* **77,** 151.

Tamir, T., ed. (1975). "Integrated Optics. "Springer-Verlag, Berlin and New York.

Taylor, H. F. (1974). Optical-waveguide connecting networks. *Electron. Lett.* **10,** 41.

Thim, H. W., Dawson, L. R., DiLorenzo, J. V., Dyment, J. C., Hwang, C. J., and Rode, D. L. (1973). subnanosecond pulse code modulation of GaAs lasers by Gunn effect switches. *Tech. Dig., IEEE Int. Solid-State Circuits Conf., 1973* p. 92.

Thompson, G. H. B., Henshall, G. D., Whiteaway, J. E. A., and Kirkby, P. A. (1976). Narrow-beam five-layer (GaAl)As/GaAs heterostructure lasers with low threshold and high peak power. *J. Appl. Phys.* **47,** 1501.

Tien, P. K. (1976). Integrated optics and new wave phenomena in optical waveguides. *Rev. Mod. Phys.* **49,** 361.

Tsang, W. T., and Logan, R. A. (1978). High-power fundamental-transverse-mode strip buried heterostructure lasers with linear light-current characteristics. *Appl. Phys. Lett.* **32,** 311.

Tsukada, T. (1974). GaAs-Al$_{1-x}$Ga$_x$As buried-heterostructure injection lasers. *J. Appl. Phys.* **45**, 4899.

Uchida, N., and Niizeki, N. (1973). Acoustooptic deflection materials and techniques. *Proc. IEEE* **61**, 1073.

Umebu, I., Abe, M., Yamoaka, T., Kotani, T., Hanano, N., Iguchi, K., and Yoshibayashi, T. (1977). GaAlAs LED's for high quality fiber-optical analog link. *Tech. Dig., Int. Conf. Integr. Opt. Opt. Fiber Commun. 1977*, p. 109.

Ward, R. B., and Peterson, D. G. (1973). Performance of a laboratory 1-Gb/s laser communications. *IEEE J. Quantum electron.* **QE-9**, 1155.

Warner, A. W., and Pinnow, D. A. (1973). Miniature Acoustooptic modulators for optical communications, *IEEE J. Quantum electron.* **QE-9**, 1155.

White, G., and Chin, G. M. (1972). Traveling wave electrooptic modulators. *Opt. Commun.* **5**, 374.

Yamada, J., Saruwatari, M., Asatani, K., Tsuchiya, H., Kawana, A., Sugiyama, K., and Kimura, A. (1978). High speed optical pulse transmission at 1.29 μm wavelength using low loss single mode fibers. *IEEE J. Quantum Electron.* **QE-14**, 791.

Yamamoto, T., Sakai, K., Akiba, S., and Suematsu, Y. (1978). In$_{1-x}$Ga$_x$As$_y$P$_{1-y}$/InP DH lasers fabricated on InP (100) substrates. *IEEE J. Quantum Electron.* **QE-14**, 95.

Yang, E. S., McMullin, P. G., Smith, A. W., Blum, J., and Shih, K. K. (1974). Degradation induced microwave oscillations in double-heterostructure injection lasers. *Appl. Phys. Lett.* **24**, 324.

Zucker, J., Lauer, R. B., and Schlafer, J. (1976). Resonse time of Ge-doped (AlGa)As-GaAs double-heterostructure LEDs. *J. Appl. Phys.* **47**, 2082.

Chapter 18

Photodetectors

TIEN PEI LEE
TINGYE LI

18.1 INTRODUCTION

The photodetector is an essential element in an optical-fiber communication system. It demodulates the optical signal, i.e., it converts the optical variations into electrical variations which are subsequently amplified and further processed. The crucial role it plays demands that it must satisfy very stringent requirements regarding performance and compatibility. The ultimate choice of a photodetector for any specific application must also involve consideration of cost.

The following are the important performance requirements:

(1) High response or sensitivity at the operating wavelengths. At present, the wavelengths of interest lie in the range 0.8–0.9 μm, where AlGaAs lasers and LEDs have their emission lines. Future operating wavelengths may shift to the 1.1- to 1.6-μm spectral region, where optical fibers have lower optical loss and minimum material dispersion.

(2) Sufficient bandwidth or speed of response to accommodate the information rate. Present systems interests extend to a few hundred megahertz, but future single-mode fiber systems may operate at multigigabit rates.

(3) Minimum additional noise introduced by the detector. Dark currents, leakage currents, and shunt conductances must be low, and, if the detector is to provide internal gain, the gain mechanism should be as noise-free as possible.

(4) Low susceptibility of performance characteristics to changes in ambient conditions. Sensitivity, noise, and internal gain of practical photodiodes all vary with ambient temperature. Compensation of temperature effects is essential in many applications.

593

Compatibility requirements involve considerations of the physical size of the detector, the coupling to the fiber and to the ensuing electronics, and the necessary power supply. The detector must be small so that it is easily packaged with other electronics and easily coupled to the fiber. It should not require excessive bias voltages or currents.

Photomultipliers can meet many of the above performance criteria, but their relative bulk and high-voltage requirement render them unsuitable for fiber system use. Solid-state photodiodes with or without internal (avalanche) gain combine good performance and compatibility with low cost. Presently available silicon photodiodes have high sensitivity near $\lambda = 0.85$ μm, adequate speed of response (to hundreds of megahertz), low dark current, negligible shunt conductance, sufficient internal gain with little excess noise, and long-term stability. Their miniature size makes coupling to fibers and interfacing with electronic circuits simple, and they require only low to moderate voltages for bias. Because the response of silicon devices falls off at wavelengths longer than 1 μm, they are not suitable for future applications near $\lambda = 1.3$ μm. Present research efforts on detectors are devoted to the investigation of semiconductor photodiode materials that have narrower bandgaps which give better response at longer wavelengths.

Several comprehensive review papers on photodetectors are found in the literature (Anderson and McMurtry, 1966; Anderson et al., 1970; Melchior et al., 1970; Melchior, 1972, 1973a, 1977; Stillman and Wolfe, 1977).

18.2 PERFORMANCE CONSIDERATIONS

In this section we discuss various factors that determine the performance characteristics of semiconductor photodiodes. The discussion includes the basic detection process, materials and their spectral response, quantum efficiency versus speed, and noise considerations. A detailed discussion of the principles of operation of PIN and avalanche photodiodes appears in the next section.

18.2.1 Basic Detection Process

The basic detection process in semiconductor photodiodes involves the photogeneration of electron-hole pairs in a region of high electric field strength, as in the depletion layer of a p–n junction. The carrier pairs are separated by the high field and collected across the reverse-biased junction. In order that the quantum efficiency η (the number of carrier-pairs generated per incident photon) be high, the depletion layer must be sufficiently thick to allow a large fraction of the incident light to be absorbed. On the other hand, since long carrier drift times limit the speed of operation of the photodiode, the depletion layer must be kept thin. There is, therefore, a tradeoff between the speed of response and quantum efficiency.

For a semiconductor with absorption coefficient α_0 at wavelength λ, the primary photocurrent produced by the absorption of incident light of optical power P_0 is given by

$$I_p = P_0 \frac{q(1-r)}{h\nu} (1 - e^{-\alpha_0 w}), \tag{18.1}$$

where q is the electronic charge, $h\nu$ is the photon energy ($h\nu = 1.24/\lambda$ eV, where λ is expressed in microns), r is the Fresnel reflection coefficient at the semiconductor–air interface, and w is the width of the absorption region. We have assumed, in the above equation, that the absorption is due to band-to-band transitions only (thus ignoring absorption due to impurity levels and traps). The quantum efficiency η is, therefore,

$$\eta = \frac{\text{(number of carrier pairs generated)}}{\text{(number of incident photons)}}$$
$$= (I_p/q)/(P_0/h\nu)$$
$$= (1 - r)(1 - e^{-\alpha_0 w}). \tag{18.2}$$

The responsivity R, often used to characterize the performance of photodiodes, is defined as

$$R = I_p/P_0 = \eta q / h\nu. \tag{18.3}$$

For an ideal photodiode ($\eta = 1$), $R = \lambda/1.24$ A/W, where λ is expressed in microns.

18.2.2 Photodiode Materials

Since the absorption coefficients of semiconductor materials are strongly dependent on the wavelength, one should choose, ideally, a photodiode material with a bandgap energy that is slightly less than the photon energy corresponding to the longest operating wavelength. Under this condition the absorption coefficient is sufficiently high to ensure good response (in terms of quantum efficiency and speed), and yet the number of thermally generated carriers can be kept small to help achieve low dark currents.

The absorption coefficients α_0 and penetration depths $1/\alpha_0$ of three commonly used semiconductor materials (silicon, germanium and gallium arsenide) are given in Fig. 18.1 for the 0.4- to 1.6-μm spectral region. The quaternary semiconductor compound, indium gallium arsenide phosphide, which is presently used to make lasers and LEDs operating near $\lambda = 1.3$ μm is also included in the figure. It may be seen that, while all

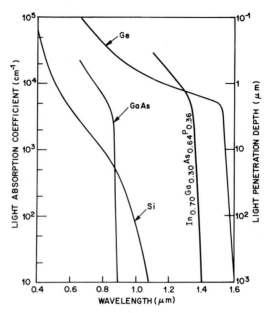

Fig. 18.1 Optical absorption coefficient or penetration depth vs wavelength for silicon, germanium, gallium arsenide, and indium gallium arsenide phosphide.

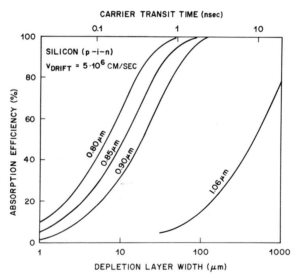

Fig. 18.2 Absorption efficiency versus depletion layer width or carrier transit time for silicon p–i–n photodiode near the absorption band edge. The curves illustrate the tradeoff between speed of response and quantum efficiency at various wavelengths of interest (Melchior, 1973b).

four materials can be used for photodiodes operating at $\lambda = 0.9$ μm, only Ge and InGaAsP (and other III–V alloys such as GaAsSb) are suitable for operation near 1.3 μm. The advantage of InGaAsP and other alloy materials is that their bandgaps can be varied by changing the relative concentrations of their constituents. Germanium photodiodes have larger dark currents because of the narrower bandgap (compared to Si).

In the spectral region from the visible to 1 μm, silicon is the preferred material since silicon technology is highly developed. Being an indirect bandgap material, its absorption coefficient changes gradually with wavelength; this allows silicon photodiodes to be optimized for particular combinations of wavelength-of-operation and speed-of-response by varying the thickness of the depletion layer. The tradeoffs between response speed (carrier transit time) and absorption efficiency for silicon p–i–n (PIN) photodiodes are shown in Fig. 18.2 for the wavelengths of interest for fiber systems (Melchior, 1973b).

18.2.3 Noise Considerations

The ultimate performance of a communication system is usually set by noise fluctuations present at the input to the receiver. Noise degrades the signal and impairs the system performance. In an optical receiver, the essential sources of noise are associated with the detection and amplification processes. Understanding the origin, characteristics, and interplay of the various noise sources is essential to the design and evaluation of any optical communication system.

The relative importance and the interplay of various noise sources in the receiver depend very much on the method of demodulation. The two basic methods of demodulating an optical signal are (1) direct detection, in which the output current of the photodetector is a linear function of the incident optical power, and (2) heterodyne detection, in which the incoming optical signal is mixed with that from a coherent local oscillator to produce a difference frequency from which information is extracted. Heterodyne detection, which is applicable only to single-mode transmission, does not appear to be practical for fiber systems at present because stable single-frequency lasers are required for both carrier generators and local oscillators. Direct detection is simple to implement; the incoming signal can be either incoherent or coherent, and the performance is independent of the polarization state or the modal content. Direct detection is therefore the preferred method of detection for optical-fiber communication systems.

Figure 18.3 depicts the various sources of noise associated with the detection and amplification processes in an optical receiver employing direct detection. The background radiation noise, which is important in an atmospheric propagation system, is negligible in a fiber system. The beat noise, generated in the detector from the various spectral components of

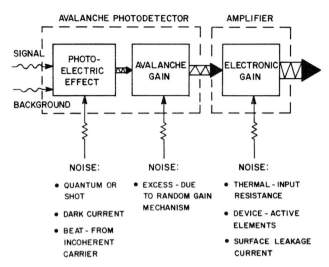

Fig. 18.3 Block diagram showing various noise sources at the front end of an optical receiver.

an incoherent carrier such as that from a LED is expected to be insignificant when a large number of modes is transmitted and received (Hubbard, 1973). The quantum noise, the dark-current noise, and the surface-leakage-current noise all manifest themselves as shot noise which is characterized by Poisson statistics. The dark-current noise and the surface-leakage-current noise can be reduced by careful design and fabrication of the detector and of the devices in the amplifier. The quantum noise, which arises from the intrinsic fluctuations in the photoexcitation of carriers, is fundamental in nature and sets the ultimate limit in the receiver sensitivity. Normally, when the photodiode is without internal avalanche gain, thermal noise arising from the detector load resistance and from the active elements of the electronic amplifier dominate. When internal gain is employed, the relative significance of thermal noise is reduced. However, carrier multiplication or avalanche gain is a random process which introduces excess noise into the receiver. The excess noise manifests itself as increased shot noise above the level that would result from only amplifying the primary shot noise. Despite the excess noise, avalanche multiplication of photocarriers does provide a useful way to improve significantly the sensitivity of optical receivers which otherwise would be limited by the thermal noise of the amplifier.

To illustrate the relative importance and the interplay of the various noise sources, consider the equivalent circuit of the front end of an optical receiver shown in Fig. 18.4a. The modulated optical power $p(t)$ is incident on the avalanche photodetector which has a junction capacitance C_d and a bias or load resistance R_L. The input capacitance and resistance of the

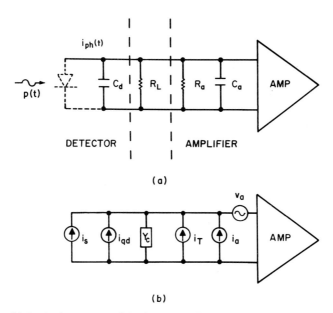

Fig. 18.4 (a) Equivalent circuit of the front end of an optical receiver. (b) Reduced equivalent circuit showing the principal signal and noise sources.

ensuing amplifier are C_a and R_a, respectively. The primary photocurrent generated in the detector

$$i_{ph}(t) = \eta q p(t) / h\nu \qquad (18.4)$$

has a dc (average) component I_p and a signal component i_p. The average value of the signal current after internal avalanche gain is

$$\langle i_s \rangle = i_p \langle m \rangle = i_p M, \qquad (18.5)$$

where m is the statistically varying multiplication factor and the symbol $\langle\ \rangle$ denotes an ensemble average. The dc component I_p determines quantum or shot noise whose mean-square current after avalanche gain is given by

$$\langle i_q^2 \rangle = 2q I_p \langle m^2 \rangle B, \qquad (18.6)$$

where B is the effective noise bandwidth and $\langle m^2 \rangle$ is the mean-square value of the internal gain. The shot noise due to the multiplied part of the dark current I_d behaves in the same way and so can be combined with the quantum noise to give a total shot-noise contribution which is*

$$\langle i_{qd}^2 \rangle = 2q(I_p + I_d)\langle m^2 \rangle B$$
$$= 2q(I_p + I_d)M^2 F(M) B. \qquad (18.7)$$

* "Mean-square current or voltage" is understood and is henceforth omitted when we refer to expressions for noise.

The noise factor $F(M) = \langle m^2 \rangle / M^2$, a measure of the degradation due to the avalanche multiplier compared to an ideal noiseless multiplier, is a function of the statistics of the avalanche multiplication process (McIntyre, 1966, 1972; Personick, 1971a,b; Conradi, 1972; Webb et al., 1974). It increases with the average gain $M = \langle m \rangle$ and is strongly dependent on the photodiode material, construction, and illumination condition. Further discussion of $F(M)$ will be given in Section 18.3.5.

The thermal noise due to the load resistance R_L is given by

$$\langle i_T{}^2 \rangle = 4K\theta B/R_L, \tag{18.8}$$

where K is Boltzmann's constant and θ is the absolute temperature. The noise sources associated with the active elements of the amplifier can be represented by a series voltage noise source $\langle v_a{}^2 \rangle$ and a shunt current noise source $\langle i_a{}^2 \rangle$. The latter includes the thermal noise associated with the input resistance of the amplifier.

The various noise sources previously discussed are shown in Fig. 18.4b, where the signal current i_s and the shunt admittance Y_c which combines the shunt resistances and capacitances of Fig. 18.4a are also shown. The amplifier is assumed to be noiseless at this point. Several important conclusions can be drawn:

(1) The thermal noise $\langle i_T{}^2 \rangle$ can be reduced by using a large load resistance R_L.* The value of R_L need not be limited by the bandwidth consideration that the time constant $R_L C_d$ be less than the reciprocal of the signaling rate, as the signal which is integrated by the large time constant at the receiver front end can be restored by differentiation at a later stage (Personick, 1973; Melchior, 1973a). This high-impedance integrating front-end approach is a commonly recognized practice in the field of nuclear-particle counters for combating thermal noise and increasing receiver sensitivity (Gillespie, 1953); thus less avalanche gain is required to achieve a given sensitivity. In practice, feedback or transimpedance amplifiers are used in optical receivers to overcome the disadvantage of limited dynamic range associated with the simple high-impedance integrate–amplify–differentiate design (Smith et al., 1978). The transimpedance amplifier is a low-noise amplifier with low input impedance. However, the usual stability problems associated with feedback amplifiers make this approach somewhat difficult to implement for large bandwidths ($\geqslant 100$ MHz).

---

* We refer here to the passive impedance which appears at the detector terminals, as distinguished from the active impedance presented at the terminals of a network containing feedback. Within Bell Laboratories credit is due to G. L. Miller for pointing out the advantage of using a high passive impedance (or an integrating amplifier) following the photodiode in the optical receiver.

(2) Referring to Fig. 18.4b and using Eqs. (18.5)–(18.8), we may write the signal-to-noise ratio at the input to the "noiseless" amplifier as

$$\frac{S}{N} = \frac{i_p^2 M^2}{2q(I_p + I_d)M^2 F(M)B + \langle i_{AMP}^2 \rangle + \frac{4K\theta B}{R_L}}, \qquad (18.9)$$

where the noise associated with the amplifier, $\langle i_{AMP}^2 \rangle$, is given by

$$\langle i_{AMP}^2 \rangle = \frac{1}{B} \int_0^B (\langle i_a^2 \rangle + \langle v_a^2 \rangle |Y_c|^2) \, df ; \qquad (18.9a)$$

$\langle i_a^2 \rangle$ and $\langle v_a^2 \rangle$ are assumed to be uncorrelated and f is the frequency. Since M^2 occurs both in the signal and quantum-noise terms and $F(M)$ increases with M, it can be readily seen that there is an optimal value of M that maximizes the signal-to-noise ratio. For th high-impedance front-end design with silicon avalanche photodiod al values of optimal M range from about 25 for a bandwid gahertz to about 100 for a bandwidth of a few gigah

(3) The required opti ceiver sensitivity increased when the sh lue, provided, of course, neither the a se is dominant.

Detailed discussior optical receiver from a systems and 20.

18.3 PRINCIPLES OF S

The discussion in this sectic neration process in a p–n junction, the relationship tion layer width and the junction capacitance, the speed o s determined by transit and diffusion times of the carriers, the avalanche multiplication process—low- and high-frequency gain, and the excess noise associated with the statistical nature of the avalanche gain process.

18.3.1 Carrier Generation in a p–n Junction

When the incident photon has energy greater than or equal to the bandgap of the semiconductor material, electron-hole pairs are generated. In a well-designed photodiode, the photocarrier-generation process occurs mainly in the depletion region of the p–n junction where the incident light is largely absorbed. As a result of the high electric field present in this region, the electrons and holes separate and drift in opposite directions as illustrated in Fig. 18.5. Carriers generated outside, but on the average within a diffusion length of either side of the depletion region, will diffuse inward and be collected across the junction. While the carriers

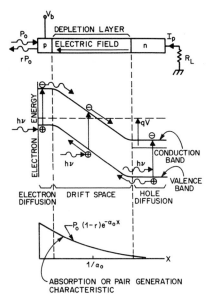

Fig. 18.5 Schematic represenation illustrating the principle of operation of a solid-state photodiode. Cross-sectional view of the p–i–n photodiode, energy band diagram under reverse bias, and optical absorption or carrier-pair generation characteristic are shown (Melchior, 1973a).

drift through the high-field depletion region, a photocurrent I_p is induced in the load, developing a voltage across the load resistor as depicted in Fig. 18.5.

The low-frequency current-voltage characteristic of the photodiode consists of a part associated with the photocurrent and another with the regular diode diffusion current (Shockley, 1949; Prince, 1955):

$$I = I_0(e^{qV/nK\theta} - 1) - I_p, \tag{18.10}$$

where I_0 is the saturation (dark) current in the absence of incident light, n is the "ideality factor" which has a value between 1 and 2 (Sze, 1967), and V is the junction voltage.

Operating in the photovoltaic mode, the photodiode is unbiased and is connected to a very high load impedance. The photogenerated carriers then induce a photovoltage across the junction:

$$V = (nK\theta/q) \log[(I_p/I_0) + 1]. \tag{18.11}$$

At sufficiently low light levels ($I_p \ll I_0$), the output voltage V will vary linearly with the light intensity.

More commonly, the diode is reverse biased so that the exponential term in Eq. (18.10) vanishes, and the output current I varies linearly with the light intensity:

$$I = -(I_0 + I_p). \tag{18.12}$$

Under the condition of reverse bias, the depletion layer widens, thus reducing the junction capacitance and allowing more photocarrier generation in the high-field region; the photodiode then can be operated with high speed and high sensitivity. In the following we limit our discussions to the reverse-biased photodiode.

18.3.2 Depletion Layer Width and Junction Capacitance

The width of the depletion layer at a p–n junction varies with the junction voltage and doping densities and profiles. The expression for the width w, assuming constant doping profiles, is given by (Sze, 1967):

$$w = \left[\frac{2\epsilon}{q}(\phi + V_a)\left(\frac{1}{N_a} + \frac{1}{N_d}\right) \right]^{1/2}, \tag{18.13}$$

where ϵ is the dielectric constant of the semiconductor material, V_a is the applied voltage, ϕ is the built-in voltage, and N_a and N_d are acceptor and donor concentrations, respectively. The value of ϕ depends on the material, and is 0.6 V for silicon and 1.1 V for GaAs. It is seen that w can be increased by having one side of the junction lightly doped (or intrinsic as in a p–i–n structure).

The junction capacitance which varies with the applied voltage is therefore,

$$C_j = \frac{\epsilon A}{w} = A \left[\frac{2}{q\epsilon}(\phi + V_a)\left(\frac{1}{N_a} + \frac{1}{N_d}\right) \right]^{-1/2} \tag{18.14}$$

where A is the diode junction area. Thus both low doping and high reverse voltage reduce the junction capacitance.

18.3.3 Speed of Response—Transit Time and Diffusion Time

The speed of response of a photodiode is ultimately limited by the time the photogenerated carriers take to sweep across the depletion region. The electric field in the depletion region is usually greater than 2×10^4 V/cm, at which the carriers attain the scattering-limited (saturated) velocity of $\sim 10^7$ cm/sec (in silicon). Hence for $w = 10 \ \mu$m, the transit time can be as short as 0.1 nsec.

In order to achieve high quantum efficiency the depletion-layer width w must be wider than $1/\alpha_0$. Also, if w is not sufficiently wide, the photocarriers generated beyond the depletion region would have to diffuse back into that region before they could be collected. Since the carrier diffusion times can be quite long (for example, the hole diffusion time through 10

μm of silicon is 40 nsec), it is important, not only from the consideration of high quantum efficiency but also of high speed, to have a diode structure with a sufficiently wide depletion layer. The tradeoffs between speed of response and absorption efficiency for silicon p–i–n photodiodes in the wavelength regions of interest for fiber systems are shown in Fig. 18.2.

From the above discussion, it is clear that the speed of response of a photodiode combined with its output circuit is dependent on the following three parameters: (i) the RC time constant of the output circuit (including the photodiode capacitance), (ii) the diffusion time of the photocarriers generated outside the depletion region, and (iii) the transit time of the photocarriers in the depletion region.

18.3.4 Current Gain—Avalanche Multiplication

When the electric field in the depletion region of a reverse-biased diode is sufficiently high (above 10^5 V/cm for Si), an electron or a hole can collide with a bound (valence) electron with sufficient energy to cause ionization, thereby creating an extra electron-hole pair. The additional carriers in turn can gain enough energy from the field to cause further impact ionization, until an avalanche of carriers has been produced (McKay and McAfee, 1953). On the average, the total number of carrier pairs created is finite and is proportional to the number of injected (primary) carriers when the diode is biased below a certain (breakdown) voltage. Very high carrier multiplication or current gain is possible through this avalanche process even at microwave frequencies (Batdorf et al., 1960; Anderson et al., 1965; Johnson, 1965; Melchior and Lynch, 1966). The ionization rate, which is the average number of electron-hole pairs created by a carrier (electron or hole) per unit distance travelled, is a strong function of the electric field. The results of the measurements of the ionization rates in several semiconductor materials are summarized in Fig. 18.6 for Si (Lee et al., 1964), Ge (Miller, 1955; Logan and Sze, 1966), GaAs (Logan and Sze, 1966; Stillman et al., 1974), $In_{0.14}Ga_{0.86}As$ (Pearsall et al., 1975; Lee et al., 1975) and $GaAs_{0.88}Sb_{0.12}$ (Pearsall et al., 1976). The measured values for silicon (Lee et al., 1964) are in good agreement with theory (Baraff, 1962). For other materials, especially III–V compounds, the measurement of ionization rates has been hindered by the lack of materials with microplasma-free junctions and of devices with good surface passivation.

It has been shown theoretically that an avalanche photodiode will have low noise and large gain-bandwidth product if only one type of carrier is capable of causing impact ionization (McIntyre, 1966; Emmons, 1967). Of the materials studied to date, only silicon exhibits a large difference between the ionization rates of holes and electrons, especially in the region of low-field intensities (see Fig. 18.6). Recently, the impact ioniza-

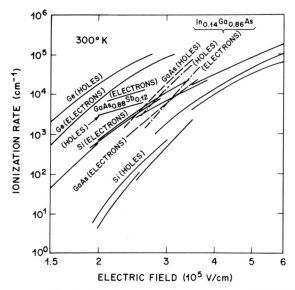

Fig. 18.6 Experimentally obtained carrier ionization rates at 300°K for silicon, germanium, gallium arsenide, gallium arsenide antimonide, and indium gallium arsenide (Melchior, 1977).

tion rates in GaAs have been found to depend on specific features of the electronic band structure of the material (Pearsall *et al.*, 1977; Capasso *et al.*, 1977). This finding will probably prompt further theoretical and experimental work on the phenomenon of impact ionization in semiconductors.

The *low-frequency gain* of the avalanche photodiode depends on the ionization rates of the carriers and on the width of the multiplication region, all of which vary with the reverse bias voltage. It is convenient to describe the low-frequency or dc multiplication factor M_0 by the empirical relation (Miller, 1955):

$$M_0 = \frac{I^M}{I^{\pi \delta}} = \frac{1}{1 - (V_j/V_B)^n} \qquad (18.15)$$

where I_M is the total multiplied current, I_{pd} is the total primary current, V_j is the effective junction voltage, V_B is the breakdown voltage at which $M_0 \to \infty$, and n is a fitting factor dependent on diode material and structure. Operation of the diode at or above V_B is undesirable because the self-sustained avalanche current tends to reduce diode sensitivity. Nevertheless, maximum gains of 100–10,000 have been observed in Si, Ge, and GaAs avalanche photodiodes (Anderson *et al.*, 1965; Melchior and Lynch, 1966; Lindley *et al.*, 1969).

At large carrier multiplication, the voltage drop due to the diode series resistance and load resistance must be taken into account, and Eq. (18.15) is then rewritten as (Melchior and Lynch, 1966):

$$M_0 = \frac{I_M}{I_{pd}} = \frac{1}{1 - \left(\frac{V_a - I_M R_M}{V_B}\right)^n},$$

(18.16)

where V_a is the applied bias voltage and $I_M R_M$ is the voltage drop. For $I_M R_M \ll V_B$ Eq. (18.16) can be approximated by:

$$M_0 = I_M/I_{pd} = V_B/n I_M R_M,$$

(18.17)

from which the maximum value of the multiplication factor is derived:

$$M_0(\text{max}) = (V_B/n R_M I_{pd})^{1/2}.$$

(18.18)

Since I_{pd} includes both the photogenerated (primary) current and the thermally excited (primary) dark current, it can be seen that the primary dark current will set a limit to the value of the maximum gain achievable. In fact, the large dark current in Ge avalanche photodiodes is responsible for the lower value of the maximum gain ($M \sim 200$) observed (Melchior and Lynch, 1966), in contrast to the much larger values ($M > 10^4$) observed for Si diodes (Goetzberger et al., 1963). Of course, by cooling the Ge diode it is possible to increase its maximum gain obtainable.

In general, the *high-frequency behavior or the bandwidth* of the avalanche diode is dependent on the carrier transit times in the high-field avalanche region (τ_n = electron transit time, τ_p = hole transit time) and on the carrier ionization rates (α = electron ionization rate, β = hole ionization rate). Figure 18.7 illustrates the avalanche of carriers initiated by electron injection in a high-field region of width W; part (a) shows the case where only one carrier type (electrons as shown) participates in impact ionization and part (b) shows the case in which both electrons and holes participate. In case (a), the injected electron produced $e^{\alpha W}$ electrons which drift toward the n-layer contact, while the holes created by ionization drift back toward the p-layer without undergoing ionizing collisions ($\beta = 0$). The induced current in the external load reaches a maximum when the last electron arrives at the n-layer contact, and the load current continues to flow until the last hole arrives at the p-layer contact. The time of response of the diode is therefore $\tau_{av} = (\tau_n + \tau_p)/2$, and is independent of the gain. Even when $\beta \neq 0$, so long as the probability of electron generation by holes is less than that by electrons, (i.e., if $\beta W e^{\alpha W} < \alpha W$, or $e^{\alpha W} = M_0 < \alpha/\beta$) the above basic transit time limitation still holds approximately (Emmons, 1967).

In case (b) where the probability of impact ionization by holes is comparable with that by electrons, the regenerative avalanche process results in the presence of a large number of carriers in the high-field region long after the primary electrons have traversed through that region. The higher

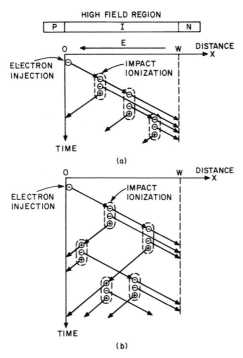

Fig. 18.7 Schematic diagrams showing impact ionization by carriers in a uniform high-field region of a semiconductor (a) only electrons undergo ionizing collisions; (b) both carriers undergo ionizing collisions (Melchior, 1972).

the multiplication is, the longer the avalanche process persists, thus implying a behavior that is set by a gain-bandwidth product.

Figure 18.8 shows the calculated bandwidth for an idealized p–i–n avalanche photodiode having an avalanche region of uniform electric field in which the drift velocities of electrons and holes are assumed to be equal (Emmons, 1967). The 3-dB bandwidth B, normalized to 2π times the average transit time τ_{av}, is plotted as a function of the low-frequency gain M_0 with the carrier-ionization-rate ratio α/β as a parameter. The dashed curve is for $M_0 = \alpha/\beta$. Above this curve where $M_0 < \alpha/\beta$, the bandwidth is largely determined by the transit time of the carriers and is approximately independent of gain. Below this curve where $M_0 > \alpha/\beta$, the curves are almost straight lines, indicating a constant gain-bandwidth product. The high-frequency multiplication factor can be approximated by

$$M(\omega) = \frac{M_0}{[1 + (\omega M_0 \tau_{eff})^2]^{1/2}}, \tag{18.19}$$

where the effective transit time τ_{eff} is approximately equal to $N(\beta/\alpha)\tau_{av}$, N

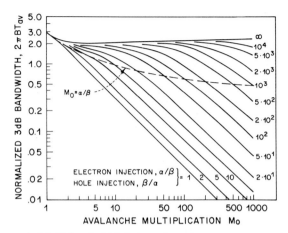

Fig. 18.8 Theoretical 3-dB bandwidth B times $2\pi\tau_{av}$ (τ_{av} = average carrier transit time) of an avalanche photodiode plotted as a function of the low-frequency multiplication factor M_0, for various values of α/β for electron injection (or β/α for hole injection). Above the dashed curve ($M_0 = \alpha/\beta$) bandwidth is nearly independent of gain; below the dashed curve a constant gain bandwidth product applies (Emmons, 1967).

is a number slowly varying from $N = \frac{1}{3}$ at $\alpha/\beta = 1$ to $N = 2$ at $\alpha/\beta = 1000$, and ω is 2π times the frequency. Equation (18.19) implies a constant gain-bandwidth product:

$$M_0 B = (\alpha/\beta)/N\tau_{av}, \qquad M_0 > \alpha/\beta \qquad (18.20)$$

In practice, τ_{eff} is also structure-dependent and this effect can be taken into account by letting β/α be k_{eff}, defined in the next section (Kaneda et al., 1976b; Goedbloed, 1977).

It is significant to note that the condition that maximizes the bandwidth, (that is, only one type of carrier causes impact ionization, or $M_0 \ll \alpha/\beta$), also minimizes the excess noise produced by the mulitiplication process (as will be discussed in the following section). In addition, because α and β are strongly dependent on the electric field intensity, the gain (M_0) is also dependent on the field intensity. The rate of increase of M_0 near the breakdown field is a function of the ratio β/α and becomes precipitously large as β/α approaches one (Webb et al., 1974). This implies that, in practice, it will be difficult to achieve uniform high gain in avalanche photodiodes made with materials in which α and β are nearly equal (such as germanium), since local variations in the electric field caused by small inhomogenieties in the doping level could lead to very large variations in gain.

18.3.5 Avalanche Multiplication Noise

As mentioned in Section 18.2.3 the avalanche process is statistical in nature because not every carrier-pair generated at a given distance x experiences the same multiplication. The statistics of the gain fluctuations are rather complicated but amenable to approximate analyses (Personick, 1971a,b; McIntyre, 1972; Conradi, 1972). Because the gain fluctuates, the mean-square value of the gain is greater than the square of the mean; the excess noise can be characterized by a noise factor $F(M) = \langle m^2 \rangle / M^2$, where $\langle\ \rangle$ denotes an ensemble average and $M = \langle m \rangle$, as defined previously. Therefore, the mean-square quantum noise current after multiplication is, as given in Section 18.2.3,

$$\langle i^2_{qd} \rangle = 2q(I_p + I_d)M^2F(M)B. \tag{18.7}$$

The excess noise factor is dependent on the ratio of the ionization rates, α/β, and on the dc multiplication factor M_0 (McIntyre, 1966). When both carrier species produce impact ionization (Fig. 18.7b), a small statistical variation in the regenerative avalanche process can cause a much larger fluctuation in gain, as compared to the case in which only one type of carrier ionizes (Fig. 18.7a). Thus either $\alpha = 0$ or $\beta = 0$ minimizes noise.

Expressions for the excess noise factor have been derived for both electron injection and hole injection (McIntyre, 1966, 1972). For electron injection alone, the simplified expression for F can be written as

$$F = kM_0 + (2 - M_0^{-1})(1 - k), \tag{18.21}$$

where $k = \beta/\alpha$ is assumed to be constant throughout the avalanche region. For hole injection alone, the above expression for F still applies if k is replaced by $k' = \alpha/\beta$.

Of particular interest are two special cases: (i) only electrons cause ionizing collisions, i.e., $\beta = 0$, and (ii) both carrier species ionize, with $\alpha = \beta$. In case (i), $F = 2$ for large M_0, and in case (ii), $F = M_0$.

In practical avalanche photodiodes the electric field in the avalanche region (of width W) is not uniform; therefore, the impact ionization rates of the carriers must be weighted accordingly to give (McIntyre, 1972)

$$k_1 = \int_0^W \beta(x)M(x)\,dx \bigg/ \int_0^W \alpha(x)M(x)\,dx \tag{18.22}$$

and

$$k_2 = \int_0^W \beta(x)M^2(x)\,dx \bigg/ \int_0^W \alpha(x)M^2(x)\,dx. \tag{18.23}$$

The excess noise factors for electron injection and hole injection become (McIntyre, 1972; Webb et al., 1974),

$$F_e = k_{eff}M_e + (2 - M_e^{-1})(1 - k_{eff}) \tag{18.24}$$

and

$$F_h = k'_{eff}M_h - (2 - M_h^{-1})(k'_{eff} - 1), \tag{18.25}$$

where $k_{eff} \simeq k_2$, $k'_{eff} = k_2/k_1^2$, and the subscripts e and h denote electron and hole, respectively.

Figure 18.9a shows F_e as a function of M_e with k_{eff} as a parameter. It is seen again that a small value of k_{eff} is desirable to minimize excess noise.

When light is absorbed on both sides of the junction so that both electrons and holes are injected into the avalanche region additional noise is introduced (Webb *et al.*, 1974; Nishida, 1977); the effective noise factor is given by (Webb *et al.*, 1974)

$$F_{eff} = \frac{fM_e^2 F_e + (1 - f)M_h^2 F_h}{[fM_e + (1 - f)M_h]^2},$$

(18.26)

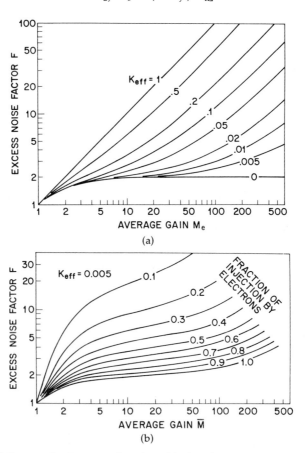

Fig. 18.9 (a) Excess noise factor as a function of the low-frequency gain for electrons, M_e, with k_{eff}, the ratio of the weighted ionization rates, as a parameter (Webb *et al.*, 1974) (b) Excess noise factor versus average gain $\overline{M}$ for mixed injection with $k_{eff} = 0.005$. The plot illustrates the importance of initiating the avalanche process by the carrier species with the higher ionization rate (Webb *et al.*, 1974).

where the injection ratio $f = I_{n0}/(I_{p0} + I_{n0})$, and I_{n0} and I_{p0} are the injected electron and hole currents, respectively. The effective excess noise factor F_{eff} for a diode having a k_{eff} of 0.005 is given in Fig. 18.9b as a function of the average gain, $\overline{M} = fM_e + (1 - f)M_h$, with the injection ratio f as a parameter. It clearly shows that the excess noise factor is both material- and structure-dependent. Hence in order to attain low noise, not only k_{eff} must be small, but also the incident light must be absorbed on the correct side of the junction.

The above discussions on avalanche photodiodes may be summed up as follows: to achieve low-noise and wide-bandwidth in an avalanche photodiode, it is necessary that the impact ionization rates of the carriers be as different as possible and that the avalanche process be initiated by the carrier species with the higher ionization rate.

18.4 STATE OF THE ART OF PHOTODIODES—MATERIALS, STRUCTURES, AND PERFORMANCE

We discuss in this section the state of the art of photodiodes for optical-fiber communication. Practical aspects and research results relating to photodetector materials, structures and performances are reviewed for various photodiodes operating in the spectral regions near 0.85, 1.06, and 1.3 μm.

18.4.1 Materials—Silicon, Germanium, and III–V Alloys

For application in the wavelength range from the visible to about 1 μm, silicon is the preferred photodiode material; it offers the desirable properties of a suitable range of absorption coefficients ($\alpha_0 \simeq 10^4$ cm^{-1} at $\lambda = 0.5$ μm to $\alpha_0 \simeq 10^2$ cm^{-1} at $\lambda = 1.0$ μm) and a large ratio of carrier ionization rates ($\alpha/\beta > 10$), thus allowing avalanche photodiodes to be made with high sensitivity, wide bandwidth, and low noise. In addition, silicon technology is already highly developed. Silicon PIN and avalanche photodiodes exhibiting good performance for operation with AlGaAs sources at $\lambda \simeq 0.85$ μm are now commercially available.

The absorption edge of germanium is near 1.6 μm at room temperature, so its absorption coefficient is large ($\gtrsim 10^4$ cm^{-1}) over the entire wavelength range of interest for optical-fiber applications. Naturally germanium is a candidate material for detectors at wavelengths beyond 1 μm where the response of silicon falls off. However, the measured value of the ratio of the carrier ionization rates is only about 2 (see Fig. 18.6), implying a relatively high excess noise factor for avalanche multiplication. ($F \simeq M/2$). In addition, because of the narrower bandgap, the bulk (diffusion-dependent) dark current that undergoes multiplication is expected to be much higher than in silicon. The dark-current problem is further aggravated by the lack of a well-developed surface-passivation technology for the material; thus surface leakage currents tend to be very high and unstable, and play an important role. These shortcomings notwithstanding, devices with usable sensitivity and fast response have been made and are available for experimental studies. Further efforts are required to develop a practical germanium device with good reliability.

Various III–V semiconductor alloys, which are being studied for use as optical-source materials, are also under active investigation for use as detector materials for longer wavelength applications. An attractive feature of these alloys is that their bandgaps depend upon composition, hence it is possible to optimize detector performance by choosing a composition which places the absorption edge just above the wavelength of operation, thereby ensuring high quantum efficiency and speed of response with low dark current. A basic reason for investigating these materials, of course, is the hope of finding one with a large difference of carrier ionization rates. However, the work associated with such a search is both tedious and difficult, for only after the formidable problems of material growth, and device processing and fabrication are overcome can meaningful measurements of ionization rates and avalanche noise be made. (The measured devices must have spatially uniform carrier multiplication that is relatively free of microplasmas, edge breakdowns, and excessive leakage currents.) Despite the difficulties, some preliminary data on ionization rates in GaAsSb and in InGaAs have been obtained, as presented in Fig. 18.6. The measured results indicate that the ratios of ionization rates in these alloy materials are not appreciably different from that of germanium. However, a recent preliminary measurement of α/β in InGaAsP has yielded a value of 3–4 (Ito et al., 1978). Other studies have shown that impact ionization rates in GaAs are dependent on specific features of the electronic band structure (Pearsall et al., 1977; Capasso et al., 1977). These findings encourage further measurements and studies which are currently underway.

Experimental photodiodes have been made using various III–V alloys, including GaAs, AlGaAs, InGaAs, GaAsSb, GaAlSb, and InGaAsP. The problem of surface passivation has yet to be solved. Considerable research interest at present is focused on ternary- and quaternary-alloy materials and photodiodes.

18.4.2 Photodiodes without Gain—Silicon and Germanium
p–n and p–i–n Photodiodes

Photodiodes without gain have the simple form of p–n or p–i–n layered structures. The junction is formed by a thin p$^+$ diffusion into a n (for p–n) or ν (for p–i–n) layer through a window in a protective SiO$_2$ film, as shown in Fig. 18.10a. The depletion-layer width in a p–n junction diode is about 1 to 3 μm and is therefore optimized for efficient detection of light in the visible wavelength range for silicon devices and near infrared for germanium devices. At shorter wavelengths (e.g., <0.5 μm for Si), the incident light is absorbed close to the surface; recombination of photocarriers at the surface and in the shallow, heavily doped p^+ or n^+ layer results in poor quantum efficiency. Thus metal-semiconductor Schottky barrier photodiodes with very thin, antireflection-coated, semitransparent metal layers are more attractive (Schneider, 1966).

At longer wavelengths where light penetrates more deeply into the semiconductor material, front-illuminated p–i–n and Schottky barrier photodiodes with wide depletion regions are preferred. Silicon p–i–n photodiodes for operation in the 0.8- to 0.9-μm wavelength range, where AlGaAs sources emit, require depletion regions as wide as 20–50 μm to attain high quantum efficiency; response times as short as 1 nsec have been achieved with such widths (Conradi and Webb, 1975; Melchior and Hartman, 1976). Various silicon p–i–n photodiodes with high quantum efficiency (>70% at $\lambda = 0.85$ μm) and fast response ($\lesssim$ few nsec) are available commercially today.

At $\lambda = 1.06$ μm, the required absorption width in silicon is 500 μm or more, which severely limits the bandwidth of a front-illuminated device. A compromise between quantum efficiency and bandwidth can be reached if the light is incident on the edge of and injected parallel to the plane of the junction, as shown in Fig. 18.10b. An experimental side-

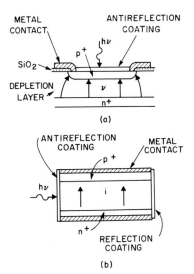

Fig. 18.10 Construction of photodiodes without gain: (a) Front-illuminated silicon p–i–n photodiode. (b) Side-illuminated p–i–n photodiode (Melchior, 1973a).

illuminated Si photodiode was observed to have a quantum efficiency of 90% at 1.0 μm and a bandwidth in the gigahertz range (Krumpholz and Maslowski, 1968). One of the disadvantages of side-illumination is the difficulty of coupling light into the relatively narrow depletion region. In an attempt to overcome this difficulty a front-illuminated silicon p–i–n photodiode employing multiple reflections within the diode has been devised and built (Müller, 1978). The device exhibited high quantum efficiency over a wide spectral range ($\eta \sim 85\%$ over 0.45–0.9 μm, and 10% at 1.06 μm), fast response (~ 100 psec), and relatively low dark current (10^{-9} A).

As mentioned before, the response of germanium photodiodes spans the entire wavelength range of interest for optical-fiber communication, but the relatively high dark-current remains a problem. Experimental p–i–n diodes with good response and large bandwidth have been built (Riesz, 1962; Conradi, 1975); both p–n and p–i–n devices are commercially available.

18.4.3 Photodiodes with Internal Gain—Silicon and Germanium Avalanche Photodiodes

The construction of an avalanche photodiode is much more sophisticated than that of a simple p–i–n photodiode (see Figs. 18.11 and 18.12). In addition to a depleted drift region where most of the primary carrier-pairs are generated, the avalanche device has a high-field region where carrier multiplication takes place. To achieve high gain, special precaution must be taken to ensure uniformity of carrier multiplication over the entire light sensitive area; microplasmas—small areas with lower break-down voltages—and excessive leakage and breakdowns at the junction edges must be avoided. In practice, microplasma-free devices are obtained through selection of defect-free materials and care in device processing and fabrication, while edge leakage and breakdowns are reduced by the use of guard-ring structures.

In order to take full advantage of the possibility offered by silicon for carrier multiplication with very little excess noise, a "reach-through" structure was proposed and implemented (Ruegg, 1967; Webb *et al.*, 1974; Conradi and Webb, 1975; Berchtold *et al.*, 1975; Melchior and Hartman, 1976; Kaneda *et al.*, 1976a; Kanbe *et al.*, 1976; Nishida *et al.*, 1977). The "reach-through" structure is composed of $p^+–\pi–p–n^+$ layers as illustrated in Fig. 18.11a. The high-field $p–n^+$ junction, where electron-initiated avalanche multiplication takes place, is formed by diffusion or ion-implantation with precise doping concentration. Under low reverse bias, most of the voltage is dropped across the $p–n^+$ junction. As the bias is increased, the depletion layer widens predominantly into the p region and, at a certain voltage V_{rt} below the breakdown voltage of the $p–n^+$ junc-

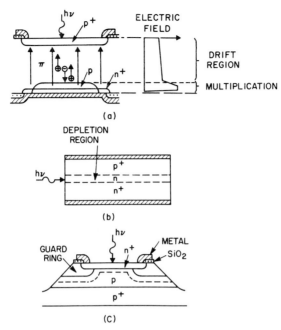

Fig. 18.11 Construction of avalanche photodiodes with internal gain: (a) Front-illuminated silicon $p^+-\pi-p-n^+$ reach-through structure—the electric fields in the drift and multiplication regions are shown on the side. (b) Side-illuminated silicon p^+-n-n^+ structure. (c) Front-illuminated germanium n^+-p mesa structure with guard ring (Melchior, 1973a).

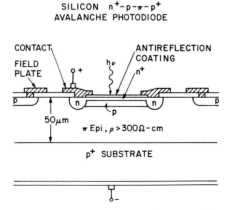

Fig. 18.12 Cross-sectional view of an epitaxial silicon reach-through avalnche photodiode (RAPD) with illumination through the n^+ contact layer of the $n^+-p-\pi-p^+$ structure. Diameter of the light-sensitive high-gain region is 100 μm (Melchior and Hartman, 1976).

tion, *reaches through* to the nearly-intrinsic π region. The applied voltage in excess of V_{rt} is dropped across the entire π region. Since the π region is much wider than the p region, the field in the multiplication region and therefore the multiplication factor will increase relatively slowly with increasing voltage above V_{rt}. In the operating range the field in the π region is substantially lower than that in the $p–n^+$ junction, but is high enough to maintain limiting carrier velocities, thus assuring a fast speed of response.

Because the light incident on the p^+ surface is almost completely absorbed in the π region, a relatively pure electron current is injected into the high-field $p–n^+$ junction where carrier multiplication takes place. Thus a nearly ideal situation is obtained where the carrier type with the higher ionization rate initiates the multiplication process, resulting in current gain with very little noise. An excess noise factor of 4 at a gain of 100 (corresponding to $k_{eff} = 0.02$) has been observed in a silicon reach-through avalanche photodiode (RAPD) produced for operation in the 0.8- to 0.9-μm wavelength range (Conradi and Webb, 1975). In order to take advantage of the larger difference in carrier ionization rates at lower fields in silicon (see Fig. 18.6), a device has been made in which the $p–n^+$ avalanche region is formed by a combination of ion implantation and epitaxy; extremely low excess noise corresponding to $k_{eff} \simeq 0.008$ to 0.014 has been obtained (Goedbloed and Smeets, 1978).

The structure shown in Fig. 18.12 is more amenable to fabrication on large silicon wafers with good control of the doping profile. The inverted construction consisting of $n^+–p–\pi–p^+$ layers is formed by ion implantation and diffusion on a π-type epitaxial material grown on a p^+ silicon substrate (Melchior and Hartman, 1976). The incident light now enters the structure through the n^+ contacting layer on the surface of the epitaxial material. The increase in excess noise caused by the resulting generation and mixed-injection of both carrier types into the avalanche region can be minimized by tailoring the field profile so that electrons injected into the region encounter a lower field than the holes generated within and in front of the region; thus fluctuations due to the *combined* electron- and hole-initiated multiplication processes are minimized. The observed excess noise factor for such a device is 5 at an average gain of 100, which compares very favorably with the value of 4 obtained under the nearly ideal light-injection conditions discussed earlier (Melchior and Hartman, 1976; Conradi and Webb, 1975).

Details of the construction of an inverted silicon RAPD designed specifically for optical-fiber application at $\lambda = 0.825$ μm are illustrated in Fig. 18.12 (Melchior and Hartman, 1976; Hartman *et al.*, 1978).[*] The lightly

---

[*] The following summary on this device is abstracted from the paper by Melchior and Hartman (1976).

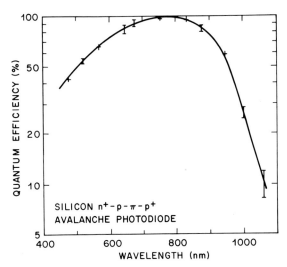

Fig. 18.13 Measured quantum efficiency of the silicon n^+–p–π–p^+ RAPD vs wavelength (Hartman *et al.*, 1978).

doped guard ring, formed by diffusion, eliminates breakdown around the periphery of the shallow p–n^+ junction, while the diffused channel stop surrounding the device prevents surface inversion and keeps the leakage current low. A precise dosage of boron is implanted into the center region and diffused to form the high-field avalanche region. A layer of Si_3N_4 is deposited to passivate the structure and to serve as antireflection coating. The field plates, which overlap the n–π and π–p metallurgical junctions, prevent surface charge accumulation which can ultimately lead to increased leakage currents and reduced breakdown voltages.

A curve of the measured quantum-efficiency (without avalanche gain) of these devices from visible wavelengths to 1.06 μm is shown in Fig. 18.13; near 0.8 μm the efficiency is close to 100%. The current gain as a function of reverse bias voltage at different temperatures is given in Fig. 18.14 for illumination at $\lambda = 0.825$ μm. In the range of bias voltage shown, the onset of gain occurs at -60 V, and avalanche breakdown occurs from about -250 to -400 V. The response has been measured using a 0.22-nsec laser pulse at $\lambda = 0.838$ μm. The observed duration (FWHM*) of the multiplied output current pulses is about 10 nsec at the onset of gain, but decreases very rapidly as the π region is depleted (complete depletion occurs at -100 V), to slightly less than 1 nsec at high gain ($M = 100$).

Avalanche carrier multiplication has been observed to be uniform over the center part of the light-sensitive area of the diodes. (At a gain of 100 the gain is uniform within $\pm 10\%$ over a diameter of 80 to 100 μm.) The

* Full-width at half-maximum.

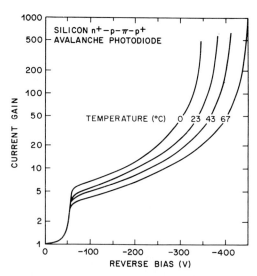

Fig. 18.14 Current-gain versus reverse bias characteristic of the silicon $n^+-p-\pi-p^+$ RAPD at different ambient temperatures measured with illumination at $\lambda = 0.825$ μm (Melchior and Hartman, 1976).

dark currents at room temperature are in the low 10^{-11} A range and depend only slightly on bias voltage. The component that is generated in the bulk and is multiplied is estimated to be in the low 10^{-13} A range. The measured variation of the excess noise factor with current gain at $\lambda = 0.8$ μm is presented in Fig. 18.15. The excess noise also varies with wavelength and is less at longer wavelengths (due to increased injection ratio for electrons).

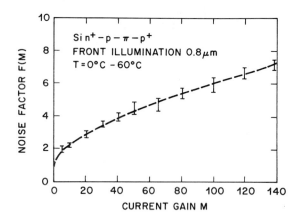

Fig. 18.15 Excess noise factor $F(M)$ of the silicon $n^+-p-\pi-p^+$ RAPD versus low-frequency current gain M for illumination at $\lambda = 0.8$ μm. The excess noise is slightly less at $\lambda = 0.9$ μm (Hartman et al., 1978).

As shown in Fig. 18.14, the current gain of the avalanche diode varies considerably with temperature, especially at high gain (Conradi, 1974). For practical operation stabilization of gain or compensation for gain variations is necessary. Many schemes are possible. A method which incorporates the required compensation in the automatic-gain-control feedback loop has been implemented in a 45 Mbit/sec optical repeater (Smith et al., 1978).

In an effort to increase the speed of response of a $n^+-p-\pi-p^+$ reach-through avalanche photodiode without increasing the operating voltage, a built-in field in the light-absorption region was provided by grading the doping profile so as to increase the carrier velocities (Kanbe et al., 1976). Another scheme involved the use of a thin-film structure with a highly reflecting back contact to increase the internal quantum efficiency of a relatively narrow depletion region (Müller and Ataman, 1976). Response times of 150–200 psec have been achieved in both cases.

For certain optical-fiber applications an array of closely packed photodiodes may be desirable. A monolithic linear array of 10 silicon avalanche photodiodes fabricated on a single chip displayed gain variations of $\pm 2\%$ from diode to diode and an isolation of -57 dB between adjacent diodes (Takahashi et al., 1977). Another array consisting of PIN photodiodes coupled to optical waveguides integrated on a single silicon chip was also demonstrated (Boyd and Chen, 1976). Using evanescent-wave coupling between waveguides and diodes, device quantum efficiencies of $\sim 80\%$ were obtained at $\lambda = 0.633\ \mu m$.

A side-illuminated silicon avalanche photodiode exhibiting large bandwidth and good response in the wavelength region from 0.4 to 1.0 μm was constructed (Maslowski, 1972). The p^+-n-n^+ structure, illustrated in Fig. 18.11b, was 300 μm long and yielded a quantum efficiency of 70%, a bandwidth of a few gigahertz and a maximum (dc) multiplication of 3000 at $\lambda = 0.9\ \mu m$.

While silicon avalanche photodiodes have been well-developed for optical-fiber applications, the development of germanium photodiodes has been hampered mainly by the lack of a good passivation technique. Nonetheless, experimental devices exhibiting very useful performance characteristics over a wide spectral range (0.8–1.65 μm) have been built. Figure 18.11c illustrates the construction of a germanium n^+-p avalanche photodiode with a low-frequency current multiplication of over 200 and a 6-GHz small-signal gain of greater than 10 at $\lambda = 1.15\ \mu m$ (Melchior and Lynch, 1966). The observed dark current was 20 nA at -16 V. A more recent device, passivated with a SiO_2 film, showed a dark current of 250 nA at a bias of -10 V (Shibata et al., 1974). Another germanium avalanche photodiode designed specifically for optical-fiber applications exhibited a gain-bandwidth product more than 5 GHz and a total dark current of ~ 200

TABLE 18.1 Avalanche Photodiodes (Si and Ge)

Diode type	Wavelength range (μm)	Light-sensitive area (mm²)	Max. current gain	Gain bandwidth product (GHz)	Excess noise factor F (M = 100)	k_{eff}	Breakdown voltage V_B (V)	Capacitance (pf)	Dark current[a] (nA)	Reference
Si n⁺-p	0.4–0.8	2×10^{-3}	10^4	100	10		23	0.8	0.05	Anderson et al. (1965)
Si RAPD (p⁺-π-p-n⁺)	0.8–0.9	4×10^{-2}	200	>100	4	0.016–0.018	200–500	2	50 (0.1)	Webb et al. (1974)
Si RAPD (n⁺-p-π-p⁺)	0.8–0.9	7.8×10^{-3}	>10³	>100	5	0.02	250–400	0.2	0.02 (0.2 pA)	Melchior and Hartman (1976)
Si RAPD (n⁺-p-π-p⁺)	0.8–0.9	7×10^{-2}	>100		6	0.032	150		2×10^{-10}	Kaneda et al. (1976a)
Si RAPD (n⁺-p-π-p⁺)	0.8–0.9		10^4	250			300	<1		Berchtold et al. (1975)
Si RAPD (n⁺-p-π-p⁺)	0.8–0.9	1.6×10^{-2}	400	150–460		0.03–0.05	150	1–1.5		Kanbe et al. (1976)
Si RAPD (n⁺-p-π-p⁺)	0.8–0.9	5.5×10^{-2}	>100		7.5		140–200			Nishida et al. (1977)
Si RAPD (n⁺-p-π-p⁺)	0.8–0.9	9×10^{-2}	>100			0.008–0.014	215			Goedbloed and Smeets (1978)
Ge n⁺-p	0.6–1.65	2×10^{-5}	200	60	10 (M = 10)		16.3	0.8	10@ −10 V	Melchior and Lynch (1966)
Ge n⁺-p	0.9–1.65	3×10^{-3}	100	>2	10 (M = 10)		28.5	1.8	250@ −10 V	Shibata et al. (1974)
Ge n⁺-p	0.9–1.65	7.8×10^{-3}	100	>5	100 (λ = 1.32 μm)	0.6–1.0	23–33	1.8	200@ 0.9 V_B (80)	Ando et al. (1978)

[a] Values in parentheses indicate the portions of the dark current that are multiplied.

nA at $0.9 \times$ breakdown voltage (about half of which is the multiplied component flowing through the n^+-p junction) (Ando *et al.*, 1978).

Salient characteristics of various Si and Ge avalanche photodiodes discussed above are summarized in Table 18.1.

18.4.4 Photodiodes of III–V Alloys

Work on photodiodes of III–V alloy materials is at a very early stage of research and development. Although experimental devices with useful performance characteristics have been produced, present efforts are directed more toward the search for materials with large differences of carrier ionization rates. As discussed earlier, the work is tedious and difficult, and lacks guidance from theory.

An experimental Pt-GaAs Schottky barrier avalanche photodiode exhibited a gain-bandwidth product greater than 50 GHz ($M = 100$) near $\lambda \approx 0.85$ μm (Lindley *et al.*, 1969). Another Schottky barrier avalanche photodiode, fabricated from epitaxially grown InGaAs on GaAs substrate gave a gain greater than 250, a rise time less than 200 psec and good quantum efficiency at $\lambda = 1.06$ μm (Stillman *et al.*, 1974).

Grown heterojunction avalanche photodiodes of various III–V alloys have been made and studied. A GaAsSb/GaAs avalanche photodiode was developed for application in a gigabit-data-rate laser communication system at 1.06 μm (Eden, 1975; Eden *et al.*, 1975). A later version of the device achieved an average gain greater than 14 with a leakage current of 150 nA (Scholl *et al.*, 1976). More recently, heterojunction photodiodes have been made from the closely lattice-matched GaAlSb/GaSb ternary material system covering the wavelength range of interest from 1.1 to 1.5 μm (Law *et al.*, 1978; Sukegawa *et al.*, 1978; Tomasetta *et al.*, 1978). The avalanche photodiode displayed a quantum efficiency greater than 50%, a high-frequency avalanche gain of 10–15, and a pulse rise time of 60 ps (Law *et al.*, 1978). A GaInAs/InP avalanche photodiode also has been made that spans the 1- to 1.6-μm wavelength range (Pearsall and Papuchon, 1978). Finally, the lattice-matched InGaAsP/InP quaternary material system, currently being used to make lasers and LEDs that emit near 1.3 μm, has been used for making photodiodes (Clawson *et al.*, 1978; Hurwitz and Hsieh, 1978; Washington *et al.*, 1978; Olsen and Kressel, 1979). The first avalanche device grown by liquid–phase epitaxy, exhibited a uniform gain of 12, rise time of 150 psec, and low-bias quantum efficiency of 45% near $\lambda = 1.2$ μm (Hurwitz and Hsieh, 1978); another, grown by vapor–phase epitaxy, showed a gain of 20 at a reverse bias of 60 V near $\lambda = 1.3$ μm (Olsen and Kressel, 1979). The observed large number of microplasma breakdowns are believed to be the cause of low gain obtained in present devices (Lee *et al.*, 1979).

18.5 PERFORMANCE OF PHOTODIODES IN REPEATERS

As discussed in Section 18.2, the performance of a photodiode in an optical communication system is determined not only by its own characteristics but also by the properties of the amplifier that follows it. Detailed consideration of optical receiver design appear in Chapter 19. We present here a summary of performance results predicted and achieved in various experimental optical repeaters.

The presentation is in the form of a chart as shown in Fig. 18.16 (Li, 1978). The ordinate is the *average* number of primary photoelectrons generated in the photodetector by the received optical signal in an interval $T = 1/(\text{bit rate})$ required to achieve a 10^{-9} error probability in a repeater; it is proportional to the optical energy per bit (pulse). The abscissa is the bit rate. The two bands represent calculated results based on current device parameters for silicon field-effect transistors (FET) and silicon bipolar transistors that are used in the preamplifiers immediately following the detectors (Personick, 1973). The upper band is for p–i–n photodiode detectors, while the lower band is for silicon avalanche photodiodes with optimal gain (ranging from about 20 at 1 Mbit/sec to about 100 at 100 Mbit/sec). The dots represent experimental results. The dashed diagonal

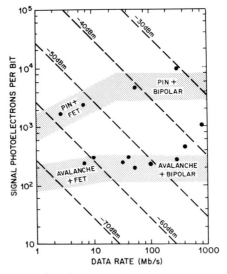

Fig. 18.16 State of the art of performance of optical repeaters using silicon photodiodes and transistors. The sensitivity of the optical receiver, measured in terms of the *average* number of signal photoelectrons per bit required to achieve an error probability of 10^{-9}, is given as a function of the bit rate. The two dotted bands represent predictions from theory. The large dots represent measured results from experimental repeaters. The dashed diagonal lines help to determine the sensitivity in terms of the input optical power at $\lambda = 0.85$ μm (Li, 1978).

lines help to determine performance values in terms of the average optical power required at the receiver ($\lambda = 0.85$ μm, $\eta = 0.80$). As can be seen, agreement between experiment and theory is rather good. At $\lambda = 1.3$ μm, the higher excess noise from a germanium avalanche photodiode would degrade the performance by about 6 dB compared to that for a silicon avalanche device.

The ultimate receiver sensitivity, limited by the fundamental quantum noise inherent in the detected signal, is represented by 11 average signal photoelectrons per bit. The best performance achieved with devices available today at $\lambda = 0.85$ μm is ~200 photoelectrons, or about 12 dB from the ultimate limit. Future devices with lower noise could bring the performance closer to the quantum noise limit, but increasingly more effort undoubtedly will be required as this limit is approached.

18.6 CONCLUSIONS

Silicon photodiodes with good sensitivity and sufficiently wide bandwidth for operation in the 0.8- to 0.9-μm wavelength region are now available commercially. They are used extensively in laboratory experiments and field trials of optical-fiber systems and their performances are found to agree well with theoretical predictions.

Although germanium photodiodes have higher dark-current and avalanche noise than their silicon counterparts, their broad spectral response covering wavelengths from the visible to 1.6 μm makes them potential candidates for application in the longer wavelength region of minimal fiber attenuation and dispersion. Indeed, germanium avalanche photodiodes are used in laboratory experiments today. Their future application in practical systems will depend on the successful development of a reliable passivation technology.

Current research efforts are directed toward the search for photodiode materials that will provide small dark currents and low excess noise for operation near $\lambda \simeq 1.3$ μm. Attention is focussed on III–V semiconductor alloys—progress is being made in the measurement of carrier ionization rates and in the fabrication of early experimental devices. Much work lies ahead in this area of research.

REFERENCES

Anderson, L. K., and McMurtry, B. J. (1966). High-speed photodetectors. *Proc. IEEE* **54**, 1335.

Anderson, L. K., McMullin, P. G., D'Asaro, L. A., and Goetzberger, A. (1965). Microwave photodiodes exhibiting microplasma-free carrier multiplication. *Appl. Phys. Lett.* **6**, 62.

Anderson, L. K., DiDomenico, J., Jr., and Fisher, M. B. (1970). High-speed photodetectors for microwave demodulation of light. *Adv. Microwaves* **5**, p. 1.

Ando, H., Kanbe, H., Kimura, T., Yamaoka, T., and Kaneda, T. (1978). Characteristics of germanium avalanche photodiodes in the wavelength region of 1.0–1.6 μm. *IEEE J. Quantum Electron.* **QE-14,** 804.

Baraff, G. A. (1962). Distribution functions and ionization rates for hot electrons in semiconductors. *Phys. Rev.* **128,** 2507.

Batdorf, R. L., Chynoweth, A. G., Dacey, G. C., and Foy, P. W. (1960). Uniform silicon p–n junctions. I. Broad areas breakdown. *J. Appl. Phys.* **31,** 1153.

Berchtold, K., Krumpholz, O., and Suri, J. (1975). Avalanche photodiodes with a gain-bandwidth product of more than 200 GHz. *Appl. Phys. Lett.* **26,** 585.

Boyd, J. T., and Chen, C. L. (1976). Integrated optical silicon photodiode array. *Appl. Opt.* **15,** 1389.

Capasso, F., Nahory, R. E., Pollack, M. A., and Pearsall, T. P. (1977). Observation of electronic bandstructure effects on impact ionization by temperature tuning. *Phys. Rev. Lett.* **39,** 723.

Clawson, A. R., Lum, W. Y., McWilliams, G. E., and Wieder, H. H. (1978). Quaternary alloy $In_xGa_{1-x}As_yP_{1-y}/InP$ photodetectors. *Appl. Phys. Lett.* **32,** 549.

Conradi, J. (1972). The distribution of gains in uniformly multiplying avalanche photodiodes: Experimental. *IEEE Trans. Electron. Devices* **ED-19,** 713.

Conradi, J. (1974). Temperature effects in silicon avalanche diodes. *Solid-State Electron.* **17,** 99.

Conradi, J. (1975). Planar germanium photodiodes. *Appl. Opt.* **14,** 1948.

Conradi, J., and Webb, P. P. (1975). Silicon reach-through avalanche photodiodes for fiber optic applications. *Proc. Eur. Conf. Opt. Fibre Commun., 1st, 1975* p. 128.

Eden, R. C. (1975). Heterojunction III–V alloy photodetectors for high-sensitivity 1.06-μm optical receivers. *Proc. IEEE* **63,** 32.

Eden, R. C., Nakano, K., Deyhimy, I., and Kim, C. K. (1975). high sensitivity gigabit data rate $GaAs_{1-x}Sb_x$ avalanche photodiode 1.06 μm optical receiver. *Tech. Dig., int. Electron. Device Meet., 1975* p. 591.

Emmons, R. B. (1967). Avalanche photodiode frequency response. *J. Appl. Phys.* **38,** 3705.

Gillespie, A. B. (1953). "Signal, Noise and Resolution in Nuclear Counter Amplifiers." Pergamon, Oxford.

Goebloed, J. J. (1977). Comments on 'Avalanche buildup time of silicon reach-through photodiodes.' *J. Appl. Phys.* **48,** 4004.

Goebloed, J. J., and Smeets, E. T. J. M. (1978). Very low noise silicon planar avalanche photodiodes. *Electron. Lett.* **14,** 67.

Goetzberger, A., McDonald, B., Haitz, R. H., and Scarlett, R. M. (1963). Avalanche effects in silicon p–n junctions. II. Structurally perfect junctions. *J. Appl. Phys.* **34,** 1591.

Hartman, A. R., Melchior, H., Schinke, D. P., and Seidel, T. E. (1978). Planar epitaxial silicon avalanche photodiode. *Bell Syst. Tech. J.* **57,** 1791.

Hubbard, W. M. (1973). Efficient utilization of optical-frequency carriers for low and moderate bandwidth channels. *Bell Syst. Tech. J.* **52,** 731.

Hurwitz, C. E., and Hsieh, J. J. (1978). GaInAsP/InP avalanche photodiodes. *Appl. Phys. Lett.* **32,** 487.

Ito, M., Kaneda, T., Nakajima, K., Toyoma, Y., Yamaoka, T., and Kotani, T. (1978). Impact ionization in $In_{0.73}Ga_{0.27}As_{0.57}P_{0.43}$. *Electron. Lett.* **14,** 418.

Johnson, K. M. (1965). High-speed photodiode signal enhancement at avalanche breakdown voltage. *IEEE Trans. Electron Devices* **ED-12,** 55.

Kanbe, H., Kimura, T., Mizushima, and Kajiyama, K. (1976). Silicon avalanche photodiodes with low multiplication noise and high-speed response. *IEEE Trans. Electron Devices* **ED-23,** 1337.

Kaneda, T., Matsumoto, H., and Yamaoka, T. (1976a). A model for reach-through avalanche photodiodes (RAPD's). *J. Appl. Phys.* **47,** 3135.

Kaneda, T., Takanashi, H., Matsumoto, H., and Yamaoka, T. (1976b). Avalanche buildup time of silicon reach-through photodiodes. *J. Appl. Phys.* **47**, 4960.

Krumpholz, O., and Maslowski, S. (1968). Schnelle Photodioden .mit Wellen langen-Unabhangigen Demodulationseigenschaften. *Z. Angew. Phys.* **25**, 156.

Law, H. D., Tomasetta, L. R., Nakano, K., and Harris, J. S. (1978). 1.0 μm to 1.4 μm high speed avalanche photodiode. *Postdeadline Pap., Top. Meet. Integr. Guided Wave Opt., 1978* Paper PD2.

Lee, C. A., Logan, R. A., Batdorf, R. L., Kleimack, J. J., and Wiegman, W. (1964). Ionization rates of holes and electrons in silicon. *Phys. Rev.* **134**, A761.

Lee, T. P., Burrus, C. A., and Dentai, A. G. (1979). InGaAsP/InP photodiodes: Microplasma-limited avalanche multiplication at 1–1.3 μm wavelength. *IEEE J. Quantum Electron.* **QE-15**, 30.

Lee, T. P., Burrus, C. A., Pollack, M. T., and Nahory, R. E. (1975). High-speed Schottky barrier photodiode in LPE In$_x$Ga$_{1-x}$As for 1.0 μm to 1.1 μm wavelength region. *IEEE Trans. Electron Devices* **ED-22**, 1062.

Li, T. (1978). Optical fiber communication—the state of the art. *IEEE Trans. Commun.* **COM-26**, 946.

Lindley, W. T., Phelan, R. J., Jr., Wolfe, C. M., and Foyt, A. G. (1969). GaAs Schottky barrier avalanche phtodiodes. *Appl. Phys. Lett.* **14**, 197.

Logan, R. A., and Sze, S. M. (1966). Avalanche multiplication in Ge and GaAs p–n junctions. *J. Phys. Soc. Jpn., Suppl.* **21**, 434.

McIntyre, R. J. (1966). Multiplication noise in uniform avalanche diodes. *IEEE Trans. Electron Devices* **ED-13**, 164.

McIntyre, R. J. (1972). The distribution of gains in uniformly multiplying avalanche photo-diodes: Theory. *IEEE Trans. Electron Devices* **ED-19**, 703.

McKay, K. G., and McAfee, K. B. (1953). Electron multiplication in silicon and germanium. *Phys. Rev.* **91**, 1079.

Maslowski, S. (1972). New kind of detector for use in communication systems with glass fiber waveguides. *Dig. Top. Meet. Integr. Opt., 1972* Paper WB6.

Melchior, H. (1972). Demodulation and photodetection techniques. *In* "Laser Handbook" (F. T. Arrecchi and E. D. Schulz-Dubois, eds.), p. 725. Elsevier, Amsterdam.

Melchior, H. (1973a). Sensitive high speed photodetectors for the demodulation of visible and near infrared light. *J. Lumin.* **7**, 390.

Melchior, H. (1973b). Semiconductor detectors for optical communications. *Conf. Laser Eng. Appl. 1973;* abstract in *IEEE J. Quantum Electron.* **QE-9**, 659.

Melchior, H. (1977). Detectors for lightwave communication. *Phys. Today* **30**, 32.

Melchior, H., and Hartman, A. R. (1976). Epitaxial silicon n$^+$–p–π–p$^+$ avalanche photodiode for optical fiber communications at 800 to 900 nanometers. *Tech. dig., Int. electron Devices Meet., 1976* p. 412.

Melchior, H., and Lynch, W. T. (1966). Signal and noise response of high speed germanium avalanche photodiodes. *IEEE Trans. Electron Devices* **ED-13**, 829.

Melchior, H., Fisher, M. B., and Arams, F. R. (1970). Photodetectors for optical communication systems. *Proc. IEEE* **58**, 1466.

Miller, S. M. (1955). Avalanche breadkdown in germanium. *Phys. Rev.* **99**, 1234.

Müller, J. (1978). Thin silicon film p–i–n photodiodes with internal reflection. *IEEE Trans. Electron Devices* **ED-25**, 247.

Müller, J., and Ataman, A. (1976). Double-mesa thin-film reach-through silicon avalanche photodiodes with large gain bandwidth product. *Tech. Dig., Int. Electron Devices Meet., 1976* p. 416.

Nishida, K. (1977). Avalanche-noise dependence on avalanche-photodiode structures. *Electron Lett.* **13**, 419.

Nishida, K., Ishii, K., Minemura, K., and Taguchi, K. (1977). Double epitaxial silicon avalanche photodiodes for optical fiber communications. *Electron. Lett.* **13**, 280.

Olsen, G. H., and Kressel, H. (1979). Vapor–grown 1.3 μm In GaAsP/InP avalanche photodiodes. *Elect. Lett.* **15**, 141.

Pearsall, T. P., and Papuchon, M. (1978). The $Ga_{0.47}In_{0.53}As$ homojunction photodiode–A
new avalanche photodetector in the near infrared between 1.0 and 1.6 μm. *Appl. Phys.
Lett.* **33**, 640.

Pearsall, T. P., Nahory, R. E., and Pollack, M. A. (1975). Impact ionization coefficients for
electrons and holes in $In_{0.14}Ga_{0.86}As$. *Appl. Phys. Lett.* **27**, 330.

Pearsall, T. P., Nahory, R. E., and Pollack, M. A. (1976). Impact ionization rates for electrons
and holes in $GaAs_{1-x}Sb_x$ alloy. *Appl. Phys. Lett.* **28**, 403.

Pearsall, T. P., Nahory, R. E., and Chelikowsky, J. R. (1977). Orientation dependence of
free-carrier impact ionization in semiconductors: GaAs. *Phys. Rev. Lett.* **39**, 295.

Personick, S. D. (1971a). New results on avalanche multiplication statistics with applications
to optical detection. *Bell Syst. Tech. J.* **50**, 167.

Personick, S. D. (1971b). Statistics of a general class of avalanche detectors with applications
to optical communication. *Bell Syst. Tech. J.* **50**, 3075.

Personick, S. D. (1973). Receiver design for digital fiber optic communication systems. Parts I
and II. *Bell Syst. Tech. J.* **52**, 843.

Prince, M. B. (1955). Silicon solar energy converters. *J. Appl. Phys.* **26**, 534.

Riesz, R. P. (1962). High speed semiconductor photodiodes. *Rev. Sci. Instrum.* **33**, 994.

Ruegg, H. W. (1967). An optimized avalanche photodiode, *IEEE Trans. Electron Devices*
ed-14, 239.

Schneider, M. V. (1966). Schottky barrier photodiodes with anti-reflection coating. *Bell Syst.
Tech. J.* **45**, 1611.

Scholl, F. W., Nakano, K., and Eden, R. C. (1976). $GaAs_{1-x}Sb_x$ 1.06 μm avalanche photodiodes. *Tech. dig., Int. Electron Devices Meet., 1976* p. 424.

Shibata, T., Igarashi, Y., and Yano, K. (1974). Passivation of germanium devices. III Fabrication and performance of germanium planar photodiodes. *Rev. Electr. Commun. Lab.* **22**,
1069.

Shockley, W. (1949). The theory of p–n junctions in semiconductors and p–n junction transistors. *Bell Syst. Tech. J.* **28**, 439.

Smith, R. G., Brackett, C. A., and Reinbold, H. W. (1978). Optical detector package. *Bell Syst.
Tech. J.* **57**, 1809.

Stillman, G. E., and Wolfe, C. M. (1977). Avalanche photodiodes. *In* "Semiconductors and
Semimetals" (P. K. Willardson and A. C. Beers, eds.), Vol. 12, p. 291. Academic Press,
New York.

Stillman, G. E., Wolfe, C. M., Foyte, A. G., and Lindley, W. T. (1974). Schottky barrier
$In_xGa_{1-x}As$ alloy avalanche photodiodes for 1.06 μm. *Appl. Phys. Lett.* **24**, 8.

Sukegawa, T., Hiraguchi, T., Tanaka, A., and Hagino, M. (1978). Highly efficient pGaSb–
$nGa_{1-x}Al_xSb$ photodiodes. *Appl. Phys. Lett.* **32**, 376.

Sze, S. M. (1967). "Physics of Semiconductor Devices," p. 104. Wiley, New York.

Takahashi, K., Takamiya, S., and Mitsui, S. (1977). A monolithic 1 × 10 array of silicon avalanche photodiodes. *Tech. Dig., Int. Conf. Integr. Opt. Opt. fiber commun., 1977* p. 37.

Tomasetta, L. R., Law, H. D., Eden, R. C., Deyhimy, I., and Nakano, K. (1978). High sensitivity optical receivers for 1.0–1.4 μm fiber-optic systems. *IEEE J. Quantum Electron.*
QE-14, 800.

Washington, M. A., Nahory, R. E., and Beebe, E. D. (1978). High-efficiency $In_{1-x}Ga_x$-
As_yP_{1-y}/InP photodetectors with selective wavelength response between 0.9 and
1.7 μm. *Appl. Phys. Lett.* **33**, 854.

Webb, P. P., McIntyre, R. J., and Conradi, J. (1974). Properties of avalanche photodiodes.
RCA Rev. **35**, 234.

Chapter 19

Receiver Design

STEWART D. PERSONICK

19.1 INTRODUCTION

The purpose of a receiver in a fiber optic communication system is to extract information from the optical carrier which impinges upon the detector. In a digital system the receiver must produce a sequence of electrical pulses ("ones" and "zeros") which faithfully reproduce the digital information which was impressed upon the light signal at the transmitter. In analog modulation systems the receiver must amplify the detector output and demodulate the analog information from the amplified waveform.

The receiver designer must model the operations of various components (e.g., amplifiers, filters) in a mathematical way in order to predict the fidelity which can be obtained by judicious use of those components. Typically, on the first "go around" fundamental fidelity limitations are studied. As the work proceeds, more and more of the practical constraints are added into the analysis. All this time, experimental verification of the predicted performance must be going on.

A fidelity criterion has been referred to in the above discussion. In some cases the criterion of fidelity is obvious. For example, in digital communication, error probability is usually the most important parameter (in some cases the frequency of occurrence of long bursts of errors or "error-free seconds" is used). In analog modulation systems, the fidelity criterion is usually mean-squared error, although in recent years more subjective criteria have been used. The fidelity criterion is used by the receiver designer to optimize the design, subject to the hardware constraints in his model.

Often it turns out that the obvious fidelity criterion, such as error rate in digital systems, is mathematically difficult to deal with. In those instances the receiver designer may invent alternate criteria (hopefully well correlated with the original criterion) which are more convenient to deal with, or which give more intuitive results.

The most fundamental limitation in the performance of a receiver is noise (Davenport, 1958). Noise is a somewhat vague term which describes unpredictable components or unwanted components of an electrical signal. Noise can come from random fluctuations of currents in electrical components (e.g., Johnson noise in resistors and shot noise in diodes). Noise can come from the fundamental randomness associated with the conversion of light power into an electrical current (quantum noise). Noise can come from interference of adjacent electrical pulses in a digital communication system or even crosstalk from other systems. Most of what will follow below will be an analysis of the effects of noise on the performance of optical fiber system receivers. To a lesser extent attention will be given to some of the practical constraints which modify the receiver design from one which is optimized on the basis of noise alone.

19.2 BASIC PRINCIPLES OF RECEIVER DESIGN

19.2.1 Performance Criteria for Digital and Analog Modulation

Figure 19.1 shows the basic structure of the optical fiber system receiver. The detector converts optical power into an electrical current. The reader will recall that optical power is a baseband quantity. That is, power varies according to the modulation and not at optical frequencies. Typically this current is very weak (nanoamperes in some cases) and it must be amplified by a low noise amplifier specifically designed to work with optical detectors. The resultant amplified signal is then processed electrically (e.g., filtered, demodulated, or regenerated) to extract the message.

19.2.1.1 Digital Systems. Figure 19.2 shows the structure of a digital system receiver often called a repeater. It consists of a detector and amplifier chain followed by a filter, a regenerator, and a timing circuit. The portion of the receiver consisting of amplifiers and filters is called the "linear channel" because all operations on the detector output current up to that point are mathematically linear. The signal falling upon the detector is a sequence of pulses of optical power of the following form

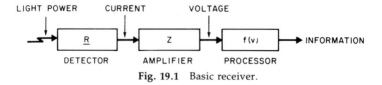

Fig. 19.1 Basic receiver.

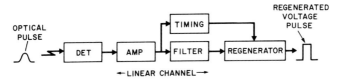

Fig. 19.2 Digital receiver.

$$p_{opt}(t) = \sum_{k=-\infty}^{\infty} a_k h_p(t - kT), \tag{19.1}$$

where T is the pulse spacing, $a_k = 0$ or 1 depending upon whether the pulse is "on" or "off" and $h_p(t)$ is the shape of an isolated pulse.

The detector converts this optical signal into an electrical current with proportionality constant R (amps/watt). The quantity R is called the detector responsivity. The resulting detector current is

$$i_s(t) = R \sum_{-\infty}^{\infty} a_k h_p(t - kT). \tag{19.2}$$

The detector output current $i_s(t)$ is amplified and filtered to produce the filter output voltage $v_{out}(t)$

$$v_{out}(t) = \sum_{-\infty}^{\infty} a_k h_{out}(t - kT), \tag{19.3}$$

where $h_{out}(t)$ is the shape of an isolated amplified and filtered pulse.

Once per time slot (pulse interval T) the waveform $v_{out}(t)$ is compared to a threshold to determine whether a pulse is present or not. The regenerator which makes this comparison is clocked by a timing circuit which is synchronized to the rate of arrival of digital pulses. Ideally, the output signal $v_{out}(t)$ would always exceed the threshold when pulses are present. In real receivers, noise and interference (from adjacent pulses or other systems) can add to or subtract from the ideal signal, causing errors to occur. The probability of errors (on the average) is called the error rate. This is the fidelity criterion of digital receivers. Typically error rates of 10^{-6} to 10^{-9} are specified for telecommunication system applications. The error rate requirements and the receiver noise and interference set a lower limit on the optical power level which must be present at the detector.

19.2.1.2 Analog Systems. Figure 19.3 shows the structure of an analog system receiver. The detector is illuminated by an optical signal $p_{opt}(t, m(t))$ where $m(t)$ is some analog message. For example, we could have simple intensity modulation where

$$p_{opt}(t, m(t)) = P_0[1 + \gamma m(t)], \tag{19.4}$$

where γ is a constant ≤ 1, and $|m(t)| \leq 1$.

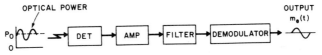

Fig. 19.3   Analog receiver (intensity modulation).

Alternatively we could have subcarrier phase modulation where

$$p_{opt}(t, m(t)) = P_0[1 + \cos[\omega t + \gamma m(t)], \qquad (19.5)$$

where ω is an i.f. frequency and γ is a constant.

The optical power is converted to an electrical current by the detector producing the current $i_s(t)$

$$i_s(t) = Rp_{opt}(t), \qquad (19.6)$$

where R is the detector responsivity (amps/watt).

This current is amplified, filtered, and demodulated. In the intensity modulation system the demodulation is not necessary since the baseband signal $i_s(t)$ is already simply proportional to the message. In the subcarrier phase modulation system the demodulator could be a phase locked loop or a discriminator. The demodulated output is ideally an exact replica of the message $m(t)$. In practice noise and other degradations result in an imperfect estimate of $m(t)$ which we shall call $m_e(t)$. If $m(t)$ is one side of a telephone conversation, a subjective evaluation of the fidelity of the system might be the evaluation of $m_e(t)$ by a series of listeners. Such subjective tests are often used to compare competing modulation and demodulation schemes. From a practical viewpoint a simple mathematical fidelity criterion is often introduced as an alternative to the subjective test. This criterion is the mean-squared error, defined as the average value of the square of the difference between the receiver output $m_e(t)$ and the actual message $m(t)$

$$\text{MSE} = \langle[m_e(t) - m(t)]^2\rangle, \qquad (19.7)$$

where $\langle \ \rangle$ signifies average value.

The mean-squared error is defined above as a statistical or ensemble average over all possible messages and degradations (noises and interferences). Since the message in an analog system is often difficult to characterize statistically, simpler but related criteria are sometimes used. We shall discuss these in the specific calculations to be made later on.

19.2.2 Signal and Noise Modeling

19.2.2.1 Signals and Noise in the Detection Process.   Detectors were discussed in Chapter 18. We shall review here some of the circuit and statistical properties of detectors (McIntyre, 1966, 1972; Personick, 1971a,b,

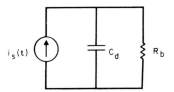

Fig. 19.4 Basic Photodiode.

1976; Personick *et al.*, 1977; Webb, 1974). The basic photodiode or avalanche photodiode as shown in Fig. 19.4 can be modeled as a current
source in parallel with a capacitor and a resistor. The capacitor represents
the sum of the junction capacitance of the detector, packaging, and lead
capacitance. The parallel resistor represents the biasing resistor which
connects the detector to the bias voltage supply.

When light is incident upon the detector, hole-electron pairs are created
which separate under the influence of the internal detector fields. This results in a displacement current represented by the current source $i_s(t)$. The
average current is related to the incident optical power by the detector
responsivity R (amps/watt).

$$\langle i_s(t) \rangle = R p_{\text{opt}}(t). \tag{19.8}$$

Equation (19.8) describes the average behavior of the detection process.
In fact, there is a more detailed description which explains certain important noises (or randomness) introduced in the detector. The holes and
electrons which are generated by the incident light are produced at
random times. For a given illumination $p_{\text{opt}}(t)$ the average number of pairs
generated in a time T has the value

$$N_{\text{av}} = \frac{\eta}{h\nu} \int_T p_{\text{opt}}(t)\, dt = \frac{\eta}{h\nu} E, \tag{19.9}$$

where η = detector quantum efficiency, $h\nu$ = energy in a photon, and
E = energy received in interval T.

On the other hand, the actual number of pairs generated can fluctuate
from this average. The probability that exactly n pairs are generated is
given by the Poisson distribution

$$p(n)_{\text{det}} = N_{\text{av}}^n e^{-N_{\text{av}}}/n!. \tag{19.10}$$

This unpredictability of the actual number of electron-hole pairs generated within the detector, in response to a known pulse of light, limits
the sensitivity (minimum allowable power level) of a receiver. In fact, for
digital systems we can quickly calculate a fundamental lower limit on the
energy which a pulse must contain if we wish to detect it with a given
probability of error. We can imagine an ideal receiver which can "see"

individual electron-hole pairs generated within the detector. (It has suffi-ciently low amplifier noise to detect the displacement current of a single pair.) In the absence of the light pulse no current will flow (we have neglected "dark current" up to this point). The receiver will decide that no pulse is present if no electron-hole pairs are generated. If even one pair is generated, a pulse is assumed present. The only way an error can be made is if the pulse is in fact present, but no pairs are generated. The probability of no pairs is given by [from (19.10)]

$$p(0) = e^{-N_{av}}. \tag{19.11}$$

Suppose an error probability of 10^{-9} is desired, leading to $e^{-N_{av}} \leq 10^{-9}$ There follows a requirement on the pulse energy of $N_{av} \geq 21$, where $N_{av} = (\eta/h\nu)E$. This minimum required pulse energy ($21/\eta$ photons per pulse) for a 10^{-9} error rate is called the quantum limit.

In many applications, the quantity of interest is the output at time t of a filter whose input is the current produced by a detector. Each electron hole pair produced in the detector causes a displacement current of total area e (1.6×10^{-19} C) to flow into its load. In this analysis the detector is as-sumed to be sufficiently fast compared to the response times of subse-quent filters that this displacement current can be idealized as an impulse. If the filter output voltage in response to an impulse current input is $h_{filt}(t)$, then the total filter response at time t due to light incident upon the detector is given by

$$v_{out}(t) = \sum_{k=-\infty}^{\infty} eh_{filt}(t - t_k) \tag{19.12}$$

where t_k is the generation time of the kth electron hole pair in response to incident light. The output voltage $v_{out}(t)$ in (19.12) shows no explicit dependence upon the incident light. The dependence is implicitly defined by the statistical relationship between the pair generation times and the light power. Standard mathematical techniques relating to Poisson pro-cesses yield expressions for the average and mean-squared deviation val-ues of the filter output (averaging over the Poisson statistics):

$$\langle v_{out}(t) \rangle = \frac{\eta e}{h\nu} \int h_{filter}(t - \tau) p_{opt}(\tau) \, dt, \tag{19.13}$$

$$\langle [v_{out}(t) - \langle v_{out}(t) \rangle]^2 \rangle = \frac{\eta e^2}{h\nu} \int h^2_{filter}(t - \tau) p_{opt}(\tau) \, d\tau,$$

where $\eta e/h\nu = R = $ detector responsivity.

The above equation (19.13) is valid for PIN detectors. For detectors with avalanche gain one must take into account the randomness of the multipli-cation process. This randomness further increases the unpredictability of the total number of electron-hole pairs which will be produced within the detector. If the detector output drives a filter with current-to-voltage impulse response $h_{\text{filt}}(t)$, then its output is given by

$$v_{\text{out}}(t) = \sum em_k h_{\text{filt}}(t - \tau_k),\tag{19.14}$$

where m_k is the number of secondary electron-hole pairs produced in the multiplication process by primary pair k. Once again the relationship between $v_{\text{out}}(t)$ and the incident light is implicitly combined in the statistics of the number of primaries generated and their generation times. One can obtain the mean value and the mean-squared deviation of the filter output as follows

$$\langle v_{\text{out}}(t) \rangle = \frac{\eta e M}{h\nu} \int h_{\text{filt}}(t - \tau) p_{\text{opt}}(\tau)\, d\tau,$$

$$\langle [v_{\text{out}}(t) - \langle v_{\text{out}}(t) \rangle]^2 \rangle = \frac{\eta e^2}{h\nu} \langle m^2 \rangle \int h_{\text{filt}}^2(t - \tau) p_{\text{opt}}(\tau)\, d\tau,\tag{19.15}$$

where M = average value of m_k and $\langle m^2 \rangle$ = average value of $m_k^2 \geq M^2$.

Once again, (19.14) and (19.15) are based on the assumption that the detection process including secondary generation is fast compared to the filter response time.

Observe that avalanche gain increases the average filter output signal by the ratio M (the mean number of secondaries per primary); and it increases the mean-squared deviation (noise) by the ratio $\langle m^2 \rangle$. An ideal amplifier (with a fixed ratio of secondaries per primary) would satisfy $\langle m^2 \rangle = M^2$. Actual detectors are characterized in some applications by the deviation from this ideal as follows

$$\langle m^2 \rangle = FM^2, \qquad F \geq 1\tag{19.16}$$

where F is called the excess noise factor.

19.2.2.2 Amplifier Noise Modeling. In many fiber system applications, the power level incident on the detector is below 10 nW. The typical detector responsivity at fiber system wavelengths (before avalanche gain) is about 0.5 A/W. Thus the detector output current is about 5 nA for p–i–n detectors and about 0.5 μA for APDs; given a 10 nW input power level. These weak currents are difficult to amplify without the introduction of large amounts of noise (Hullett, 1976; Personick, 1973, 1976a). Often the noise added in the amplification process will dominate the noise associated with the randomness of the detection process (discussed above). Thus the receiver designer must choose an amplifier which introduces as little added noise as possible.

Figure 19.5 shows the circuit model of a typical amplifier. The amplifier has an input impedance represented as the parallel combination of a re-

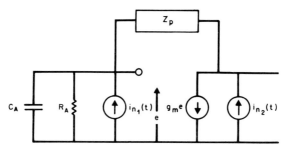

Fig. 19.5 Typical amplifier.

sistor and a capacitor. Voltages across this impedance cause current to flow in the amplifier output. The voltage controlled current source representing this relationship is characterized by a transconductance g_m (amps/volt). The amplifier has two current noise sources $i_{n1}(t)$ and $i_{n2}(t)$ whose origins will be described below. Also shown is an impedance between the input and output of the amplifier. This may represent parasitic capacitance or an intentional feedback element.

The noise sources of the amplifier will be assumed independent and Gaussian in their statistics. Gaussian noise sources are completely characterized by their noise spectral densities. If the noise source is white, then the output of a filter excited by the noise source has mean-squared value given by

$$\sigma^2 = v_{out}^2(t) = S_I \int_{-\infty}^{\infty} h_{filt}^2(t)\, dt = S_I \int_{-\infty}^{\infty} |H_{filt}(f)|^2\, df, \tag{19.17}$$

where S_I is the noise spectral density (A²/Hz), $h_{filt}(t)$ is the filter impulse response (V/A) and $H_{filt}(f) = \mathscr{F}\{h_{filt}(t)\}$ = Fourier transform of $h_{filt}(t)$.

Furthermore, the output voltage of a filter excited by Gaussian noise is a Gaussian random variable. The probability that the voltage takes on any given value v is given by

$$\operatorname*{prob}_{v\,out} (v) = [1/(2\pi)^{1/2}\sigma]\exp -\{v^2/2\sigma^2\}. \tag{19.18}$$

From Fig. 19.6 we can calculate the noise at the output of an amplifier–filter combination. First, observe that current from the detector

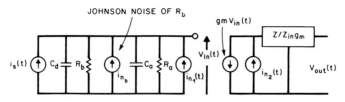

Fig. 19.6 Amplifier–filter combination (neglect feedback).

produces a voltage at the amplifier input according to the (frequency domain) relationship

$$V_{in}(f) = I_s(f)Z_{in}(f),$$ (19.19)

where

$$Z_{in}(f) = \left[\frac{1}{R_b} + \frac{1}{R_a} + j2\pi f(C_d + C_a)\right]^{-1}.$$ (19.20)

Thus as defined, the filter output due to the detector current is not dependent upon the details of the amplifier. Thus a variation of the amplifier parameters to minimize the amplifier noise at the filter output does not vary the detector contribution.

From Fig. 19.6 we can calculate the noise at the filter output due to $i_{n1}(t)$ and $i_{n2}(t)$, and due to the Johnson noise of the detector load resistor i_n (t).

$$\langle v_{out}^2 \rangle \Big|_{\substack{\text{due to amplifier} \\ \text{and bias resistor}}} = 2S_{I_1}Z^2B + (4kTBZ^2/R_b)$$

$$+ S_{I_2} \int_{-\infty}^{\infty} \frac{Z^2}{g\,m^2|Z_{in}(f)|^2}\, df,$$ (19.21)

where S_{I_1} = spectral density of $i_{n1}(t)$, S_{I_2} = spectral density of $i_{n2}(t)$, and kT = Boltzmann's constant · absolute temperature.

We can now evaluate (19.21) for specific amplifiers.

19.2.2.2.1 FET Amplifiers. For optical fiber receivers operating below 25 Mbit/sec the lowest noise amplifier device (widely available) is the silicon FET. Figure 19.7a and b show simplified models of the grounded source FET configuration. Typical FETs have very large input impedances. For most applications we can set $R_a = \infty$ and $S_{I_1} = 0$. The input capacitance including header and lead capacitance is a few picofarads. The output noise source can be associated with the channel resistance and has value $S_{I_2} = 1.4kTg_m$. The transconductance g_m of a good silicon FET is

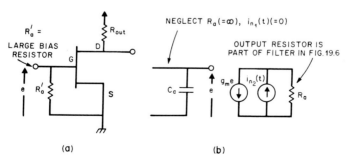

Fig. 19.7 (a) FET amplifier. (b) FET amplifier equivalent circuit.

around 5000 microSiemens. Using these values for R_a, S_{I_1}, and S_{I_2} we can obtain a simplified version of (19.21)

$$\langle v^2_{out} \rangle = \frac{4kTBZ^2}{R_b} + \frac{2.8kTBZ^2}{R_b^2 g_m} + \frac{2.8kTB^3Z^2}{3g_m} [2\pi(C_a + C_d)]^2. \quad (19.22)$$

Equation (19.22) results from substituting the expression for Z_{in} of (19.19) and performing the simple integration.

Note that it is desirable to make the detector biasing resistor very large. This will increase the low frequency impedance of the detector load and tend to integrate the signal $i_s(t)$. This integration can be compensated by differentiation in the filter following the amplifier provided nonlinearity due to overload of the amplifier does not occur. At high frequencies the amplifier produces a noise at the filter output which has mean-squared value proportional to the cube of the filter bandwidth.

Observe also that at high frequencies, $g_m/(C_a + C_d)^2$ is a figure of merit of the amplifier noise performance. For a fixed material (say silicon) the FET geometry can be varied with a fixed ratio of g_m/C_a, determined by the carrier mobility. The optimized value of $C_a = C_d$; however, in practice this device optimization may not be economical.

19.2.2.2.2 *Bipolar Amplifiers.* At high frequencies, the current gain of the FET drops to values near unity because g_m is fixed and the input impedance is decreasing. In the frequency range around 25 to 50 MHz it is necessary to switch from FET to bipolar amplifiers. A typical bipolar grounded emitter amplifier is shown in Figs. 19.8a and b. The input resistance of the bipolar amplifier is determined by the base bias current through the relationship

$$R_a = kT/eI_{base}. \quad (19.23)$$

The transconductance of the bipolar amplifier is its fixed current gain, β, divided by its input resistance R_a. Thus the transconductance is adjust-

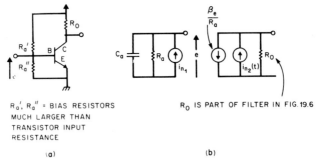

R_a', R_a'' = BIAS RESISTORS
MUCH LARGER THAN
TRANSISTOR INPUT
RESISTANCE

R_0 IS PART OF FILTER IN FIG. 19.6

(a)   (b)

Fig. 19.8   (a) Bipolar amplifier. (b) Bipolar amplifier equivalent circuit.

able for bipolar transistors. Input capacitances of typical bipolar transistors are a few picofarad. The input current noise source of the bipolar is the shot noise of the base bias current. The output current noise is the shot noise of the collector bias current. Thus, both of these noises are also adjustable.

$$S_{I_1} = eI_{\text{base}}, \qquad S_{I_2} = eI_{\text{collector}} = e\beta I_{\text{base}} \qquad (19.24)$$

By substituting (19.23) and (19.24) into (19.21) one obtains

$$\langle v_{\text{out}}^2 \rangle = \frac{4kTBZ^2}{R_b} + \frac{2kTZ^2B}{R_a} + \frac{2kTR_aZ^2B}{\beta[R_aR_b/(R_a + R_b)]^2}$$

$$+ \frac{2kTR_aB^3Z^2}{3\beta} \, 2\pi(C_a + C_d)^2. \qquad (19.25)$$

With the assumption that the biasing resistor R_b is large compared to R_a (which it should be for low noise), the following simple expressions for the optimal input resistance and minimized noise are obtained:

$$R_{a \text{optimal}} = \frac{1}{2\pi(C_a + C_d)B} \, (3\beta)^{1/2} \qquad \text{for} \quad \beta \gg 1, \qquad (19.26)$$

$$\langle v_{\text{out}}^2 \rangle_{\text{minimum}} = \frac{4kT(2\pi(C_a + C_d))}{(3\beta)^{1/2}} \, B^2Z^2. \qquad (19.27)$$

Observe that for bipolar transistors at optimum bias the noise is proportional to the square of the filter bandwidth. Observe also that the optimal input resistance R_a has a large value compared to the impedance of the input capacitance C_a in parallel with the detector capacitance C_d at the upper frequency B. Thus, the impedance loading the detector at optimum bias tends to integrate the detector input current. This integration is compensated by differentiation in the output filter.

For bipolar transistors $\beta/(C_a + C_d)^2$ is a figure of merit of amplifier performance. As with FETs, the bipolar transistor can in principle be optimized to work with a given detector, with $C_a = C_d$.

19.3 PERFORMANCE CALCULATIONS FOR DIGITAL SYSTEMS

19.3.1 Exact Calculations

Having reviewed the sources of noise in detectors and amplifiers, we can discuss briefly the complicated problem of performance analysis for digital systems (Dogliotti, 1976; Personick, 1971a,b, 1973; Personick et al., 1977).

In digital communication the fidelity criterion is the error rate as discussed in Section 19.2.1.1. After amplification and filtering, the resulting linearly processed detector output is compared to a threshold once per time slot to determine whether or not a pulse of light was present at the detector in that time slot. The error probability is the probability that the filter output voltage at the sampling time exceeds the threshold when a pulse of light is not present, plus the probability that it is below threshold when a pulse is present, all divided by two (for equally likely pulse present and pulse absent). To calculate these error probabilities one needs to know the probability distribution of the signal at the filter output. This voltage can be expressed as follows (see Fig. 19.9.)

$$v_{\text{out}}(t) = \sum_{-\infty}^{\infty} em_k h_{lc}(t - t_k) + n_{\text{amp}}(t), \tag{19.28}$$

where $h_{lc}(t)$ is the linear channel impulse response (volts/amp), e = electron charge, m_k is the number of secondaries produced by primary k (generated at time t_k), and $n_{\text{amp}}(t)$ is the amplifier noise. The average value of $v_{\text{out}}(t)$, given the sequence of data digits is

$$\langle v_{\text{out}}(t) \rangle = \sum_l q_l h_{\text{out}}[t - lT], \tag{19.29}$$

where $\{q_l\}$ are the data digits

$$h_{\text{out}}(t) = h_p(t)R * h_{lc}(t),$$

$h_p(t)$ is the shape of an isolated optical power pulse, R = detector responsivity, and the asterisk (*) signifies convolution.

Since the isolated filter output pulses may overlap (referred to as intersymbol interference) the error probabilities in general depend upon the values of adjacent data digits. We can assume that the filter is chosen to shape the output pulses (for known input power pulse shape) to have zero intersymbol interference at the sampling times as shown in Fig. 19.10. Thus the error probabilities are determined by the statistical variations in the detection process (number of electron-hole pairs generated and their

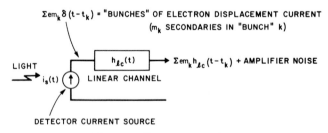

Fig. 19.9 Detector statistics.

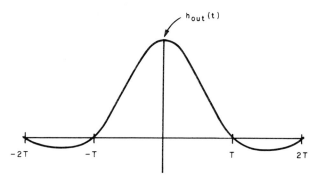

Fig. 19.10   Zero forcing equalization.

generation times) and by the amplifier noise. Because of the difficulty in working with the exact statistics of avalanche gain, various approximations are used to calculate the error probabilities. To illustrate these calculations we shall consider a somewhat simplified situation where the receiver linear channel output at the sampling time for data digit m is proportional to the sum of the secondary electrons emitted by the detector during the interval occupied by optical pulse m (plus amplifier noise). In practice, the receiver output is a weighted sum of the secondaries produced by the optical pulse m and its adjacent optical pulses, as given by (19.28).

The simplified receiver output at the sampling time is then

$$\left(\sum_{1}^{N} m_k\right) + n_{\text{amp}} = X, \tag{19.30}$$

where N is the total number of primary electron-hole pairs produced by optical pulse m; m_k is the number of secondaries produced through avalanche multiplication by primary k and n_{amp} is amplifier noise. (If there is no avalanche gain $m_k = 1$ for all k.)

Decisions are made by comparing X to a threshold. In (19.30) N, m_k (for all N values of k), and n_{amp} are random. To calculate error probabilities we need to know the probability distribution of X for the optical-pulse-present and optical-pulse-absent conditions. Up to now we have assumed that in the absence of an optical pulse, no primary electrons are produced within the detector. We have also assumed that the optical power is completely extinguished in the pulse "off" condition. In practice there is some dark current (given in primary dark-current electron-hole pairs per time slot) and some optical signal in the pulse "off" condition (due to practical transmitter constraints). To calculate the probability distribution of X we need to know the average number of signal primary electron-hole pairs for the pulse "on" condition [given by (19.9)]; the statistics of primary pair generation [given by (19.10)], the variance of n_{amp} (can be calculated from the results of Section 19.3.2.2, the number of primary dark current counts per time slot, the extinction ratio (energy in an "off" pulse/energy in an "on" pulse), and the statistics of the avalanche gain m_k. Avalanche gain statistics have been derived for various types of APDs, and using these statistics, required optical signal levels for a fixed error probability have been calculated. The calculation is done using a general purpose digital computer. A typical calculation is shown in Fig. 19.11 where the required number of photons per optical pulse, for a receiver with a somewhat high noise amplifier and an APD of moderate quality, is plotted versus avalanche gain M. We see that about 1200 photons per optical pulse must be detected for a 10^{-9} error probability with this receiver. A state of the art APD receiver could achieve this fidelity with 400 photons/pulse. This can be compared to the quantum limit of 21 photons per pulse in the "on" state, or half that number on the average per pulse interval if the pulses are 50% "on" and 50% "off."

19.3.2 Approximations

Since the exact calculation alluded to above is tedious to perform (almost impossible for some parameter situations), various approximate techniques have been formulated. One approach is to approximate X of (19.30) as a Gaussian random variable. One need only calculate its mean and variance for optical pulse "on" and pulse "off" conditions to determine the error probabilities, e.g.,

Prob $(X > \text{Threshold} = \gamma/\text{given optical pulse "off"})$

$$= \text{erfc*} \left(\frac{\gamma - \langle X \rangle \text{ off}}{\sigma_{\text{off}}} \right), \quad (19.31)$$

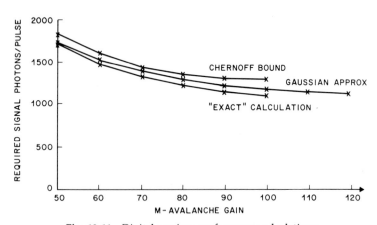

Fig. 19.11 Digital receiver performance calculations.

where $\langle X \rangle_{\text{off}}$ = mean value of X, given the optical pulse is "off," σ_{off} = standard deviation of X, given the optical pulse is "off," and

$$\text{erfc}^*(y) = \frac{1}{(2\pi)^{1/2}} \int_y^\infty \exp(-y^2/2) \, dy$$

Equation (19.31) and its analog for the optical pulse "on" condition is easy to evaluate, and yields the result shown in Figure 19.11; along with the exact calculation. The advantages of the Gaussian approximation are its ease of computation (even for receivers with intersymbol interference) and its ability to give analytical expressions for the effects of various receiver parameters on the required power for a fixed error rate. For example, if dark current is negligible, and if the transmitter is completely extinguished in the "off" state, then the required average number of primary photoelectrons, $\langle N \rangle$, in the "on" state (for a 10^{-9} error rate) is given by $\langle N \rangle = 36F + (12\langle n^2_{\text{amp}} \rangle^{1/2}/M)$ where F is the detector excess noise factor, $\langle n^2_{\text{amp}} \rangle^{1/2}$ is the rms thermal noise (in electrons) and M is the average avalanche gain. For equally likely "on" and "off" states we obtain from this expression an equation for the required detected average optical power $\eta \bar{p}$

$$\eta \bar{p} / h \nu = 18FB + [6\langle i^2_{\text{amp}} \rangle^{1/2}/(eM)], \qquad (19.31a)$$

where η is the detector quantum efficiency, $h\nu$ = energy in a photon, $\bar{p}$ is the received average optical power, B = bit rate = $1/T = 1/\text{pulse}$ spacing, $\langle i^2_{\text{amp}} \rangle^{1/2}$ = rms amplifier equivalent input noise (amperes), and e = electron charge.

One can relate $\langle i^2_{\text{amp}} \rangle^{1/2}$, $\langle n^2_{\text{amp}} \rangle^{1/2}$, and $\langle v^2_{\text{out}} \rangle^{1/2}$ of (19.21)–(19.27) as follows

$$\langle n^2_{\text{amp}} \rangle = \left\langle \frac{i^2_{\text{amp}}}{(eB)^2} \right\rangle = \langle i^2_{\text{amp}} \rangle \frac{T^2}{e^2} = \left\langle \frac{v^2_{\text{out}}}{e^2 B^2 Z^2} \right\rangle \qquad (19.31b)$$

where Z = transimpedance defined in (19.21). The disadvantage of the Gaussian approximation is its underestimation of the effect of the tail of the avalanche gain distribution on the probability of threshold crossings from below in the optical pulse "off" condition. Thus using the Gaussian approximation one tends to underestimate the required threshold setting and to underestimate the effect of dark current and imperfect extinction on the error rate. On the other hand, these distribution tails contribute to the variance of the distribution in the pulse "on" condition which gives an overestimate in the pulse "on" condition in the required separation between the threshold and the average pulse "on" output. These two errors partially cancel, yielding a fairly good estimate of the required signal level as illustrated.

Other approximations such as the Chernoff bound have also been applied to this problem of calculating the error probability with a reasonable tradeoff of accuracy and computational simplicity. A review of the entire error probability calculation problem for fiber receivers is provided in Personick *et al.* (1977).

19.3.3 Effects of System Parameters on Performance and Receiver Design

19.3.3.1 Bit Rate.   As one increases the bit rate, the number of pulses per second increases in proportion (for two-level systems). For a fixed required energy per pulse, the average power required at the receiver would grow in proportion to the bit rate. For frequencies below 25 to 50 Mbit/sec where FET receivers are used, the average energy required per pulse grows with the maximum frequency passed by the receiver filter (bandwidth) (due to the B^3 dependence of the FET amplifier noise). Therefore, the average power required at the receiver grows faster than the bit rate for frequencies up to about 25 Mbit/sec. Beyond this bit rate, bipolar amplifiers are used, and the required power grows linearly with the bit rate (due to the B^2 dependence of bipolar amplifier noise) up to about 200 to 300 Mbit/sec where speed limitations of devices becomes severe. The required-power variation with bit rate for typical p–i–n and avalanche detector receivers is shown in Fig. 19.12. The crosses indicate experimental values.

19.3.3.2 Pulse Overlap.   All of the above assume that there is no pulse overlap in the optical signal. If the optical pulses overlap, they can in principle be separated by equalization (high frequency enhancement) in the receiver linear channel. However this high frequency enhancement also

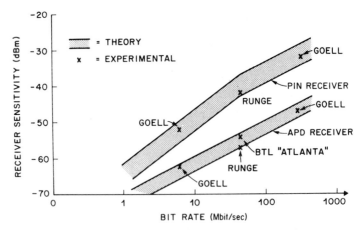

Fig. 19.12 Theory and experiment for digital receiver performance.

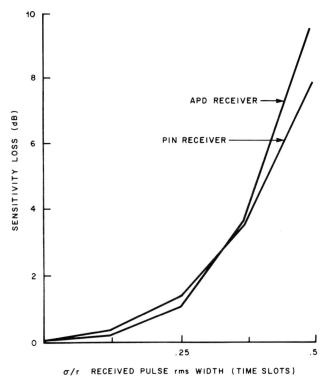

Fig. 19.13 Effects of fiber dispersion on digital receiver performance.

enhances receiver noises, reducing the receiver sensitivity relative to the nonoverlapping case. As an alternative to equalization, the pulses can be allowed to interfere; which also reduces the receiver sensitivity. Figure 19.13 shows a typical plot of receiver sensitivity loss in decibels versus received rms optical pulse width (for Gaussian-shaped pulses and an equalizing receiver). We see that for sensitivity losses below 1 dB (relative to infinitely narrow received optical pulses) the rms optical pulse width must be kept below 0.25 time slots. In practice, since optical sources are peak power limited rather than average power limited there is a tradeoff between transmitted pulse width and pulse energy. In any event, it is important to keep the rms fiber delay distortion below 0.25 time slots to prevent delay distortion from limiting the receiver sensitivity.

19.3.3.3 Amplifier Noise. In receivers with PIN detectors, the required optical power for a fixed error rate grows in proportion to the rms amplifier noise at the linear channel output. Thus for pin receivers amplifier noise strongly affects the sensitivity. For APD receivers, increased amplifier noise can be partially compensated by increased avalanche gain. Thus the effect of increased amplifier noise is mostly to make the required APD gain go up; which in turn makes the APDs more difficult to fabricate and bias.

In practice the design of the amplifier may involve a trade between amplifier noise and dynamic range. The so-called transimpedance (feedback) amplifier design (Hullett, 1976) is often used in practical repeaters because it has more range of input signal levels than the (integrate–differentiate) high impedance design. On the other hand, noise from the feedback resistor of transimpedance amplifiers often dominates the other amplifier noise sources.

19.3.3.4 Transmission Codes (Takasaki *et al.*, 1976). Amplifiers for optical fiber receivers are usually ac coupled.

As a result each optical pulse which impinges upon the detector produces a linear channel output response with a long duration low amplitude negative tail. Tails from a sequence of pulses can accumulate causing a condition known as baseline wander. If the number of pulses "on" and "off" can be kept roughly balanced for periods short compared to the tail length, then the effect of ac coupling is merely to introduce a constant offset in the receiver linear channel output which can be compensated by adjusting the threshold of the regenerator. One approach to this problem is to "scramble" the data digits at the transmitter. This results in a very low probability of long sequences of unbalanced pulses. The difficulty with this scheme is the cost of the scrambling and descrambling operation.

Another approach to balance is illustrated in Fig. 19.14. Here an "on" pulse is replaced by a half-width pulse in the left half of the time slot and an "off" pulse is replaced by a half-width pulse in the right half of the time slot. Since each time slot contains a pulse, dc balance is no problem. On the other hand, this binary pulse position modulation scheme increases the bandwidth requirements on the fiber and the receiver.

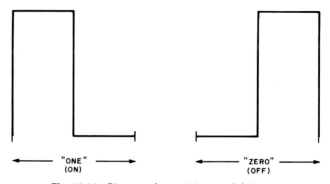

"ONE"   "ZERO"
(ON)   (OFF)

Fig. 19.14 Binary pulse position modulation.

In addition to the problem of baseline wander, coding also is required to aid in timing recovery. Typical fiber repeaters derive the clock signal for the regenerator directly from the linear channel output using a rectifier followed by a high Q tank circuit or a phase-locked loop. In order for this timing recovery process to work properly, a sufficient frequency of transitions from pulse "off" to pulse "on" must be provided. This can be accomplished by coding.

19.4 PERFORMANCE CALCULATIONS FOR ANALOG SYSTEMS (Hubbard, 1973)

19.4.1 Intensity Modulation

The simplest analog modulation scheme is intensity modulation. Here the message $m(t)$ is used to vary the light power in the following manner

$$p_{opt}(t) = P_0[1 + \gamma m(t)], \tag{19.32}$$

where γ is a constant referred to as the modulation index and the message $m(t)$ is assumed to satisfy a peak value constraint.

$$|m(t)| \leq 1. \tag{19.33}$$

The receiver consists of a detector which converts $p_{opt}(t)$ into a current with proportionality constant R (amps/watt), an amplifier, and a filter. Using the receiver model shown in Fig. 19.15 we obtain the average filter output as

$$\langle v_{out}(t) \rangle = \gamma R Z P_0 m(t), \tag{19.34}$$

where $R = (\eta e / h\nu)M$.

In addition to the average filter output, there are noise components due to the randomness of the detection process; and due to the amplifier and bias resistor noise sources. The mean-squared noise at the filter output is given by

$$\langle [v_{out} - \langle v_{out}(t) \rangle]^2 \rangle = \sigma^2 = (2P_0/h\nu)\eta e^2 \langle m^2 \rangle B + N_{th}, \tag{19.35}$$

where M = average avalanche gain of the detector, e = electron charge, B = bandwidth of the filter, $\langle m^2 \rangle$ = mean-squared value of the avalanche

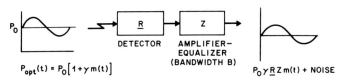

Fig. 19.15   Analog receiver circuit.

gain, N_{th} = noise output contribution from the amplifier and biasing resistors, $h\nu$ = photon energy, and η = detector quantum efficiency.

We can characterize the fidelity of the receiver output by a peak signal-to-rms noise ratio (which is related to the mean-squared error fidelity criterion)

$$S/N = [\langle v_{out}(t)\rangle_{peak}]^2/\sigma^2 \tag{19.36}$$
$$= (\gamma\eta e/h\nu)^2 P_0^2 M^2 / \left(\frac{2P_0}{h\nu} \eta e^2\langle m^2\rangle B + N_{th}\right).$$

For good APDs the relationship between $\langle m^2\rangle$ and M is given by

$$\langle m^2\rangle = M^2\left[kM + \left(2 - \frac{1}{M}\right)(1-k)\right], \tag{19.37}$$

where k = ratio of ionization probabilities for holes and electrons (k_{ideal} = 0, k for silicon APDs ~0.025–0.1) as shown in Fig. 19.16.

From (19.36) and (19.37) we see that for a given desired signal-to-noise ratio and a given amplifier noise there is an optimal value of avalanche gain, which results in a mininum required optical power level P_0. At high signal-to-noise ratios this optimal gain is so close to unity that there is no advantage to using the APD rather than a PIN detector. Figures 19.17a and b shows calculated values of required optical power for an APD with detector parameter $k = 0.1$, various values of modulation index γ, and signal-to-noise ratio; for a typical fiber system with a 4-kHz bandwidth and for a typical fiber system with a 1-MHz bandwidth. Also shown are performance curves for PIN detectors ($M = 1$).

At high signal-to-noise ratios this required power can be approximated well by the quantum limit. This limit is obtained by setting N_{th} equal to zero and $M = 1$ (no avalanche gain). One obtains

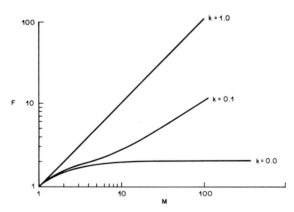

Fig. 19.16   Avalanche detector excess noise F versus Gain M.

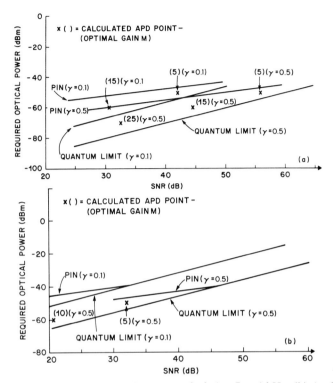

Fig. 19.17 (a) Analog receiver performance calculation $B = 4$ kHz; (b) Analog receiver performance calculation $B = 1$ MHz.

$$P_0 \bigg|_{\text{quantum limit}} = 2 \frac{S}{N} \frac{h\nu}{\eta} \frac{B}{\gamma^2}. \tag{19.38}$$

The use of analog intensity modulation in optical systems presents two difficulties. First, it is difficult to obtain linear modulation of optical sources, particularly lasers. Second, the high signal-to-noise ratio requirements of typical analog systems combined with the low modulation index required for linearity results in a small allowable loss between the transmitter and the receiver. To get around these problems various alternative analog modulation schemes have been proposed.

19.4.2 Sampled Systems—Pulse Position Modulation

One possible approach to analog modulation (other than digital encoding) is to first sample the information signal $m(t)$ and then to use the samples of $m(t)$ to modulate the position, width, or duration of a sequence of pulses. The message $m(t)$ is assumed to be band limited to some maximum frequency B. It is then sampled at the rate $2B$ using any of a number

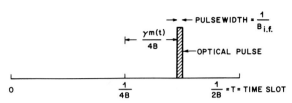

Fig. 19.18   Pulse position modulation schematic.

of sampling schemes. The resulting sequence of voltage samples $\{m_k\}$ are assumed to take on analog values between -1 and 1.

In pulse position modulation, the transmitted optical power waveform has the form

$$p_{\text{opt}}(t) = \sum_{-\infty}^{\infty} h_p \left[t - \frac{k}{2B} - \frac{\gamma m_k}{4B} \right], \tag{19.39}$$

where γ is a modulation index, and $1/2B$ = time slot width as shown in Figure 19.18. Each narrow optical pulse $h_p(t)$ is positioned in its time slot with a delay relative to the time slot center given by $\gamma m_k/4B$ where m_k is the kth sample of $m(t)$. In this scheme, linearity of the delay versus the value of m_k is important rather than linearity of the optical source. The scheme is ideally suited to sources which work best in a high peak power, low duty cycle mode.

At the receiver, the optical pulses are detected, amplified and filtered to produce a sequence of voltage pulses of the form (neglecting noise)

$$v_{\text{out}}(t) = \sum_{-\infty}^{\infty} h_{\text{out}} \left(t - \frac{k}{2B} - \frac{\gamma m_k}{4B} \right). \tag{19.40}$$

The filter output is applied to a threshold crossing detector which is triggered on at the beginning of each pulse interval $1/(2B)$ and which is triggered off when the pulse at the filter output crosses a fixed threshold. Ideally the resulting pulse from the threshold crossing detector would have a width proportional to the delay of the optical pulse in its time slot; and therefore proportional to the message sample m_k at the transmitter. In practice various errors occur. Noise can add to the ideal signal at the filter output [given in (19.40)] resulting in a perturbation of the threshold crossing time. In addition, noise can cause an early false threshold crossing, or prevent a threshold crossing from occurring at all. The fidelity criterion for the PPM system is the rms error in the estimate of the signal $m(t)$ (which has unity peak value).

A typical variation of the peak signal-to-rms noise ratio with the level of received optical power is shown in Figure 19.19. We see that at low power levels, the performance is very poor due to missed threshold crossings and false threshold crossings (sometimes called global errors). At some thresh-

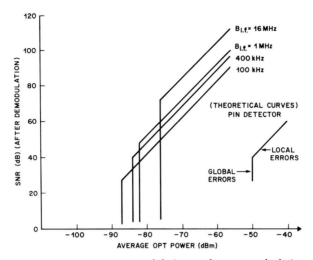

Fig. 19.19 Pulse position modulation performance calculation.

old level of optical power, the global errors are negligible compared to the local perturbations of the true threshold crossing. Beyond this point the SNR improves gradually with increasing optical power. We see that as the width of the optical pulses decreases (and the required fiber and receiver bandwidth increases) the signal-to-noise ratio for a fixed optical power level (average power) increases, provided this power remains above the knee of the evolving performance curve (negligible global errors).

19.4.3 Subcarrier Modulation Systems

A disadvantage of the ppm scheme described above is the requirement of waveform sampling and therefore careful waveform band limiting. An alternative approach to analog modulation with nonlinear sources is subcarrier FM. In this scheme the message $m(t)$ is used to control the frequency of a sinewave which in turn intensity modulates the optical power

$$p_{opt}(t) = P_0 \left[1 + \cos\left[\omega t + \gamma \int_{-\infty}^{t} m(\tau)\, d\tau\right]\right], \qquad (19.41)$$

where $\gamma m(t)|_{max}$ is the peak frequency deviation.

At the receiver (shown in Fig. 19.20) the detected, amplified, and filtered signal is applied to a discriminator or a phase-locked loop. Through this FM scheme two benefits are derived. Since source linearity is unimportant, large indices of power modulation can be used. (In fact, the source can be modulated to produce a square wave of varying frequency.) Second, since FM is a bandwidth expanding modulation scheme (as is PPM) one obtains an additional signal-to-noise ratio improvement

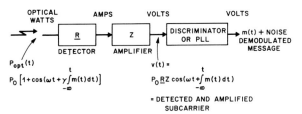

Fig. 19.20 Subcarrier FM schematic.

approximately equal to the square of the ratio of peak frequency deviation to modulation bandwidth. When using FM one must be careful to use a sufficiently large subcarrier frequency to allow for the bandwidth of the FM signal. Exact performance calculations for FM systems with APD receivers are difficult to carry out. However, as mentioned, one gets roughly an improvement in SNR proportional to the square of the bandwidth expansion. One disadvantage of FM over intensity modulation is the increased bandwidth required of the optical fiber. This may be important, for example, in television transmission using subcarrier FM.

19.5 CONCLUSIONS

This chapter has reviewed briefly some of the important concepts in optical fiber system receiver design. Because of limited space and the broad subject matter of this book, no attempt has been made to give a complete review of the complex subject of statistical receiver analysis. However, it is hoped that the reader has come away with a rough understanding of what is involved.

REFERENCES

Davenport, W., and Root, W. (1958). "Random Signals and Noise." McGraw-Hill, New York.

Dogliotti, R., Guardincerri, A., Luvinson, A., and Pirani, G. (1976). Error probability in optical fiber transmission systems. *Proc. Natl. Telecommun. Conf., 1976* pp. 37.6–1 to 37.6–5, IEEE Cat. No. 76CH1149–4CSCB.

Hubbard, W. M. (1973). Efficient utilization of optical frequency carriers for low and moderate bit rate channels. *Bell Syst. Tech. J.* **52,** 731–765.

Hullett, J. L., and Moui, T. V. (1976). A feedback receive amplifier for optical transmission systems. *IEEE Trans. Commun.* **COM-24,** 1180–1185.

McIntyre, R. J. (1966). Multiplication noise in uniform avalanche photodiodes. *IEEE Trans. Electron Devices* **ED-31,** 164–168.

McIntyre, R. J., and Conradi, J. (1972). The distribution of gains in uniformly multiplying avalanche photodiodes. *IEEE Trans. Electron Devices* **ED-19,** 713–718.

Personick, S. D. (1971a). New results on the statistics of avalanche multiplication with applications to optical detection. *Bell Syst. Tech. J.* **50,** 167–189.

Personick, S. D. (1971b). Statistics of a general class of avalanche detectors with applications to optical communication. *Bell Syst. Tech. J.* **50,** 3075–3095.

Personick, S. D. (1973). Receiver design for digital fiber optic communication systems. *Bell Syst. Tech. J.* **52,** 843–886.

Personick, S. D. (1976a). Receiver design. *In* "Fundamentals of Optical Fiber Communications" (M. K. Barnoski, ed.), pp. 183–204. Academic Press, New York.

Personick, S. D. 1976b). Photodetectors for fiber systems. *In* "Fundamentals of Optical Fiber Communications" (M. K. Barnoski, ed.), pp. 155–181. Academic Press, New York.

Personick, S. D., Balaban, P., Bobsin, J. H., and Kumar, P. (1977). A detailed comparison of four approaches to the calculation of the sensitivity of optical fiber system receiver. *IEEE Trans. Commun.* **COM-25,** 541–549.

Takasaki, Y., *et al.* (1976). Optical pulse formats for fiber optic digital communications. *IEEE Trans. Commun.* **COM-24,** 404–413.

Webb, P. P., Conradi, J., and McIntyre, R. J. (1974). Properties of avalanche photodiodes. *RCA Rev.* 234–278.

Chapter 20

Transmission System Design

STEWART E. MILLER

The preceding chapters have given the fundamentals associated with the parts of a fiberguide transmission system and in some instances have touched on the tradeoffs that are available in settling on design parameters. We seek here to view directly the system problem of providing a telecommunications service and will try to clarify how lightwave transmission via fiberguide cables relates to other alternatives available.

We will be concerned primarily with fundamental relations, and less with engineering practices; the latter are very important but require a depth of treatment of individual systems that is beyond the intended scope of this book.

20.1 GENERAL FORM OF TRANSMISSION CONFIGURATION

One might consider an all-optical transmission system in which a laser is modulated with the voice signal, with this wave carried all the way to the intended listener on fiberguides. At that point an optical wave to acoustical wave transducer might convert the signal to an intelligible form (Kleinman and Nelson, 1978). However, as described in Chapter 5, there are lightwave power limitations due to nonlinear effects which are most pronounced for single-mode fibers but which also prevent the use of multimode fibers at power levels above a few watts. Therefore, only in very limited circumstances could one transport optical power to the end of a fiberguide path for direct conversion to an audio signal or to provide power for baseband circuits.

The configuration which is most attractive is to use the optical wave as a carrier, modulated in one of a variety of alternative configurations (pulse code modulation, pulse-position modulation, intensity modulation, etc.)

653

in a manner analogous to familiar microwave carrier systems. The length of the fiberguide link may be very short (a few meters) or may be very long (many kilometers). The optimum degree of complexity is very different for these "systems" and we will explore this in subsequent paragraphs. The essentials, however, are a modulated carrier generator (LED or laser), a fiberguide link, and a detector that converts photons to electrons. This simple subsystem can be integrated into the transmission network in a great many ways, all essentially transparent to the operation of the communications network.

One might question why a heterodyne detection arrangement is not employed. It is indeed possible to construct an optical heterodyne receiver (DeLange and Dietrich, 1968) but this requires two very high-quality (single-frequency) lasers, one as the transmitter and another as a local oscillator. The resulting signal-to-noise is only 5–10 dB better than that attainable with a high-quality avalanching photodiode detector; the latter is more feasible at the present state of the art.

In the initial use of fiberguide transmission, switching is done by the same electrical switching methods and using the same switching "crosspoints" as in the absence of fiberguide. However, fiberguide transmission links can be used as subassemblies in the switching machine (data buses, for example), and may indeed induce significant functional differences in the switching-transmission network's operation when the full impact of fiberguide advantages has been felt.

This is not intended as dismissal of possible use of lasers and fibers as the heart of the "crosspoint" function in a switching machine. However it does not now seem attractive to build a switcher using optical crosspoints to be a one-for-one replacement for existing switching machines. The impact of lasers and fibers on switching is more likely to be through a more complex transformation of the transmission and switching functions, a change that has been vaguely anticipated for many years and which is still evolving. These changes may be the subject of some future book when the art has matured.

20.2 ADVANTAGES OF FIBERGUIDE TRANSMISSION

When undertaking a design to provide a service or to accomplish some given transmission objective, the designer may engage in an iterative process of tabulating the required or desired features and comparing them to the corresponding characteristics of a series of alternative technologies that may be potentially available for use. With that process in mind, we can summarize the advantages of fiberguide transmission.

(1) *Low transmission loss.* As compared with wire pairs or coaxial conductors, fiberguides have far less loss for signal frequencies above a few megahertz. This is very important to system economics, because it increases the allowed distance between repeaters.

(2) *Wide bandwidth.* Either graded-index multimode fibers or single-mode fibers can now carry digital signals at a bit-rate-distance product of 2000 Mbit·km with the same low losses as for low-frequency signals; graded-index fibers have a theoretical ·capability an order of magnitude higher and single-mode fibers have a theoretical capability nearly two orders of magnitude higher.

(3) *Small bending radius.* With proper cable design, a bending radius of the order of a few centimeters can have negligible effect on transmission.

(4) *Nonradiative, noninductive, and nonconductive.* These features avoid radiative interference, ground-loop problems and, at least in some cases, lightening-induced interference.

(5) *Small crosstalk.* The anticipated low fiber-to-fiber crosstalk has been verified experimentally, made possible by designing the cladding and the splices properly.

(6) *Graceful growth.* One might install a fiberguide cable for a relatively low bit-rate system using the advantageous low-loss property, and later convert the associated electronics for a higher bit-rate system. Some wire-pair and coaxial systems have evolved this way in the past, and fiberguide systems may do likewise.

Early thinking predicted low cost for glass fiberguides, based on the abundance of the raw material. Costs of obtaining the required material purity and fiber fabrication costs have thus far prevented this prediction from becoming true. Fiberguide cost differences on the order of 50% have major impact on the economic viability of some fiberguide systems, and changes in manufacturing volume can easily change fiber costs by more than 50%. We now view fiber cost as neither an advantage nor a disadvantage for fiberguides generically—but the prospect for advantageous cost values in selected applications should eventually be realized.

However, there are some real limitations for fiberguides. In wire-pair systems, dc and low frequencies are used to supply repeater power or to signal (ring) in the subscriber loop. These techniques are not available with fiberguides, but carrier-signal alternatives for subscriber-loop signaling appear quite feasible. Use of one wire pair as a power source or to carry signaling currents for many fibers is another example of an alternative design approach to carry out needed functions within the capabilities of fibers.

20.3 SYSTEM DESIGN CHOICES

The system designer has many options, in theory at least, when selecting design center values of the component parameters. We list a few to orient the following discussion:

(1) Fiberguide parameters: (a) single- or multimode, (b) index difference Δ, (c) core-index profile, (d) mode mixing (intentional or inadvertent);

(2) Source type—LED or laser;

(3) Carrier wavelength;

(4) Detector type—simple PN (or PIN) junction versus avalanching type;

(5) Modulation method.

There is of course a strong direct interdependence between these choices. In addition, other important system criteria are also strongly coupled to each other—crosstalk, splicing effects, cabling process and the resulting cable strength and handling properties, and, importantly, cost. The preceding chapters have given detailed relations useful for evaluating new design proposals as well as illustrating attractive prospects. We discuss here some of possible combinations that present or foreseeable technology makes possible.

Viewing the potential fiberguide application as anywhere that pairs of wires or coaxials are now used, it is clear that a wide variety of design choices will be used. The system combination that is optimum for an on-premises application with maximum length perhaps 500 m will differ greatly from that for an intercity or undersea cable system where maximum repeaters spacing and/or maximum information transmission on each fiber usually gives lowest system costs. In these two cases the desired fiber type, source, detector, and modulation format may differ significantly. The economic viability of each proposed system is evaluated separately and a decision to use or not to use is made; these decisions are made more or less independently—the on-premises use is not directly related to an intercity or undersea cable use. Indirectly the economics of many use proposals are coupled, as a consequence of the relationship between cost and production-volume for fibers, for LEDs, etc. This will be touched on more fully in the following chapter.

20.4 DIGITAL TRANSMISSION SYSTEMS

PCM is well-suited to transmission over fiberguide transmission systems, and we use that case to illustrate the interrelations between the system design choices.

The most important overall system performance number is the repeater spacing which is feasible. One makes this calculation by putting together the following component performances:

(1) End-of-life average transmitter power.
(2) Required receiver input power for low error rate (10^{-9} for example), taking into account component deterioration during life.
(3) Installed fiber cable loss, including splice loss and the effects of aging and normal environmental changes.

There are very many contributions to the above numbers, and one approach to handling the complexity is to use Monte Carlo calculations based on known or assumed distributions of the performance to determine the repeater spacing at which a large percentage (perhaps 98%) of the component combinations would meet the error rate specification. We will not pursue this detail here—this design process is similar to that needed for other transmission systems.

Looking more simply at the problem we can block out a first-order system layout by using nominal component performance figures. Some of the functional relations can be seen in the following expression for required receiver input power (Personick, 1973; Goell, 1974):

$$p_i = \frac{h\nu}{2\eta} Q \left\{ QBF(m) + \frac{2\langle i^2_{eq}\rangle^{1/2}}{e\langle m\rangle} \right\}, \tag{20.1}$$

where $h\nu$ is Planck's constant times frequency, η is the quantum efficiency of the detector, B the bit rate, m the avalanche detector's instant multiplication factor, $F(m) = \langle m^2\rangle/\langle m\rangle^2$ the excess noise factor of the detector, e the electronic charge, $\langle i^2_{eq}\rangle$ the equivalent mean-square noise current representing all amplifier noise components referred back to the detector output, and Q the ratio of required baseband signal to rms noise at the sampling time. This expression corresponds to Eq. (19.31a) in Chapter 19 which contains a discussion of the limitations in its use.

The factor Q has a value 6 for 10^{-9} error rate, and drops to only 5 for 10^{-7} error rate (Personick, 1973). The value of $\langle i^2_{eq}\rangle$ can be approximately calculated, for an FET preamplifier (Smith, 1978; Personick, 1976):

$$\langle i^2_{eq}\rangle = \frac{4KTBI_2}{R_i} + 0.7(4kT)(2\pi)^2 \left(\frac{C_T^2}{g_m}\right) B^3I_3 \tag{20.2}$$

in which kT is the Boltzmann energy, R_i is the input shunt resistance, C_T the total input capacitance, g_m the transconductance of FET, and B the bit rate. Alternatively, for a bipolar transistor preamplifier one can approximate $\langle i^2_{eq}\rangle$ by (Smith, 1978; Personick, 1976)

$$\langle i^2_{eq}\rangle = \left\{\frac{4KT}{R_i} + \frac{2eI_c}{\beta}\right\} BI_2 + 2(2\pi KTC_T)^2 \frac{B^3I_3}{eI_c} \tag{20.3}$$

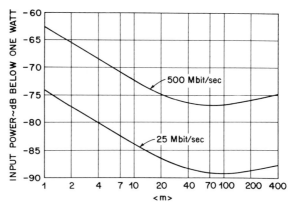

Fig. 20.1 Required receiver input power versus detector gain $\langle m \rangle$ for 10^{-9} error rate.

in which β is the transistor beta and I_c is the collector current. In both cases I_2 and I_3 are functions only of the input and output pulse shapes and are not very sensitive to those shapes; reasonable values are $I_2 \simeq 0.5$ and $I_3 \simeq 0.04$ (Personick, 1973). The bipolar amplifier has an optimum collector current to minimize Eq. (20.3), given by

$$I_{c_{\text{optimum}}} = \left(\frac{2\pi KT}{e}\right) C_T \beta^{1/2} \left(\frac{I_3}{I_2}\right)^{1/2} B \tag{20.4}$$

Amplifier noise is minimized by using a large R_i^* and by minimizing C_T. The principal variation in required receiver input power comes from the bit rate B, with secondary influence from the amplifier quality parameters and avalanching detector gain $\langle m \rangle$ if employed. Figure 19.12 of Chapter 19 illustrates the receiver input power required for 10^{-9} error rate when using silicon detectors in the 0.8- to 0.9-μm wavelength region. Below 20 Mbit/sec the FET receiver technology is most advantageous and the required receiver power (with no detector gain) varies as $B^{3/2}$. Above 30 Mbit/sec bipolar receiver technology is more advantageous and the required receiver input varies directly with B. The optimum avalanching gain appears from a balancing of the excess detector noise $F(m)$ which increases with $\langle m \rangle$ and the suppression of receiver noise through the second term in Eq. (20.1). Figure 20.1 illustrates how the required receiver input power varies with detector gain for a 25 Mbit/sec and a 500 Mbit/sec receiver; the element values assumed are given in Table 20.1.

In either case, about 15 dB maximum improvement due to avalanching gain can be realized. For small detector gain $\langle m \rangle$ the required receiver input power varies inversely as $\langle m \rangle$.

---

* Alternatively, feedback can be used around the first amplifier stage to produce a high active input impedance and correspondingly low noise.

TABLE 20.1
Receiver Element Values Used for Fig. 20.1

Transistor type	25 Mbit/sec receiver silicon FET	500 Mbit/sec receiver silicon bipolar
C_T	8 pF	4 pF
g_m	6×10^{-3} mho	
β		160
R_i	8000 Ω	800 Ω
I_c		0.63×10^{-3} A

Using modest margin and assuming a typical GaAlAs injection laser output of 0.5 mW one calculates the nominal repeater spacing versus bit rate shown in Fig. 20.2; this plot is applicable to systems using single-mode fibers and laser sources with sufficiently narrow spectral width so that pulse dispersion effects are negligible. In the Atlanta system experiment (Kerdock and Wolaver, 1978) installed fiber-cable losses in the 6 dB/km region were realized; additionally, experimental repeaters have been demonstrated at bit rates to 300 Mbit/sec (See Fig. 19.12), so the 6 dB/km line on Fig. 20.1 is well-supported out to 300 Mbit/sec. Better base-band transistors are needed to achieve 1 Gigabit rate repeaters, but this seems to be nearly within reach (Nawata *et al.*, 1978). The 1 dB/km line on Fig. 20.2 is less well-documented. However, individual fiber losses below 1 dB/km at wavelengths near 1.1 to 1.3 μm have been reported (Horiguchi and Osanai, 1976), and other work has succeeded in producing fiber cables with added cabling losses of $\frac{1}{4}$ dB/km or less. The economic value of the large repeater spacings indicated for 1 dB/km in Fig. 20.2 lend great impetus to work aimed at achieving such low losses and to solving the problems of single-mode fibers.

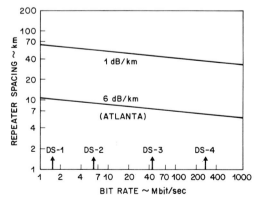

Fig. 20.2 Nominal repeater spacing versus bit rate for a loss-limited system. (Figures 20.5 and 20.6 show how dispersion effects limit the attainable repeater spacing).

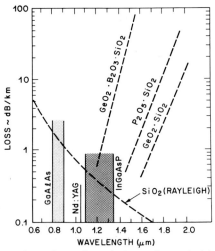

Fig. 20.3 Approximate lower boundaries on losses in high-silica fibers.

The boundaries on the achievable losses are approximately as indicated in Fig. 20.3 (Osanai *et al.*, 1976). Rayleigh scattering sets a lower limit in the wavelength region shorter than about 1 μm, and above 1 μm the tails of vibrational absorption peaks give different minimum values depending on the choice of dopant used to raise or lower the index of refraction. The trends of these boundaries are also shown in the measured loss for a germanium- and boron-doped fiber (Fig. 20.4). The additional loss peaks at 0.95, 1.2+ and 1.4+ μm are due to OH⁻ ions, which cause added loss at the rate 1.25 dB/km per part per million at the 0.95-μm wavelength. This shows the importance of keeping very low OH⁻ levels, and of avoiding the

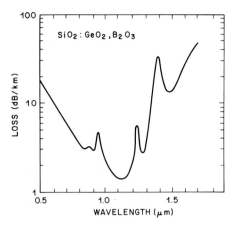

Fig. 20.4 Measured loss versus wavelength for a germanium- and boron-doped silica fiber.

system choice of carrier wavelength at those wavelengths where OH⁻ overtone losses peak. To use carriers at those wavelengths would be a need-less risk of disaster due to minor impurity increases during manufacture.

The above discussion of fiber loss effects is adequate reason to support a major effort to get sources and detectors in the 1.1- to 1.3-μm region. There is an additional reason—minimization of pulse dispersion.

When one takes account of the inherent spectral width of lasers and LEDs one finds that the attainable repeater spacing may be limited by pulse broadening (yielding pulse-to-pulse interference) rather than by thermal- or shot-noise effects. The fundamentals of this situation are as follows. In an infinite medium of fused silica, signal components at dif-ferent wavelengths travel with different group velocities and develop a delay difference as drawn in Fig. 20.5. For wavelengths near 0.82 μm the delay difference is near 100 psec/km-nm; present-day LEDs and injection lasers made in the AlGaAs material system have typical spectral widths of 35 and 2 nm, respectively. To assess the effect on repeater spacing we find the rms impulse response σ for the fiber excited by an impulse of the LED or laser and transform this into an allowed bit rate for 1 dB loss of signal-to-noise ratio (Personick, 1973). The result is

$$B = 1/4\sigma. \qquad (20.5)$$

For material dispersion, σ is directly proportional to fiber length, leading to a bit-rate limitation expressed as a product of bit-rate and distance. For the AlGaAs LED and laser the material-dispersion-limited bit rates are 140 and 2500 Mbit·km. Figure 20.6 shows as dotted lines these wavelength-

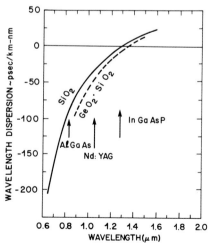

Fig. 20.5 Wavelength dispersion versus wavelength for an infinite medium of fused sil-ica and germania-doped fused silica.

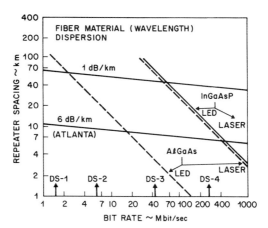

Fig. 20.6 Nominal repeater spacing versus bit rate showing the limitations imposed by dispersion of source spectral components.

dispersion limits; for the 6 dB/km cable loss the AlGaAs LED system is dispersion limited above about 18 Mbit/sec and the laser system is dispersion limited above about 400⁺ Mbit/sec. Wavelength dispersion in fused silica goes through a minimum at wavelengths near 1.3 μm (see Fig. 20.5), the minimum being about 2.5×10^{-2} psec/km·nm² due to higher order effects (Kapron, 1977). Consequently, sources at longer wavelengths are desired to reduce dispersion as well as to realize the advantage of lower fiber loss. Chapter 16 gives more information on this work. We note here that InGaAsP shows great promise for use in LEDs and lasers in the 1.3-μm region.

There is another cause for pulse dispersion in multimode fibers—group velocity differences among the propagating modes. For equal excitation of all modes by an input impulse, one can find a root-mean-square output pulse width σ; then the allowable system bit-rate for 1 dB signal-to-noise degradation is again given by Eq. (20.5). For fiber cables that are uniform longitudinally (no mode mixing) the bit-rate-distance product again characterizes the system effect and this again appears as a 45° boundary line on the repeater spacing versus bit rate plot. Figure 20.7 shows as dotted lines the boundaries due to this modal dispersion for a $\Delta = 0.01$. At bit rates where the dotted line lies below the loss-limit boundary, the system repeater spacing is modal dispersion limited. The best graded-index fibers have less modal dispersion than step-index fibers by a factor of about 100; another factor of about 10 is theoretically achievable [see Chapter 4, Eq. (4.33)].

Although intentional mode mixing should be effective in reducing modal dispersion [see Chapters 3 and 4, Eq. (3.170) and (4.47)], no information on practical realization has yet been published. Unintentional

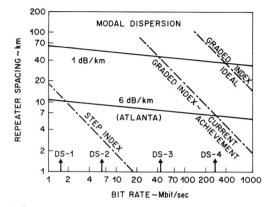

Fig. 20.7 Nominal repeater spacing versus bit rate showing the limitations due to pulse spreading caused by group-velocity differences among the propagating modes in multimode fibers ($\Delta = 0.01$).

mode mixing is common, and frequently causes pulse dispersion to grow more nearly as the square root of fiber-cable length, rather than linearly with fiber length as would happen in uniform structures. Mode mixing can be introduced not only by longitudinal variations in the fiber cable itself, but also by splices which join dissimilar or imperfectly aligned fibers. From the system design standpoint, these effects introduce some loss and reduce the accumulated pulse dispersion, the latter being a performance improvement. However, the normal objectives of striving for uniformity and accuracy in the fiber-cable manufacture and in the splicing operations tend to *eliminate* the unintentional mode mixing. It is therefore crucial to monitor the performance of the cable during installation if the system performance is dependent on dispersion reduction of unintentional mode mixing. By way of contrast, intentional mode mixing built into the fiber during manufacture would be a fully reliable system-performance parameter.

20.5 SELECTION OF SYSTEM COMPONENTS

Let us consider the choices of design-center values to resolve some of the open questions listed in Section 20.3.

20.5.1 Choice of Carrier Wavelength

It is essential that the carrier wavelength be one at which a satisfactory detector is available and one where the fiber cable has satisfactory loss and dispersion. At the present writing most all lightwave systems use sources emitting in the 0.82- to 0.9-μm region, where either GaAs or AlGaAs junction devices emit. An overtone loss due to the OH$^-$ ion peaks at 0.95 μm

(Fig. 20.4) but has a linewidth that extends ±0.05 μm from that wave-length leading to a choice in the 0.82- to 0.85-μm region. The detector need is filled by a silicon device (which can be used at wavelengths up to about 1.08 μm). A longer wavelength source would give less loss and dispersion (Figs. 20.4 and 20.5); an appropriate source compatible in the sense of size and complexity is the Nd:YAG fiber laser (see Chapter 16.5.2) but this at present has not been fully developed. Shorter wave-length sources are unattractive from the viewpoint of fiber loss and dispersion.

Even longer wavelength sources, i.e., near 1.3 μm, are most attractive from the viewpoint of fiber loss and dispersion, but at present there is available no really satisfactory detector and the LED–laser art has just produced a few research models of suitable devices. For systems wherein fiber loss and dispersion is economically important, development of longer wavelength sources and detectors will be important. For other shorter length systems like data-bus on-premises applications, fiber loss and dispersion are not very critical and the wavelength choice will be made to minimize the costs of the LED, laser, and detector.

20.5.2 The LED-Laser Choice

In addition to wavelength-related factors discussed above, the factors influencing carrier generator choice include system modulation method, modulation bandwidth, and cost. The cost factor includes not only the cost of the LED or laser itself, but also the cost of the associated driving circuits. Because the laser is a threshold device, because its threshold is significantly influenced by ambient temperature and because it is necessary (from the viewpoint of device life) to operate at driving currents only 10 to 30% above threshold, the laser driving circuits are intrinsically more expensive than those for an LED.

In the 0.82- to 0.9-μm region where sources and detectors are now available, the LED spectral width in combination with the wavelength dispersion of silica fibers yields a bit-rate-distance product of about 140 Mbit-km, as illustrated in Fig. 20.6. This is very adequate for bit rates beyond 100 Mbit/sec in on-premises applications, and for spans of 5 km or so at bit rates out to around 20 Mbit/sec. In the future, when InGaAsP LEDs (or some other 1.1–1.3 μm emitters are available the LED systems will be cable loss limited out to around 1500–2000 Mbit-km.

The available power into the fiber from an LED runs about 10–15 dB below that available from a laser so this fact alone gives preference to the laser in an application where receiver noise is limiting. Furthermore for high-bit-rate long-span systems the narrower spectral width of the laser is needed to avoid wavelength dispersion. Figure 20.6 illustrates this with a

2500 Mbit-km 45° line for existing AlGaAs injection lasers, and with further refinement a wavelength dispersion figure as high as 25,000 Mbit-km could theoretically be attainable with AlGaAs lasers and much higher with InGaAsP lasers.

20.5.3 The Detector Choice

For carrier wavelengths between 0.8 and 1.08 μm, silicon detectors provided useful quantum efficiency (see Chapter 18, Fig. 18.1), and there are commercially available both PIN and avalanching types. Germanium detectors are also available for the longer wavelength region, but dark current and leakage currents make their noise excessive. There is now considerable research interest in new material systems and in means for quieting the germanium detectors with the objective of producing a satisfactory PIN and avalanching detector for the 1.1- to 1.3-μm region.

The choice between PIN and avalanching (APD) types is principally one of cost. The PIN–APD cost situation is analogous to that for the laser–LED. The avalanching process in the APD has a sharp threshold which is sensitive to ambient temperature, which may lead to the requirement of dynamic control of a relatively high bias voltage. It follows that the APD control and driver circuits are much more expensive than those for the PIN detector, and the APD itself is more expensive than the PIN device. Therefore low-cost systems will be more readily achievable with PIN detectors. The penalty in receiver sensitivity is shown in Fig. 20.1; optimum APD gives gain about 15 dB more receiver sensitivity than for the PIN— $\langle m \rangle = 1$. There may be a place for a system using low APD gain ($\langle m \rangle$ in the 5 to 10 region) with a simpler design of APD and associated control circuits.

20.5.4 The Fiberguide Choice

In selection of the fiberguide design, we choose between single mode and multimode, we prescribe an index difference Δ, we put requirements on the core's index profile, and we bear in mind the effects of mode mixing whether intentional or inadvertent. These choices react strongly on system performance and on the cable design, in which materials choices and structural alternatives of differing costs are traded off against the nominal choice of Δ and mode mixing and microbending susceptibility. These latter relations are discussed more fully in Chapters 13 and 6.

If an LED source is used, a multimode fiber must be used in order to carry any significant amount of power. With increasing values of core–cladding index difference Δ the power collected from the LED and the fiber's modal dispersion both increase linearly.* A Δ of at least 0.01 (corresponding to a numerical aperture NA of 0.21) is typically used, and this yields a tolerable bit-rate-distance product of 18 Mbit-km in step-index fibers. By stretching the fiber technology one can make germania-doped silica fibers with a Δ of 0.03 (NA $\simeq$ 0.36); step-index fibers with about the same Δ can be made by using pure silica core and plastic cladding. The bit-rate-distance product for a Δ of 0.03 is 6 Mbit-km, which is good for 10 Mbit/sec or more in on-premises applications since these are usually well under 1 km in length. If an LED system is to be pushed toward its high-speed limit, grading of the profile is desirable and perhaps necessary. For GaAlAs LEDs with wavelength dispersion of 140 Mbit-km, a $\Delta = 0.01$ fiber must be graded for a factor of 10 modal dispersion reduction (corresponding to a profile exponent slightly less than 3) in order to make use at the 140 Mbit-km rate possible. For a $\Delta = 0.03$ fiber, a factor of 20–25 modal dispersion reduction is needed to bring modal dispersion down to the level of wavelength dispersion, and this corresponds to a profile exponent of about 2.2. By using 1.3-μm LEDs (InGaAsP, for example), the wavelength dispersion limit is about 2500 Mbit-km; to reduce the modal dispersion to this region requires a graded-index fiber with less than 1% of the modal dispersion of the step-index fiber, which is the best that has yet been done. Unintentional mode mixing is very small in the lowest loss fibers, and should not be counted on for dispersion reduction in spans of less than 1 km. For longer spans, splices may introduce some mode mixing and minor reductions from linear accumulations of dispersion may be experienced. The full potential of intentional mode mixing for dispersion reduction has yet to be explored.

For systems using a laser source, either a single-mode or multimode fiber may be used. The small transverse size of the guided-wave field in single-mode fibers (5 to 10 μm diameter) makes the splicing operation more critical and more expensive than for multimode fibers (typical core diameter 50 μm). This is a strong reason for using multimode fibers, even with lasers which usually radiate in only one or a few modes. Wavelength dispersion for typical AlGaAs lasers in silica fibers is in the order of 2500 Mbit-km which (as noted above) requires the best available graded-index fibers to bring modal dispersion down to wavelength dispersion. Further refinement could reduce the GaAlAs spectral width by a factor of 10, and theory shows a factor of 10 still available also in modal dispersion reduction (see Eq. (4.33)). The dispersion limits that one may go to with lasers in multimode graded-index fibers fall in the range 1500 to 15,000 Mbit-km; feasible manufacturing tolerances will set this boundary.

---

* Ideally, graded-index fibers yield modal dispersion proportional to Δ^2, but (as discussed in Section 4.4) in the presence of profile imperfections a linear-Δ relationship is also appropriate.

For the ultimate in high-bit-rate systems, the laser with a single-mode fiber is best. By refining the laser to deliver its power in a single mode and confining its spectral width to 1 or 2 Å, a bit-rate-distance product of about 30 Gbit-km/sec could be achievable. A 1 Gbit/sec pulse rate with a repeater span of 30 km is shown by Fig. 20.2 to be within physical noise limits, so this goal might be realizable.

20.5.5 Sample System Layouts

We can illustrate the results of the preceding discussion with a few sample systems.

System #1
  Signal: 25 Mbit/sec binary PCM
  Source: GaAlAs LED
  Carrier wavelength: 0.85 μm
  Fiber Type: plastic-coated silica

$$\Delta \simeq 0.03, \qquad NA \simeq 0.35$$

Transmitter power into fiber: 50 μW (average)
Detector: silicon PIN
Required Receiver Input Signal: -44 dBm (44 dB below 1 mW)
Fiber cable and splicing loss: 15 dB/km.

This combination has about a 10-dB margin against noise at 1.4-km repeater spacing. The components should all be achievable at low cost.

System #2
  Signal: 25 Mb/s binary PCM
  Source: GaAlAs LED
  Carrier wavelength: 0.85 μm
  Fiber type: high-silica graded index

$$\Delta \simeq 0.02, \qquad NA \simeq 0.3$$

Transmitter power into fiber: 200 μW (average)
Detector: silicon avalanching photodiode (APD)

$$APD \text{ gain } \simeq 20^*$$

Required receiver input signal: -56 dBm
Fiber cable and splicing loss: 4 dB/km.

This combination has a 9-dB margin against noise at a 10-km repeater span. The fiber grading must be good enough for a bit-rate-distance product of 250 Mbit-km, which is a factor of 28 better than step index. The components would be of intermediate cost.

System #3

   Signal: 500 Mbit/sec binary PCM
   Source: GaAlAs laser, single-mode output
   Carrier wavelength: 0.85 μm
   Fiber type: single-mode high silica
   Transmitter power into fiber: 1 mW (average)
   Detector: silicon avalanching photodiode

$$\text{APD gain} \simeq 20^*$$

   Required receiver input signal: -44 dBm
   Fiber cable and splicing loss: 2 dB/km.

This combination has an 8-dB margin against noise at an 18-km repeater span. The laser spectral width must be no more than 5 Å to keep wavelength dispersion sufficiently small. These component performances have been achieved individually under laboratory conditions, but are not now on-line as manufacturable products.

20.6 ANALOG FIBERGUIDE TRANSMISSION

In Sections 20.4 and 20.5 we have reviewed some fiberguide system design choices using binary PCM as the modulation method. We will now address the question of modulation choice a little more broadly.

If a fiberguide transmission link is a subsection of a larger communication network there will frequently be strong influence favoring a particular modulation format. Avoidance of coding equipment costs and avoidance of coding noise would suggest using analog transmission on a fiberguide link that is part of a larger analog network—for example, a microwave radio relay network using narrow-band FM as the modulation method. It is indeed feasible to do so. However, there are limitations on lightwave fiberguide links that are generic to all analog applications and we will give an indication of these problems with an example.

Consider an analog channel in the baseband region, with a 4-MHz bandwidth potentially suitable for a video signal. There are readily available transistors with good linearity at 100 mW output as a transmitter signal P_s. The fundamental limiting noie is KTB, with K as Boltzmann's constant, T absolute temperature, and B the equivalent flat noise bandwidth of the channel. The quantity KTB at room temperature is 1.65×10^{-14} W, so with the 100-mW signal power P_s the maximum ratio of carrier to noise is

---

* The APD gain is below optimum, but only by <3 dB in required input power (see Fig. 20.1). This choice may ease tolerances and reduce cost.

$$P_s/KTB = 6 \times 10^{12} \quad \text{or} \quad 128 \text{ dB}.$$

The comparable lightwave channel has available a signal carrier power P_s of about 1 mW. The fundamental noise is $h\nu B$, with h the Planck's constant, ν the frequency, and B the equivalent flat noise bandwidth of the channel (Gordon, 1962). The quantity $h\nu B$ is about 8×10^{-13} in our 4-MHz band at 1-μm wavelength. With a detector quantum efficiency η of 0.8, the maximum ratio of carrier to noise ratio is

$$\eta P_s/h\nu B = 1 \times 10^9 \quad \text{or} \quad 90 \text{ dB}.$$

Thus the fiberguide channel has about 38 dB less carrier-to-noise ratio available than has the baseband transistor alternative. This is 38 dB less loss that can be tolerated in the transmission medium—the fiberguide cable or the coaxial cable as the case may be. This deficit comes both from $h\nu B$ being larger than KTB and from less available transmitter carrier power.

The total equivalent input noise in the lightwave analog receiver includes both the quantum-noise term $h\nu B$ and thermal noise from the first amplifiers following the detector.* We now show how these two noise contributors influence the carrier-to-noise ratio at different received signal power levels P_i; this is a significant relation, because the quantum noise term varies with received signal power whereas the thermal noise level is a constant.

The detected carrier signal current i_c is

$$i_c^2 = \left(\frac{\eta P_i e}{h\nu}\right)^2 \langle m \rangle^2. \tag{20.6}$$

The mean-square value of the thermal noise current $\langle i_{TH}^2 \rangle$, referred back to the detector output, can be taken as Eq. (20.2) (with the approximation that the value B is the equivalent noise bandwidth). The mean-square value of the quantum noise may be written

$$\langle i_Q^2 \rangle = \frac{2e^2}{h\nu} \eta P_i B \langle m \rangle^2 \langle m \rangle^x. \tag{20.7}$$

with x in the range 0.3 to 0.5 and with other symbols defined under Eq. (20.1). The carrier to noise ratio is then

$$\frac{i_c^2}{\langle i_{TH}^2 \rangle + \langle i_Q^2 \rangle}. \tag{20.8}$$

This ratio has been calculated for an illustrative case: an FET first amplifier

---

* A less ideal detector might also exhibit dark current or leakage currents, as described in Chapter 18, see particularly Section 18.2.3 and Fig. 18.2.

with C_T of 8 pf, g_m of 6×10^{-3}, R_{in} of 5000 Ω and the 4-MHz bandwidth. Figure 20.8 shows the results for no detector gain and for an average gain of 20. With no detector gain the carrier to noise ratio is dominated by quantum noise above input powers of 10^{-2} mW and by thermal noise below input powers of 10^{-3} mW. With detector gain of 20 the transition from quantum noise limiting to thermal noise limiting is dropped down to received powers near 10^{-4} mW; however, the excess noise introduced by the avalanching detector [represented by $\langle m \rangle^x$ in (20.7)] *reduces* the carrier-to-noise ratio compared with that for no detector gain at large received signal power levels. In the region where quantum noise dominates the carrier-to-noise ratio would not be improved by increasing detector gain further, since both i_c^2 and $\langle i_q^2 \rangle$ vary as $\langle m \rangle^2$, and excess detector noise will eventually cause a decrease in carrier to noise ratio. There is of course an optimum value for $\langle m \rangle$ in a particular application.

Figure 20.9 shows for the same case the total noise $\langle i^2_{TH} \rangle + \langle i_q^2 \rangle$ as a function of received power. This illustrates more dramatically the way the system output noise varies with received signal power level.

The above example assumed a 4-MHz bandwidth, and we may inquire how the result is affected by changing B. Referring to Eq. (20.7) we observe that $\langle i_q^2 \rangle$ varies as B. For an optimized bipolar amplifier it is found that $\langle i^2_{TH} \rangle$ varies as B^2 (Personick, 1976) Eq. 6.8), and for an FET amplifier $\langle i^2_{TH} \rangle$ varies as $(k_1 B + K_2 B^3)$ (see Eq. 20.3). Hence in Fig. 20.8 the region which is thermal noise limited (below $P_i = 10^{-2}$ mW for no detector gain), the carrier-to-noise ratio drops 1 to 3 dB for 1 dB increase in B; in the region which is quantum noise limited (above $P_i = 10^{-2}$ mW for no detector gain), the carrier-to-noise ratio drops 1 dB for each 1 dB increase in B. Naturally the transition from thermal noise limiting to quantum noise limiting shifts to higher received power levels as B increases. With de-

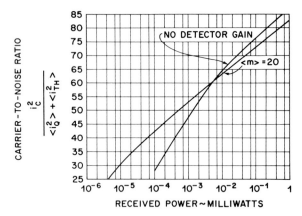

Fig. 20.8 Carrier-to-noise ratio versus received power for a 4-MHz bandwidth receiver.

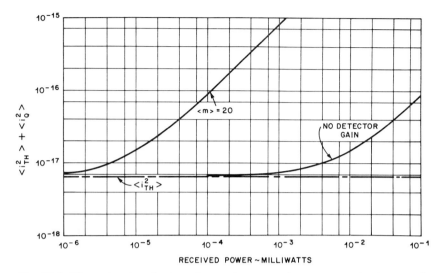

Fig. 20.9 Thermal noise plus quantum noise on carrier versus received power for a 4-MHz bandwidth receiver.

tector gain, quantum noise usually dominates and the carrier-to-noise ratio varies inversely as B.

This discussion has been framed in terms of carrier-to-noise ratio. Similar relations will be found for the signal-to-noise ratio in various analog transmission schemes (Hubbard, 1973). For intensity modulation the signal-to-noise ratio will usually be about 5 to 10 dB less than the carrier-to-noise ratio, depending on the modulation index permissible with acceptable linearity. For some other analog arrangements (such as using an FM subcarrier) one can get a larger signal-to-noise ratio at the expense of some bandwidth expansion. These cases must each be analyzed to get a precise answer but will not be pursued further here.

Because many analog systems require 40 dB or more signal-to-noise ratio at the end terminal, and because frequently the subsystem links are required to be much better than this as part of the budgeting of the acceptable noise contributions, analog transmission on fiberguide is not attractive in many applications where analog coaxial cable systems are able to perform with ease.

20.7 MORE SOPHISTICATED SYSTEMS

The preceding discussion has centered on using a single lightwave carrier on each fiberguide. As time passes there are certain to be situations in which a more efficient usage of the fiberguide transmission potential is feasible and economically desirable. The straight forward approach is

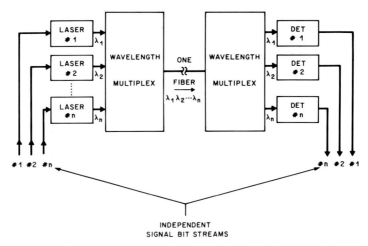

Fig. 20.10 Block diagram of wavelength multiplexing of "*n*" independent bit streams onto a single-fiber transmission line.

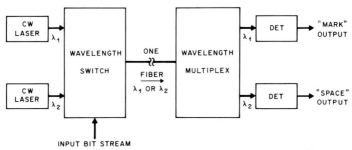

Fig. 20.11 Use of a wavelength switch as modulator to send a digital bit stream over a fiber.

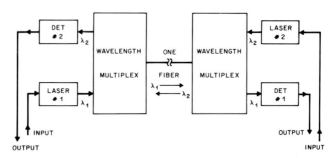

Fig. 20.12 Block diagram showing method for two-way transmission on one fiber by employing wavelength multiplexing.

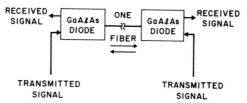

Fig. 20.13 A single GaAlAs diode can act as an LED transmitter or as a detector on a time-division basis, thus providing a two-way link over a single fiber.

frequency-division multiplexing, conceptually the same as has been done in microwave radio systems. This is illustrated in Fig. 20.10. Each laser (or LED) is independently modulated, either with a digital or analog signal, and wavelength multiplexers make possible the combining of the different carrier wavelengths at the transmitter and separating them at the detector. There are a number of possible forms of optical filters potentially applicable (Tomlinson, 1977) and suitable filters have been demonstrated in the laboratory (Tomlinson and Aumiller, 1977; Sugimoto *et al.*, 1978; Alferness and Schmidt, 1978). Bearing in mind that 1-nm bandwidth channels can carry a gigabit or more and noting the large number of nanometers of low-loss transmission in the best fibers, one can extrapolate to a very large total information capacity on a single fiber. The penalty (with existing technology) comes in reduced repeater spacings, for the filters have very significant loss. Additionally there is added complexity and dollar cost in the filters themselves.

One of the wavelength combining filters (Alferness and Schmidt, 1978) has the capability of rapid switching, leading to a hypothetical arrangement wherein "marks" are sent with a pulse at one wavelength and "spaces" are sent with a pulse at another wavelength (Fig. 20.11). Error correction can be added, using the redundancy involved.

Another variation involves sending one direction on the fiber using a λ_1 carrier, and sending in the opposite direction on the same fiber using a λ_2 carrier (Fig. 20.12) (Sugimoto *et al.*, 1978).

Finally, a proposal advanced by E. L. Chinnock and K. Ogawa of Bell Laboratories involves using a single GaAlAs diode as both LED transmitter and PIN-diode detector, with appropriate forward- or reverse-biasing of the junction, respectively. Illustrated in Fig. 20.13, this provides a time-division two-way channel with one fiber and one diode at each end of the fiber.

The above paragraphs may give a clue to the potential for novelty in future lightwave fiberguide systems. Lightwave configurations very different from those used in lower frequency systems are certainly feasible and will probably prove attractive.

REFERENCES

Alferness, R. C., and Schmidt, R. V., (1978). Tunable optical waveguide directional coupler type filter. *Tech. Dig.*, O.S.A. and IEEE *Top. Meet. Integr. Guided Wave Opt., 1978* p. TuA3.

DeLange, O. E., and Dietrich, A. F. (1968). Optical heterodyne experiments with enclosed transmission paths. *Bell Syst. Tech. J.* **47**,.161–178.

Goell, J. E. (1974). Input amplifiers for optical PCM receivers. *Bell Syst. Tech. J.* **53**, 1771–1793; An optical repeater with high impedance input amplifier. *ibid* **53**, 629–643.

Gordon, J. P. (1962). Quantum effects in communications. *Proc. IRE* **50**, 1898–1908.

Horiguchi, M., and Osarai, H. (1976). Spectral losses of low-OH-content optical fibers. *Electron. Lett.* **12**, 310.

Hubbard, W. M. (1973). Efficient utilization of optical-frequency carriers for low and moderate bandwidth channels. *Bell Syst. Tech. J.* **52**, 731–765.

Kapron, F. P. (1977). *Electron. Lett.* **13**, 96–97.

Kawana, A., Miyashita, T., Nakahara, M., and Hosaka, T. (1977). Fabrication of low-loss single-mode fibers. *Electron. Lett.* **13**, 188.

Kerdock, R. S., and Wolaver, D. H. (1978). Results of the Atlanta lightwave system experiment. *Bell Syst. Tech. J.* **57**, 6, 1857–1879.

Kleinman, D. A., and Nelson, D. F. (1978). Nonlinear analysis of a photo voltaic optical telephone receiver. *Bell Syst. Tech. J.* **57**, 1569–1596.

Nawata, K., Machida, S., and Ito, T. (1978). An 800 Mbit/s optical transmission experiment using a single-mode fiber. *IEEE J. Quantum Electron.* **QE-14**, 98.

Osanai, H., Shioda, T., Moriyami, T., Araki, S., Horiguchi, M., Izawa, T., and Takata, H. (1976a). Effect of dopants on transmission loss of low-OH-content optical fibers. *Electron. Lett.* **12**, 549.

Personick, S. D. (1973). Receiver design for digital fiber optic communication systems. Parts I and II. *Bell Syst. Tech. J.* **52**, 843–886.

Personick, S. D. (1976). Design of repeaters for fiber systems. *In* "Fundamentals of Optical Fiber Communications" (M. K. Barnoski, ed.). Chapter 6. Academic Press, New York.

Ravenscraft, I. A. (1977). Feasibility trial of optical-fiber transmission system. *P. O. Electr. Eng. J.* **70**, Part 2, 119.

Smith, R. G. Detectors and receivers for optical fiber applications. *Proc. IEEE Int. Symp. Circuits Syst., 1978* p. 20.

Sugimoto, S., Minemura, K., Kobayashi, K., Shikada, M., Nomura, H., Kaede, K., Ueki, A., and Matsushita, S. (1978). Wavelength division two-way fiber-optic transmission experiments using micro-optic duplexers. *Electron. Lett.* **14**, 15–17.

Tomlinson, W. J. (1977). Wavelength multiplexing in multimode optical fibers. *Appl. Opt.* **16**, 2180–2194.

Tomlinson, W. J., and Aumiller, G. D. (1977). Optical multiplexer for multimode fiber transmission systems. *Appl. Phys. Lett.* **31**, 169–171.

Chapter 21

Potential Applications

STEWART E. MILLER

21.1 INTRODUCTION

This concluding chapter calls attention to the anticipated applications of lightwave transmission via fiberguides. The preliminary listing of the advantages of fiberguides given in Section 20.2 should be reviewed, and from there it may be concluded that lightwave transmission on fibers has not a *single* future, but *many* futures. This is a consequence of the large variety of potential applications that fiber transmission has—on-premises transmission of data or pictures, local loop transmission between the customer and the central office, trunking between central offices, intercity trunks, and undersea cables in the civilian sphere, and for shipboard communications, aircraft data bus and battlefield portable communication lines in the military sphere (Miller *et al.*, 1973; Miller, 1977). Many writers have noted that fibers are potentially applicable wherever wire pairs or coaxials have been found useful. It is more typical for a new art to be potentially applicable to a more limited market—such as intercity trunks *or* local loops. The technologic competitor to fiberguide will be very different in the various proposed applications—for example, satellites or millimeter waveguide for the intercity trunks and a 1.5 or 3 Mbit/sec wire-pair-cable system for the interoffice trunk or local-loop "wideband" facility. The futures for the various lightwave possibilities are evolving in parallel, and success in one will not necessarily be accompanied by success in another. Nevertheless there is some interrelation between the various fiber application prospects through commonality of components, such as sources, fibers, and detectors or through commonality in the manufacturing processes and know-how for differing specific designs of the components. There follows a more specific discussion of proposed applications, which should sharpen this picture.

675

As implied, commercial usage seems to be on the threshold and the next several years should show whether this art will explode into the telecommunications field or just gradually infiltrate. There are threshold effects in the economics which relate to production volume and which are regenerative. At this writing the art is not above these thresholds.

The initial approach to evaluating the attractiveness of fiberguides is to compare them with the alternatives in presently perceived functions: fiberguide versus coaxial cable in undersea or intercity systems. Fiberguide versus wire-pair carrier systems in the interoffice trunking function. These evaluations are reasonable and necessary, but probably miss some of the major impact that fiberguide will—with benefit of hindsight 20 years from now—be found to have produced. What fibers offer, beyond valuable conveniences such as small size and freedom from radiative interactions with their environment, is very small loss in bandwidths that range from one to one-hundreds or even one-thousand megahertz. Fibers do cost at this writing more than an order of magnitude more than wire pairs (per meter); although presently comparable in cost to coaxials, fibers are expected to cost an order of magnitude less than coaxials. This sets the stage for the future. For one voice circuit a fiber is not much of a competitor. For broadband applications it should compete very well and it will open up the possibilities to new configurations in which relatively inexpensive broadband transmission plays a role. Broadband services to the customer is a frequently cited example. Just as important may be some less obvious links in a future communication network which capitalizes on the low-loss broadband characteristic of fibers. For example, the switching apparatus itself may become more distributed geographically with high-speed links interconnecting its parts.

21.2 ON-PREMISES APPLICATIONS

The on-premises category can include all applications for which the fiber cable length is up to 1 km approximately and which involve transmission between terminals located entirely within a single building or other enclosure. The military groups have shown interest in fiberguide for transmission of data and other communication between parts of an aircraft or between parts of a ship and for battlefield communications (Taylor, 1977). Trade journals are beginning to carry accounts of this type of work (Markstein, 1977). It is reported that several measuring equipment manufacturers will soon offer fiberguide links as auxiliary apparatus to mate with data accumulation and data processing equipment. Computer manufacturers are also reported to have designs in process.

In the telephone system a pioneering demonstration was carried out in

1973 at Bell Laboratories by J. E. Goell, W. M. Muska, and Tingye Li, who built and installed a 16 Mbit/sec fiberguide link in a laboratory model No. 4 ESS toll digital switching machine. Numerous digital links of length 40 to 400 m are an integral part of the switching machine, and the small size, low radiation properties, freedom from ground loops and low loss of the fiber version are attractive. Extremely high reliability is also necessary in all switching machine elements and this had not been proven for light-wave components at the time No. 4 ESS went into manufacture. Interest continues (Hackett *et al.*, 1977).

Digital data buses in general seem likely applications both in telecommunications equipment and in other computer and control applications. As noted in Chapter 20, analog transmission is feasible in limited circumstances, but most uses appear to be digital (Albanese and Lenzing, 1978).

Typically these fiberguide subsystems involve LED sources accompanied by multimode fibers and either PIN or APD detectors. Although early work employed fiber bundles to carry the output of a single large-area LED, this inefficient approach is being replaced by single-fiber-per-channel systems with high-brightness LEDs. The bit rate is usually 20 Mbit/sec or lower, but of course the art will handle in excess of 100 Mbit/sec for up to 1 km. Current work provides integrated driver circuitry and LED in one compact assembly, with a similar integrated baseband amplifier and detector at the receiving end. New work will soon show the feasibility of integrating an array of LEDs, lasers, or detectors with their associated electronics onto a single assembly to mate with an array of fibers for inexpensive multichannel applications (Albanese and Holden, 1979). The combination of LSI baseband circuitry with the above elements seems certain to become feasible.

21.3 POWER COMPANY COMMUNICATIONS

Fiberguide transmission provides electrical isolation inexpensively, a feature that can be important in service to power companies. The required service may take the form of simple voice or voiceband-data service provided over the normal switched telephone network, or may take the form of a private monitoring and control facility that is part of the basic operation of generating and distributing the power.

In the case of normal switched telephone service, it is necessary with wire-pair customer connections to provide high-voltage isolation transformers on the telephone line so that high voltages induced or connected to the subscriber line during accidental fault conditions are prevented from appearing on the telephone network remote from the power company location. Isolation equipment of this kind can cost thousands of dollars, thereby providing an opportunity for fiberguide. By using a fiberguide cable with no conducting members, the electrical isolation is inherently obtained. An experimental subsystem which provides the necessary customer ringing and signaling functions on a carrier basis has been designed and demonstrated by W. M. Hubbard and E. L. Chinnock at Bell Laboratories.

For the type of system used to monitor and control the power distribution system itself, rather extensive work has been carried out in Japan. In one field trial (Aoki *et al.*, 1977) a differential PCM signal at 6.3 Mbit/sec was employed on a 3-km route near Tokyo, associated with a 275-kV power cable. Freedom from induced interference as well as electrical isolation is advantageous.

Both of these exploratory investigations were successful and revealed no basic problems. Neither one is a large-volume application, so the costs of manufacturing the needed lightwave components for this application alone would be prohibitive. Economic prove-in may occur as the art matures and other applications develop.

21.4 INTEROFFICE TRUNKS

At present there are many 1.5 Mbit/sec digital trunks in use between central offices in U.S. cities using one pair of wires for each direction of transmission. Digital carrier systems are utilized in rapidly growing links or where large capacity is needed. Conversion to a digital carrier system is also standard practice for obtaining additional voice circuits when an existing wire-pair cable no longer has adequate capacity if used with a single two-direction voice circuit per pair. The 1.5 Mbit/sec system carries 24 channels, and recently several other systems have become available providing 48 channels using pulse rates in the 2–3 Mbit/sec range.

The advantages fiberguide systems offer are several: low fiber losses permit spanning central offices spaced as far as 7 or 10 km with no amplifiers in between, whereas wire-pair systems require a repeater at about 1 km from each central office (for crosstalk reasons) and at about 2-km intervals thereafter because of cable attenuation. It is expensive to provide these manhole-located repeaters. Furthermore, the small size of fiberguide cables provides far more channels per duct (see Fig. 1.2) thereby saving duct cost. The other advantages of fiberguide (see Section 20.2) also apply.

Interoffice trunking appeared likely to be an early commercial application in the Bell System and an effort to achieve feasibility led first to an experimental installation at the Western Electric location near Atlanta, Georgia (Jacobs, 1976) and then in 1977 (Jacobs and Miller, 1977) to a field evaluation carrying commercial voice, data, and video traffic in Chicago (Schwartz *et al.*, 1978). This system is engineered to run at 44.7 Mbit/sec carrying 672 voice channels, with permitted repeater spacing at least 7 km.

In the U.S. the standardized digital hierarchy uses 1.544 Mbit/sec for 24 channels, 3.152 Mbit/sec for 48 channels, 6.312 Mbit/sec for 96 channels, 44.736 Mbit/sec for 672 channels, and 274.176 Mbit/sec for 4032 channels. In Europe the standards are somewhat different: 2.048 Mbit/sec for 30 channels, 8.448 Mbit/sec for 120 channels, 34.368 Mbit/sec for 480 channels, 139.264 Mbit/sec for 1920 channels, and 565 Mbit/sec for 7680 channels (Turner, 1975). In both the U.S. and Europe there is interest in low-bit-rate as well as in high-bit-rate systems. The high-bit-rate light-wave systems appear more economical because the repeater spacing and repeater cost are relatively insensitive to bit rate (see Figs. 20.2 and 20.6), but this economy is only applicable where there are customers to use the available capacity. The fact is that there are perhaps an order of magnitude more links where a 24- or 48-channel system is needed. However, the low-bit-rate application is sensitive to fiber cable costs and LED-laser costs, because the competing wire-pair systems are in volume production at low cost.

The Bell System's 44.7 Mbit/sec system uses AlGaAs lasers and silicon avalanche photodiodes to achieve the maximum repeater spacing. For lower bit rates or short spans an LED source would be usable. In the near future multimode graded-index fibers will be preferred over single-mode fibers to accommodate multimode lasers and LEDs; this preference is likely to persist in the interoffice-trunk application to hold down the splicing costs.

This art is evolving and will be a candidate for longer wavelength sources and detectors when they become feasible.

21.5 SUBSCRIBER LOOPS

Between the central office and the subscriber most all of the links are simple voice-frequency wire pairs. There is, particularly on long rural loops, some use of digital carrier on wire pairs as a feeder to an interface located near the subscriber. If Picturephone or some other wideband service appears, the economic feasibility of a digital carrier system will be expedited and fiberguide systems should become attractive. Even now digital carrier feeders are beginning to prove in (subscriber-loop carrier) using wire pair lines. The most probable future is an evolutionary one in which the fibers could provide broadband transmission to those customers who want it, and the wire pairs could provide power, signaling, and simple voice service.

21.6 INTERCITY

The intercity trunk application could follow the general format of a conventional digital coaxial system. Low fiber losses lead to repeater spacings far longer than the one mile of the analog L-5 coaxial system and could provide 4000 channels at 274 Mbit/sec per fiber (compared with 10,800 channels per $\frac{3}{8}$ in. coaxial). As noted in earlier chapters, higher bit rates will become feasible. In this application splices could be made less frequently and the higher costs per splice of single-mode fibers would be tolerable. Therefore, component choices are likely to be single-mode fibers, or very precisely graded-index fibers, in combination with laser sources operated at the highest bit rate the art will allow (Bask *et al.*, 1977; Shimodaira *et al.*, 1977; Abbott *et al.*, 1978).

Alternatives to lightwave systems in this application are millimeter waveguide and satellite radio, after single-sideband microwave radio has been used to its full capacity. The advantages of fiberguide over millimeter waveguide are greater flexibility in route layout (tolerable bend radius of a few feet) and greater adaptability to branching into smaller cross sections of capacity; advantages for millimeter waveguide include far longer repeater spacing (>25 mi) and ruggedness. Most new guided wave systems must pay for the right of way; satellite systems have an appreciable launching cost. For intercity distances below some value (probably under 1000 mi) it seems clear that radio relay or cable systems will prove-in over satellite systems. Total system cost differences as small as 25% are large in dollars and are a basis for a firm choice between alternatives. A comprehensive cost evaluation is needed to approach that kind of accuracy of comparison and is beyond the scope of this work.

21.7 UNDERSEA SYSTEMS

Low losses and broad bandwidth are especially attractive in undersea cable systems, which again suggests fiberguide systems. This application is a close relative of the domestic intercity trunk, but with some significant differences. Repeaters must have even longer life since it is very expensive to pull up an undersea cable for repairs. Furthermore, power is supplied only from the ends and may be of limited availability. The most attractive configuration is likely to be a laser source, a single-mode or precisely graded-index fiber, and a digital rate as high as component reliability will support. The alternative is satellite radio, and the choice is both economic and political due to the number of countries and interests involved. Five years ago, lightwave device lifetime was so low that an undersea cable system seemed academic. Now, with 10^8 hr LEDs and 10^6 hr lasers being quoted from hard data, one can plan a system with a high probability of success.

21.8 THE IMPACT OF VIDEO

The preceding discussion has not reflected the full impact that video of various kinds will have on future transmission configurations. We add a few words here to call attention to some of the interactions that will occur.

It is familiar that even a small fraction of subscribers using a video service of standard broadcast quality would vastly increase the total transmission bandwidth employed, because conventional video uses 1000 times the bandwidth of a voice channel. However, bandwidth-saving techniques are coming down in cost and picture-storage techniques are becoming practical. Thus it is likely that at least some forms of video will evolve without quite so large a bandwidth being required.

As described in Section 20.6, lightwave transmission on fiberguide is not well adapted to analog transmission, and this is particularly true for carrier-type analog systems of the type sometimes used for CATV service over coaxials or wire pairs. Fiberguide can be used for very short analog video circuits, but for large-scale video distribution a digital form is preferable. The bandwidth is available on fibers, and when combined with storage and band-saving techniques in the terminals, this may become feasible in time.

It seems evident that a major impact on subscriber loop would be necessary if video usage grew, since the needed bandwidth is not available presently. Interoffice trunks and intercity trunks would also need to grow.

Major questions exist on the form of video service needed to fill a practical function. There is ahead of us an innovative period in which social customs may evolve and entrepreneurs find ways of using what technology makes possible.

CATV, providing entertainment video on cable to the homes, is expanding and typically costs $7 to $8 per month for basic broadcast service, plus additional charges for closed-circuit video presentations. There are major policy questions being considered in Washington as to how this service should be related to common-carrier services. The CATV industry itself is evaluating the prospects for fiber optics in future CATV and concludes that the driving factor is cost (Hollis and Ecker, 1977).

In addition, forms of video-telephone common-carrier service are being explored both for intercity conferences and for more personal communications. To this writer, it seems probable that a video service for home subscribers would take the form of one-way video into the home, switchable at the central office to a variety of sources, at the initiation of either the home subscriber or the entrepreneur originating the video.

At this writing the timing and form of large-scale video transmission to the subscriber is unclear. The capabilities of fiberguide transmission are on call, and may prove essential to a viable system configuration.

21.9 ECONOMIC EVALUATIONS

In the introduction of this chapter we referred to component costs and to some prospective interrelations between the various prospective applications through component cost. One large-volume fiberguide system could carry along dozens of small-volume subsystems as a by-product.

Let us illustrate the cost versus volume relationship with a few examples. Hypothetically, let us assume a source that costs $500 each if initially sold in a quantity of 1000. From analogy with other "learning curves" let us assume a cost decrease of 25% every time the integrated production volume doubles. Then at 2000 devices the cost would be $375, at 4000 devices $280, at 128,000 devices $66, and at 10^6 devices $28. For order of magnitude this should be correct. If a device costs $500 in the initial 1000 it requires an accumulated production of around 10^6 units to bring the cost down to around $25. Sharing this volume among a number of systems could enable all to realize reduced costs. It also follows that if the initial cost for 1000 devices is $50 each, the 10^6 unit cost can be in the $2.50 vicinity.

Let us take a similar example in fiber cost. Let us assume a cost of $1 per fiber meter at a volume of 20,000 km; then at a total volume of 40,000 km the cost is 75¢/fiber-m, at 80,000 km 56¢, and at 1.28×10^6 km it reaches 18¢/fiber-m. From this one concludes that millions of kilometers of total production volume are needed to bring fiber costs down to really attractive values. Sharing the integrated volume among many users surely will help, but short-length systems will have little effect on cost as compared with long-distance systems.

It seems highly probable that LED systems will have lower cost devices available, by virtue of enjoying larger total production volume as well as by virtue of being a somewhat simpler device. However, fiber costs for short-length systems (which many LED systems will be) may stay rather high (per meter) until a large production-run system exists to bring down the cost.

Each prospective lightwave system use is usually evaluated on an individual basis—economic comparison with the alternatives. The picture outlined above shows that a broader view could be useful—a view of several prospective uses all taken at once, against the alternatives. This may be difficult to do, but there may be evidence that the organizations who are potential users of the above systems are sensing the future in just this more broad way. They want a sample of a lightwave system to show and to get experience with, more or less without traditional economic prove-in documentation. Once above some threshold, fiberguide usage should proliferate, and many groups are anxious to be first at that point.

REFERENCES

Abbott, S. M., Muska, W. M., Lee, T. P., Dentai, A. G., and Burrus, C. A. (1978). 1.1 Gb/s pseudorandom pulse-code modulation of 1.27 μm wavelength C. W. InGaAs/InP d.h. lasers. *Electron. Lett.* **14**, 349–350.

Albanese, A., and Lenzing, H. F. (1978). Video transmission tests performed on intermediate-frequency lightwave entrance links. *Society of Motion Pictures and Television Engineers Journal* **87**, 821–824.

Albanese, A., and Holden, W. S. (1979). LED array package for optical data links. *Bell Syst. Tech. J.* **58**, 713–720.

Aoki, F., Ando, K., Ueno, Y., Kajitani, M., Tsukada, K., and Shiraishi, S. (1977). Field trial of optical fiber communication for power system. *Proc. Top. Meet. Opt. Fiber Transm., 2nd, 1977* p. ThB4.

Bask, C., Elze, G., Wolf, G., Gliemeroth, G., Krause, P., and Neuroth, N. (1977). Optical transmission experiment at 1.12 G Bit/s using a graded-index fibre with a length of 1.652 KM. *Electron. Lett.* **13**, 452–453.

Hackett, W. H., Jr., Brackett, C. -A., Dombrowski, L. C., Howarth, L. E., Shumate, P. W., Smith, R. G., Warner, A. W., Riggs, R. S., and Jones, J. R. (1977) Optical data links for short-haul high level performance at 16 and 32 Mb/s. *Proc. Top. Meet. Opt. Fiber Transm., 2nd, 1977* p. ThB2.

Hollis, S., and Ecker, A. (1977). Fiber Optics for CATV in perspective. *IEEE Trans. Cable Television* **CATV-2**, 154–157.

Jacobs, I. (1976). Lightwave communications passes its first test. *Bell Lab. Rec.* pp. 291–297.

Jacobs, I., and Miller, S. E. (1977). Optical transmission of voice and data. *IEEE Spectrum* **14**, 33–41.

Kawachi, M., Kawana, A., and Miyashita, T. (1977). Low-loss single-mode fibre at the material-dispersion—free wavelength of 1.27 μm. *Electron. Lett.* **13**, 442–443.

Markstein, H. W. (1977). Fiber optics for electronic interconnection. *Electron. Packag. Prod.* pp. 34–43.

Miller, S. E. (1977). Photons in fibers for telecommunications. *Science* **195**, 1211–1216.

Miller, S. E., Marcatili, E. A. J., and Li, T. (1973). Research toward optical-fiber transmission systems. *Proc. IEEE* **61**, 1703–1751.

Schwartz, M. I., Reenstra, W. A., Mullins, J. H., and Cook, J. S. (1978). Chicago lightwave communications project. *Bell. Syst. Tech. J.* **57**, 1881.

Shimodaira, M., Hirimatsu, T., and Honma, K. (1977). An experimental 400 Mb/s optical transmission system using SELFOC fibers. *Radio Sci.* **12**, 511–517.

Taylor, H. F. (1977). Review and assessment of fiber optics for military applications. *AGARD Conf. Proc.* **219** (available from National Technical Information Service, 5285 Port Royal Road, Springfield, Virginia).

Turner, R. J. (1975). Preliminary engineering design of digital transmission systems using optical fibre. *P. O. Eng. J.* **68**, Part I, 7.

Index